THE SCIENCE OF COOKING

THE SCIENCE OF COOKING

Understanding the Biology and Chemistry Behind Food and Cooking

JOSEPH J. PROVOST
KERI L. COLABROY
BRENDA S. KELLY
MARK A. WALLERT

WILEY

Published by John Wiley & Sons, Inc., Hoboken, New Jersey
Published simultaneously in Canada

For general information on our other products and services or for technical support, please contact our Customer Care Department within the United States at (800) 762-2974, outside the United States at (317) 572-3993 or fax (317) 572-4002.

Wiley also publishes its books in a variety of electronic formats. Some content that appears in print may not be available in electronic formats. For more information about Wiley products, visit our web site at www.wiley.com.

Library of Congress Cataloging-in-Publication Data

Names: Provost, Joseph J., author. | Colabroy, Keri L., author. | Kelly, Brenda S., author. |
    Wallert, Mark A., author.
Title: The science of cooking : understanding the biology and chemistry
    behind food and cooking / Joseph J. Provost, Brenda S. Kelly, Mark Wallert,
    Keri L. Colabroy.
Description: Hoboken, New Jersey : John Wiley & Sons, 2016. | Includes
    bibliographical references and index.
Identifiers: LCCN 2015041520 (print) | LCCN 2015044584 (ebook) |
    ISBN 9781118674208 (pbk.) | ISBN 9781119210320 (pdf) |
    ISBN 9781119210337 (epub)
Subjects: LCSH: Food–Analysis. | Biochemistry. | Food–Composition. |
    Food–Biotechnology.
Classification: LCC TX545 .P76 2016 (print) | LCC TX545 (ebook) | DDC
    664/.07–dc23
LC record available at http://lccn.loc.gov/2015041520

Printed in the United States of America

SKY10033177_021422

CONTENTS

9 Eggs, Custards, and Foams   311

10 Bread, Cakes, and Pastry   343

PREFACE

Interest in cooking, baking, and food has risen tremendously over the past few years. In fact, the popularity of food and cooking within the 18–34-year-old demographic group draws more than 50 million viewers to food- and cooking-based cable shows and websites each month. Many faculty members have tapped into this interest, creating unique and interesting courses about science, food, and/or cooking. This aim of *The Science of Cooking: Understanding the Biology and Chemistry Behind Food and Cooking* is to teach fundamental concepts from biology and chemistry within the context of food and cooking. Thus, the primary audience for the text is nonscience majors, who are fulfilling a science curricular graduation requirement. However, we anticipate that there may be instructors and students with a more significant interest in science who may utilize the book as a catalyst to fuel further study in the area. We hope that this book helps reduce the barriers to teach courses related to science, food, and cooking and opens up new opportunities for those already teaching about food and cooking.

We also recognize that there are important pedagogical approaches to learning that are well beyond the scope of a textbook. The companion website has over 35 guided inquiry activities covering science basics such as chemical bonding, protein structure, and cell theory and such food-focused topics as meat, vegetables, spices, chocolate, and dairy. These are carefully crafted and classroom-tested activities designed for student teams to work on under the guidance of an instructor. The activities introduce the scientific concepts in a way that complements the text while giving students practice in critical thinking about the relevant foundational principles of chemistry and biology. We have also created a series of food- and cooking-based laboratories. These experiential learning opportunities involve hypothesis design and help teach the scientific process and critical concepts while engaging students in fermentation, cheesemaking, analyzing food components, and other hands-on exercises. The laboratories have been designed to minimize cost and hazardous materials; some are even appropriate to assign as homework to be done in a student's home kitchen.

The Science of Cooking: Understanding the Biology and Chemistry Behind Food and Cooking is food centered while including several chapters that introduce fundamental concepts in biology and chemistry that are essential in the kitchen. In the first few chapters of the book, students will learn about molecular structure, chemical bonding, cell theory, signaling, and biological molecule structure. These concepts are drawn upon in later chapters; for example, students will learn the science behind cheesemaking, meat browning, and fermentation processes. The chapters are also full of interesting facts about the history of the food, ailments, or cures associated with the food, all guided by in-depth discussions of the science behind the food.

Of course, there is a rich history of literature on and the science of food and cooking. We have taken some space to acknowledge those who helped build and grow modernist cooking. Special thanks go to Harold McGee and Shirley O. Corriher for their pioneering work, inspiration, and kind words as we developed this work. We hope to add to the scientific culture that they and others have created in the kitchen.

Inquire, Learn, Investigate, and Eat Well!

ABOUT THE AUTHORS

Dr. Joseph J. Provost is a professor of chemistry and biochemistry at the University of San Diego. He has helped create and teach a science of cooking class and taught to small and large classes. Provost has served on educational and professional development committees for the American Society for Biochemistry and Molecular Biology, Council on Undergraduate Research, and the American Chemical Society while teaching biochemistry, biotechnology, and introductory chemistry laboratories. For the past 18 years, he has partnered with Dr. Mark Wallert as they research non-small cell lung cancer focusing on processes involved with tumor cell migration and invasion. When not in the lab or class, Provost can be found making wine and cheese, grilling, and then playing or coaching hockey.

Dr. Keri L. Colabroy is an associate professor of chemistry at Muhlenberg College in Allentown, Pennsylvania, where she created and teaches a course on kitchen chemistry for nonscience majors. When she isn't evangelizing nonscience majors with her love of chemistry, Colabroy is teaching organic chemistry, biochemistry courses, and a first-year writing course on coffee while also serving as codirector of the biochemistry program. Her scholarly research is in the area of bacterial antibiotic biosynthesis with a focus on metalloenzymes and actively involves undergraduates. Colabroy serves as coordinator for undergraduate research at the college and participates on the Council on Undergraduate Research in the Division of Chemistry. When not in the lab or class, Colabroy can be found chasing her two small children or singing in the choir.

Dr. Brenda S. Kelly is an associate professor of biology and chemistry at Gustavus Adolphus College in St. Peter, Minnesota. Kelly's immersion into teaching about science and cooking began in 1997 when she cotaught a January term course, The Chemistry of Cooking, that enrolled science majors who knew little about cooking and nonscience majors who were excellent cooks. The immense number of resources that she used to gather information for the course, as well as the diverse student population who would have benefited from a single resource, suggested a need for an undergraduate textbook for such a course. In addition to talking with her students about cooking as one big science experiment, Kelly teaches courses in biochemistry and organic chemistry and has an active undergraduate research lab where she engages her students in research questions related to protein structure and function. When she is not busy in her current interim role as associate provost and dean of the Sciences and Education at Gustavus, Kelly enjoys cooking, baking, and running (not at the same time) and spending time with her family.

Dr. Mark A. Wallert is an associate professor of biology at Bemidji State University in Bemidji, Minnesota. Mark was an inaugural member of Project Kaleidoscope Faculty for the twenty-first century in 1994 and has worked to integrate inquiry-driven, research-based laboratories into all of his courses. For the past 18 years, he has maintained a research partnership with Dr. Joseph Provost where they investigate the role of the sodium–hydrogen exchanger in cancer development and progression. Mark is the Northwest Regional Director for the American Society of Biochemistry and Molecular Biology Student Chapters Steering Committee where he has  helped organize the Undergraduate Research in the Molecular Sciences annual meeting held in Moorhead, Minnesota, for the past 10 years. In 2005, Mark was recognized as the Council for Advancement and Support of Education/Carnegie Foundation for the Advancement of Teaching Minnesota College Professor of the Year. When not engaged in campus and research activities, Wallert can be found spending time with his family and enjoying the abundance of nature in northern Minnesota.

ABOUT THE COMPANION WEBSITE

This book is accompanied by a companion website:

www.wiley.com/go/provost/science_of_cooking

The website includes:

- Guided Inquiry Activities
- Inquiry and Scientific Method based Laboratory Experiments
- Color Infographics with Recipe and Science Behind the Food
- Powerpoint files with all chapter images
- Powerpoint files for teaching
- Learning Objectives for a course and each chapter
- Practice Questions

1

THE SCIENCE OF FOOD AND COOKING: MACROMOLECULES

Guided Inquiry Activities (Web): 1, Elements, Compounds, and Molecules; 2, Bonding; 3, Mixtures and States of Matter; 4, Water; 5, Amino Acids and Proteins; 6, Protein Structure; 7, Carbohydrates; 8, pH; 9, Fat Structure and Properties; 10, Fat Intermolecular Forces; 11, Smoking Point and Rancidity of Fats

1.1 INTRODUCTION

The process of cooking, baking, and preparing food is essentially an applied science. Anthropologists and historians venture that cooking originated when a pen holding pigs or other livestock caught fire or a piece of the day's catch of mammoth fell into the fire pit. The smell of roasted meat must have enticed early people to "try it"; the curious consumers found culinary and nutritional benefits to this new discovery. The molecular changes that occurred during cooking made the meat more digestible and the protein and carbohydrates more readily available as nutrients. Contaminating microbes were eliminated during cooking, which made the consumers more healthy and able to survive. Moreover, the food was tastier due to the heat-induced chemical reactions between the oxygen in the air and the fat, proteins, and sugar in the meat. Harnessing the knowledge of what is happening to our food at the molecular level is something that good scientists and chefs use to create new appetizing food and cooking techniques.

We are all born curious. Science and cooking are natural partners where curiosity and experimentation can lead to exhilarating and tasty new inventions. Scientific

The Science of Cooking: Understanding the Biology and Chemistry Behind Food and Cooking, First Edition. Joseph J. Provost, Keri L. Colabroy, Brenda S. Kelly, and Mark A. Wallert.
© 2016 John Wiley & Sons, Inc. Published 2016 by John Wiley & Sons, Inc.
Companion website: www.wiley.com/go/provost/science_of_cooking

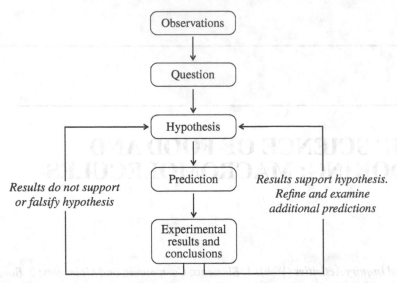

FIGURE 1.1 The scientific method. Scientists use a testable method originating from observations to generate a testable hypothesis to conduct their work. A cook or baker can also use this method to create a more interesting food.

discovery is driven by hypothesis (see Fig. 1.1 for a model of the scientific method). An observation of an event creates a question and/or a statement that explains the observation or phenomenon: the hypothesis. The hypothesis can then be tested by a series of experiments and controls that supports or falsifies the hypothesis, starting the cycle over again. For example, a scientist might observe that the growth rate of cancer cells in a petri dish slows when the cells are exposed to a sea sponge. The scientist may then hypothesize that a molecule found in the sponge binds to a protein in cancer cells. After adding the compound to a tumor, its growth slowed and the cells die. Looking at how all of the individual molecules found in the sea sponge affect the growth of cancer cells can test this hypothesis. These experiments can lead to a more advanced hypothesis, testing and eventually finding a new compound that can be used to fight cancer.

Cooking can also be a hypothesis-driven process that utilizes biology, chemistry, and physics. As you cook, you use biology, chemistry, and physics to create hypotheses in the kitchen, even if you weren't aware of being a scientist. Each time you try a recipe, you make observations. You may ask yourself questions about what you added to the concoction or how the food was baked or cooked. This creates a hypothesis or a statement/prediction that you can test through experimentation (your next attempt at the dish). A nonscientific idea is often approached as something to prove. That is different from hypothesis testing. A hypothesis is falsified rather than proven by testing. Cooking does just this; it will falsify your test rather than prove it. Tasting, smelling, and visualizing your results tell you if your hypothesis was supported or falsified. If wrong, you may create a new hypothesis that might be generated by the time you have washed the dishes from your first experiment! Learning more of the basic science behind food and cooking will help you appreciate the world around you and become a better scientist and a better cook, baker, and consumer.

1.2 FUNDAMENTALS OF FOOD AND COOKING

Bread baking provides a great example of the importance of having a scientific understanding of cooking and baking. Take a close look at bread. Notice that it is made of large and small caves surrounded by a solid wall (Fig. 1.2).

The key to bread is making a way to trap expanding gases in the dough. Adding water to flour and sugar allows for the hydration and mixing of proteins and carbohydrates. Kneading the dough stretches a protein called gluten, which allows for an interconnected network of protein ready to trap gas that is generated by the yeast. During the proofing step of making bread, the yeast converts sugar into energy-filled molecules, ethanol, carbon dioxide gas, and other flavorful by-products. The heat applied during baking allows the water to escape as steam, which expands the bread, links the gluten protein molecules further, and traps carbon dioxide gas. While this is happening, the heat catalyzes chemical reactions between proteins and sugars, creating a beautiful brown color, a dense texture, and over 500 new aromatic compounds that waft to your nose. Clearly there is a lot of science that goes into making a loaf of bread.

FIGURE 1.2 Structure of bread. A close look at bread demonstrates the requirement of proteins and carbohydrates needed to trap expanding gases.

Preparing food and drink is mostly a process of changing the chemical and physical nature of the food. Molecules react to form new compounds; heat changes the nature of how food molecules function and interact with each other, and physical change brings about new textures and flavors to what we eat. To gain a better appreciation for these chemical and physical processes, a fundamental understanding of the building blocks of food and cooking must first be understood. In the following two chapters we will study the basic biological principles of cooking, tasting, and smelling.

One of the most important building blocks of food is water; our bodies, food, and environment are dependent on the unique chemistry and biology of this molecule. Large biological molecules such as proteins, carbohydrates, and fats comprise the basic building blocks of food. Smaller molecules, including vitamins, salts, and organic molecules, add important components to cooking and the taste of food. Finally, the basics of plant and animal cells and cellular organization are key to understanding the nature of food and cooking processes. However, before we get into some of the science fundamentals, it is important to recognize and acknowledge the origins of and the chefs who first embraced the science behind their profession.

1.2.1   Science, Food, and Cooking

Many chefs and bakers embrace the collaboration of science and food. Historically, one means whereby science has been utilized in the kitchen is in the area of food technology—the discipline in which biology, physical sciences, and engineering are used to study the nature of foods, the causes of their deterioration, and the principles underlying food processing. This area of food science is very important in ensuring the safety and quality of food preparation, processing of raw food into packaged materials, and formulation of stable and edible food. College undergraduates can major in "food science" or attend graduate studies in this area, working for a food production company where they might look at the formulation and packaging of cereals, rice, or canned vegetables. Recently a new marriage of science and food, coined molecular gastronomy, has grown to influence popular culture that extends far beyond the historical definition of food science. A physicist at Oxford, Dr. Nicholas Kurti's interest in food led him to meld his passion for understanding the nature of matter and cooking. In 1984 Harold McGee, an astronomist with a doctorate in literature from Yale University, wrote the first edition of the influential and comprehensive book *On Food and Cooking: The Science and Lore of the Kitchen* [1]. This fascinating book is the basis for much of the molecular gastronomy movement and describes the scientific and historic details behind most common (and even uncommon) culinary techniques. Together with cooking instructor Elizabeth Cawdry Thomas, McGee and Kurti held a scientific workshop/meeting to bring together the physical sciences with cooking in 1992 in Erice, Italy. While there were more scientists than chefs attending, with a five to one ratio, the impact of the meeting was significant. It was at Erice that the beginnings of what was then called molecular and physical gastronomy became the catalyst for an unseen growth in science and cooking. Hervé This, a chemist who studies the atomic and subatomic nature of chemistry, attended the workshop and has been a key player in the growth of molecular gastronomy. Dr. This blames a failed cheese soufflé for sparking his interest in culinary precisions and has since transformed into a career in molecular gastronomy. Other participants of the meetings include chef Heston Blumenthal and physicist Peter Barham, who have collaborated and influenced many molecular-based recipes and projects. Finally another scientist, biochemist Shirley O. Corriher, was present at these early meetings (Box 1.1). Shirley found her love of cooking as she helped her husband run a school in Nashville in nearby Vanderbilt Medical School where she worked as a biochemist. Her influence on science and cooking includes a friendship and advisory role with Julia Child and the many informative, science approach-based cookbooks (Ms. Corriher, personal communications, June 2012). The impact on popular culture and influence on modernist cooking are immense. For 13 years, Alton Brown brought the scientific approach to culinary arts in the series *Good Eats*. Through the work of all of these scientist chefs, use of liquid nitrogen, a specialized pressure cooking called sous vide, and unique presentation and mixtures of flavors are now more commonplace and creating new options for the daring foodie.

BOX 1.1 SHIRLEY CORRIHER

Shirley Corriher has long been one of the original scientists/cooks to influence the new approach to cooking and baking. Using everyday language as a way to explain food science, Shirley has authored unique books on becoming a successful cook and baker with her books *CookWise* [2] and *BakeWise* [3]. Her influence on popular acceptance of science on cooking and baking includes a friendship with Julia Child, appearances on several of Alton Brown's *Good Eats* episodes, and her involvement in the growth of the science and cooking. Shirley earned a degree in biochemistry from Vanderbilt University where she worked in the medical school in a biomedical research laboratory while her husband ran a school for boys. She recalls her early attempt to cook for the large number of boys. Little did she know this experience would be the beginning of a new career. Shirley describes how she struggled with the eggs sticking to the pan and worrying that there would be no food for the students. Eventually she learned to heat the pan before adding the eggs. The reason was that the small micropores and crevices of the pan would fill and solidify in the pan. This sparked the connection between science and cooking for her. After a divorce Ms. Corriher and her sons were forced into a financial struggle, where they had to use a paper route as a source of income, a friend, Elizabeth Cawdry Thomas, who ran a cooking school in Berkeley, California, asked her to work for her cooking school where she learned formal French cooking while on the job. Later Shirley found herself mixing with a group of scientists and chefs who appreciated the yet to be studied mix of science and cooking. In 1992, the group including Thomas, Kurti, and Harold McGee obtained funds to bring scientists and chefs together to support workshops on nonnuclear proliferation in Erice, Sicily. Shirley was a presenter at that first meeting leading discussions on emulsifiers and sauces and continued as a participant in each of these early workshops (Ms. Corriher, personal communications, June 2012). Corriher recalls that the term molecular gastronomy was voted on by the core group to reflect both the science and culinary aspects of the meeting. Shirley talks of a respect and friendship between herself and leading food scientist Harold McGee. Shirley recalls reading his book and called him to ask him where had he been all this time? She said, "You don't know me, but I and many other ladies in Atlanta are going to bed with you every night!" Her books using science to explain how to become a better cook and baker are extremely popular. Her approach to trust in yourself and understanding the science of kitchen work is certainly an inspiration given by a person with a unique route to her spot in American culinary society.

1.3   THE REAL SHAPE OF FOOD: MOLECULAR BASICS

What are the fundamental units of all food and cooking processes? Atoms and molecules! All living systems (animals, microbes, and smaller life forms) are made of atoms and molecules. How these atoms and molecules are organized, interact, and react provides the building blocks and chemistry of life. It makes sense that to best understand cooking and baking at the molecular level, you must first appreciate how atoms and compounds are put together and function. Let's start with the basics and ask, what is the difference between an atom and molecule? The answer is simple: an atom is the smallest basic building block of all matter, while molecules are made when two or more atoms are connected to one another.

An atom consists of three main components also known as subatomic particles. These subatomic particles are called protons, neutrons, and electrons. A simple description of what and where these particles are located is that protons and neutrons are found in the center or nucleus of the atom, while electrons orbit the core of the atom (Fig. 1.3). Protons are positively charged particles with an atomic mass of one atomic mass unit. Neutrons essentially also have an atomic mass of one, but do not have an electrical charge. Electrons have almost no mass and have an electrical charge of −1.

The elements of the periodic table are arranged and defined by the number of protons present within an atom of a given element. The number of protons defines an atom, not the electrons or neutrons. A quick examination of a periodic table shows that their proton number organizes atoms: from the smallest atom, hydrogen, to the largest atom, ununoctium. As stated, the number of protons in an atom defines that atom. Any atom with six protons is a carbon; any atom with seven protons is nitrogen. Thus, if a carbon atom gains a proton, it becomes a nitrogen atom. However, if a carbon atom gains or loses an electron, it still is a carbon, but now has a charge associated with it. The total number of protons and electrons defines the charge of an atom. An atom of any element with an equal number of protons and electrons will have a net neutral charge; atoms that have gained an electron will have a negative charge, and those that have lost an electron will have a positive charge. Most of the atoms of the elements on the periodic table can gain or lose one or more electrons. The numbers of neutrons within a given type of atom can also vary. Isotopes are

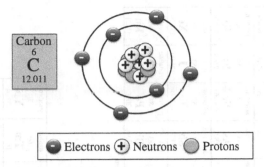

FIGURE 1.3 Atomic structure. Atoms are made of electrons in orbitals around the nucleus where protons and neutrons are found. The identity of an atom is the number of protons.

atoms that have the same number of protons but differ in the number of neutrons. Carbon 12 and carbon 13 both have six protons (thus they are carbon), but carbon 12 has six neutrons for a total atomic mass of 12, while carbon 13 has seven neutrons and when including the mass of the protons has an atomic mass of 13 (6 protons + 7 neutrons = 13 atomic mass) (Fig. 1.4).

What about a compound or a molecule? How does a molecule differ from an atom or compound? A molecule is a substance of two or more atoms connected by sharing electrons (covalent bonds). A compound is a chemical substance made of different atoms. Compounds can be made of atoms held together by ionic or covalent bonds where molecules are made only of covalently bonded atoms. Thus all molecules are compounds, but not all compounds are molecules. Molecules are often categorized further into organic (those molecules containing mostly carbon atoms) and inorganic molecules (everything else).

Most of the compounds found in living things contain carbon, hydrogen, nitrogen, or hydrogen atoms. A group of other elements, including sulfur, magnesium, and iron, make up less than 1% of the atoms in most living systems. Trace elements, such as copper, zinc, chromium, and even arsenic, although necessary for biological function, only make up a minute portion of an organism, less than 0.01% of all atoms. Due to their complexity and impact on their behavior in cooking, let's talk a little bit more about the bonds that connect atoms together.

1.3.1 Ionic and Covalent Compounds

There are two types of bonds that connect two atoms to yield a molecule or compound: ionic and covalent. Ionic bonds form between atoms that have opposite charge due to the loss or gain of electrons (Fig. 1.5). Atoms that have become charged have their own name—ions. Ionic bonds form when an ion with a positive charge (a cation) is bonded to an ion with a negative charge (an anion). The resulting molecule is called an ionic compound or a salt. This terminology is apropos because the salt that you sprinkle on your popcorn, NaCl, is an ionic compound consisting of a positively charge sodium atom or ion (Na^+) and a negatively charged chlorine atom or ion (Cl^-).

1	2	3	4	5	6	7	8	9	10	11	12	13	14	15	16	17	18
hydrogen 1 **H** 1.0079																	helium 2 **He** 4.0026
lithium 3 **Li** 6.941	beryllium 4 **Be** 9.0122											boron 5 **B** 10.811	carbon 6 **C** 12.011	nitrogen 7 **N** 14.007	oxygen 8 **O** 15.999	fluorine 9 **F** 18.998	neon 10 **Ne** 20.180
sodium 11 **Na** 22.990	magnesium 12 **Mg** 24.305											aluminium 13 **Al** 26.982	silicon 14 **Si** 28.086	phosphorus 15 **P** 30.974	sulfur 16 **S** 32.065	chlorine 17 **Cl** 35.453	argon 18 **Ar** 39.948
potassium 19 **K** 39.098	calcium 20 **Ca** 40.078	scandium 21 **Sc** 44.956	titanium 22 **Ti** 47.867	vanadium 23 **V** 50.942	chromium 24 **Cr** 51.996	manganese 25 **Mn** 54.938	iron 26 **Fe** 55.845	cobalt 27 **Co** 58.933	nickel 28 **Ni** 58.693	copper 29 **Cu** 63.546	zinc 30 **Zn** 65.39	gallium 31 **Ga** 69.723	germanium 32 **Ge** 72.61	arsenic 33 **As** 74.922	selenium 34 **Se** 78.96	bromine 35 **Br** 79.904	krypton 36 **Kr** 83.80
rubidium 37 **Rb** 85.468	strontium 38 **Sr** 87.62	yttrium 39 **Y** 88.906	zirconium 40 **Zr** 91.224	niobium 41 **Nb** 92.906	molybdenum 42 **Mo** 95.94	technetium 43 **Tc** [98]	ruthenium 44 **Ru** 101.07	rhodium 45 **Rh** 102.91	palladium 46 **Pd** 106.42	silver 47 **Ag** 107.87	cadmium 48 **Cd** 112.41	indium 49 **In** 114.82	tin 50 **Sn** 118.71	antimony 51 **Sb** 121.76	tellurium 52 **Te** 127.60	iodine 53 **I** 126.90	xenon 54 **Xe** 131.29
caesium 55 **Cs** 132.91	barium 56 **Ba** 137.33	lutetium 71 **Lu** 174.97	hafnium 72 **Hf** 178.49	tantalum 73 **Ta** 180.95	tungsten 74 **W** 183.84	rhenium 75 **Re** 186.21	osmium 76 **Os** 190.23	iridium 77 **Ir** 192.21	platinum 78 **Pt** 195.08	gold 79 **Au** 196.97	mercury 80 **Hg** 200.59	thallium 81 **Tl** 204.38	lead 82 **Pb** 207.2	bismuth 83 **Bi** 208.98	polonium 84 **Po** [209]	astatine 85 **At** [210]	radon 86 **Rn** [222]
francium 87 **Fr** [223]	radium 88 **Ra** [226]	lawrencium 103 **Lr** [262]	rutherfordium 104 **Rf** [261]	dubnium 105 **Db** [262]	seaborgium 106 **Sg** [266]	bohrium 107 **Bh** [264]	hassium 108 **Hs** [269]	meitnerium 109 **Mt** [268]	ununnilium 110 **Uun** [271]	unununium 111 **Uuu** [272]	ununbium 112 **Uub** [277]		ununquadium 114 **Uuq** [289]				

*Lanthanide series 57–70:

lanthanum 57 **La** 138.91	cerium 58 **Ce** 140.12	praseodymium 59 **Pr** 140.91	neodymium 60 **Nd** 144.24	promethium 61 **Pm** [145]	samarium 62 **Sm** 150.36	europium 63 **Eu** 151.96	gadolinium 64 **Gd** 157.25	terbium 65 **Tb** 158.93	dysprosium 66 **Dy** 162.50	holmium 67 **Ho** 164.93	erbium 68 **Er** 167.26	thulium 69 **Tm** 168.93	ytterbium 70 **Yb** 173.04

**Actinide series 89–102:

actinium 89 **Ac** [227]	thorium 90 **Th** 232.04	protactinium 91 **Pa** 231.04	uranium 92 **U** 238.03	neptunium 93 **Np** [237]	plutonium 94 **Pu** [244]	americium 95 **Am** [243]	curium 96 **Cm** [247]	berkelium 97 **Bk** [247]	californium 98 **Cf** [251]	einsteinium 99 **Es** [252]	fermium 100 **Fm** [257]	mendelevium 101 **Md** [258]	nobelium 102 **No** [259]

FIGURE 1.4 Periodic table. Each atom is arraigned based on the number of proton (elemental number) increasing from left to right and top to bottom. Scientists use the periodic table to understand the physical characteristics. LeVanHan, https://commons.wikimedia.org/wiki/File:Periodic-table.jpg. Used under CC-BY-SA 3.0 Unported https://creativecommons.org/licenses/by-sa/3.0/deed.en, 2.5 Generic https://creativecommons.org/licenses/by-sa/2.5/deed.en, 2.0 Generic https://creativecommons.org/licenses/by-sa/2.0/deed.en and 1.0 Generic license https://creativecommons.org/licenses/by-sa/1.0/deed.en.

FIGURE 1.5 Ionic compound (sodium chloride). A positively charged cation (Na^+) forms an ion bond to a negatively charged anion (Cl^-) to form an ionic compound.

Thus compounds are divided into molecules that have a charge or those without a charge. Ionic compounds are molecules that have somehow lost or gained an electron resulting in a compound with two parts; one atom or group will be positive charged and bonded to another atom or group of atoms with a negative charge. One of the atoms in an ionic compound will have at least one metal element (Na, K, Ca, Al, etc.). Metal atoms more readily give or accept electrons transforming the atoms into charged ionic elements. The simplest ionic compounds are formed from monoatomic ions, where two ions of opposite charge act as the functional unit. A good example is table salt, or sodium chloride (NaCl). In addition to single atom ions, a group of covalently bound atoms can also possess an overall charge called polyatomic ions. Polyatomic ions are made of several atoms bonded as a group, which is charged. Potassium nitrate, commonly called saltpeter and used in curing meat, is a complex polyatomic ion with the chemical formula KNO_3, where the potassium ion (K^+) provides the positive charge and the nitrate ion provides the negative charge (NO_3^-). Nitrate compounds have been historically used to preserve meats and fish. The nitrate dries the meat by drawing the water out of the muscle tissue leaving an inhospitable environment for bacteria to grow.

As a solid, ionic atoms are tightly held together by opposite charges in large networks called a lattice. In water, however, the attractive force between cation and anion components of the ionic compound is shielded by water and separate from one another. You can see this phenomenon with your very own eyes as you watch a teaspoon of salt dissolve in a glass of water. What is happening at the molecular level? Water is a polar covalent molecule with a positive and negative partial charge. The hydrogens have a partial positive charge, while the oxygen has a partial negative charge. Water molecules align with the charge of the ion forming a solvating shell of water (Fig. 1.6). This coating of water acts to shield the attraction between the ions, which can then separate from one another, dissolving in the water.

Salts are a very important aspect of foods, cooking, and taste and are often key to the demise of success of a given dish. Thus, when we refer to salts throughout the rest of this text, we will specify whether we are using the scientific definition of salt (an ionic compound made up of a cation and anion) or the common definition of salt (meaning table salt, or NaCl).

Can you have molecules that are made up of uncharged atoms? Yes, these molecules are called covalent or molecular compounds (as opposed to the ionic compounds or salts referred to earlier). In covalent compounds, sharing electrons holds atoms

Sodium chloride (i.e., table salt) is an ionic compound. It is made of two different types of atoms that are held together by a positive to negative attraction called an *ionic bond*

Na⁺ Cl⁻

+

H₂O

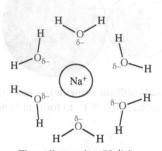

The sodium cation (Na⁺) is surrounded by a cloud of water molecules that are oriented to present their *slightly* negative oxygens toward the positively charged sodium

The chloride anion (Cl⁻) is surrounded by a cloud of water molecules that are oriented to present their *slightly* positive hydrogens toward the negatively charged chloride

FIGURE 1.6   Salt dissolves in water. In water, the polar nature of water surrounds and reduces the attractive force between ionic compounds dissolving each ion into the water solution.

A line between two atoms indicates they are joined by a *covalent bond*

Two lines represent a *double covalent bond*

The molecule *glycine*

The two electrons being *shared*. This joins the atoms together

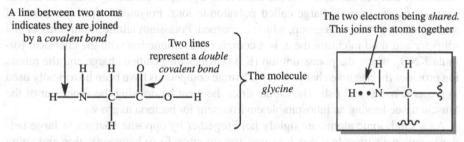

FIGURE 1.7   Covalent bonds have shared electrons. The sharing of two electrons forms a covalent bond. The straight line between atoms represents these electrons. Electrons are very tiny particles with negative charge. Every atom of each unique element has a specific number of electrons. For example, every hydrogen atom has one electron.

together; the force that ties the atoms together is called a covalent bond. The amino acid glycine is a great example of a covalent compound (Fig. 1.7). In a molecule of glycine, each nitrogen, carbon, oxygen, and hydrogen atom shares electrons with neighboring atoms forming a bond. The sharing of electrons that creates these covalent bonds has a particular order. Sharing of one set of electrons between atoms creates a single bond often shown by a single line drawn between the atoms. A double or triple bond is created when two or three pairs of electrons are shared between atoms (Fig. 1.8). Covalent compounds are made up of nonmetal atoms and are typically much more diverse (i.e., different arrangements of atoms) and larger (i.e., more atoms) than ionic compounds. The main difference between ionic and covalent compounds is that covalent compounds are not held together by charges, but atoms are bonded

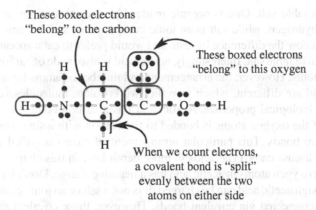

These boxed electrons "belong" to the carbon

These boxed electrons "belong" to this oxygen

When we count electrons, a covalent bond is "split" evenly between the two atoms on either side

FIGURE 1.8   Counting electrons with covalent bonds. Shared electrons making a covalent bond are often drawn as pairs of dots. However most molecular structures use single lines to represent the shared pairs of electrons.

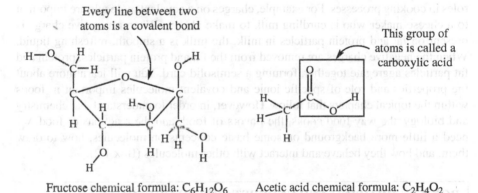

Every line between two atoms is a covalent bond

This group of atoms is called a carboxylic acid

Fructose chemical formula: $C_6H_{12}O_6$ Acetic acid chemical formula: $C_2H_4O_2$

FIGURE 1.9   Structure of fructose and acetic acid. The organization, shape, and chemical nature of the bonds and atoms create very different tastes and biological roles for these simple compounds.

together by sharing electrons in what is called a covalent bond. Molecular compounds make up the bulk of our food and include water, sugars, fats, proteins, and most vitamins. Sugars, fat, protein, and most vitamins are covalent compounds. Given their importance in food and cooking, let's look at two detailed examples of covalent compounds, fructose, and acetic acid.

Fructose is a sweet tasting sugar found in fruit and honey, while acetic acid is responsible for the sour taste in vinegar (Fig. 1.9). Looking at its molecular structure, the six carbon atoms are bonded (shown by the lines connecting atoms) to the 12 hydrogen or six oxygen atoms. Because of its atomic components, the molecular formula of fructose is $C_6H_{12}O_6$. This molecule is relatively large and has no overall charge, and all of the atoms are nonmetals. Clearly at the molecular level, fructose is different from table salt. One is organic made of a special arraignment of carbon, oxygen, and hydrogen, while salt is an ionic compound of sodium and chloride. Of course we all know the difference by taste and would prefer to eat a spoonful of honey over a spoonful of table salt. Interestingly, acetic acid is also made of carbon, hydrogen, and oxygen atoms. However, the arrangement and number of atoms between fructose and acetic acid are different, which give the two covalent molecules very different chemical and biological properties. Acetic acid has a carbon bonded to two oxygen atoms. One of the oxygen atoms is bonded to the carbon with a single bond, and the second has two bonds. This particular arrangement of atoms is called a carboxylic acid; we will discuss carboxylic acids in more depth later in this chapter. Notice also that one of the oxygen atoms in acetic acid has a negative charge. Don't let this confuse you! Even though acetic acid can be charged, it is not a salt or an ionic compound since its atoms are connected via covalent bonds. However, these covalent molecules do behave very differently than those covalent molecules that are uncharged. Specifically, these "charged" covalent molecules have acidic or basic properties. You have heard of acids and bases and likely have surmised that acetic acid is, in fact, an acid. Covalent molecules that are acidic or basic (and their corresponding charges) play many key roles in cooking processes. For example, charges on a covalent molecule are important to a cheese maker who is curdling milk to make cheese. When a negative charge is present on fat and protein particles in milk, the milk is a smooth, refreshing liquid. When the negative charges are removed from the fat and protein particles, protein and fat particles aggregate together, forming a semisolid curd. You will learn more about the properties and role of specific ionic and covalent molecules important in foods within the topical chapters that follow. However, in order to understand that chemistry and biology, the way food cooks, the flavors of food, and the reactions of food, we need a little more background on some basic concepts on molecules, how to draw them, and how they behave and interact with other molecules (Box 1.2).

BOX 1.2   DRAWING AND UNDERSTANDING CHEMICAL STRUCTURES

Scientists use a number of ways to represent chemical compounds. The simplest way to represent a molecule is the molecular formula. This is simply a count of each kind of atom in a molecule. The subscript describes the number of atoms in the molecule for the preceding element. While simple, it does not describe very much about the way the atoms are joined together. For example, both glucose and fructose can be described by the molecular formula $C_6H_{12}O_6$, as both a molecule of glucose and a molecule of fructose contain 6 carbons, 12 hydrogens, and 6 oxygen atoms. However, a molecular formula is often used for simple molecules to show how they react. For example, to produce caramel from table sugar ($C_{12}H_{22}O_{11}$), the applied heat results in a loss of water and a decomposition of sucrose to yield caramelen ($C_{36}H_{50}O_{25}$):

$$C_{12}H_{22}O_{11} \rightarrow 8H_2O + C_{36}H_{50}O_{25}$$

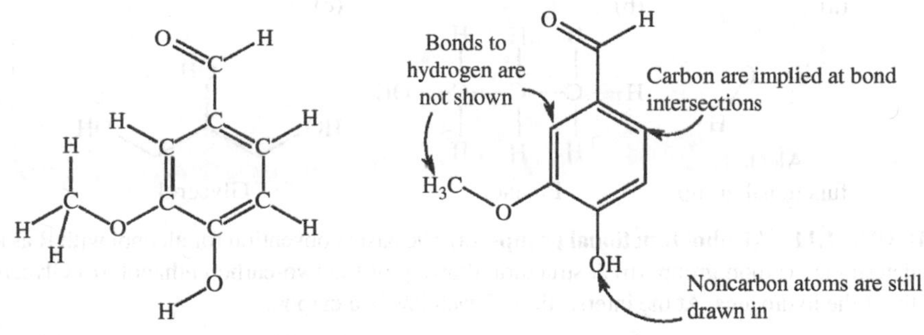

Vanillin—*complete structural formula*   Vanillin—*skeletal structure*

FIGURE 1.10   Structure of vanillin. On the left: structural formula of vanillin. Each atom is drawn and each bond is clearly marked—notice the single and double bonds and carbon atoms bonded to H, O, and other C atoms. On the right: Skeletal formula of vanillin. Note the implied carbons at the intersection and end of each line. Groups of atoms are explicitly drawn. Double and single bonds are drawn the same as shown in a structural formula.

A complete structural formula is used to depict the way atoms are bonded together and show every atom and every bond. Covalent bonds are illustrated as a line between atoms. For example, C—C shows that there is one (—) bond between two carbon atoms, where each bond is a pair of (i.e., two) shared electrons. Some compounds have two bonding sets of electrons, a double bond shown as =. Some molecules even have triple bonds, which involve six shared electrons (≡). Let's use vanillin, the molecule responsible for artificial vanilla extract odor and flavor, as an example of different bonding arrangements. The molecular formula of vanillin is $C_8H_8O_3$. Vanillin contains both single and double bonds between atoms (Fig. 1.10).

A third common way of depicting molecules with lots of carbon atoms is to draw a line structure formula, sometimes referred to as a skeletal formula. Skeletal formulas are useful in that they provide the information contained within a complete structural formula, but they are drawn in a shortcut manner. In a skeletal formula, a carbon–carbon bond is drawn without specifically showing the carbons and hydrogens, but all of the other atoms or groups of atoms are included. In these drawings, a carbon is implied at each bend and end of a line (the line represents the bond of a carbon atom); if any carbon atom doesn't have four covalent bonds, then there are hydrogens present to ensure that each carbon atom is involved in four covalent bonds.

1.3.2   Properties of Covalent Molecules

1.3.2.1   *Functional Groups*   The structure of a molecule defines how it functions in a cell and how a food may taste or react when cooking or baking. Special groups of molecules called functional groups define the behavior of molecules. Functional groups are arrangements of atoms that have specific chemical and biochemical behavior. These groups of atoms are useful to predict and understand properties of

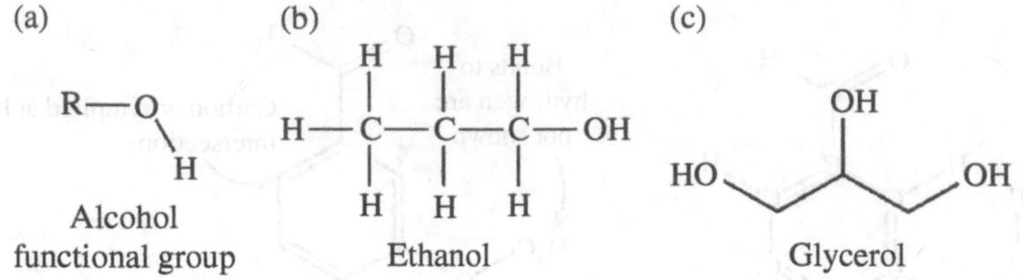

(a)   (b)   (c)

R—O   H   H   H   OH
      \ | | |
       H   H— C— C— C— OH   HO   OH
                       | | |
   Alcohol   H   H   H
functional group   Ethanol   Glycerol

FIGURE 1.11   Alcohol functional groups. (a) The basic convention for alcohol with R as an undetermined carbon group. (b) A structural drawing of the two-carbon ethanol. (c) Glycerol without the hydrogens. At the intersection of each line is a carbon.

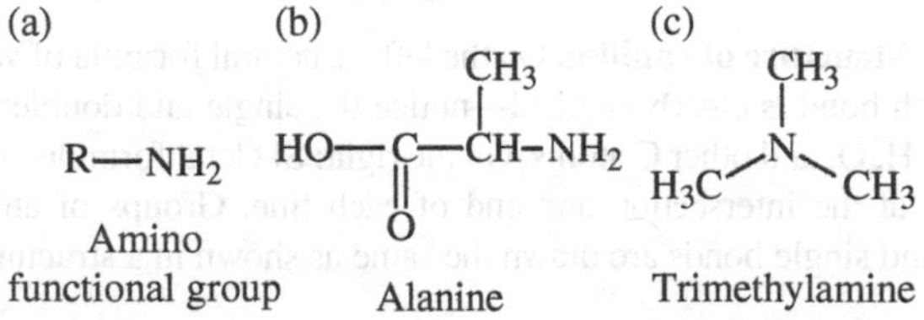

(a)   (b)   (c)

                      CH₃   CH₃
                      | |
     R—NH₂   HO— C—CH–NH₂   H₃C   N   CH₃
     Amino   ‖
functional group   O
                   Alanine   Trimethylamine

FIGURE 1.12   Amino functional groups. (a) The basic convention for amino with R as an undetermined carbon group. (b) L-Alanine, one of the common 20 amino acids used to make proteins. (c) Three CH₃ (methyl) groups bound to central nitrogen to make trimethylamine.

organic molecules and molecules important in food and cooking. Specific functional groups and examples of molecules that are important in food and cooking are shown throughout the book.

Alcohol —OH   An alcohol is the simplest of all functional groups. It is an oxygen atom covalently bonded to a hydrogen atom, often designated as —OH (Fig. 1.11). Sugars, like fructose, have many alcohol groups. Molecules of ethanol and glycerol both contain alcohol functional groups. The —OH plays key roles in allowing these molecules to interact with and dissolve in water. You likely already know a little about or have experienced the use of fructose (honey) and ethanol in food or drinks. Glycerol is a sweet, sticky, and thick compound that is often added to bread, cookies, and cakes to keep them moist. A glycerol molecule also provides the molecular framework for fat molecules.

Amino —NH₂ and —NH₃⁺   A group of atoms containing a nitrogen covalently bonded to hydrogen is called an amine or amino group (Fig. 1.12). Two or three hydrogen atoms can bond to the nitrogen, creating a neutral (—NH₂) or positively charged (—NH₃⁺) group. Amino acids, which combine to make proteins, contain an amine functional group. The molecule trimethylamine provides the unique odor associated with fish.

Saltwater fish contain high amounts of trimethylamine oxide in their muscle cells to counter the high salt content in water balancing the resulting osmotic pressure in the cells of the fish.

Carboxylic Acid —COOH and —COO⁻   The tangy taste associated with a nice cool glass of lemonade or a sour citrus hard candy is provided by carboxylic acids (Fig. 1.13). This functional group consists of a carbon bound to two oxygen atoms, where one of the

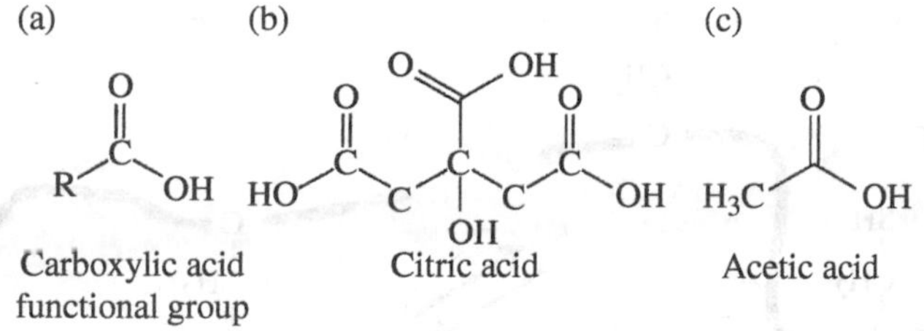

FIGURE 1.13 Carboxylic acid functional groups. (a) The basic convention for carboxylic acid with R as an undetermined carbon group. (b) The sour tasting weak acid citrate with three carboxylic groups. (c) The molecular structure of acetic acid whose household name is vinegar.

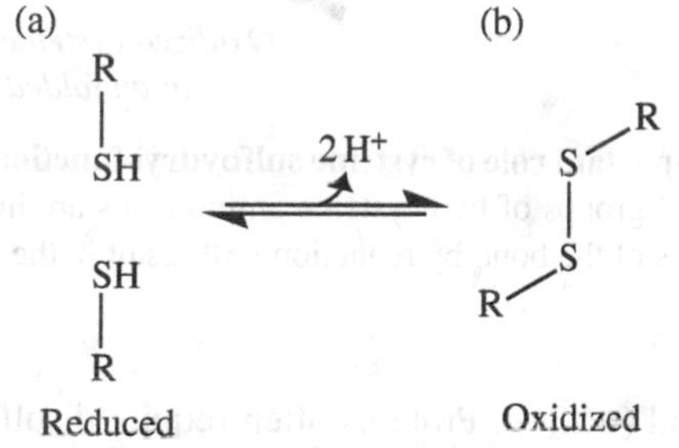

FIGURE 1.14 Thiol functional groups. (a) The basic convention for a reduced sulfhydryl with R as an undetermined carbon group. (b) The change in oxidation state of a sulfhydryl group from reduced (R—SH) to oxidized (R—S—S—R).

oxygen atoms may also be bonded to a hydrogen ion. Thus, it is designated as R—COOH or R—COO⁻. Why is the hydrogen sometimes absent? Due to oxygen's affinity for electrons and hydrogen's lack of affinity electrons, the bond between the hydrogen and oxygen in carboxyl groups is easily broken, yet the oxygen keeps the electron from the previously shared covalent bond, yielding a carboxyl group that lacks a hydrogen ion (H⁺) and maintains a negative charge (R—COO⁻). The R—COO⁻ is a weak organic acid, hence the name carboxylic acid. Carboxylic acids are found throughout food and cooking, most notably in citrus fruits (citric acid) and vinegar (acetic acid). The acid component of these foods stimulates the sour taste receptor on our tongues giving these foods a sour taste. An example is malic acid. Malic acid is an organic acid that is found in unripe fruit like green apples and gives the food a sour green apple flavor.

Sulfhydryl (Thiol) Group —SH Sulfur atoms that are contained within a molecule have a very important and diverse role in cooking and baking, depending upon its bonding partners. The amino acid cysteine has an —SH group. When sulfur is bonded to a hydrogen atom, we call the functional group a sulfhydryl or thiol group and designate it as —SH (Fig. 1.14). Most proteins found in plant and animal tissues have various amounts of cysteine and therefore sulfhydryl groups. However, the sulfur in cysteine does not have to remain bonded to a hydrogen; it can also be bonded to another sulfur atom (often found in a different cysteine amino acid) when a chemical reaction, called an oxidation/reduction reaction, occurs, resulting in the formation of a

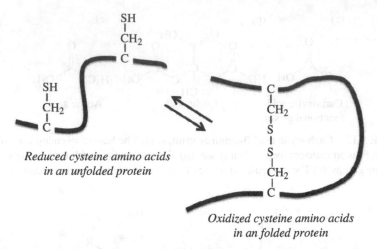

*Reduced cysteine amino acids
in an unfolded protein*

*Oxidized cysteine amino acids
in an folded protein*

FIGURE 1.15   The important role of cysteine sulfhydryl functional groups. When proteins are folded, the sulfhydryl groups of two cysteine amino acids are involved in maintaining the shape of the protein. Loss of the bond by reduction will result in the loss or denaturation of the shape of the protein.

covalent disulfide bond (—SS—). Proteins often require disulfide bonds to be present to keep the protein folded in a functional, native state, and in solution (Fig. 1.15). However, because an S—S bond is weaker than a C—C bond, heat can break disulfide bonds. The more disulfide bonds, the more heat that is required to break them and unfold the proteins. Some compounds will change the "oxidation state" of disulfide bonds and will contribute to the denaturing of the protein. In cooking, we visualize this process of protein unfolding when we cook eggs. Eggs have several different kinds of proteins. Those found in egg whites have relatively few disulfide bonds, and low levels of heat cause the proteins to denature. You observe this when the egg whites change from clear to white and "cook" in your warm skillet. In contrast, proteins found in the egg yolk have more disulfides and require a higher temperature to unfold and "cook" these proteins. Disulfides also play an important role in baking and wheat.

1.3.3   Gluten, Fumaric Acid, and Tortillas

A handmade tortilla is a simple food made from wheat flour, water, shortening, and salt. Wheat flour has two gluten proteins that include large numbers of cysteine (sulfhydryl-containing) amino acids. Once processed, the gluten proteins link together via disulfide bonds providing an elastic, chewy texture to the tortilla. Unfortunately, machine processing of tortillas creates excess links between the proteins resulting in a rubbery, less than satisfying tortilla. The molecule fumaric acid has a carboxylic acid functional group used to overcome this problem. Fumaric acid is naturally made in tissues of plant and animals. Fumaric acid acts as a reducing agent, keeping the —SH groups from forming disulfide bonds (S—S) and decreases the pH level of the flour dough, defeating the toughening disulfide bonds of gluten (Fig. 1.16).

(a) (b)

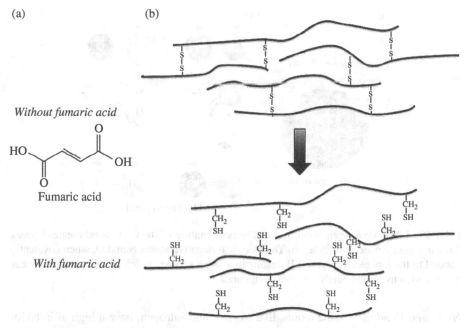

Without fumaric acid

Fumaric acid

With fumaric acid

FIGURE 1.16 Soft tortillas. (a) The molecular structure of fumaric acid. (b) The impact of fumaric acid on disulfide formation in gluten found in tortillas. The addition of the acid keeps the disulfide bonds in the reduced state limiting the cross-linking of gluten for a softer chewier food.

This results in softer more machinable tortilla dough. In addition to limiting the gluten cross-linking, fumaric acid also acts as an antimicrobial agent and does not easily bind to water from the atmosphere increasing the shelf life of the food from a few days to over 2 months.

Now you know something about the individual molecular components of food molecules; however that doesn't provide the full picture of what happens when a protein clumps when eggs are cooked, when fat globules curdle together when making cheese, or when flour is added to broth to make a thick gravy. In all of these processes and many others, it is the interaction of different molecules that causes the cooking or baking process to take place.

1.3.3.1 Interaction of Food Molecules: Intermolecular Forces Forces that attract or repel two different molecules are called intermolecular forces. There are a number of different kinds of these forces, with different strengths and properties, but a key concept is that intermolecular forces are not bonds that hold atoms together. Intermolecular forces are weaker interactions that bring molecules together or keep them apart. Once you know some details about intermolecular forces, you will have a better understanding of how to make a foam or emulsion, why adding lime juice slows down the browning of avocado, and why destroying the structure of protein makes a solid in your cooked scrambled eggs.

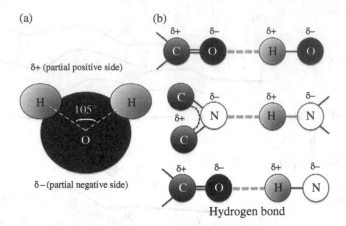

FIGURE 1.17 Hydrogen bonding. (a) The polar nature of the O—H bond creates a weak dipolar across a water molecule. (b) The electrical negative atoms N and O, when covalently bonded to the less electronegative H, results in a weak charge, which will form special weak bonds called hydrogen bonds between compounds.

Hydrogen Bonding Some atoms, like oxygen and nitrogen, have a high affinity for electrons, while other atoms, like hydrogen, have a low affinity for electrons. When atoms with differing affinities for electrons are bonded to one another, the high electron affinity atom (i.e., nitrogen or oxygen) pulls on the shared electrons more than the low electron affinity atom. Since electrons are negatively charged, the oxygen or nitrogen atoms become slightly negative, indicated by a partial charge ($\delta-$). At the same time, the hydrogen atom that has "lost" some of the shared electrons has a very weak positive charge ($\delta+$). The resulting partial positive and negative portions of the atoms can become attracted to and attract partially positive and negative atoms from nearby molecules or even within the same molecule. The resulting interaction between a partial positive component of one molecule and a partial negative component of another molecule is called a hydrogen bond (Fig. 1.17). It is called a hydrogen bond because of the involvement of hydrogen as the low electron affinity atom; the high electron affinity atom is typically nitrogen or oxygen in foods and cooking (Fig. 1.18).

Given the hydrogen bonding potential for water, as H_2O, and the presence of water in many foods and cooking processes, hydrogen bonding is a very important inter-molecular interaction. Let's look at the example of starch. Anyone who has made gravy with cornstarch has experienced the frustration of adding hot water to dried starch and the resulting blob at the bottom of the dish. As we will learn later, starch is a long polymer of glucose molecules (from hundreds to thousands of glucose molecules) resulting in tens of thousands of —OH groups (Fig. 1.18). That is a *lot* of alcohol functional groups! In fact, you may be thinking this is good because then water molecules can interact with the starch via hydrogen bonding.

As predicted, when water is added to dried starch, the water molecules form inter-molecular interactions with the many alcohol groups (—OH) on starch. However, there are so many —OH functional groups on the surface of starch granules that the

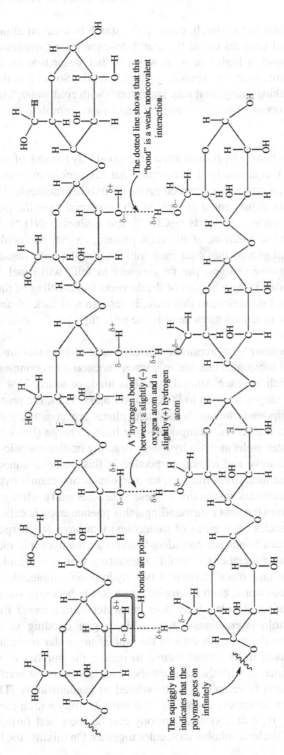

The dotted line shows that this "bond" is a weak, noncovalent interaction.

A "hydrogen bond" between a slightly (−) oxygen atom and a slightly (+) hydrogen atom

O—H bonds are polar

The squiggly line indicates that the polymer goes on infinitely

FIGURE 1.18 **Hydrogen bonding in starch.** Long glucose polymers of starch form tangles of hydrogen-bonded strands, which serve to thicken a gravy. Think of a tangle of yarn.

water binds too tightly to the starch, causing the starch to form an almost solid gel. Additional structural changes cause the starch to expand and eventually contract, which happens at such a high rate with warm or hot water that an impenetrable blanket of water forms over the expanding starch granule. So what is the take-home message? When making gravy, first mix your starch with cold water. The cold water slows down this process to allow a controlled and more complete hydration of the starch granules.

Electrostatic Interactions   Opposites attract is a good way to think of the interaction between molecules that are charged. Molecules that have one or more charged atoms will be attracted to an oppositely charged group on another molecule. Proteins have many different kinds of functional groups, in which several have the potential to be charged, including carboxylic acids ($-COO^-$) and amines ($-NH_3^+$). Electrostatic interactions govern the behavior of the milk protein, casein. Molecules of casein have carboxylic acid groups that coat each milk fat droplet with negative charges. Because of the negative charges, the fat droplets in milk will repel one another, reducing the possibility of aggregation of the droplets and curdling of the milk. Thus the key electrostatic interaction, in this case, is repulsion or lack of an interaction, which allows the fat to remain suspended in the milk liquid.

Hydrophobic Interactions   Hydrophobic interactions are forces that are of particular importance for food molecules that are in a water (aqueous) environment. Plant and animal tissues are rich in water. Animal muscle is made of nearly 70% water, while plant water content ranges from 75 to 90% of total mass. Thus, the proteins, sugars, fats, and other compounds in our bodies and plants are constantly exposed and surrounded by water molecules. Compounds that have a charge (full or partial) will interact with the water molecules via hydrogen bonding or electrostatic-like interactions; they easily dissolve and remain suspended in this water or aqueous environment. However, some molecules, like fats, have no charge and cannot hydrogen-bond or be involved in electrostatic interactions. These molecules tend to clump or aggregate together to "hide" from the water surroundings; this phenomenon is called the hydrophobic effect. Molecules (or regions of molecules) that have no charge and do not participate in hydrogen bonds are considered nonpolar; the hydrophobic interaction brings these molecules together to "avoid" interacting with water molecules. Why does this interaction take place? Consider two hydrophobic molecules (Fig. 1.19). When first placed into water, each hydrophobic molecule becomes surrounded by a shell or cage of water molecules. Why does the water form a cage? Because there are minimal favorable interactions (such as hydrogen bonding or electrostatic interactions) between the hydrophobe and the water, any water molecule that does interact organizes itself in the caged format to reduce the number of water molecules that have to interact with the hydrophobe. This allows more water molecules (in the entire solution) to remain in a disordered or random array. The scientific term for disorder or randomness is entropy. The more entropy within the system, the better. Thus, in this type of a system, entropy can be increased further through a "clumping" of all of the hydrophobic molecules together. On mixing, the hydrophobic

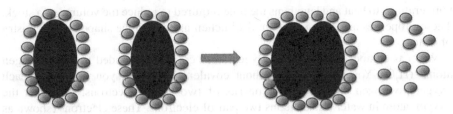

FIGURE 1.19 Hydrophobic effect is driven by entropy. Hydrophobic compounds shown in black are separate on the left. Because water (shown as small gray globes) cannot hydrogen-bond to the water-fearing compound, they are forced to form a rigid shell of water where the water hydrogen bonds to itself. If the two hydrophobic compounds come together, there are less is a smaller surface area and less water needed to form the cage around the combined molecules. There is more order on the right than on the left. Entropy drives the hydrophobic molecular interaction.

molecules begin to interact with one another (via a hydrophobic interaction), thus reducing the number of water molecules that are caged around the hydrophobes. Overall then, there is more randomness or disorder within the system. Thus, the hydrophobic interaction isn't so much about an attraction of the hydrophobes (although there is a transient attractive force called a van der Waals force that can occur); it is more about the increase in disorder or entropy of the system when the hydrophobic molecules come together.

We have been using a lot of scientific terminology in our discussion to this point. It might be useful to clarify some of that terminology and summarize our discussion of intermolecular interactions here. The terms hydrophobic and nonpolar, often used interchangeably, describe molecules or components of molecules that possess only carbon and hydrogen atoms. Polar molecules or regions possess partial charges because they contain nitrogen and/or oxygen atoms. Charged molecules or regions possess full charges, because of their charged nature; they behave similarly to polar molecules. In which types of intermolecular interactions do each of these species participate? The phrase "like dissolves like" is useful here. Nonpolar molecules interact with other nonpolar molecules in hydrophobic interactions. Nonpolar molecules (like olive oil) do not interact well with water (a polar molecule); thus olive oil and water do not mix. Given that water is polar, polar molecules will interact with water via hydrogen bonding and electrostatic-like interactions and will dissolve. A good example of this is sugar. The alcohol functional groups (—OH) are able to hydrogen-bond and will easily dissolve in water.

1.3.4 Molecules in Motion: Water

As already mentioned, water is a major component of food: a cucumber is 95% water, an avocado is 73% water, and a chicken breast is 69% water. Thus, the properties of water must be considered when thinking about food and cooking. For example, water expands when freezing, which impacts the texture of and types of foods that can be frozen and thawed without damage, and water has a high heat of vaporization and fusion, which makes water-rich foods take longer and higher temperatures to heat and lengthens the time required to reduce the volume of a stock. These properties and their impact in the kitchen are due to the shape and chemistry of water.

Water is composed of a single oxygen atom covalently bonded to two hydrogen atoms (H_2O). You already know about covalent bonds, so you know that each oxygen–hydrogen covalent bond consists of two shared electrons. However, the oxygen atom in water also contains two pair of electrons. These electrons (shown as a pair of "dots" or ".." in Figs. 1.7 and 1.8) cause the hydrogens to bend away from the electrons. Thus, these elements give the water molecule a bent rather than straight structure (Fig. 1.20). How does this shape impact the properties of water? You already know that the —OH bonds make every water molecule polar, where the hydrogen atoms have a partial positive and the oxygen atom has a partial negative charge. The "v" shape actually adds to the polar nature of water, as it allows other polar molecules (up to four) to more easily access and interact with water via hydrogen bonding; oxygen has a strong affinity for electrons, called electronegativity, while hydrogen does not have as high of an electronegative hold on these shared molecules. This results in an unequaled sharing of the common set of electrons between oxygen and hydrogen. The result leaves one atom (hydrogen) with a slightly positive charge from the protons in the nucleus and the oxygen with the unbounded pair of electrons on the oxygen atom with a slightly negative charge. The shape and arraignment of electrons (free and those in bonds) make water a "polar" molecule. These are not the full charges found in ions, but the charges are strong enough to attract other partial or fully charged compounds.

How do you get four hydrogen bonds from one water molecule? The partial positive charge on each hydrogen atom attracts a partial negative charge oxygen atom in a different water molecule (this makes two). The partial negative charge oxygen in the same water molecule can attract partial positive charge hydrogens in two different water molecules. Thus, each water molecule can hydrogen-bond to four different water molecules at one time! This occurs when water freezes, forming a solid lattice and hydrogen-bonded network of water molecules. Water molecules can also interact with fully charged ions. In this case, the water molecule surrounds the charged ion with the appropriate partial charge of the water molecule, forming a jacket or hydration shell, around the ion; this allows the ion to dissolve in the aqueous solution (think about dissolving a teaspoon of table salt in water). Other polar covalent molecules can hydrogen-bond and interact with ions in a similar manner. However, when other molecules are mixed into water, these "nonwater" interactions interfere with water's ability to interact with itself. Thus dissolved particles in water depress the point at which water freezes.

Another important property of water is that, in food, some of the water is tied up in solid hydration shells (and is held rigidly in its hydrogen bonding interactions), while other water molecules are not close enough to be held in place by this water and is considered bulk or free water. Ice cream has sections of both liquid (free) and solid (rigid) water molecules. The liquid water is filled with dissolved proteins, sugars, and salts, while the solid water is frozen in the form of ice crystals. The moisture content of food is a measure of free water; it is measured in a term called water activity (a_w)

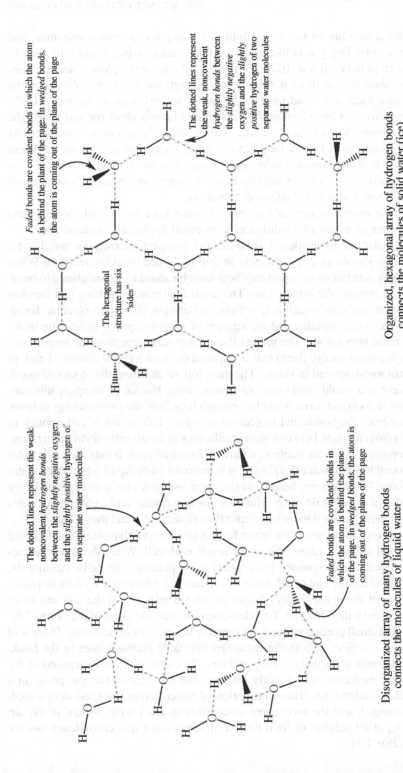

Faded bonds are covalent bonds in which the atom is behind the plant of the page. In *wedged* bonds, the atom is coming out of the plane of the page

The dotted lines represent the weak, noncovalent *hydrogen bonds* between the *slightly negative* oxygen and the *slightly positive* hydrogen of two separate water molecules

The hexagonal structure has six "sides"

Organized, hexagonal array of hydrogen bonds connects the molecules of solid water (ice)

The dotted lines represent the weak, noncovalent *hydrogen bonds* between the *slightly negative* oxygen and the *slightly positive* hydrogen of two separate water molecules

Faded bonds are covalent bonds in which the atom is behind the plane of the page. In *wedged* bonds, the atom is coming out of the plane of the page

Disorganized array of many hydrogen bonds connects the molecules of liquid water

FIGURE 1.20 Water molecule interactions in liquid and solid form. The hydrogen bonds in liquid and solid form help it both form a cohesive liquid and a solid that is highly ordered and less dense than liquid water.

and provides a measure of the accessibility of water for microbes, enzymes, and chemical reactions. Dry goods like ground coffee beans, or powdered milk, and or even potato chips have a low a_w (0.2–0.08). Foods like those with low water activities have a long shelf life, as there is little or not enough water for microbe microbial survival. Fresh meat and bread have a high water activity with a_w values of 0.95–0.9765, respectively. Correspondingly, these foods have a short (or no) shelf life. Interestingly honey has a long shelf life, with a relatively moderate water activity of 0.6. In honey, which lasts a long time at room temperature, most of the water molecules are hydrogen bonded to the high concentration of sugar molecules (like fructose). This results in a water activity of 0.6 a_w which means there is not enough free or bulk water available for microbes to survive.

What do the intermolecular interactions of water have to do with cooking and foods? The state of matter of a substance is governed by the intermolecular interactions of the individual molecules. A solid has the strongest interactions. Solid water molecules, also known as ice, participate in four hydrogen bond interactions. When the ice changes state (melts or vaporizes), heat must be added to the substance to break some of the intermolecular interactions. The transition between melting and freezing or vaporization and condensation is a balance between the intermolecular forces holding the molecules together and the amount of energy required to free the molecules from these interactions. The greater the number and strength of the intermolecular forces, the more energy (heat) that is required to cause a phase change. Think of two blocks of wood covered in Velcro. The more Velcro attracting the blocks of wood, the more force you would need to use to separate these blocks. If we apply this concept to water, in its liquid form, water has enough heat from the surroundings to bend, rotate, and vibrate the bonds and molecules of water. This action is just enough to cause the hydrogen bonds between water molecules to continually form and reform, thereby preventing the most stable structure of four hydrogen bonds for each water molecule found in solid water (ice). As heat is removed from liquid water, each water molecule forms four hydrogen bonds and the water molecules form an expanded cage or lattice relative to the liquid water. Thus, ice has fewer water molecules per unit area than liquid water, causing it to be less dense than liquid water and thus float.

The expansion that occurs when water freezes to ice is problematic when freezing food. Plant and animal tissues have water inside each cell. When the water freezes and forms crystals, the expanded ice crystals often puncture the cells, causing the tissue to become mushy and expel water when thawing. Moreover, with large pieces of food, some of the water doesn't freeze because it mixes with the salt and other molecules present within the cell. This decreases the freezing point of that area of the food wherein a small portion of the food remains liquid, leading to mush. There will be more on this subject in both the meat and vegetable chapters later in the book. Freezer burn is another equally vexing problem. When frozen food is exposed to dry cold air, water molecules can directly escape from solid ice to the gas phase in a process called sublimation. The sublimation of water leaves the food dehydrated, the cells damaged, and the food more susceptible to react with oxygen in the air. The resulting dried patches of food have a different color and consistency but are safe to eat (Box 1.3).

BOX 1.3 MODERNIST COOKING: SOUS VIDE

It takes a lot of heat to turn liquid water to gaseous water (steam), again because of the strength of hydrogen bonding. Heat adds energy to the water molecules, giving them more kinetic activity until the molecules are free from all or almost all of the hydrogen bonds, allowing for escape into the gas phase (steam water). The energy required to vaporize liquid water is called the latent heat of vaporization; this is what allows foods to cook at or near 100°C (212°F) as long as there is considerable water content in the tissue. Simmering large pieces of meat in water or stock allows the cook to keep a constant temperature; the water stays near boiling point because any excess energy (heat) is being spent as water molecules escape the water phase into the vapor phase. Modernist cooks also called molecular gastronomists use a method called sous vide, French for cooking under vacuum. Imagine a fish with a large midsection and long tapering thin end. The thinner parts of the fish would reach a higher temperature earlier while cooking than would the thicker middle portion of the fish. Sous vide-style cooking allows for a precise, even temperature maintained throughout the food. This means the food is cooked to the same temperature in the core of the food without excess temperature at the surface of the food. Essentially, sous vide cooking involves placing a food item and seasoning into an airtight sealable plastic bag, removing the air, immersing the bag in a controlled water bath, and precisely heating the food to a controlled temperature (Fig. 1.21). This style of cooking allows the entire food item, regardless the shape or different size, to be

FIGURE 1.21 Sous vide cooking. Using a controller as shown here to maintain a narrow temperature range while circulating water, food in a bag can be cooked to a precise heat throughout the food. Photo credit Jeff Rogers.

cooked to a specific temperature. Sous vide is more about controlling the temperature in a water bath than removing (vacuum) the air in the bag. Today many sous vide cooks do not use a vacuum but instead limit air using ziplock bags. This technique results in a less dry, more consistently cooked food. Most foods are cooked in the bag to include juice or liquids. The result is a constant exposure to a less dry consistency and less oxidation in heated air conditions.

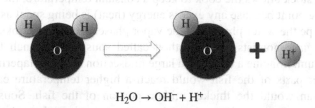

$$H_2O \rightarrow OH^- + H^+$$

FIGURE 1.22 Dissociation of water. Will dissociate into the OH$^-$ (hydroxyl) and H$^+$ (proton) but only a small amount of the water molecules will do this.

1.3.5 Acids, Bases and pH

Water has another important feature that impacts cooking. One of the hydrogen atoms of a water molecule can separate (also known as dissociate) from the rest of the atoms. This produces two new molecules that are charged, a positively charged hydrogen ion called a proton (H$^+$) and a hydroxide ion (OH$^-$) (see Fig. 1.22).

HINT! Don't confuse the hydroxide ion (OH$^-$) with the alcohol functional group (—OH). A hydroxide ion is charged and is not covalently bonded to any other atoms, while an alcohol is bonded to a carbon atom and is uncharged.

However, in pure water, only a very small fraction of the water molecules are dissociated into protons and hydroxide ions, approximately two out of every billion (10^9) molecules. However, even with this small fraction, this dissociation plays a significant role in our lives and our food.

The concentration of protons in a solution is a measure of acidity or basicity. A Danish biochemist, Søren Sørensen working at the Carlsberg Laboratories in Copenhagen, invented a scale to measure acid levels/proton concentration while he was studying proteins, enzymes, and yeast involved in making Carlsberg beer. This scale is called pH: "p" stands for *puissance* (French) or *potenz* (German), both words translating to power. Thus, pH stands for the power of hydrogen and is used to determine the acid and base content of a substance (Fig. 1.23).

The pH scale is a measure of the balance of both protons and hydroxide ions; it ranges from the most acidic (pH = 0; 10^1 H$^+$ and 10^{-14} OH$^-$ mol/l concentration) to the most basic (pH = 14; 10^{-14} H$^+$ and 10^1 OH$^-$ mol/l concentration). A pH of 7 is considered neutral and occurs when H$^+$ and OH$^-$ ions are at the same concentration (neutral solutions are not acid-free, but have an equal number of protons and hydroxide ions). Substances with a pH higher than 7 have more hydroxide ions relative to protons;

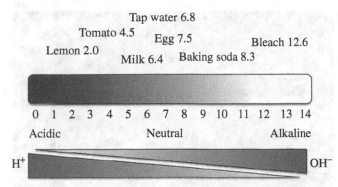

FIGURE 1.23 pH scale and foods. The pH scale from low pH (High H^+ concentration) to high pH (low concentration of H^+) and a few common food and household items.

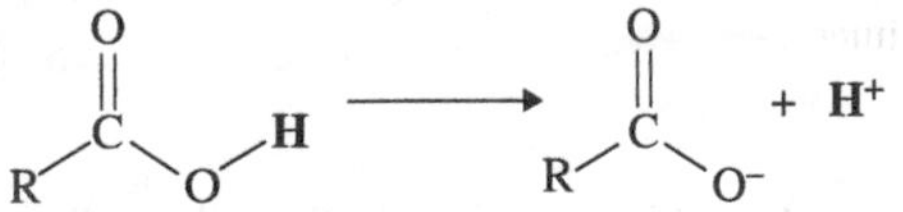

FIGURE 1.24 Weak acids. A weak acid such as vinegar containing a carboxyl functional group will weakly ionize to form a charged COO– group and protons (hydrogen ions). Unlike a strong acid such as HCl, only a fraction of the weak organic acids will react to generate a free proton.

these solutions are considered basic or alkaline. Foods or drinks with a pH less than 7 are considered acidic and have a higher concentration of protons relative to hydroxide. Acidic foods typically range in pH from 2 to 4, while the pH of more basic foods may approach 8 or even 9; most foods are neutral to acidic. Although pure water has a neutral pH, tap water contains minerals and gases that can cause the pH to range between 6 and 8. The pH of most living cells is 7.2 and the pH of human blood ranges between 7.35 and 7.41.

While pH is a measure of how many protons and hydroxide ions are in solution, in foods, the definition of acids and bases is expanded to include consideration of pH or proton concentration, taste, and feel. Acidic food and drinks have those compounds that have a sour taste. The term *acid* is Latin for *acidus*, meaning "sour or tart." Bacterial contamination of food reduces pH and produces vinegar making the food taste sour. The Latin term for vinegar, *acetum*, is related to the *acidus* and has been around since antiquity to describe our food and drink. A more technical scientific definition of an acid is a compound that causes an increase in proton concentration in water. The increase in proton concentration is due to release of a proton or by reducing the concentration of hydroxide ion ($H_2O^- \rightarrow H^+ + OH^-$). For many acidic foods and drinks, the acidity comes from the presence of weak organic acids such as citric, malic, tartaric, and acetic acids. These are considered weak acids because only a fraction of the compounds will dissociate (also called ionize) to generate a free proton (Fig. 1.24). Citric fruits include many kinds of organic acid molecules containing one, two, or even three carboxylic functional

FIGURE 1.25   Using weak acids to make ceviche. Raw shrimp is made tender and tangy by weak acids in citrus juices.

groups; each of these carboxylic groups can dissociate to generate a proton into solution increasing the acidity (by increasing the H^+ concentration). There are many examples of these weak organic acids in food. Sour hard candies are coated with two organic acids: citric and malic acids. Grapes, lemons, and limes include tartaric and citric acids. Lactic acid, another carboxylic acid containing organic weak acid that is produced by yeast, helps create a tart flavor in cheese. Ceviche is a Spanish and South American dish that contains raw or partially cooked seafood treated with carboxylic acid-containing molecules found in lemon or lime juice (Fig. 1.25). Acids from the juices provide flavor, partially break down the seafood protein that tenderizes the meat, and slow the growth of some harmful microbes. Thus, acids in foods, in addition to their role in flavor, also act as preservatives (because microbes don't survive well in acidic environments) and tenderizing and hygroscopic agents. What are hygroscopic agents? Acids help keep dry goods free flowing because many carboxylic acids have a low attraction for water (they are hydroscopic) and inclusion of the acid limits clumping of dry goods like flour, sugar, and other components due to moisture.

A basic solution is something that has a slippery feel or a bitter taste. The slippery phenomenon is due to the fact that strong bases dissolve the oils and fats naturally covering your skin creating a slippery feel. Chemically, bases add hydroxide ions to a system or have different functional groups that react with protons; in both cases, the relative number of protons is reduced within the solution, so the pH increases. The traditional description of a base as a compound that tastes bitter is somewhat misleading. Many compounds that taste bitter, like coffee, unsweetened cocoa, or beer hops have a pH greater than 7 and are basic. However, as you will learn, the bitter taste is not directly due to high pH of the food or drink, but is due to the presence of a compound (often a base) that binds to a taste receptor that signals a bitter flavor to the brain. Many plant materials have a basic or alkali pH and contain toxins. Recognition of a bitter taste (which might stop an organism from eating the toxic plant), although an evolutionary benefit, is not necessarily an appropriate modern definition or descriptor for a base.

1.3.6 Macromolecules (Proteins, Sugars, and Fats)

You have likely read about some of the different types of food molecules, like carbohydrates, proteins, and fats, on a nutritional food label. These large biological molecules, proteins, carbohydrates (sugars), and fats are the functional units of a cell and are key components of food and drink. Each is comprised of simple starting building blocks (amino acids, simple sugars, individual fatty acids) that are chemically combined to make a larger, more complex molecule that plays numerous functions within the cell and food.

How are these large molecules made? The chemistry of assembly and disassembly of the complex molecules is perhaps surprisingly similar, even if the details of the molecules are distinct. Polymerization is a process where smaller molecules, called monomers, are chemically combined to produce a larger chain known as a polymer (Fig. 1.26). Starch, a carbohydrate polymer, consists of hundreds or thousands of individual sugar molecules that are connected to produce the final product. Chemically,

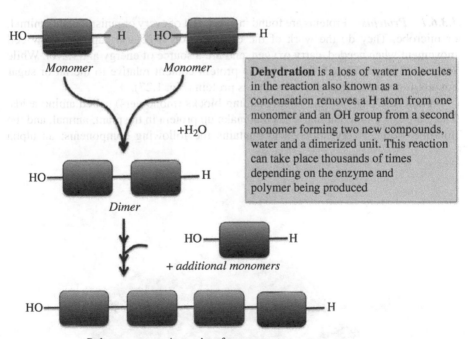

Dehydration is a loss of water molecules in the reaction also known as a condensation removes a H atom from one monomer and an OH group from a second monomer forming two new compounds, water and a dimerized unit. This reaction can take place thousands of times depending on the enzyme and polymer being produced

Polymer—repeating units of monomer

FIGURE 1.26 Growing a biological polymer. Adding individual building blocks called monomers into long strands of monomers creates polymers such as proteins, starches, or DNA molecules. Enzymes that dehydrate the monomers linking them together while generating a water molecule create most biological polymers.

two individual units are combined in a process called condensation or dehydration. In this example (Fig. 1.26), a hydrogen atom is lost from one of the units, and an —OH group is removed from another unit, allowing the two units to combine or condense together in formation of a new covalent bond, resulting in a larger linked growing polymer. The "lost" H and OH reform to generate a molecule of water as a side product. Depending on the molecule being created, this dehydration reaction can be repeated thousands of times making a larger and more complex new molecule. When discussing polymers, we often give the polymer a name that indicates how many monomeric units have been condensed together. When two monomers are linked together, the growing polymer is called a dimer; the sugars lactose, maltose, and sucrose are dimeric polymers, also known as disaccharides.

As you can imagine, polymers can also be broken down, and individual monomeric units can be chemically removed. The process of removing a monomer from a polymer is called hydrolysis; this terms means breaking (lysing) bonds through the addition of water. Hydrolysis reactions happen naturally in living cells, but in cooking, the presence of acid and heat often promote hydrolysis reactions and the breakdown or degradation of complex molecules like proteins, lipids, and sugars. You will see the theme of degradation throughout the book as we describe the production and breakdown of foodstuffs.

1.3.6.1 Proteins   Proteins are found in every cell of every organism, plant, animal, or microbe. They do the work of the cell, provide structural support, allow cell movement when needed, carry oxygen, and are a source of energy and flavor. While some food is considered high or low in protein content relative to the fat or sugar content found in a food, all food contains protein (Fig. 1.27).

Proteins are made of individual building blocks (monomers) called amino acids. There are 20 common amino acids that make up protein in the plant, animal, and the microbe world. Every amino acid contains the following components: an alpha

FIGURE 1.27   Protein-rich foods. Many foods are good sources of proteins including milk, cheese, meat, and fish.

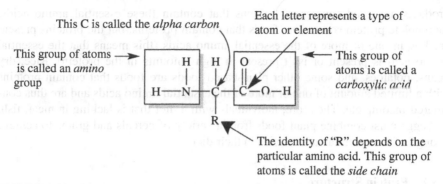

FIGURE 1.28 **The anatomy of an amino acid.** An amino acid has four main components; the R group is the portion unique for each of the 20 common amino acids.

TABLE 1.1 **Essential Amino Acids and Some Food Sources.**

Food	Missing Essential Amino Acid
Eggs, fish, meat	None
Beans	Methionine, tryptophan
Corn	Lysine, tryptophan
Wheat and rice	Lysine
Peas	Methionine
Almonds and walnuts	Lysine, tryptophan

carbon (the center carbon in the structure), an amino functional group, a carboxylic acid functional group, and another group called the side chain (Fig. 1.28). Each of the 20 different amino acids has a unique side chain group; thus it is this group (often chemically notated as an "R group") that makes each of the 20 common amino acids different from the others. Chemically the side groups can be organized by their chemical reactivity/properties. Some are hydrophobic or nonpolar, while others are polar or charged (the term hydrophilic might be used to describe these amino acids because of the side chain interaction with water). Some amino acids can also be described as acidic or basic, while several have other unique chemical qualities. In order to make a protein, individual amino acids are linked together via a covalent bond in a dehydration/condensation reaction that involves the carboxylic group of one amino acid and the amino group of an other amino acid. The resulting covalent bond between the two amino acids is called a peptide bond (Table 1.1).

The human genetic makeup codes for about 20,000–25,000 different proteins. Each of these proteins is made from combinations of the same 20 amino acids. A typical protein is between 500 and 900 amino acids in length. Humans can make some of our amino acids from scratch; however, we need to obtain nine amino acids from our diet since we cannot synthesize them and they are required to make the proteins that our cells need to support life. Proteins that are high in these nine "essential" amino acids are called high-quality proteins. Meat, fish, eggs, and dairy products are good sources of proteins that contain these essential amino acids. Incomplete protein sources are foods that contain proteins, but the proteins present are low in one or more of the essential amino acids (this means that the essential amino acid is present or isn't present in high amounts in the protein). Rice, dry beans, potatoes, and some other plant-based foods are foods that contain proteins with a limited amount of one or more of the essential amino acids and are thus considered incomplete. Therefore, individuals with a diet that is lacking in meat, fish, and eggs must combine plant foods from a variety of cereals and grains to achieve enough of the essential amino acids in their diet.

1.3.7 Protein Structure

Once synthesized from amino acids, each protein molecule folds into a unique and special shape, a shape that is influenced by the order and number of amino acids present. How a protein folds and stays in this folded shape is based on the intermolecular forces imparted by the amino acids and their interactions of the side chains (Fig. 1.29). Many of the hydrophobic amino acids are found clumped or aggregated together on the inside of the protein structure; here, they can avoid the water that fills the cells. Positive and negative charged amino acids attract each other in electrostatic interactions or ionic bonds. Polar amino acids tend to be involved in hydrogen bonding interactions with other polar amino acids or water. The amino acid called cysteine contains sulfur in the side group. If you recall from our discussion of functional groups, two cysteines can come together in a folded protein to form a sulfur–sulfur covalent bond that is known as a disulfide bond. If the amino acids on the surface of a protein are charged, depending on the overall charge (positive or negative charge), proteins can attract other proteins of the opposite charge or repel proteins of the same charge.

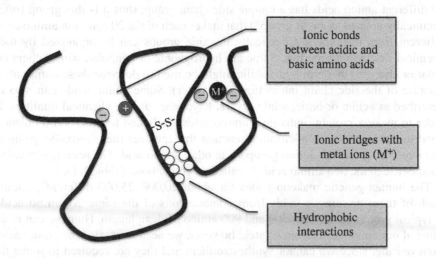

Ionic bonds between acidic and basic amino acids

Ionic bridges with metal ions (M^+)

Hydrophobic interactions

FIGURE 1.29   Forces maintaining the protein structure. The backbone of a protein (shown as a black line) is folded in its native state by the chemical interactions of the side chains.

1.3.8 Protein Denaturation

A protein folded into its functional shape is called a native folded protein. When proteins are subjected to heat, acids, or bases, the intermolecular forces holding the protein in its native structure are broken and the protein unravels; we call these proteins denatured (Fig. 1.30). Ovalbumin, one of the proteins in egg whites, in its native structure is suspended/dissolved in the water present in the egg white, so light can pass through the liquid egg white and it looks transparent or clear. Heat denatures the proteins (ovalbumin and others) in the white, causing the individual protein molecules to aggregate or clump together. As a result, the egg white solidifies, and light is reflected off the egg white, creating a solid white appearance. In contrast, the proteins in yolks denature more slowly (or require more heat to be denatured) due to differences in the amino acids and intermolecular interactions that maintain the protein structure. As an example of this diversity, let's look at the three egg proteins: ovalbumin, conalbumin, and ovomucin. Ovalbumin, the major protein component of an egg at approximately 54%, denatures at 80°C. Conalbumin has fewer intermolecular forces that hold the protein in its native structure and denatures at 63°C. By contrast, ovomucin has a larger number of cysteine amino acids, which can form disulfide bonds that stabilize the protein's native structure and denatures at a higher temperature. The thin part of an egg white has very little ovomucin; you can observe this by watching the thin egg white solidify first when frying an egg.

Acids and bases can change the charge state on the side chain groups of amino acids; a negatively charged carboxylic acid side chain might become neutral (uncharged) in the presence of an acid, while a positively charged amine might become neutral in the presence of a base. This change in charge state also leads to an unraveling or denaturation of protein molecules. As an example, the acid produced

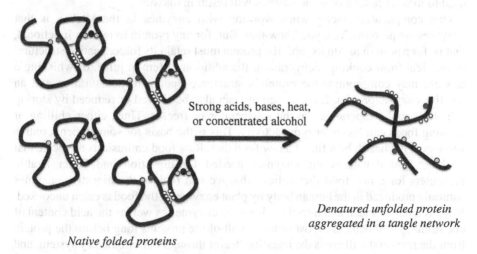

Strong acids, bases, heat, or concentrated alcohol

Native folded proteins

Denatured unfolded protein aggregated in a tangle network

FIGURE 1.30 **Denaturing protein.** Extreme conditions can lead to protein unfolding causing a mesh network of insoluble protein. Cheese curds (from acid) and egg whites (from heat) are two classic examples of denaturing proteins and food.

by lactobacillus bacteria causes negatively charged milk proteins to become neutral. In turn, the milk proteins aggregate and curdle into yogurt or soft cheese.

Proteins can also denature due to agitation or stress. In the kitchen, you can agitate proteins with a whisk (this is what you do when you make meringue). The mechanical agitation of the egg white proteins introduces air into the proteins, causing the proteins to distort, stretch, and denature. The long tangles of denatured proteins then ravel together forming a cage around the air bubbles; the whites expand creating a foam.

1.3.9   Protein Function

Proteins play a diverse role in living cells. One special group of proteins called enzymes aids in chemical reactions. Enzymes reduce the energy needed for a chemical reaction, thus increasing the rate of the reaction. In other words, enzymes serve as biological catalysts. Some enzymes are involved in making new molecules needed for a growing cell, others are involved in breaking down molecules for cellular energy, and others are released from digestive organs to help break down food. Enzymes also play an important role in cooking and baking. Some enzymes chemically cut (or cleave) proteins into smaller pieces. Other enzymes chemically modify sugars or fats, converting them into new compounds with different flavors. For example, aged steaks are enhanced in flavor due, in part, to the enzymes released by dying cells that break down some of the connective tissue, which makes a tenderized meat, and generate flavorful amino acids. Aged cheese contains enzymes that chemically alter the protein, sugars, and fats in the cheese, creating a more mature and stronger tasting food. However, enzymes can also create havoc during cooking and baking. Lysozyme, an enzyme found in egg white, degrades large carbohydrate sugar polymers into much smaller sugar molecules. Thus, the presence of egg white in an egg yolk that is being used to make a custard or some pastries will result in disaster.

One complicating factor when working with enzymes in the kitchen is that enzymes are proteins. Yes, you knew that. But, for any protein to remain functional, that is, for a protein to "do its job," the protein must retain its folded, native structural form. Heat from cooking, dehydration, the addition of lemon juice, or whisking a mixture may compromise the protein's structure, leading to denaturation and an inactive enzyme/protein. Enzyme activity can also be slowed or reduced by storing a food at cool temperatures in the refrigerator or freezer. Thus, either chilling or cooking foods can lessen enzyme activity. This is the basis for some recent controversy over the health benefits of a raw food diet. Raw food enthusiasts advocate that cooking food denatures the enzymes needed for digestion and better health. Promoters for a raw food diet believe that we can replace the digestive enzymes naturally produced in the human body by plant enzymes if the food is eaten uncooked. However, our stomach and intestinal digestive enzymes as well as the acid content of our stomach will denature most or almost all of the proteins long before the protein from the raw food will cross the intestine, travel through our circulatory system, and get to our organs.

Another important role of proteins in cooking is their capacity to hold water. As described earlier, the native structure of most proteins keeps the hydrophobic amino acids tucked away inside the structure where water is excluded, leaving the charged and polar amino acids on the surface of the protein free to form hydrogen bonds with water. When a protein denatures, the hydrophobic portions of the unraveled protein are exposed, driving the protein to interact with other protein molecules instead of water. Thus, native proteins help to retain water in foods, while denatured proteins allow water to be released. This phenomenon plays a critical role in food texture in meat, milk, plant products, and baked goods. Native proteins in meat allow the tissue to remain moist after cutting or grinding. However, once heated, the meat proteins denature and have fewer interactions with water. The juices leaked from a cooked steak are mostly water that is no longer retained by the myosin proteins of meat (colored with other components from the tissue). Resting a steak after cooking allows some of the water to find new interactions with proteins; thus less juice leaks out upon cutting. Later, we will see that starch plays a similar role in holding water-baked goods, keeping them moist.

It may seem like proteins are the only molecular players in food chemistry; however carbohydrates and fats play equally important roles. Let's take a look at the molecular components and cooking characteristics of carbohydrates next.

1.3.9.1 Sugars are Carbohydrates   Sugar, saccharides, polysaccharides, complex carbohydrates, simple sugars, starches, pectins, fiber, and gums all refer to the same family of biomolecules called carbohydrates. Carbohydrates, as the name suggests, contain carbon, hydrogen, and oxygen atoms, often arranged in a manner in which a chain of bonded carbon atoms is each bonded to —OH and —H groups or a single oxygen via a double bond (C=O). A carbon that has a double bond to an oxygen atom is a functional group called carbonyl. Because of the presence of multiple alcohol groups on carbohydrates, carbohydrates are soluble in water; the alcohol groups are also important in making the carbohydrate sweet. The carbonyl group is key for the browning action that occurs when sugars are combined with protein, as observed when a cooked pie crust that was coated with an egg wash.

The simplest carbohydrates are the monosaccharides (Fig. 1.31). Monosaccharides are linear or single ring structures. Any monosaccharide can be described structurally by a linear or ring structure the carbonyl group has reacted with an —OH from another carbon forming a ring structure. There are a multitude of monosaccharide sugars; however only a few are involved in day-to-day cooking and baking.

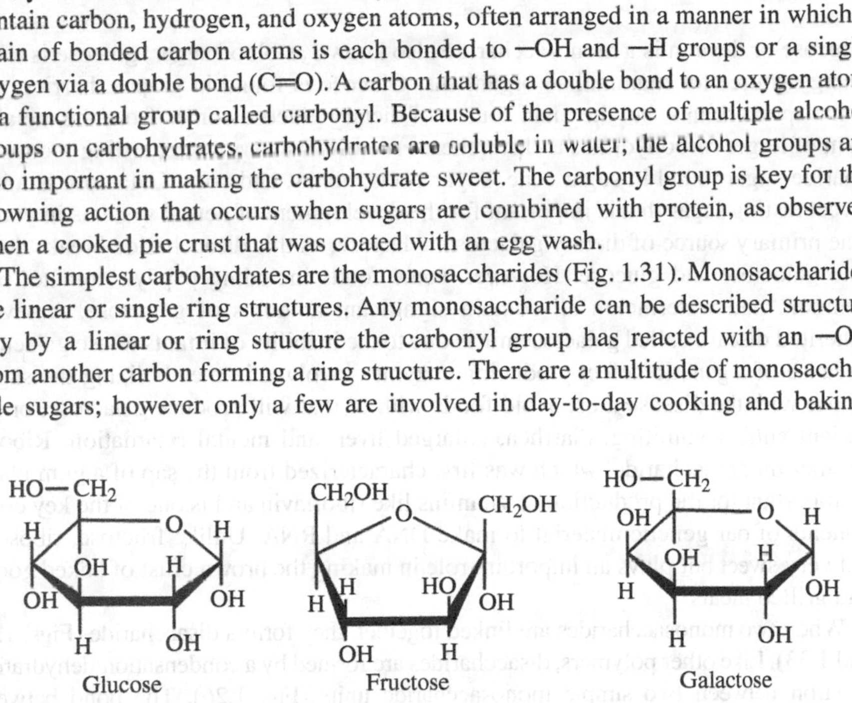

FIGURE 1.31 Common monosaccharides. Simple sugars (monosaccharides) found in food and drink.

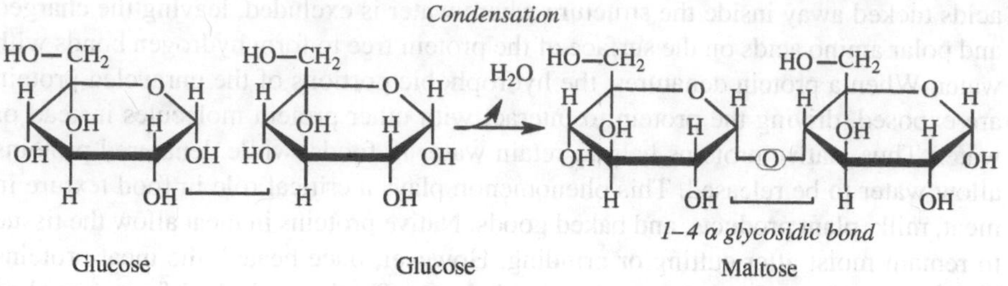

FIGURE 1.32 **Disaccharides and glycosidic bonds.** Maltose shown on the right is a disaccharide made of two glucose molecules held together by a special bond called a glycosidic bond.

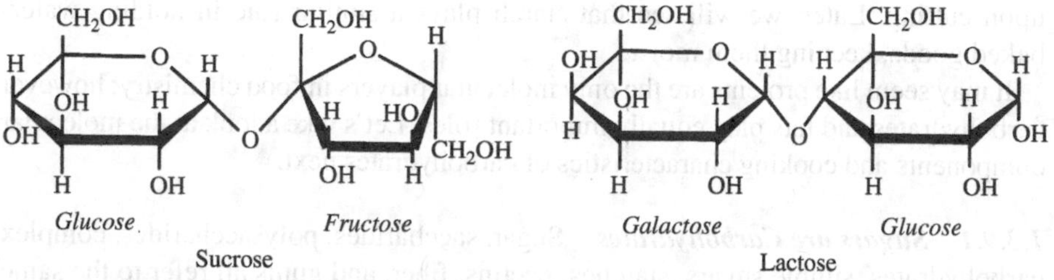

FIGURE 1.33 **Common disaccharides.** Disaccharide sugars found in food and drink.

Glucose or dextran is a monosaccharide also known as blood sugar; glucose is a key energy source for mammalian organisms. Glucose is found in grapes, berries, and some sports drinks, but the main source of dietary glucose comes from the metabolism of larger carbohydrates and starches. Fructose is a very sweet tasting monosaccharide sugar found in sugar cane, sugar beets, honey, and corn. Galactose is a less sweet tasting sugar that is important for the development of neural systems in youth. The primary source of dietary galactose is lactose, as this disaccharide breaks down into galactose and glucose. However, some foods including papaya, tomato, persimmon, and watermelon all contain significant amounts of galactose. A severe inherited disease called galactosemia is due to the inability of infants to use galactose because of a genetic defect producing enzymes involved in metabolizing the sugar. Those with the disease must avoid foods with or that will produce galactose, or the patient suffers vomiting, diarrhea, enlarged liver, and mental retardation. Ribose, another monosaccharide, which was first characterized from the sap of a gum plant, is important for the production of vitamins like riboflavin and is one of the key components of our genetic material to make DNA and RNA. Unlike fructose, ribose is not very sweet but plays an important role in making the brown crust of baked goods and grilled meats.

When two monosaccharides are linked together, they form a disaccharide (Figs. 1.32 and 1.33). Like other polymers, disaccharides are formed by a condensation/dehydration reaction between two simple monosaccharide units (Fig. 1.26). The bond between the linked sugars is called a glycosidic bond. The bond can be formed in two configurations: an α-glycosidic bond and a β-glycosidic bond. α-Glycosidic bonds are formed when the oxygen atom between the monosaccharides falls below the carbon atoms. β-Glycosidic bonds happen when one of the linking carbon atoms falls above the oxygen atom in the bond. The α- or β-character of the glycosidic link is important to the function and structure of the saccharide polymer. Disaccharides cannot be used by the body in the disaccharide form for energy; thus they must be metabolized into individual monomer sugars or monosaccharide units for biological use. However, they do play an important role as key ingredients in many cooking and baking recipes. Dietary examples of disaccharides include lactose and sucrose. Joining glucose and galactose together via a β-glycosidic bond makes lactose, often called milk sugar. You will learn more about lactose in the chapters on milk, cheese, and metabolism/fermentation. Sugar beet and sugar cane plants produce sucrose, often called table sugar, as the plant cells trap energy from the sun by chemically combining carbon dioxide and water. A molecule of sucrose is formed when glucose and fructose monosaccharides are linked together in an α-glycosidic bond. If purified, there is no difference in sucrose between the two sources. Maple syrup and sorghum also contain sucrose. The conversion of sucrose into its single sugar components, glucose, and fructose, is catalyzed by enzymes secreted into human saliva. The enzyme invertase splits the indigestible sucrose into the usable fructose and glucose in the mouth. Boiling sucrose for an extended time can also hydrolyze or break apart the two simple sugars from sucrose. This process is enhanced in the presence of acids like lemon juice or tartaric acid. Inverted sugar is just sucrose that has been reduced into monomer sugars and should no longer be considered sucrose. This is a trick some winemakers use when fortifying their grapes with an additional boost of carbohydrates. Bakers sometimes use inverted sucrose (called invert syrup) because of the increased sweetness of fructose compared with sucrose.

Once you begin to link more than two sugars together, the terminology used to describe a sugar or carbohydrate may be unclear to the novice. What exactly do we mean at a molecular level when using the terms simple sugars, oligosaccharides, or complex carbohydrates? Now that you know a little more about the molecular structure of mono- and disaccharides, these terms are easily clarified by examining the number of monomer saccharides incorporated into the molecule. Both monosaccharides and disaccharides are considered "simple sugars"; these sugars are easily absorbed by the body and are readily available as an energy source or building block for other biomolecules. Carbohydrates that contain less than a hundred monosaccharide units, called oligosaccharides, are found in dried beans, peas, and lentils. These molecules are poorly digested by the human body and typically pass through your digestive system unaltered. However, once in the gut, your intestinal bacteria metabolize the oligosaccharides and produce gases. Complex carbohydrates consist of even longer chains of monosaccharide units (on the order of hundreds or thousands of the monosaccharide building blocks). The individual monosaccharide units in complex carbohydrates often participate in multiple glycosidic bonds, which gives them a structure that is much more complex than the shorter-chain carbohydrates. Moreover, like proteins, these long polymers bind tightly to water through hydrogen bonds; thus they readily absorb water. Complex carbohydrates can be placed into two major classes. Dietary or nutritionally unavailable complex carbohydrates include gums, fibers, and pectins. These complex carbs are certainly important in cooking, even though they lack nutritional value (think about the pectin that you might add to fruit that thickens a jelly). Other complex carbohydrates are nutritionally important, such as starches. Let's look more in depth at both types of complex carbohydrate.

Starches are long polymers of glucose and serve as a source of glucose storage that naturally occurs in plants and animals. There are three main forms of starch: glycogen, amylose, and amylopectin. Glycogen is the form of starch that is used in animals as our glucose reserve; it is made and stored in the liver and red muscle tissue. Glycogen is a good example of a branched polymer; this means that approximately every 10-glucose monomer has two glucose monosaccharide units linked by a glycosidic bond. The "second" link serves as a branch point to grow another polymeric chain. The energy storage molecule in plants is called starch; it also consists of long polymers of glucose molecules that serve as energy stores of glucose in seeds and the roots of rice, corn, wheat, potatoes, beans, and cereals. Plant starch consists of two types of molecules: amylose and amylopectin. Amylose, which makes up about 20% of plant starch, consists of unbranched chains of 200–4000 glucose molecules that structurally forms a coil. The remaining 80% of most plant starches are amylopectin. Amylopectin is branched, like glycogen, but the branch points occur less frequently, approximately every 25 glucose units. In both glycogen and plant starch, each glycosidic bond that connects one glucose to another is an α-glycosidic bond. Enzymes in our saliva can break down (hydrolyze) the α-glycosidic bonds in these large complex carbohydrates so that we can use the remaining smaller pieces as an energy source. By contrast, cellulose is also a complex carbohydrate made of glucose, where the glucose molecules are linked via a β-glycosidic bond. Cellulose is used by plants to provide rigid strength in cell walls in wood and fibrous plants and has a linear, extended (noncoiled) structure, which contributes to its role as a structural protein. Furthermore, the linear arrangement of the carbohydrate allows for hydrogen bonding between different cellulose chains, thus creating a strong cross-linked fiber. Our salivary enzymes do not break down cellulose because of the β-glycosidic bonds, so cellulose is unavailable to humans as an energy source. However, cows, goats, and termites (to name a few) have symbiotic bacteria living in their gut that can break down the cellulose to yield glucose for use as an energy source; thus cows and goats can survive by eating grasses and termites thrive on wood (Fig. 1.34 and Box 1.4).

Another class complex carbohydrate is dietary fiber. Dietary fibers, also called roughage or just fiber, are poorly digested plant polymers that contain a diverse mixture of monosaccharide components, many of which are chemically modified. Fiber comes in two forms, soluble and insoluble. Fiber that readily dissolves in water is soluble. Soluble dietary fiber tightly binds water through hydrogen bonds, swells, and turns into a gel. The thick water-soluble fiber gel slows down digestion; thus foods that contain soluble fiber create a feeling of being "full." Moreover, some soluble fibers bind cholesterol and aid in carrying it through the intestinal system. Good sources of soluble fiber include oatmeal, lentils, apples, pears, celery, and carrots. Insoluble fiber does not dissolve in water and speeds digestion and transit of molecules through the digestive system. Sources of insoluble fiber include whole wheat, whole grains, bran, seeds and nuts, dark leafy vegetables, grapes, and tomatoes. Both soluble and insoluble fibers are important to food, nutrition, cooking, and flavor. Let's look at some examples.

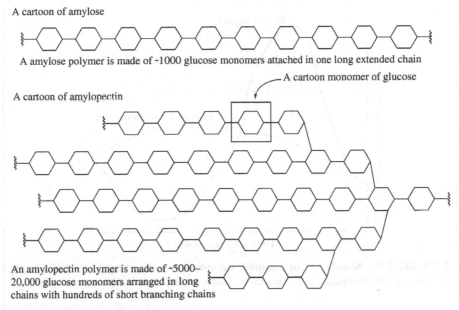

A cartoon of amylose

A amylose polymer is made of ~1000 glucose monomers attached in one long extended chain

A cartoon monomer of glucose

A cartoon of amylopectin

An amylopectin polymer is made of ~5000–20,000 glucose monomers arranged in long chains with hundreds of short branching chains

FIGURE 1.34 Plant starches. Two glucose polymers amylose and amylopectin are depicted. Amylopectin is a coiled unbranched polymer, while amylopectin is branched with a tree-like structure.

BOX 1.4   SIMPLE VERSUS COMPLEX SUGARS: WHY ARE THEY BAD FOR YOU?

Sugar or "refined sugar" has gained a reputation as being "bad" for you. Refined sugar is simply purified sucrose, fructose, or glucose that comes from sugar cane, sugar beets, or other plants such as corn. There is nothing intrinsically bad about the sugar molecule when it is present in food or used in cooking. In fact, consuming 1 g of sucrose or fructose that is added to a food in cooking will have the same effect as eating 1 g of the same sugar from fruit or other plant sources. As an example, one banana contains about the same amount of simple sugar as a prepared food that contains four tablespoons of granulated sugar per serving! However, excess consumption of sugar, like any food, results in poor health consequences. Spikes in glucose (or blood sugar) occur after you consume a supersized candy bar that contains 40 g of carbohydrates (in the form of simple sugars). These sugars are easily absorbed into the body, causing a spike in blood glucose levels from 70–100 milligram per deciliter of blood (mg/dl) to 120–200 mg/dl depending on the individual. Having chronic high blood sugar leads to a number of negative health consequences including glaucoma and nerve damage in the extremities. Why do you get a "crash" after eating a large quantity of sugar? Once glucose levels rise, the body releases a hormone called insulin from the

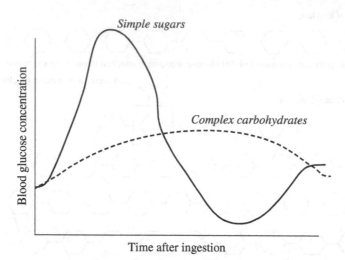

FIGURE 1.35 Simple versus complex carbohydrate. The impact on blood sugar (glucose) levels after eating simple mono- or disaccharides versus complex carbohydrates.

pancreas. Insulin allows the glucose to move from the circulatory system into the muscle tissue. The amount of insulin released is directly proportion to the levels of glucose in the blood. Thus, abnormally high amounts of sugar, due to mega candy bar consumption, cause abnormally large amounts of insulin to be secreted by the pancreas, resulting in high amounts of glucose to be transported into the muscle tissue. This phenomenon results in a serious reduction in blood sugar to a level that is much lower than a premeal level. Because the brain gets most of its energy from blood glucose, low levels of blood glucose cause you to feel tired, weak, confused, and even dizzy (Fig. 1.35). Interestingly, eating the same mass of complex carbohydrates avoids this spike and crash in blood sugar levels because the starch/glycogen polymers have to be broken down by the digestive enzymes in the gut before the individual glucose molecules can be transported from the intestine into the bloodstream. This takes time and allows for the gradual release of glucose into the circulatory system. The result is a slow, gradual increase and decline in blood sugar levels due to the modest and controlled release of insulin. The rate at which a food spikes the blood glucose level is called the glycemic index. Foods that contain high amounts of simple sugars have a high glycemic index. In contrast, complex carbohydrates including starches and fibers have a much lower glycemic index.

Because of the gel-like character that is imparted by soluble fiber in water, soluble fiber is often used as a thickening agent in cooking. Pectin, a soluble fiber found in plant cell walls, is made of the monosaccharide galacturonic acid (Fig. 1.36). An immature apple contains chemically unmodified pectin, which gives the fruit a hard rigid consistency. As the fruit ripens, enzymes modify pectin by adding methyl

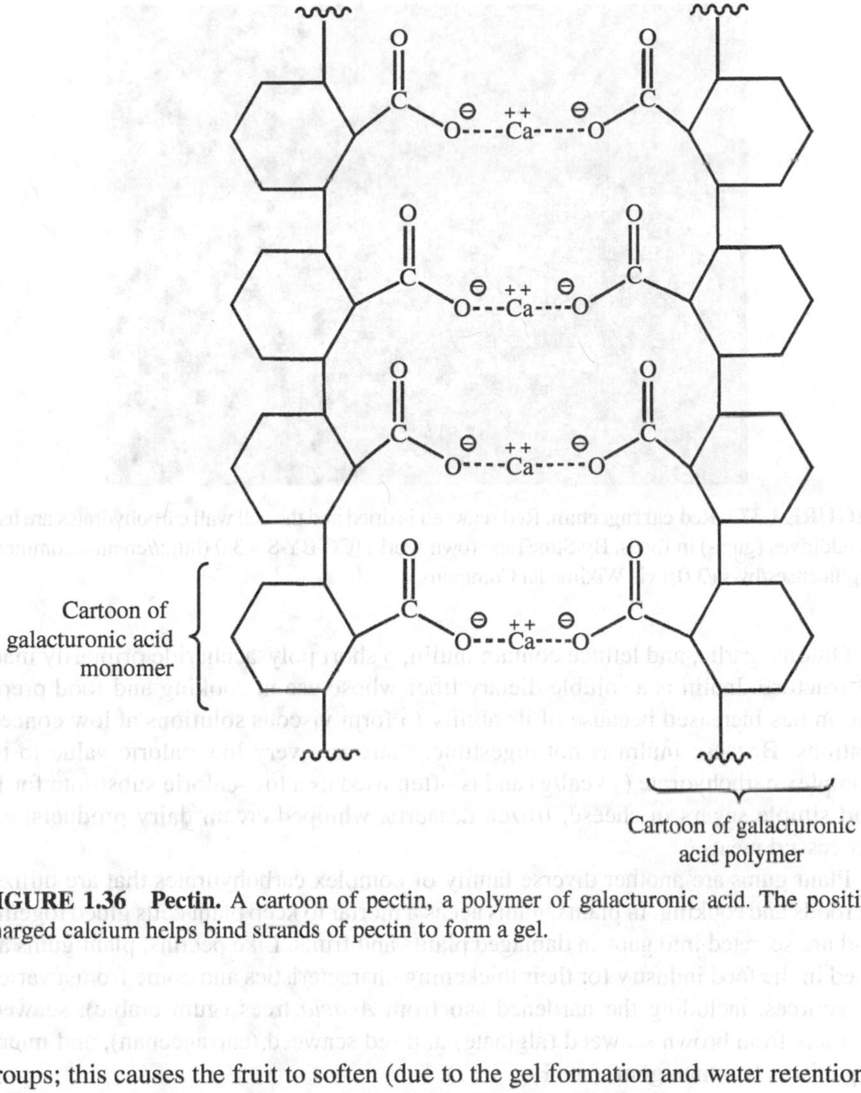

Cartoon of galacturonic acid monomer

Cartoon of galacturonic acid polymer

FIGURE 1.36 Pectin. A cartoon of pectin, a polymer of galacturonic acid. The positive charged calcium helps bind strands of pectin to form a gel.

groups; this causes the fruit to soften (due to the gel formation and water retention). Moreover, in cooking, a more stable gel can be formed during the preparation of applesauce or apple pie. If you have ever tried to make an apple pie with unripened apples, the resulting watery mess was due to the lack of gel-forming pectin found in the fruit. Natural (from within the fruit) or added pectin is key to form an effective gel in jellies and jams. The pectin "gel" consists of long tangles of carbohydrate polymer that is tangled together into a rigid mass that tightly retains water. A great example is the gel in jelly. In acidic conditions when citric acids are included, they interact with mono- and disaccharides to form a stable gel that swells several times its dried volume. Mature fruits provide pectin from the cell walls to form an effective gel in making jellies and jams. The gel-like nature of pectin is also used to stabilize yogurts (Fig. 1.36).

FIGURE 1.37   Red carrageenan. Red seaweed is dried and the cell wall carbohydrates are used as additives (gums) in foods. By StinaTano (own work) (CC BY-SA 3.0 (http://creativecommons. org/licenses/by-sa/3.0)) via Wikimedia Commons.

Onions, garlic, and lettuce contain inulin, a short polysaccharide primarily made of fructose. Inulin is a soluble dietary fiber whose use in cooking and food preparation has increased because of its ability to form viscous solutions at low concentrations. Because inulin is not digestible, there is a very low caloric value to the complex carbohydrate (1 kcal/g) and is often used as a low-calorie substitute for fat and simple sugars in cheese, frozen desserts, whipped cream dairy products, and processed meats.

Plant gums are another diverse family of complex carbohydrates that are utilized in foods and cooking. In plants, gums act as a mortar to keep plant cells glued together and are secreted into gaps in damaged plants and fruits. Like pectins, plant gums are used in the food industry for their thickening characteristics and come from a variety of sources, including the hardened sap from *Acacia* trees (gum arabic), seaweed extracts from brown seaweed (alginate) and red seaweed (carrageenan), and microorganisms (xanthan gum) (see Fig. 1.37).

What makes a gum different from a fiber? Gums are specifically defined as a hydrocolloid; this means that a gum can form a gel or solid depending on the amount of water present. How are they used in food preparation? Gums are used to improve mouthfeel, emulsify liquids, and trap or encapsulate flavor molecules. Mouthfeel is the way the food or drink feels thick or thin as it is eaten. When someone says a sauce is too thin, this is mouthfeel. The gum arabic present in a can of Mountain Dew acts as an emulsifier, helps to keep the oil in suspension, and improves mouthfeel. Some gums produce very smooth textures in processed foods and tightly hold on to water and reduce ice crystal formation during freezing/thawing. Because of these properties, gums are often added to process frozen prepared microwave meals from separating into water and solids. Agar agar is a gum used by molecular gastronomy-inspired cooks and chefs. Agar agar (also just called agar) is an extract of algae that has been used historically for many years by microbiologists as a solid gel infused with growth compounds to culture bacteria. This use of course was influenced by cooks who had used agar in Java, as well as in other Asian countries, for centuries. Agar agar is an efficient gelling agent that readily forms a semisolid consistency, can be easily infused with flavors (or can infuse flavors), can provide a firm solid shape and mouthfeel, and can suspend or encapsulate liquids and solid food particles. Unlike other gels, once formed, agar agar will retain its solid consistency even when heated at 90°C! Some creative cooks use the gel to create unusual dishes with unique shapes and properties (Box 1.5).

BOX 1.5   THICKENERS

The ability to thicken a soup, gravy, sauce, or beverage is a critical tool for any cook. The trick to creating a successful thickened sauce, soup, or gravy is through incorporation of a complex carbohydrate and/or protein that binds water, yielding a more viscous liquid. The feeling that comes with tasting and eating a thick gravy or salad dressing is called mouthfeel. In order to understand how thickening works, you need to learn some physics about liquids. Water is considered a Newtonian fluid because there is a proportional relationship between viscosity and the force applied to the fluid. The more viscous the Newtonian fluid, the more force required to get it to flow. Gravies and other thickened foods are non-Newtonian fluids; these fluids require a larger force to start movement, but once in motion non-Newtonian fluids move with a greater ease. Most of you have experience with ketchup. What a great example of a fluid that does not easily start to move but, once flowing, flows much faster than you often desire! What does a non-Newtonian fluid have to do with mouthfeel? The stickiness and viscosity of a non-Newtonian fluid are amazingly detected by the human mouth and give foods containing these fluids a distinctive, pleasurable character. Think of a soup that seems watery versus one that has been thickened with starch or gum. The thickened soup tastes better even when the flavor of the soups is identical.

One cooking technique used to thicken a recipe is called reduction. Heating a mixture provides enough energy to the water molecules to allow them to rotate, vibrate, and escape the intermolecular forces that hold the water molecules in place. As the water "boils off," the remaining molecules of complex carbohydrates or proteins are forced to interact with each other, increasing the viscosity of the fluid. Unfortunately, this preparation that requires high heat or long cooking times often alters the flavor and nature of the fluid in unintended ways. The starches found in flour can form pastes and gelatinize for use in gravies and stews. Thick flour pastes are difficult to form, as the starch solidifies in a low water environment. Cooks will use fats to coat and interact with the starch. Such a mixture, called a roux, is the base for pasta gravies and some thick stews.

How do you decide what thickening agent to use—a protein, starch, or gum? It depends upon the flavor and final presentation of the desired food. Is the food to be served hot or cold? Proteins coagulate upon cooling (making a solid mess), while gums and starches are more effective thickeners when cold. Starch thickeners quickly clump when reheated. If you want a clear thickened soup, some gums (like xanthan gum) can be used in a small amount to avoid clumping and cloudiness. Both starch and xanthan gums work well for more viscous preparations. pH can also impact the behavior of a thickening agent. Acidic foods cause some gums to be less effective; thus cornstarch or arrowroot starch is a better choice. Clear Jel is a chemically modified cornstarch that is specially made for baking and freezing acidic foods; it is a good choice for acid-containing fruit pies that need thickening.

1.3.9.2 Lipids (Fats, Oils, Waxes, Phospholipids, Fatty Acids) Besides the well publicized nutritional role fats and oils play in our food, this third biological macromolecule is important in cooking and the taste of the foods we eat. Lipids are a class of molecules composed mostly of carbon and hydrogen atoms and are poorly insoluble in water but soluble in solvents like chloroform or ether. Fats, fatty acids, oils, waxes, cholesterol, membrane lipids, and other molecules all belong in the lipid family. Lipids, unlike other biological macromolecules, are not long polymers (repeating units of monomers); however many of the lipid molecules are composed of several smaller molecules bonded together to form a larger functional compound. We will look at the structure and chemical nature of several important lipids and investigate how lipids impact cooking and baking.

The simplest lipids are the fatty acids. Fatty acids are long chains of carbon atoms bonded to hydrogen atoms ending with a carboxyl group (Fig. 1.38). The carbons in fatty acids will be bonded to one, two, or three hydrogen atoms, making this a very nonpolar, water-insoluble molecule. The (COO−) carboxyl groups on the end gives the molecule the ability to hydrogen-bond to water or ionic bond to positive charged ions and that portion of the fatty acid is water soluble or hydrophilic. By definition, *lipids* are insoluble in water, so that means triglycerides are *insoluble* in water. To be *soluble* means that two molecules will dissolve in one another to form a homogeneous mixture. When compounds are *insoluble*, the combination forms a heterogeneous mixture. When a lipid (e.g., oil) is mixed with water, you will see boundaries form between the two *phases*—literally, the two cannot mix. Polar compounds can mix with or *dissolve/are soluble in* water (*hydrophilic*) to form homogeneous mixtures (i.e., sugar dissolving in water, lemon juice dissolving in water, vinegar dissolving in water). Nonpolar compounds can mix with or *dissolve/are soluble in* oils (*hydrophobic*) to form homogeneous mixtures (e.g., vanilla extract dissolving in oil, melted butter mixing with olive oil). Figure 1.39 highlights the differences between polar and nonpolar bonds. These facts are described by the principle *like dissolves like.*

Molecules with regions that are both hydrophilic and hydrophobic are considered "amphipathic." The amphipathic nature of fatty acids drives the hydrophilic and hydrophobic regions to align together creating globules or complicated micelles and even into sheets of lipids called membranes. Fatty acids are naturally found in plants

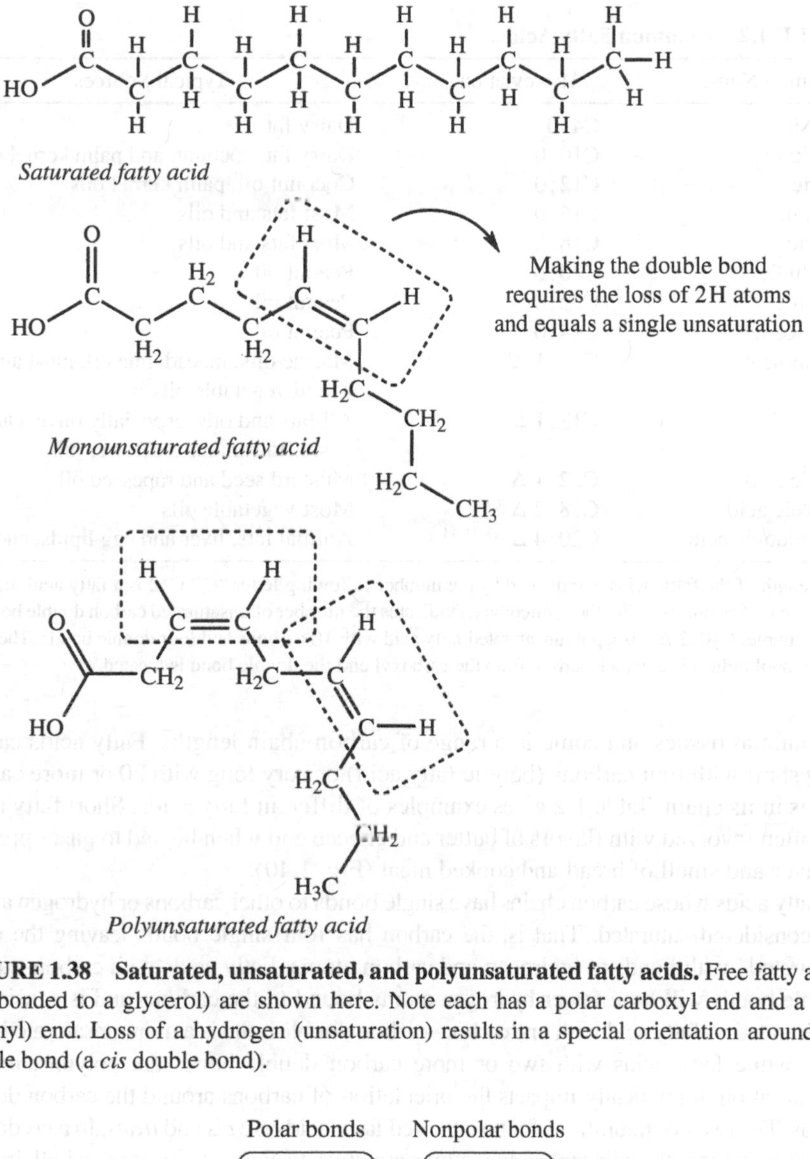

Saturated fatty acid

Monounsaturated fatty acid

Making the double bond
requires the loss of 2 H atoms
and equals a single unsaturation

Polyunsaturated fatty acid

FIGURE 1.38   Saturated, unsaturated, and polyunsaturated fatty acids. Free fatty acids (not bonded to a glycerol) are shown here. Note each has a polar carboxyl end and a CH_3 (methyl) end. Loss of a hydrogen (unsaturation) results in a special orientation around the double bond (a *cis* double bond).

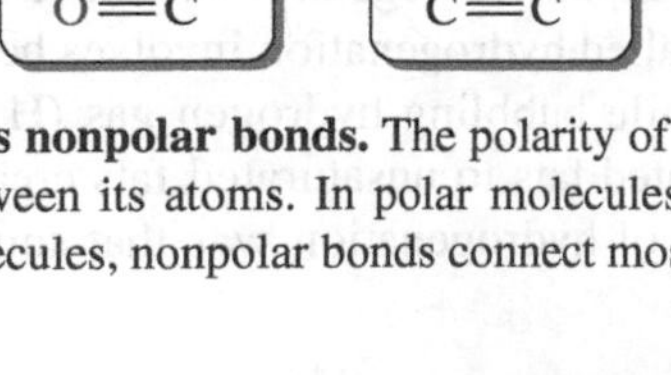

Polar bonds   Nonpolar bonds

δ− δ+	C—H
O—H	C—C
N—H	
O=C	C=C

FIGURE 1.39   Polar versus nonpolar bonds. The polarity of a molecule is determined by the separation of charge between its atoms. In polar molecules, most atoms are connected polar bonds. In nonpolar molecules, nonpolar bonds connect most atoms.

TABLE 1.2   Common Fatty Acids.

Common Name	Abbreviation	Typical Sources
Butyric	C4:0	Dairy fat
Capric	C10:0	Dairy fat, coconut, and palm kernel oils
Lauric	C12:0	Coconut oil, palm kernel oils
Palmitic	C16:0	Most fats and oils
Stearic	C18:0	Most fats and oils
Arachidic	C20:0	Peanut oil
Behenic	C22:0	Peanut oil
Lignoceric	C24:0	Peanut oil
Palmitoleic	C16:1 Δ^9	Marine oils, macadamia oil, most animal and vegetable oils
Oleic	C18:1 Δ^9	All fats and oils, especially olive, canola, sunflower, and safflower oils
Erucic acid	C22:1 Δ^{13}	Mustard seed and rapeseed oil
Linoleic acid	C18:2 $\Delta^{9,12}$	Most vegetable oils
Arachidonic acid	C20:4 $\Delta^{5,8,11,14}$	Animal fats, liver and egg lipids, and fish

The length of the fatty acids is indicated by the number following letter "C." C12 is a fatty acid made of 12 carbons. The number after the semicolon (:) indicates the number of unsaturated carbon double bond(s). For example, C10:2 $\Delta^{2,4}$ is a polyunsaturated fatty acid with 10 carbons and four double bonds. The delta (Δ) symbol indicates at which carbon from the carboxyl end the double bond is located.

and animal tissues and come in a range of carbon chain lengths. Fatty acids can be very short with four carbons (butyric fatty acid) or very long with 20 or more carbon atoms in its chain. Table 1.2 gives examples of different fatty acids. Short fatty acids are often involved with flavors of butter and cheese and when heated to gases provide a flavor and smell of bread and cooked meat (Fig. 1.40).

Fatty acids whose carbon chains have single bonds to other carbons or hydrogen atoms are considered saturated. That is, the carbon has four single bonds leaving the atom "saturated" with bonds to hydrogen and carbon atoms. Fatty acids with carbon–carbon double bonds will have fewer hydrogen atoms bound to the carbons and is considered unsaturated. A fatty acid with one carbon–carbon double bond is a monounsaturated fatty acid, while fatty acids with two or more carbon double bonds are polyunsaturated. Unsaturation dramatically impacts the orientation of carbons around the carbon double bonds. The two configurations for unsaturated fatty acids are *cis* and *trans*. In a *cis* double bond orientation, the adjacent carbon atoms are close to each other leaving a kink in fatty acid chain at the unsaturation. *Trans* configuration leaves the carbon atoms on either side of the double-bonded carbons farther away (*trans*) from each other, resulting in a fatty acid with a straight chain shape. The *cis* or *trans* shape to the fatty acid can have a significant impact on the melting point and chemical reactivity (Fig. 1.41).

In the early 1900s less expensive vegetable oil was used to create margarine and shortening. The process called hydrogenation involves heating fats with high heat in the presence of nickel while bubbling hydrogen gas (H_2 (g)) through the mixture. Changing the *cis* unsaturated fats to unsaturated fats created an inexpensive way to make food. A by-product of hydrogenation was that some of the *cis* double bonds

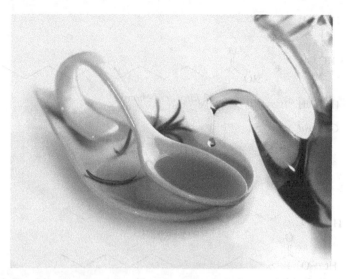

FIGURE 1.40 Oil. Unsaturated and polyunsaturated fatty acids are liquid at room temperature.

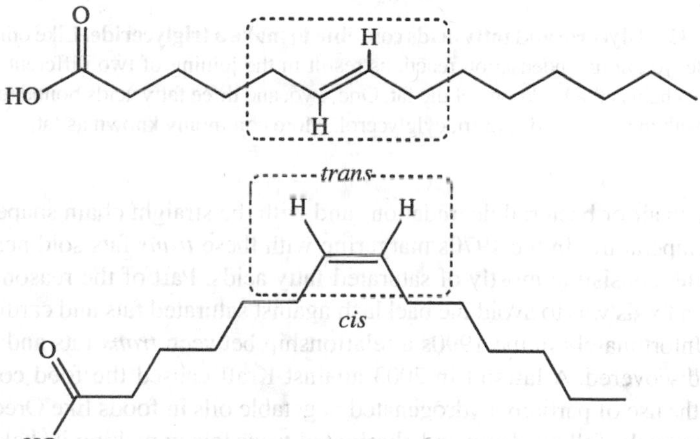

FIGURE 1.41 *Cis* **and** *trans* **fatty acids.** Loss of two hydrogens in a fatty acid chain results in the sharing of electrons forming a double bond. The arraignment of the atoms around the double results in the hydrogens being on the same (*cis*) or opposite (*trans*) side of the double bond. Note the impact of a *cis* and *trans* desaturation on the shape of the fatty acid chain.

were converted to *trans* double bonds. The unintentional creation of *trans* fatty acids during the production of margarine and shortening was initially considered a bonus effect of hydrogenation. This was because *trans* fatty acids, due to their shape, are more resistant to reaction with oxygen and are a poor food for bacteria. These *trans* fats have a much longer shelf life and less likely to become rancid (a product of

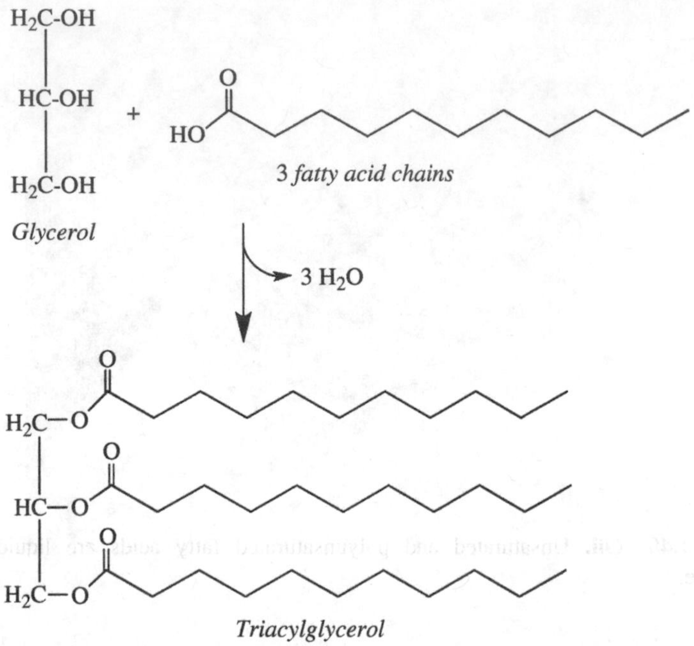

FIGURE 1.42   **Glycerol and fatty acids combine to make a triglyceride.** Like other polymer reactions, dehydration/condensation reactions result in the joining of two different molecules. The glycerol chain is the backbone of the fat. One, two, and three fatty acids bound to a glycerol molecule result in a mono-, di-, or triacylglycerol. More commonly known as fat.

reaction with air or bacterial degradation) and with the straight chain shape are solid at room temperature. In the 1970s margarine with these *trans* fats sold nearly twice that of butter consisting mostly of saturated fatty acids. Part of the reason for using *trans* fats in foods was to avoid the backlash against saturated fats and cardiovascular disease. Unfortunately in the 1990s a relationship between *trans* fats and heart disease was discovered. A lawsuit in 2003 against Kraft caused the food company to eliminate the use of partially hydrogenated vegetable oils in foods like Oreo cookies. In 2006 Starbucks followed suit and eliminated *trans* fats in making its baked goods. While low levels of *trans* fatty acids are found in some foods like pork, lamb, and milk, the problem with *trans* fatty acids is the higher incidence of coronary disease. *Trans* fats raise the levels of LDL that can contribute to blood flow blocking fatty deposits in arteries and can cause one type of white blood cells to change, collect cholesterol from LDL, and harden, blocking flow to the heart.

Fats and oils are essentially the same molecule, a triacylglycerol. Triacylglycerol is the major form of energy for animals and is stored in specialized fat cells called adipose cells. To create a triacylglycerol, three fatty acids are bonded to a short organic molecule called glycerol through the same dehydration reaction seen with proteins and carbohydrates (Fig. 1.42). Overall the new molecule is very hydrophobic and is nonpolar. A triglyceride is made from the combination of fatty acids and

TABLE 1.3   Smoke Point of Common Cooking Oils.

Fat	Oil Grade/Quality	Smoke Point
Vegetable shortening	Hydrogenated	360°F/180°C
Butter	Unsalted	350°F/175°C
Clarified butter	Removed sugars, water, and protein	450°F/230°C
Canola oil	Refined	400°F/205°C
Corn oil	Refined	450°F/230°C
Olive oil	Extra virgin	350–410°F/177–210°C
Sesame oil	Refined	350–410°F/175–210°C
Peanut oil	Refined	450°F/230°C
Safflower	Refined	510°F/265°C

glycerol; it has the basic structure shown in Figure 1.42 and Table 1.3. Fatty acids can be from 4 to 35 carbons long, but 14–20 carbon fatty acids are most common in food. When fatty acids and glycerol combine, bonds are broken and formed in a chemical reaction to produce a triglyceride and three molecules of water. In that process, a new group of atoms is formed called an *ester*. The properties of a given triglyceride depend upon the chemical structure of the three fatty acids it contains, and the properties of a lipid depend upon the particular mixture of triglycerides it contains.

A fat is a triacylglycerol that is solid at room temperature and found in fats in meat and milk. Oil is a triacylglycerol with fatty acid chains usually from plants and is liquid at room temperature. Essentially, the only difference between fat and oil is the length and number of double bonds on the fatty acids. However, these are significant differences both chemically and for cooking.

Melting of solid fat or oil like butter is a change in states of matter. Melting butter takes place when heat is applied, and the state of matter moves from solid to liquid. Fat molecules will stack and interact with each other with weak but numerous bonds. The more contact between fats, the more bonds can be formed and the more stable the fat will be in a solid state. Heat supplies enough energy to shift, move, and vibrate the molecules far enough apart from each other that they feel very little attraction. Now the fat molecules behave like a liquid and, with weak interactions attracting each other, are free to move. The melting point is the temperature at which molecules shift from a solid configuration to a liquid state. Fats that melt at a higher melting point require more heat to melt than a fat with a low melting point. The reason for this is the number of contacts between fats. Saturated fats pack very efficiently, densely packed with lots of interactions. Longer fatty acid chain results in more interactions between fatty acid chains. Thus it takes more energy to separate longer fatty acid chains than shorter fatty acid chains. *Cis* unsaturated fatty acids create the kinked fatty acid shape discussed earlier, keeping the fatty acid carbon chains from efficiently packing, reducing the heat needed to separate and melt a solid fat. One the other hand, *trans* fatty acids have a shape similar to unsaturated fatty acids and pack like the unsaturated fats. A simple approach is to remember that the longer the fatty acid chain, the higher the melting point and the more *cis* unsaturated double bonds, the lower the melting point.

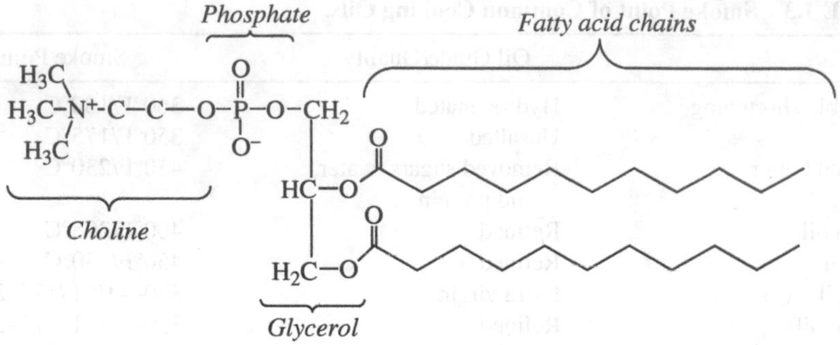

FIGURE 1.43   **Phospholipid.** A phospholipid similar to a diacylglycerol has a very different function and includes a polar head group (charged compound) linked to the glycerol backbone via a phosphate group.

Another lipid important in biology and cooking is the phospholipid. Like the triacylglycerol, phospholipids have a glycerol backbone, but, instead of three fatty acids, phospholipids contain two fatty acid chains. The third —OH of glycerol is bonded to a range of phosphate-containing molecules. The phosphate group is called the polar head group of the phospholipid. These molecules are now polar with the fatty acids creating a hydrophobic region, and the phosphate-containing portion usually is charged and hydrophilic. Phospholipids make up the major component of cell membranes—which surround the cell containing the contents of the cell. Phospholipid membranes are tough and form into globules in milk fat to provide a durable enclosure for the fatty acids and triacylglycerol. Lecithin, formally called phosphatidylcholine (Fig. 1.43), is a special phospholipid with a choline head group. Lecithin is found in high amounts in soybean and egg yolk and is used for cooking. The nonstick cooking spray PAM is primarily lecithin and water and utilized for its oil-like properties. A quick search of your pantry and refrigerator will find lecithin in a number of items including salad dressing, chocolate, and a number of interesting molecular gastronomy-inspired dishes and preparations. A curious dessert is liquid popcorn with caramel froth. Popcorn, sugar, and water are used with syrup and dry lecithin to create a two-phase drink where lecithin plays a role in keeping caramel cohesive.

Fats provide special flavors and characteristics to foods. The lubricating properties of lipids create a slippery smooth mouthfeel to foods and make some foods seem moist. Fats are often reported to tenderize food. Think of the difference between a lean and marbled steak. The cut marbled with fat is easier to chew and more tender. In baking, fat acts as a shortening agent. That is, fat coats some of the proteins and starches in flour, limiting the network and keeping a crust together. Adding fat (shortening) to a pie crust recipe creates a softer smaller crumb with thin layers of pie crust rather than a solid thick sheet of baked dough (Fig. 1.44). Short-chain fatty acids have a special flavor, creating a complex flavor to several foods. Microorganisms used to make some cheeses create sharp cheese flavors and some of the odors of cheeses. When cooked at high heats, fats of meat will change chemically to volatile

FIGURE 1.44   Role of shortening on pastry crust. The Greek pastry, baklava. Notice the sheets of flaky crust due to the addition of shortening. Caused by fat blocking or shortening the gluten interactions.

molecules that impart the flavor and smell of meat. Raw meat has little aroma and a very simple, blood-like taste. Phospholipids and fatty acids in meat react with oxygen to form smaller molecules called aldehydes, unsaturated alcohols, ketones, and lactones. Each provides a single note of flavor and smell of cooked meat. Certain fats like molecules derived from plants called terpenoids provide strong flavors to cooked foods. Examples include cinnamon, cloves, and mint.

Cooking with fats and oils provide a unique challenge. Heating fats and oils can create a smelly, smoky mess that leaves food tasting bitter. Fats will melt into oils when warmed but do not boil. Before the fat can reach the boiling point, it smokes and breaks down. The breakdown of fat at high temperatures is due to several factors. At high temperatures, oxygen in the air will oxidize the double bonds of unsaturated fatty acids creating a rancid and smelly product. Impurities in the oil including sugars, water, and proteins will burn in the oil, producing dark colors and off-tasting molecules. Free fatty acids are naturally present in fats and oils in very small amounts, but the amount of free fatty acids increases as the fats/oils are heated. The smoke point of oil is the temperature at which overheating causes a fat to give off smoke. At this point the fat decomposes where the fatty acids are released from glycerol and form into longer polymers that are very unpleasing to taste. The released glycerol further reacts with oxygen and heat to form acrolein and water. Acrolein is a toxic compound that irritates soft tissues. The fatty acid component and purity of oil influence the smoke point, and choosing the right oil or fat is critical to avoid ruining your food. More saturated fats decompose quickly, and butter, made mostly of saturated fatty acids, will smoke much more quickly than canola oil, which is a mixture of mono- and polyunsaturated fats. Safflower oil (12% monounsaturated and 75% polyunsaturated) has one of the highest smoke points. Impurities also decrease the temperature at which oil will smoke. Butter is a mixture of fats, water, proteins, and lactose. The latter components readily burn and are the reason why sautéing with butter leads to a mess. Professional chefs will separate the fat from protein and water (clarified butter) to sauté food. To sauté in a higher temperature, use a vegetable oil-like canola or saffron oil.

Olive oil poses a special challenge when cooking. There are several grades of olive oil, each depending on the way the oil is extracted and stored. One way of making olive oil is to simply squeeze or press the fruit. The resulting oil is separated from large solids by filtering or centrifugation. This is cold-pressed oil. Sometimes hot water is then added and the oil extracted from the remaining paste to recover more oil. Another method involves dissolving the ground olive in a solvent to extract the oil. To be called a "virgin" olive oil, no chemical purification can be used in its production. Extra virgin olive oil is free of acids found in some species of olives and retains a good flavor. This olive oil is often used to add to foods for its flavor and not cooking. Refined olive oil is the extra virgin olive oil that does not qualify for the extra virgin status and has been chemically treated to remove impurities. Some oils, simply labeled "olive oil," are a blend of refined and virgin olive oil and will have some of the flavors but will contain less of the impurities. Unrefined olive oil has a very low smoke point with a mix of impurities (320°F/160°C). Extra virgin olive oil has a slightly higher smoke point of 185°C/365°F, while processed purified, refined olive oil without the particles and acids from the olive paste will have a much higher smoke point of 450°F/232°C and serves as a good cooking medium.

1.3.9.3 *Emulsions* Proteins, lipids, and complex carbohydrates can each be used to prepare a critical technique for food and cooking called an emulsion. Emulsions are evenly mixed (homogeneous) dispersions of two components that repel each other. Examples of food emulsions include sauces, dressings, or mayonnaises. In chemical terms, the two phases are immiscible. A common culinary emulsion is a mixture of oil and water. If there is more water than oil, the emulsion is called oil-in-water emulsion, and if there is more oil than water, the emulsion is a water-in-oil emulsion. Mayonnaise is an oil-in-water emulsion, while butter is a water-in-oil emulsion. Either way, the basic concept of cooking with an emulsion is to create small enough droplets or fragments of the oil or water to be evenly dispersed through the mixture. Small droplets are created by a physical force called sheering is most commonly performed using blenders, whisking, or food processors. Industrial kitchens use fine-gauge strainers and force the fluids to create an emulsion. Consider shaking a mixture of oil and water. Shake or mix hard enough and the tiny spheres of oil will seemingly mix within the water. However, the two phases quickly separate from each other. That is the challenge of making an emulsion, keeping the solutions mixed.

Emulsifiers are food additive compounds that help create emulsions from separating back into two distinct layers. For oil-in-water emulsions, an emulsifier must have both a water-loving and water-fearing (oil-loving) region. Emulsifiers surround oil droplets with the hydrophobic portion of the molecule and align the hydrophobic region of the molecule to face the water. A second additive called a stabilizer is also often used in making emulsions. Stabilizers help keep the small droplets prepared in the emulsion from reuniting into a larger mass and finally fully separating into two layers. Some molecules can act as both an emulsifier and a stabilizer. Common ingredients used for emulsifiers and stabilizers are egg yolks. The key compound in egg yolks used as an emulsifier is the phospholipid lecithin. The fatty acid component of the lipid buries itself into the oil droplets with the charged choline head group facing outward where it can interact with water by hydrogen bonds. This coats the newly formed oil droplet with a negative charged compound that allows the droplet to stabilize interactions with water and will avoid other oil droplets. Sugars that have been modified by adding short fatty acids are often used. The base sugar–fat compound used as emulsifiers is called sorbate. The carbon chain fatty acid interacts with the lipid, while the carbohydrate end of the molecule interacts with water. Polysorbate 80 is an emulsifier used in ice creams to keep the fats separated and ice cream smooth and slows down the "liquefaction" or appearance. Some proteins also serve as effective emulsifiers. Casein, one of the main proteins found in milk, coats the globules of fat in a similar way as lecithin coats oil droplets. The hydrophobic portions of casein interact nicely with the oil, and the charged component of the protein faces water to hydrogen bond, keeping the droplet intact.

Once the emulsion is formed, even in the presence of emulsifiers, the mixture is doomed and will eventually coalesce and form into distinct layers creating a watery, runny, and oily sauce. Complex carbohydrates like starches and gums and proteins interact and bind to emulsifiers and are often used as emulsion stabilizers and thicken the solution as well. The key job of a stabilizer is to help keep the droplets of oil or water apart so they cannot collect to form larger droplets where oil and water will eventually separate into two different phases. Starches and gums from tomatoes are released during preparation and act as stabilizers in pastes and ketchup. Dried ground mustard seed contains a considerable amount of gum carbohydrates, which serve well as stabilizers in hollandaise sauce. Agar and carrageenans are used as stabilizers of emulsions in prepared food. A quick examination of Velveeta processed cheese will find alginate and whey proteins—both used to emulsify and stabilize the cheese. Thickening the emulsification limits the diffusion or movement of the oil or water droplets so they cannot come together.

In making a good emulsion (sauce, mayonnaise, etc.), Harold McGee, the influential author of *On Food and Cooking* [1], writes that "the cook has made one of three mistakes: he has added the liquid to be dispersed too quickly to the continuous liquid, or added too much of the dispersed liquid, or allowed the sauce to get either too hot or too cold." Key rules to making an effective emulsion are to start with the largest volume liquid (oil or water) in the bowl or mixer first. Add the lesser liquid to the greater liquid and add slowly to allow the droplets to form and become surrounded by the emulsifier. Keep the proportions correct; the larger phase should be three times that of the dispersed liquid. Ensure there is enough emulsifier for the dispersed liquid. If there isn't enough emulsifier, the droplets will not be covered. High temperatures can cause proteins to fully denature, and emulsifiers and stabilizers will stop working and begin to curdle. At colder temperatures the oils will begin to solidify. Commercial mayonnaise is refrigerated and remains stable because more polyunsaturated oils are used that remain fluid at lower temperatures.

REFERENCES

[1] McGee, H., ed. (2004) *On Food and Cooking*. Simon and Schuster, Inc., New York.

[2] Corriher, S., ed. (1997) *Cookwise: The Secrets of Cooking Revealed*. HarperCollins Publishers Inc., New York.

[3] Corriher, S., ed. (2008) *Bakewise: The Hows and Whys of Successful Baking with Over 200 Magnificent Recipes*. Simon and Schuster Inc., New York.

2

THE SCIENCE OF TASTE AND SMELL

Guided Inquiry Activities (Web): 6, Protein Structure and Function; 7, Fats, Intermolecular Forces; 8, pH; 13, Flavor

2.1 INTRODUCTION

How you taste and smell impact the flavor of our food and drink. Understanding the molecular nature (biology and chemistry) of how we perceive our food as well as the nature of the molecules that signal to our senses can help one appreciate food and drink and give a chef, cook, or baker a leg up on creating appealing dishes.

2.2 THE PHYSIOLOGY OF TASTE, SMELL, AND FLAVOR

You walk through the door and it hits you. You immediately recognize the smell of your favorite holiday foods. Almost as quickly your body responds. Your mouth begins creating saliva and your stomach rumbles. The smell of food not only allows you to recognize it is time to eat, but it lets your body begin to prepare to digest your food. While smell is a very strong memory stimulus for all humans, it has a special relationship with taste and our concept of flavor.

Scientifically, smell (olfaction) and taste (gustation) are part of the human body's sensory system. Specifically, they function through chemoreception, which is the ability of a body to respond to chemicals in the immediate environment. In gustation your body is responding to chemicals referred to as tastants, chemicals that can stimulate taste receptors. Similarly, in olfaction, receptors in your nasal cavity are

The Science of Cooking: Understanding the Biology and Chemistry Behind Food and Cooking,
First Edition. Joseph J. Provost, Keri L. Colabroy, Brenda S. Kelly, and Mark A. Wallert.
© 2016 John Wiley & Sons, Inc. Published 2016 by John Wiley & Sons, Inc.
Companion website: www.wiley.com/go/provost/science_of_cooking

responding to chemicals called odorants that stimulate smell receptors. Typically when we comment on how great a specific food or beverage tastes, what we are really commenting on are is the flavor of that item. Flavor is the blend of taste and smell sensations you perceive while eating and drinking. The term flavorant can be used as a compound that uses both senses to impart a flavor. Another common usage of tastant is a compound that mimics a flavor or gives a food or drink its flavor. We will use the term tastant throughout this chapter to focus on how the compounds bind to their receptors and signal to the brain. We will also discuss how many of the sensations we describe as taste are actually provided by the olfactory system, so to truly enjoy the flavor of our favorite foods, both systems must participate in forming sensations.

Sensory receptors are specialized tissues that initiate electrical signals that carry information in your central nervous system (CNS; brain and spinal cord) and allow you to respond to your environment. We have a variety of senses each tuned to different environmental energies. These sensory modalities include touch, temperature, pain, proprioception (body position), taste, smell, sight, and sound. Each of the sensory systems responds to a specific sensory input called an adequate stimulus and sends electrical impulses to the CNS in the process of sensory transduction. When you put a hot pack or ice pack on your skin, your touch receptors inform you that something is on your skin, but the temperature sensors tell you whether it is hot or cold. Each sensory receptor type responds to one physiological adequate stimulus. To understand how this works for taste and smell, we will first look at the basic tastants and how taste receptors function, and then do the same for odorants and the olfactory system.

Jean Anthelme Brillat-Savarin, an eighteenth-century gastronomist wrote "smell and taste are in fact but a single sense, whose laboratory is the both and whose chimney is the nose" Gustation, or how we perceive taste, is a combination of both smell and taste. To understand gustation, one must first recognize the relationship between taste buds, the olfactory system, and how we perceive taste and smell in the brain and to understand there is a difference between taste and flavor. Taste is the sensation originating from the oral cavity, while flavor is the combination of odor and taste. In addition to the molecules that signal to our gustation and olfaction senses, flavor is influenced by other inputs called "somatosensory sensations." Somatosensory sensation is a complex contribution of sensations including temperature, pain, and the density of food (pressure sensors). Thus how we perceive food and drink is much more complicated than what food hits our tongue.

To best understand the process of flavor, we should follow the path from food flavor molecule to the brain. Flavor formally initiates when a small molecule (tastant) escaping from our food and drink binds to a protein receptor on the surface of one of two specialized cells: a taste receptor cell (TRC) (Fig. 2.1) or olfactory nerve fibers (Fig. 2.2) sending the signal perceiving flavor to our brain. A careful tracing of the route of taste and smell to and from TRC and olfactory nerve cells can involve three pathways. First, an initial sniffing of food brings the odorant and tastants directly to the olfactory nerves (orthonasal route) where small volatile components of food and drink will bind to the receptor nerves of the olfactory nerve fibers in the nasal cavity. Second, the process of chewing (masticating) food

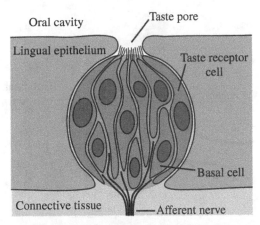

FIGURE 2.1 **Taste bud and taste receptor cells.** Taste buds filled with specialized taste receptor cells. At the surface adjacent to the taste bud, the membrane of the taste receptor cell possess proteins that bind the flavor molecules and transmit the signal to afferent nerves located deep within the bottom of the taste bud. This is how you taste via the gustatory system. Taken from the web. "Taste bud" by NEUROtiker—Own work. Licensed under CC BY-SA 3.0 via Wikimedia Commons—https://commons.wikimedia.org/wiki/File:Taste_bud.svg#/media/File:Taste_bud.svg

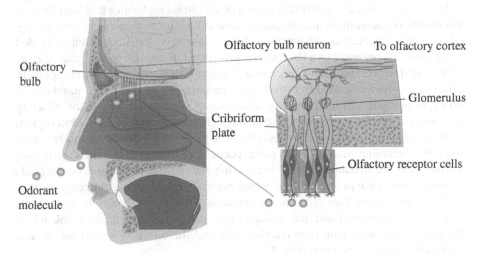

FIGURE 2.2 **Tasting by smell.** The olfactory system is able to identify a diverse range of odors and through this system many tastants. Located at the posterior of the nasal cavity, the olfactory bulb is connected to receptor cells exposed to receive volatile compounds.

will break open vegetative and animal material adding to the existing volatile compounds. As the food is swallowed, the mouth is closed and exhaled air brings the new mix of compounds back through the nasal cavity, providing a second round of smells and flavors to be experienced in what is called retronasal olfaction. This pathway is perceived as a flavor coming from the mouth but is actually a second

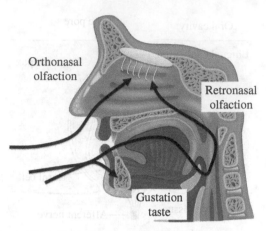

FIGURE 2.3 How we taste our food. Food and drink are sensed in the mouth by smell and through the retronasal pathway odors released by chewing.

form of smell. The third pathway of flavor perception is the tastant molecules binding to specific taste receptors in the TRC cells found in the taste buds of our tongue (Fig. 2.3).

As we will learn later in this chapter, the small tastant molecules that bind the protein receptors found in the membrane of these cells initiate a cascade of events signaling within the receptor cell, continue the signal to nerve cells, and eventually alert the brain of the presence of a tastant. The perception of tastants by the brain is quite complicated as the brain must process and distinguish a diverse set of compounds. Olfactory nerves embedded in the olfactory mucosa are located in the nasal cavity. Olfactory neurons will send their signal to a set of specialized cells in the olfactory bulb where a second set of nerves signal through the thalamus and different regions of the cerebral cortex called the neocortex. The neocortex is where the higher order functions including sensory perception occur. The nerves associated with taste receptor cells are transmitted to the brain using one of three different nerves (chorda tympani, glossopharyngeal, and the vagus nerves). These nerves converge through the brain stem where they signal to the thalamus. From the thalamus the taste signal is carried to neocortex. Later in this chapter we will continue our detailed look at how the nerves associated with taste receptor cells and the olfactory nerves initiate and send their signals to the brain (Fig. 2.4).

2.3 GUSTATION: THE BASICS OF TASTE

While a human can perceive thousands of different flavors, the gustatory system basically responds to five distinct classes of tastants. The stimulation of these five basic classes to taste receptors in different intensities and different combination provides these thousands of different tastes.

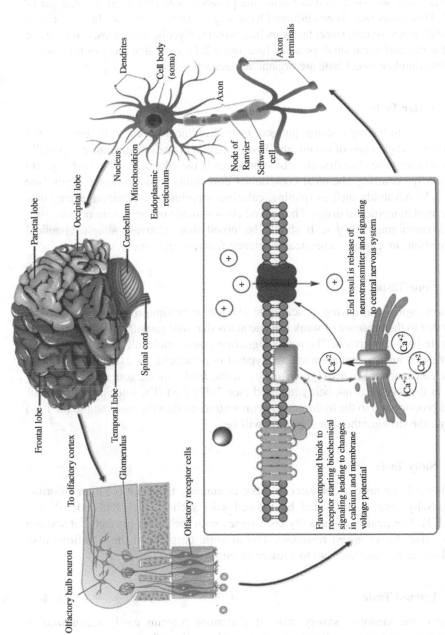

FIGURE 2.4 From mouth and nose to the brain. Odorants and tastants stimulate receptors in the olfactory and taste bud receptors. Depending on the type of receptor activated, biochemical changes lead to membrane depolarization and release of neurotransmitter to the sensory nerves and central nervous system. These signals are then carried to the brain where flavor and smell are perceived.

Labels within figure:

Parietal lobe
Occipital lobe
Cerebellum
Frontal lobe
Temporal lobe
Spinal cord
Glomerulus
To olfactory cortex
Olfactory bulb neuron
Olfactory receptor cells

Dendrites
Cell body (soma)
Nucleus
Mitochondrion
Endoplasmic reticulum
Axon
Node of Ranvier
Schwann cell
Axon terminals

Flavor compound binds to receptor starting biochemical signaling leading to changes in calcium and membrane voltage potential

End result is release of neurotransmitter and signaling to central nervous system

Ca^{+2}

2.3.1   Sweet Taste

Sweet is commonly described as having the pleasant taste characteristic of sugar or honey. This sweet taste is not initiated by a single classification of chemicals. The chemicals that stimulate sweet tastes include sugars, glycols, aldehydes, esters, some amino acids, and some small proteins (see Table 2.1). Note that most of the chemicals that stimulate sweet taste are organic molecules.

2.3.2   Bitter Taste

Bitter relates to having a sharp, pungent taste. Similar to sweet taste, bitter is not caused by a single class of chemicals. Two particular classes of substances typically induce a bitter taste. The first class of substances is the alkaloids, which are a group of naturally occurring chemical compounds containing basic nitrogen atoms (see Table 2.2). Alkaloids such as quinine, caffeine, strychnine, and nicotine are commonly used in medicinal drugs. The second class is made of long-chain organic substances containing nitrogen. It should be noted that relatively simple chemical modification can change a chemical substance from sweet to bitter.

2.3.3   Sour Taste

Sour is recognized as having an acid taste like lemon or vinegar. A sour taste sensation is due to the presence of weak organic acids that will partially dissociate into its conjugate base and proton. The actual signaling tastant molecule is caused by both the hydrogen ion concentration and the type of organic acid, for example, citric acid (citrus fruit sour), malic acid (green apple), acetic acid (vinegar sour), or other weak acids, in the food or drink being ingested (see Table 2.3). The intensity of the sour taste is proportional to the hydrogen ion concentration; thus the more acidic the food or drink, the stronger the sour sensation will be.

2.3.4   Salty Taste

Salty is the basic taste of seawater or more commonly table salt (sodium chloride, NaCl). Salty taste is stimulated by ionized salts such as Na^+, K^+, and Li^+ (see Table 2.4). The quality of the salty taste varies with the cation present, but sodium cations give the strongest response. The anionic partners to these cations also contribute to the salty taste but to a lesser extent.

2.3.5   Umami Taste

Umami is the meaty or savory taste of glutamate proteins easily recognized in monosodium glutamate or Parmesan cheese. Umami is a Japanese word meaning delicious. Most people describe it as a pleasant taste that is qualitatively distinct from the other four taste sensations. Umami is the predominant taste of the amino acid L-glutamine (see Table 2.5).

TABLE 2.1   Taste Classifications, Tastant Structure, and Taste Index for Sweet Compounds.

Taste Classification	Substance	Taste Index	Structure
Sweet	Sucrose	1	
	Glucose	0.8	
	Fructose	1.7	
	Chloroform	40	
	Aspartame (Equal)	200	
	Sucralose (Splenda)	600	
	Saccharin	675	

TABLE 2.2 Taste Classifications, Tastant Structure, and Taste Index for Bitter Compounds.

Taste Classification	Substance	Taste Index	Structure
Bitter	Quinine	1	
	Caffeine	0.4	
	Nicotine	1.3	
	Strychnine	3.1	
	Brucine	11	

TABLE 2.3 Taste Classifications, Tastant Structure, and Taste Index for Sour Compounds.

Taste Classification	Substance	Taste Index	Structure
Sour	Hydrochloric acid	1	H—Cl
	Carbonic acid	0.06	
	Citric acid	0.46	
	Acetic acid	0.55	
	Formic acid	1.1	

TABLE 2.4 Taste Classifications, Tastant Structure, and Taste Index for Salty Compounds.

Taste Classification	Substance	Taste Index	Structure
Salty	NaCl	1	Na^+ Cl^-
	KCl	0.6	K^+ Cl^-
	$CaCl_2$	1	Ca^{2+} Cl^-
	NaF	2	Na^+ F^-
	NH_4Cl	2.5	NH_4^+ Cl^-

TABLE 2.5 Taste Classifications and Tastant Structure for Umami Compounds.

Taste Classification	Substance	Structure
Umami	L-Glutamine	
	L-Glutamic acid	
	Monosodium glutamate	
	Inositol monophosphate	

2.4 WHY DO WE TASTE?

The value of this system of five basic tastes comes in part through its use as a basic survival mechanism. Taste allows us to evaluate the nutritious content of food while preventing us from eating toxic substances. Salty taste allows us to identify foods and liquids that contain essential salts necessary for the maintenance of electrolyte balance. Sweet taste allows us to identify food that is high in energy-rich nutrients. Umami allows recognition of amino acids and proteins necessary as building blocks for proteins in our bodies. On the other hand, bitter and sour tastes give us a warning against the intake of potentially noxious or poisonous substances. This taste discrimination becomes even more important during pregnancy. To ensure both the growth of the fetus and the well-being of the mother, specific metabolic adaptations occur. The most predominant change is a need for additional food calories to account for the energy needs of both the mother and the developing fetus. In terms of taste sensation, the two most consistent changes identified in pregnant women come in the area salty and bitter tastes. Pregnant women have a decreased threshold for salty taste, which means that they show an increased preference for salty food as compared with women who are not pregnant. This increased preference for salty food coincides with an increase in salt requirements that a pregnant woman faces as the fetus develops. Pregnant women also display an increased sensitivity to bitter tastes. Since one role for bitter taste is to prevent ingestion of potentially toxic substances, an increased sensitivity to bitter tastes in pregnancy would protect the developing fetus, which would have a much lower tolerance for these same noxious or poisonous substances than the mother [1].

2.5   THE DIVERSITY OF TASTANTS

To better understand taste distinctions, Tables 2.1, 2.2, 2.3, 2.4, and 2.5 identify different tastants by the sensory system they activate and the intensity of that activation. We will start by looking at the tastants for sweet taste. We are all familiar with sucrose, which is common table sugar. It is the sweetener we most frequently use in baking. The taste index sets the sweetness level of sucrose at 1. Sucrose is a disaccharide composed of the two monosaccharides, glucose, and fructose. Notice that in terms of sweetness, glucose tastes less sweet than sucrose while fructose tastes sweeter than sucrose. This is one reason that high fructose corn syrup is used as a commercial sweetener. You get the same level of sweetness with a lower amount of sugar. On the sweetness scale, however, these natural sugars are not leaders. For example, the artificial sweeteners, Equal (aspartame), Splenda (sucralose), and saccharine, taste 200, 600, and 675 times as sweet as sucrose. Artificial sweeteners are designed to trick our taste buds into believing we are eating high-calorie sweet food without the calories. This in theory allows people to enjoy their favorite sweet-tasting beverages without the additional calories provided by natural sugars. Experiments have found that sweet taste, regardless of its caloric content, enhances your appetite. This means that drinking diet beverages containing artificial sweeteners can lead to an increased food intake, thus undermining a person's effort to reduce calorie intake or lose weight. Aspartame has been found to have the most pronounced effect in this area, but the same applies for other artificial sweeteners, such as acesulfame potassium and saccharin [2].

The taste intensity of different bitter substances do not vary as widely as they do for sweet tastes, but the response to these bitter substances can be quite dramatic (Table 2.2). In this category, quinine is used to set the baseline of one. Quinine is a natural white-colored crystalline alkaloid with a variety of therapeutic properties. Quinine has been used as a fever reducer (antipyretic) and an anti-inflammatory agent. The most common use for quinine however is in the prevention and treatment of malaria. Both the Union and Confederate armies understood the value of quinine as a treatment for ague or intermittent fever that frequently occurred in soldiers

FIGURE 2.5 From medicines to poisons. The Bitter Truth, early 1800s strychnine bottle.

sleeping in swampy wet areas. The symptoms of what later became known as malaria are chills, shakes, fever, and headache. The Union army alone purchased 595,544 four-ounce tins of quinine sulfate (Fig. 2.5) during the American Civil War [3].

You may have also heard of the presence of quinine in tonic water. As initially developed in India, tonic water was a carbonated soft drink with a significant amount of quinine dissolved in it. The original purpose was as a prophylactic preventative of malaria. Tonic water today still contains quinine but at a much lower concentration. The US Food and Drug Administration limits the amount of quinine in tonic water to 83 mg/l. This is substantially lower than the therapeutic range for quinine that is 167–333 mg daily. Many of the medicines we take are accompanied by a bitter taste similar to that of quinine.

Strychnine is also a crystalline alkaloid. These colorless crystals are highly toxic in human and were first used as poisons for killing small vertebrates such as mice and rats. Strychnine poisoning causes muscular convulsions and eventually death through asphyxia or lack of oxygen from not being able to breathe. Brucine is a white alkaloid that is closely related to strychnine. While it is poisonous to humans, it requires a substantially higher dose than strychnine does. Medicinally, brucine has been used to treat high blood pressure. Brucine in a concentrated sulfuric acid solution can be used to test for nitrates or nitric acid as the mixture gives off a red color.

The presentation of sour and salty tastes is much more direct. A simple but not totally complete understanding of sour taste is directly related to the acid content of the food or drink you are ingesting (Table 2.3). The intensity and a main portion of the sour taste come from the presence of the concentration of H^+ in the food or drink. The functional groups discussed in Chapter One play a role in the concentration of hydrogen ion in solutions. Most weak acids found in food and drink have one or more carboxyl group that depending on the acid will produce various amounts of H^+ in our food depending on its pK_a. We use the pH scale to measure proton concentration in solution (Fig. 2.6). An increase in the proton concentration in solution makes the solution more acidic, which is a decrease in pH. Solutions with a pH less than 7 will have a sour taste. The lower the pH, the more H^+ in solution and the more intense the

(a)   (b)

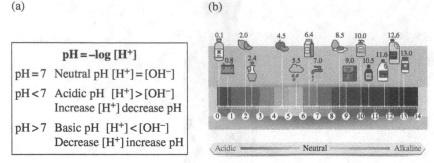

$$pH = -\log [H^+]$$

pH = 7   Neutral pH $[H^+] = [OH^-]$

pH < 7   Acidic pH $[H^+] > [OH^-]$
         Increase $[H^+]$ decrease pH

pH > 7   Basic pH $[H^+] < [OH^-]$
         Decrease $[H^+]$ increase pH

FIGURE 2.6   The pH scale. (a) Equation for calculating pH and pH concepts. (b) The pH scale and the pH of some different common household items, foods, and drink.

sourness of food. Sour candies are often coated with citric acid, and as the solid acid dissolves in our mouth, a high local concentration gives the strong sour taste.

On the taste index, hydrochloric acid (HCl) is used to set the baseline of 1. Hydrochloric acid is a strong acid and is the most common acid in stomach acid, which is very acidic. The other acids on the list are organic acids, which contain a carboxylic acid group. Typically, organic acids are weaker acids and therefore do not release as many protons into solution as HCl and thus have a lower sour taste rating. Formic acid is one organic acid that has a higher taste index than HCl.

Salty taste is primarily controlled by cations in salts. Sodium chloride or common table salt is used to set the taste index (Table 2.4). The most commonly occurring natural salts, NaCl, KCl, and $CaCl_2$, give relatively similar taste sensations. In cooking you will occasionally see recipes that call for sea salt, which is simply salt produced by evaporating seawater. Some cooks believe that it tastes better and since taste is a personal opinion that is justified. There are cooks that talk about sea salt and use the term low sodium. This is inaccurate and misleading. While sea salt is a mixture of salts, it is still approximately 98% sodium chloride by weight. Thus it does have lower sodium but certainly not low enough to help lower sodium levels in your body.

Umami taste is defined by the savory taste of the naturally occurring amino acid L-glutamine. In this case, glutamine has a very similar structure to another amino acid glutamic acid. Table 2.5 shows that glutamic acid is the same structure as glutamine with the terminal amino group replaced with a hydroxyl group. Another common descriptor for umami taste is monosodium glutamate, the sodium salt of glutamic acid.

2.6   GUSTATION: SIGNALING—RECEPTORS, CELLS, AND TISSUE

Each tastant will diffuse through the fluids of the mouth and bind to protein receptors embedded in the surface of the taste receptor cells clustered in our taste buds. Taste buds are small groups of taste receptor cells found throughout the tongue. The taste receptor cells are located primarily on the top or dorsal surface of the tongue. The taste receptors are localized into tissues projections called papillae. The surface of the tongue has three distinct types of papillae that are located in three different areas. The papillae are small

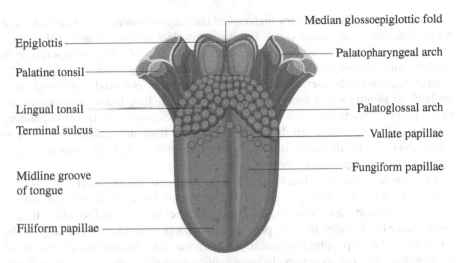

FIGURE 2.7   Anatomy of the tongue.

TABLE 2.6   Numbers of Papillae and Taste Buds per Papillae.

Papillae	Papillae/Tongue (Average)	Taste Buds/Papillae	Total Taste Buds
Fungiform papillae	200	1–18 Front of tongue 1–9 Middle of tongue	1120
Foliate papillae	11	117	1280
Circumvallate papillae	8	252	2200

pimple-, ridge-, or mushroom-shaped structures each a couple of millimeters in size. The three types of papillae are directly involved in taste sensations (Fig. 2.7). Fungiform papillae are pimple-shaped structures located on the surface primarily down the sides of the tongue. As you move down the side of the tongue, the next set of papillae is foliate papillae that are a series of ridges with one distinct set on each side of the tongue. Finally, in the back center of the tongue are the circumvallate papillae that present functionally as a row of mushroom-shape structures at the back of the tongue. Each of these types of papillae contains multiple taste buds (Table 2.6).

Each taste bud is composed of 50–150 taste receptor cells, a substantial series of basal and support cells that surround the receptor cells, and a set of sensory afferent neurons that carry sensory inputs from the receptor to the brain. The number of each type of papillae and the number of taste buds in each type of papillae vary. Table 2.6 gives average values for these numbers. A fourth type of papillae called the filiform papillae, which are long cone-shape structures, are present on the tongue. The filiform papillae are the most numerous papillae on the tongue. The filiform papillae don't participate in gustation and serve a mechanical function.

On average, people have 2000–5000 taste buds with the extremes of the range being from 500 to 20,000. As stated earlier, each of the taste buds has 50–150 taste receptor cells. People at the very high end of the range of number of taste buds have been identified as supertasters. It is estimated that 35% of women and 15% of men are categorized as supertasters. The evolutionary value of being a supertaster is unclear. Supertasters would have an advantage in avoiding noxious or toxic substances because of the increased ability to taste them in potential food items. It is difficult to identify other evolutionary advantages. In fact, many supertasters are rather picky eaters as the tastes of many foods and beverages can be overwhelming to them. A few bites of a rich desert or a salty main dish are typically sufficient to satiate the desire for that taste. But that does not mean that all picky eaters are supertasters or that all supertasters are picky eaters.

The structure of the taste bud lies beneath the surface of the tongue (Fig. 2.7). Thus contrary to popular belief, you cannot see your taste buds. What you see are the papillae on the surface of your tongue. The opening between the surface of the tongue and the taste bud is called the taste pore. The taste pores are not located directly on the top of the gustatory papillae but typically located near the sides and in the crevasses of the papillae. The taste receptor cells are modified epithelial cells with microvilli that project into the taste pore. The microvilli contain the membrane receptors that bind the tastants to initiate the taste process. The taste cells are surrounded by basal cells that support the structure of the taste bud and taste cells but do not directly participate in gustation. The taste cells form a chemical synapse with the sensory neurons leaving the taste buds. When activated, the taste cells release neurotransmitter activating the sensory neuron that carries electrical impulses into the brain.

As scientists learn more about the nature of the taste receptors, taste receptor cells, and the anatomy of taste buds, they have found the old model of taste map of the tongue (where different regions of the tongue were thought to be responsible for a specific taste) is wrong (Fig. 2.8). This map did not include umami or account for all sorts of types of taste receptor cells for each of the flavor types. As stated earlier, each taste bud contains 50–100 taste receptor cells. And each taste receptor cell is unique and a taste bud is comprised of a diverse collection of taste receptor cells with several kinds of taste receptors. Each taste receptor cell will only have one kind of taste receptor and is associated with its own afferent nerve that signals to the thalamus and neocortex. This means that each taste bud is wired and can recognize a variety of different tastes and flavors. The makeup of each bud and density of the buds can vary leading to the incorrect mythological taste map of the tongue.

Complicating matters is that some of the taste receptor cells express only one type of taste receptor (e.g., the sweet taste receptors) and bitter taste receptor cells have many different receptors produced on a single taste receptor cell. Thus for each cell with only one type of receptor, those cells with its associated nerve signal to the neocortex can distinguish between different types of flavors such as sweet. While the several bitter compounds will each bind and activate the same taste cell and its pathway to the brain. The bottom line is that we can tell the difference between different sweet flavors but even though there are thousands of different bitter compounds, we cannot tell the differences between them (Box 2.1). They are all signaling through the same taste receptor cell and nerve to the same place in our brain!

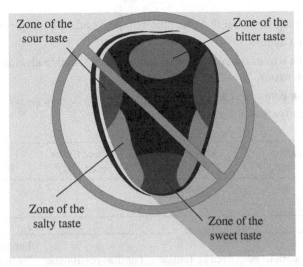

Zone of the sour taste

Zone of the bitter taste

Zone of the salty taste

Zone of the sweet taste

FIGURE 2.8 The mythical tongue map.

BOX 2.1 ARE YOU A SUPERTASTER?

The first step to determining whether you are a supertaster is to evaluate the number of fungiform papillae present on your tongue. The fungiform papillae are small pink bumps on the surface of the tongue. The following process will make it easier to count the papillae and determine whether you are a super taster.

Supplies

- Vial of blue food coloring
- 2 paper towels per person
- 2 cotton swabs per person
- 2 cotton balls per person
- 2 sticky notebook hole reinforcement circles per student
- A magnifying glass
- A ruler with a millimeter scale

Procedure

- Work with a partner.
- Use the cotton swab as an applicator to paint the surface of the tip of your partner's tongue with the blue dye.
  - Put a few drops of the food coloring on the cotton swab.
  - Paint the tip of the tongue with the dyed cotton swab.
- Move your tongue around in your mouth to make sure the dye covers the entire tongue. The dye is safe to swallow.
- Pat your tongue dry with one paper towel.
- Place the reinforcement circle on the tip of your partner's blue-dyed tongue. This circle defines the sample area on the tongue.

- Count the fungiform papillae (the pink bumps) inside the sample area.
- Record your data.
- Measure the size of the person's tongue (Measure length and width of the front 1/3 of the tongue).
- Record your data.
- Repeat the experiment for the partner.

Partner	Papillae Counted (P)	Diameter of Circle (mm)	Area of Circle $A_c = \pi(0.5\,D)^2$	Papillae per mm² (PPM = P/A)
1				
2				

Partner	Dimensions of Tongue	Area of Tongue A_T (mm²)	Papillae per Tongue (PPT = PPM × A_T)	Fungiform Papillae Taste Buds per Tongue (PPT × 8)
1				
2				

Questions

- How does the number of fungiform papillae on your tongue compare with the average in your data table?
- From the calculation, how does your total number of fungiform papillae taste buds compare with the average in your data table?
- Are your numbers high enough to be considered a supertaster? Explain.

2.7 GUSTATION: MEMBRANE PROTEINS, MEMBRANE POTENTIAL, AND SENSORY TRANSDUCTION

As we have seen, taste sensations are categorized as five distinct sensations. With these five distinct flavor sensation, humans can distinguish between 4000 and 10,000 different chemical sensations. The sensory transduction that permits these thousands of different sensations is broken down into two classes. Salty and sour tastes function through the direct use of ion channels, while sweet, umami, and bitter tastes use a family of membrane receptors calls G protein-coupled receptors (GPCRs).

2.7.1 Taste Signaling

The basic order of taste or gustatory signaling is for the tastant to bind to its specific receptor on one of the cluster of taste receptor cells found in the taste bud (Fig. 2.9). Regardless of the type of tastant, the taste receptor cell will cause the release of a neurotransmitter stored in the taste cell. The neurotransmitter will bind to the afferent

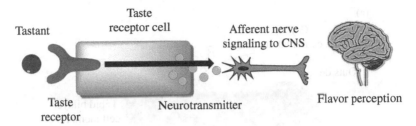

Tastant   Taste
           receptor cell   Afferent nerve
                            signaling to CNS

Taste   Neurotransmitter   Flavor perception
receptor

FIGURE 2.9   Tastant signal transmission. The basic flow of information from a taste receptor to the central nervous system begins with a tastant binding to its specific receptor.

nerve and start a wave of membrane potential to ultimately end in other nerves in the brain. This differs from olfactory signaling where the tastant binds directly to the nerve starting the signal, bypassing the taste bud altogether.

2.7.1.1 *Plasma Membrane Basics*   Cells are the structural, functional, and reproductive units of living systems. All cells are enclosed by a plasma membrane that separates the cell from its surrounding environment. The fluid mosaic model describes the plasma membrane as being composed of a bilayer of amphipathic lipids and associated proteins (Fig. 2.10). The primary lipids in the membrane are phospholipids and cholesterol. The hydrophobic tail of the amphipathic lipids align forming the nonpolar center of the bilayer while the polar head groups face the aqueous environment inside and outside the cell. The proteins embedded in the membrane provide the primary functional elements of the membrane but are defined in the fluid mosaic model based upon how they are associated with the membrane.

Proteins are associated with the membrane in one of two ways. Integral membrane proteins span both sides of the membrane possessing hydrophobic amino acids that come in contact with the hydrophobic core of the membrane. Peripheral proteins are associated with the polar head groups or bind directly to the integral proteins of the membrane. One common type of integral protein that is important for our understanding of taste and smell are transmembrane proteins. These proteins span the entire membrane having part of their structure in the extracellular fluid, part contained in the hydrophobic core of the membrane, and part of their structure in the intracellular fluid. Figure 2.10 shows the basic structure of a plasma membrane with integral and peripheral proteins.

There are two types of integral proteins present in the plasma membrane essential to taste perception: receptor proteins and channel proteins. Channel proteins are transmembrane ion pores (Fig. 2.10b). When the channels are open, they allow ions to move into or out of cells changing the electrical potential difference across the membrane. The second type of integral transmembrane proteins involved with taste reception is the receptor protein (Fig. 2.10c). Receptor proteins bind to specific molecules in the extracellular fluid, in this case tastants that ultimately transmit information into the cell without allowing the tastants to cross the membrane. The receptors involved in taste perception are receptors that interact with a special class of proteins that also bind to nucleotides called guanosine triphosphate (GTP).

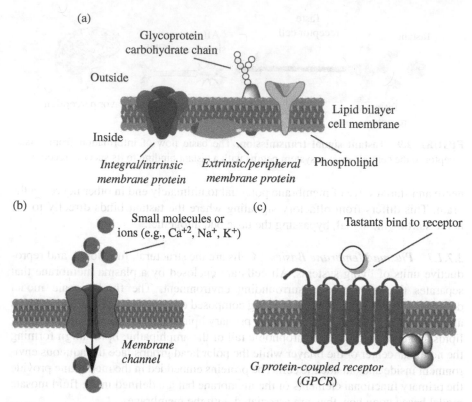

FIGURE 2.10   Plasma membrane structure and function. (a) Fluid mosaic model of cell structure. (b) Transmembrane integral protein that functions as a channel. (c) Transmembrane integral protein that functions as a receptor.

Such receptors that bind signaling molecule on the extracellular portion of the protein (tastant) and interact with GTP-binding proteins inside the cell are called G protein-coupled receptors (GPCR). Sometimes these receptors are called 7-pass transmembrane proteins because of the number of times the protein passes through the membrane (Figs. 2.10c and 2.11).

Once activated by tastant, both transporter/channel and receptor signaling lead to the release of neurotransmitters, which initiate a cascade of events. Neurotransmitter activates the afferent nerve signaling to the brain stem through the thalamus and finally ending in the cerebral cortex. There are three main cranial nerves that bundle all the afferent nerves into the brain stem (VII, IX, and X). The signal is carried from the dendrites through the cell body and axon ending at the synapse. Membrane potential is responsible for the release of neurotransmitters in the receptor cell and for the propagation of signal through neurons.

2.7.1.2   An In-depth Look at Membrane Potential and Neuronal Transmission   The basis for the membrane potential is the fact that both the intracellular and extracellular fluids contain cations and anions. When the distribution of anions and cations

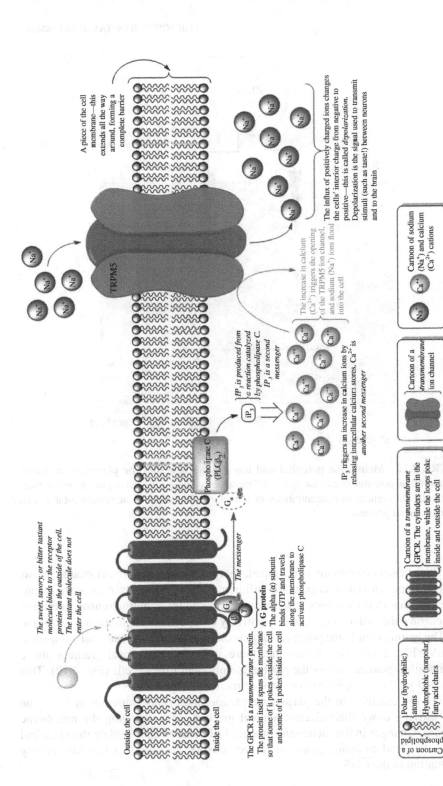

FIGURE 2.11 **Transmembrane proteins that bind and signal the presence of taste and flavor.** Receptors bind and activate integral proteins which in turn will activate GTP-binding proteins inside the cell leading to activation of signaling proteins including phospholipase C. Membrane channel proteins allow ions such as the sodium ion to enter the cell, which results in a change in charge potential across the membrane.

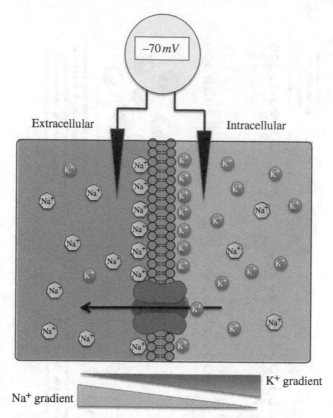

FIGURE 2.12   Membrane potential and ion gradients across the plasma membrane.
The membrane potential is measured as a difference in voltage between the inside and outside
of the cell. Differences in concentrations of ions across the plasma membrane establish an
electrochemical gradient.

across the plasma membrane is not equal, there is a difference in net charge across
the membrane, an electrical potential difference, which is referred to as the membrane potential (Vm). For most cells in the human body, including sensory receptors
and neurons, the inside of the cell is negatively charged relative to the outside of the
cell. The resting membrane potential for neurons and sensory receptor cells is around
−70 mV (Fig. 2.12). That means that there is both an electrical gradient and a
concentration gradient across the plasma membrane of these cells (Fig. 2.12). This
combined gradient is referred to as an electrochemical gradient.

When ion channels in the plasma membrane open, ions can enter or leave the
cells, moving down their electrochemical gradient thus changing the membrane
potential. Changes in the membrane potential of sensory cells initiate the electrical
events that lead to action potential in the sensory neurons that provide sensory
information to the CNS.

(a)   (b)

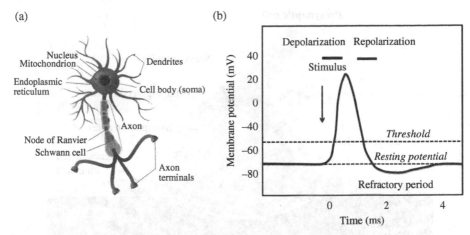

FIGURE 2.13 Neuron structure and action potentials. (a) The basic structure of neurons showing both presynaptic and postsynaptic neurons. (b) An action potential showing the three distinct phases.

2.7.1.3 Neuron Structure, Action Potentials, and Chemical Synapses   Nerve cells or neurons are cells specifically designed to transmit electrical signals from one location to another in the body. All neurons consist of a soma or cell body, dendrites, which are structures that receive neural inputs, and an axon, which sends neural inputs to the next location in the body (Fig. 2.13). The junction between two neurons or between a receptor cell and a neuron is called a synapse. There are electrical synapses where the two cells are in direct contact with each other and there are chemical synapses where the presynaptic cell releases a chemical called a neurotransmitter that stimulates the postsynaptic cell (Fig. 2.14). The synapses involved in gustation are chemical synapses. The chemical synapse is activated when an action potential arrives at the axon terminal. Figure 2.13 shows the basic structure of the action potential. Neurons have a resting membrane potential of −70 mV. When signals enter the dendrites of a neuron, the membrane potential becomes more positive until it reaches threshold. Once the membrane reaches threshold, an all-or-none event known as an action potential is triggered. An action potential involves a rapid depolarization phase when the Vm becomes positive, a repolarization phase when the Vm returns toward normal, a hyperpolarization where Vm drops below normal, and finally a return to resting. Functionally, action potentials move down the axon delivering an electrical signal to another part of the body.

When an action potential reaches the axon terminal at a chemical synapse, a distinct series of events occur that ultimately lead to the initiation of an action potential in the postsynaptic neuron. These events are:

1. Voltage-gated Ca^{2+} channels open.
2. Ca^{2+} enters the axon terminal increasing the Ca^{2+} concentration inside the cell.

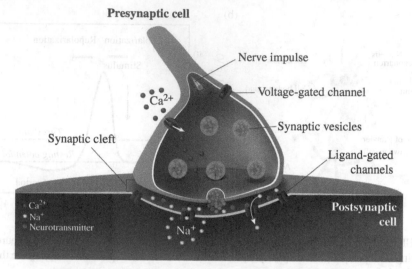

FIGURE 2.14   Signal transmission at a chemical synapse.

3. The Ca^{2+} stimulates the binding of vesicles containing neurotransmitter to the plasma membrane.
4. Neurotransmitter is released into the chemical synapse.
5. The neurotransmitter binds to receptors on the postsynaptic cell.
6. Ion channels in the postsynaptic cell plasma membrane open causing an increase in membrane potential to threshold.
7. An action potential is initiated in the postsynaptic cell.

2.7.1.4   Tasting through Receptors or Transporters   Sweet, bitter, and umami tastants signal differently than sour or salty tastants. Sweet, bitter, and umami signals act in a similar manners to hormones in that both trigger a cellular response without entering the cell. The tastants are small molecules that bind to the receptor proteins found on the surface of the taste receptor cell. These molecules bind the protein receptors in a very specific "lock and key" fashion. The lock (receptor protein) and key (tastant molecule) are very specific in the shape and set of interactions for the two to bind. In fact, most receptors will only bind one or a few very similar shaped tastants. Salty and sour tastes signal to their receptor cells in a very different fashion. In this case, ions and other compounds can bind and open transporters or channels allowing ions to enter into the taste receptor cell. Like receptor signaling, neurotransmitters are ultimately released by the sour and salty receptor cells activating afferent nerves to the brain. Now we will focus on the specifics for each taste receptor cell and tastant.

2.7.1.5   Sensory Transduction of Salty Tastes   The primary salty taste in our diets comes from Na^+. We have evolved to enjoy the taste, and this helps ensure that we take this essential component into our diet. The Na^+ channel found in taste receptors

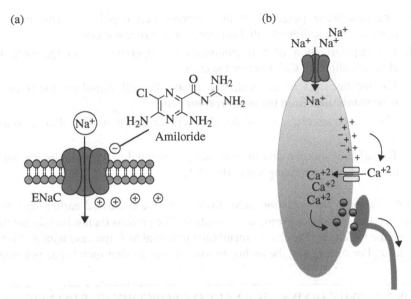

FIGURE 2.15 Salty taste transmission. (a) The epithelial sodium channel (ENaC) transports sodium into the taste receptor cells and is inhibited by amiloride. (b) Influx of sodium leads to a series of signaling events involving calcium and results in neurotransmitter release.

is called epithelial sodium channel (ENaC). As sodium channels function, ENaC is somewhat unique in that it is open to conduct sodium ions most of the time. This means that when Na^+ is ingested, whether in food or drinks, dissolved sodium ions in the fluids of your mouth diffuse to taste receptor cells. When the Na^+ containing fluid surrounds the taste receptor cells, the extracellular concentration of Na^+ increases allowing more sodium to move down its electrochemical gradient and enter the cells. The process driven by entropy does not require additional energy and is very efficient. The electrochemical gradient is the combination of electrical and chemical forces that cause ions to move across a membrane. In the case of Na^+, both forces drive the ion to enter the cell. The inside of the cell is negatively charged ($Vm = -70\,mV$), which attracts cations like Na^+ into the cell. Simultaneously, the Na^+ concentration outside the cell is higher than inside the cell. Thus, the chemical gradient uses entropy to drive Na^+ into the cell. The combined impact of the two components of the electrochemical gradient drive Na^+ into the cell and triggers a series of events that leads to action potentials being sent to the brain indicating you are ingesting salty food. Figure 2.15 shows the events involved in the signal transduction process which are:

1. Na^+ increases in the extracellular fluid around the taste receptor cells.
2. The increase in Na^+ concentration outside the cells increases the electrochemical gradient across the membrane for Na^+.
3. Na^+ enters the taste receptor cell.

4. The membrane potential of the receptor cell depolarizes (becomes less negative). No action potential is formed in the receptor cell.

5. The depolarization of Vm stimulates the opening of voltage-gated Ca^{2+} channels allowing Ca^{2+} to enter the cells.

6. The increase in Ca^{2+} concentration inside the cell stimulates the release of neurotransmitter from the taste receptor cell.

7. The neurotransmitter stimulates the depolarization of the afferent sensory neuron.

8. The afferent sensory neuron sends action potential to the brain indicating you have ingested something salty (Box 2.2).

2.7.1.6 Sensory Transduction Sour Tastes As we discussed earlier, sour tastes are induced by changes in proton concentration. The protons themselves do not move through the ion channel to change membrane potential as is the case with Na^+ ions in salty taste. For sour tastes, the hydrogen ions appear to alter membrane potential in

---

BOX 2.2   WHY DO WE LIKE SALT ON POPCORN BUT DO NOT DRINK OCEAN WATER?

While the previous information is our most current understanding of how salt is "tasted," there are some discrepancies. Amiloride is a drug that binds very effectively to ENaC channels and blocks the sodium taste receptor (Fig. 2.15a). In fact, amiloride is commonly used to block sodium channels in biology and is a key way to tell a sodium channel from other channels. However, amiloride does not block all of the sodium tastant signaling in taste cells. Other salt ions and high levels of sodium are not affected by amiloride. This indicates that there must be another receptor–channel for salty tastants. At low concentrations (<100 mM NaCl) the ENaC receptor is activated in an amiloride-sensitive manner and stimulated a positive taste or appetite for salty foods. At higher concentrations of sodium, typically greater than 300 mM NaCl, a yet undetermined receptor is stimulated. Scientists working at the University of California, San Diego, and Columbia University found that high levels of salt activate both the bitter and sour tastant receptor cells helping to create an aversion for high salt concentrations. In fact, in mice where bitter and sour taste receptor cells have been genetically knocked out, the mice showed a much higher level of salt intake than the control mice. These mice did not avoid high salt solutions, even salt found at ocean water levels. Such results suggest that humans have evolved a trigger to control salt intake. Low levels of salt bind and activate one receptor (ENaC) that is attractive to our senses, and another undiscovered receptor–channel seems to lead to activation of a sour and bitter reaction helping us recognize and avoid potentially dangerous high salt solutions. It is too early to make too many conclusions, but it is suggested that a better understanding would help modulate and control our high salt intake, which leads to hypertension and heart health issues [4].

one of two ways. The protons can bind to and activate the TRPP3 Na^+ channel or bind to and inactivate K^+ channels. The activation of Na^+ channels causes a depolarization due to the increase in positively charged sodium ions entering the cytoplasm of the sensory cell. Blocking the K^+ channel also leads to a depolarization of the membrane. Under resting conditions in the sensory cells, K^+ channels are typically open, allowing positively charged potassium ions to leave the cell. Positive charges leaving the cell help keep the membrane potential more negative. Blocking the efflux of K^+ through the potassium channels retains the positive charge causing a depolarization in the sensory cell. Whether the depolarization is caused by opening Na^+ channels or closing K^+ channels, the end result is the same as we saw with salty taste. Depolarization leads to the opening of Ca^{2+} channels causing an increase in calcium concentration in the cytoplasm and the release of neurotransmitter stimulating the postsynaptic sensory neuron.

As with our understanding of salt gustation, the complete story of how we perceive sour taste is only partly complete. As described previously, protons have a major role in activating channels to drive sour signals to the brain. However it seems that each acid is not the same. Organic acids such as acetic acid and citric acid have a more sour taste than HCl at the same pH (number of protons in solution). Because of the number of total molecules needed to achieve the same pH, there are fewer total HCl molecules in a solution with the same pH as acetic or other organic acids. Over the past 30 years, many scientists have tried to understand the role of both the proton and the anion (acid donor) with many different and confusing results. One current hypothesis is that there is likely more than one receptor–channel and that both the proton and the anion have a role in triggering the sour taste. This could explain why at equal molar concentrations weak acids have different perceived sourness.

There are several new and interesting receptor–channels possibly responsible for sour taste. One study found a proton-stimulated (H^+ gated) channel involved in sour detection is inhibited by amiloride, the same compound that inhibits the sodium channel. This observation indicates that sodium is involved in a new taste receptor–channel, and H^+ alone is not enough to stimulate sour taste. Further showing the role for sodium is that this channel protein, called the acid-sensing ion channel 2a (ASIC2a), is produced in sour taste receptor cells. While we have yet to understand the role of sodium in sour taste, there is a second very interesting question about how we sense sourness. The organic acid that produces H^+ may also play a role. Evidence for the organic acid anion was found when comparing various H^+-producing organic acids. When cells are exposed to the same H^+ concentrations and were held at the same concentration, acetic acid produced a larger signal (taste receptor depolarization) than HCl, indicating a role for both the H^+ and the organic acid anion (acetate) in sour taste. As both solutions were held to the same hydrogen ion concentration, the two solutions should have given the same result, but did not. A new set of genes has been found to respond to different types of acids. This protein, originally discovered in kidney called polycystic kidney disease protein (PKD are ion channel proteins), has also been found in taste receptor cells. The role of these types of ion channels is not clear nor are the specific ions they transport. PKDs are a family of

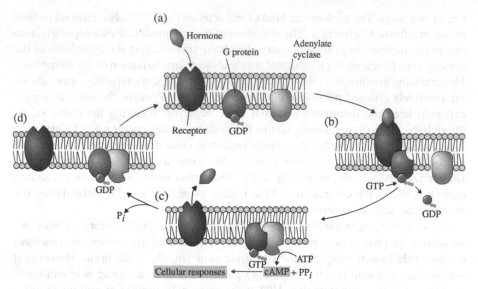

FIGURE 2.16   G protein activation and inactivation. From Voet and Voet.

similar related proteins (PKD1L1, PKD2L1, and PKD1L3) where two of the three
proteins form membrane complexes to allow ions into the taste receptor cells. Mice
which were unable to make this protein could detect sweet and bitter tastes but not
sour. PKD2L1 and LKD1L3 combine to form functional acid sensing receptor–
channels in mouse taste receptor cells. Unfortunately the complete mechanism
describing how these receptors function in taste cells has yet to be elucidated. It is
interesting that several of the PKD acid-sensitive receptors are also found in other
tissue, raising another topic that many of the taste receptors are produced in tissues
that have nothing to do with taste but perhaps allow those cells to respond to sweet,
bitter, or acid conditions (Fig. 2.16).

2.7.1.7   Sweet, Umami, and Bitter Tastes use G Protein Coupled Receptors   As
we briefly presented earlier, the remaining three taste sensations use a family of
receptors called GCPRs. GCPRs are involved in the regulation of a large array of
body functions. These receptors are involved in the regulation of heart rate and
strength of cardiac contraction, stimulate smooth muscle contraction and regulate
blood pressure, and function as the light receptors in the eye that allow vision; and
these are only a few examples. As the name implies all GPCRs are coupled to an
intracellular signaling molecule called a heterotrimeric G protein. These G proteins
have three distinct subunits α, β, and γ (Fig. 2.17). The G proteins get their name
because the α subunit is a guanosine triphosphate (GTP)-binding protein. In the inactive state, the α subunit is bound to guanosine diphosphate (GDP) and is associated
with the β and γ subunits. For G proteins to become active, they must bind to a molecule called a guanine nucleotide exchange factor that allows them to release GDP
and bind GTP. When the α subunit is bound to GTP, it dissociates from the

Sucrose

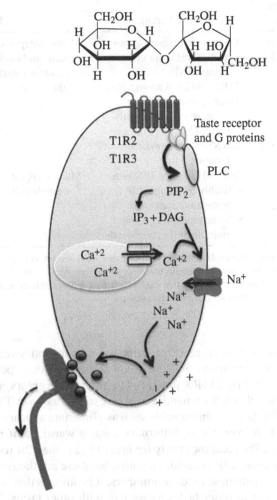

FIGURE 2.17 Sweet taste transmission.

$\beta\gamma$ subunits. The α-GTP can then bind to other proteins activating them, leading to cell-specific responses in the stimulated cells.

Two families of GPCRs-termed taste receptor proteins called the T1R and T2R families account for sweet, umami, and bitter tastes. There are multiple members of each family that bind to different classes of tastants. See Table 2.7 for a listing of variants in these receptors. For example, the T1R receptor family has three family members named T1R1, T1R2, and T1R3. The G protein involved in this system is named gustducin because of its role in modulating gustation. For each type of taste perception, there is a different combination of T1R and T2R receptor

TABLE 2.7   GPCR and Gustation Perception.

Tastant	Receptor Type	Genetic Variation	Misc
Umami	T1R1 and T1R3	T1R1—six different genes with 30 different effective known SNPs T1R3—442 unknown SNPs Three known mutations increase umami sensitivity two fold or more	Mouse receptors bind most amino acids and nucleotides—human receptor mostly nucleotides and glutamate
Sweet	T1R2 and T1R3	706 SNPs with unknown function	
Bitter	T2R	Over 30 different T2R genes • Highly evolving with many mutations • Some receptors can bind multiple bitter compounds	Many receptors are expressed on a single cell

SNPs—Single nucleotide polymorphism: single base changes in the gene of a protein. Some changes may alter the coded protein. SNPs provide genetic variation among people. Known SNP effects are genetic changes that lead to a protein (receptor in this example) with a different function. Such changes could increase or decrease a receptor's sensitivity for its tastant.

subunits involved. The taste receptor cells for umami and sweet taste present specific receptor combinations and thus only respond to one specific taste. The sweet receptors are T1R2/T1R3 dimers. The umami receptors are T1R1/T1R3 dimers. This means that both umami and sweet receptors have T1R3 subunits in their receptors. The taste of amino acids such as glutamate appears to depend upon the T1R1 subunit. As you recall, bitterness usually warns us of noxious or poisonous substances. The receptor family for bitter taste is the T2R receptor. Humans have about 30 different T2R receptor subunits, but there are thousands of different bitter compounds! With these kinds of numbers, it is obvious that we cannot distinguish between different bitter tastes as we can with other tastes. This is because many if not all of the 30 different T2R bitter taste receptors are produced on *each* bitter receptor cell. Thus if one or 20 bitter compounds are in a food, the same signal will be sent to the brain: "stop don't eat!"

The diversity of how we taste and respond to the thousands of different tastants is due to variations of the gene coding for each receptor (Table 2.7). Single amino acid changes can enhance how a tastant binds to a receptor, increasing or blocking binding altogether. There are three different amino acid variations in the T1R3 gene that increase the sensitivity for umami flavors 2- to 10-fold. One of many ways in which domesticated cats differ from dogs is in their lack of interest in sweets. Cats do not express the gene (make protein) for the T1R2 receptor and thus have no ability to distinguish sweet from other tastes. A quick survey of people around you will find a great variation of what is thought to be sweet, bitter, or savory. The T1R receptors (involved with both sweet and umami tastes) have 47 known variants. One bitter receptor gene for the T2R38 receptor (the 38th different bitter receptor) has three different variants. People with different amino acids at position 49-valine, 262-alanine, or 296-valine will not taste the highly bitter phenylthiocarbamide. Such mutations are inheritable and thus families, and, over long periods of times, regions may lose sensitivity to some bitter compounds or tastants. Similar mutations in other receptors are implicated in altering food and affinity for alcohol intake.

GPCR signaling for taste and smell involves the use of second messengers. Second messengers are intracellular signaling molecules produced when a receptor is activated by a primary messenger, in this case a tastant or an odorant. The role of the second messenger is to trigger a series of intracellular events that ultimately lead in the cell changing its function in response to the primary messenger. For taste we will see the second messengers inositol trisphosphate (IP_3) and diacylglycerol (DAG). For smell reception the second messenger used in cyclic AMP (cAMP).

Once the appropriate tastant is bound to its receptor on the surface of a taste receptor cell the activation pathway that stimulates the afferent sensory neuron is the same for each of the sensory receptor cell types. It is the receptors that distinguish the type of signal the sensory receptor cell will send to the brain.

For example, sweet taste perception includes the following events (Fig. 2.17):

1. A sugar compound binds to the sweet T1R GPCR receptor leading to its activation.

2. The activated receptor binds to an inactive G protein.

3. The inactive G protein releases GDP and binds GTP to become active.

4. The active α-GTP dissociates from the $\beta\gamma$ subunits and binds to an enzyme called phospholipase C (PLC).

5. PLC catalyzed the formation of two second messengers, IP_3 and diacylglycerol.

6. The IP_3 stimulates the release of Ca^{2+} from intracellular stores.

7. This increase in calcium ions leads to the activation of another type of Na^+ channel called TRPM5.

8. TRPM5 activation causes a depolarization of the Vm.

9. This depolarization opens voltage-gated Ca^{2+} causing a further increase in the intracellular calcium ion concentration.

10. Neurotransmitter is released stimulating the afferent sensory neuron that sends action potentials to your brain indicating you have eaten something sweet.

The exact same intracellular pathway activating PLC and forming IP_3 is used by both umami and bitter taste receptor cells. The only difference is the GPCR dimers these taste receptor cells express on their surface. Remember that each taste receptor cell contains one type of receptor that defines the adequate stimulus for that receptor cell.

84

THE SCIENCE OF TASTE AND SMELL

To get a true sense of the flavor of food and drink, you need to have a combination of taste and smell receptors working together. That means we need to investigate the olfactory system and its role in taste perception. In other words, as Dwayne Johnson used to say on WWE, "Can you smell what the Rock is cooking?"

2.7.1.8 The Sixth Taste: Fat Increasing acceptance in science and culinary circles is a sixth taste for fat. For most, our understanding of the sensation of fat has been in the texture and smell of lipids or fats. However, people who produce a protein called CD36 are more likely to pick out fat in taste samples than those without the protein. This protein, first found in rodents, seem to ensure a fat-seeking behavior to encourage rats to seek a high-calorie diet. When this gene was knocked out in rats, the preference for fatty foods was lost. Thus far, three receptors for free fatty acids (see Chapter 1) have been identified, CD36, GPR120, and GPR40. Each bind and is activated by different size (chain length) of fatty acids and signals in a similar fashion to other taste receptors, activating calcium and TRPM3 in taste receptor cells.

CD36 is more specific to bind for long-chain fatty acids than GPR120 and GPR40. The latter two receptors are produced in rodents but seem to be at very low levels in the human tongue. Thus so far in humans, CD36 is produced in larger quantities in taste receptor cells of the circumvallate papillae and is implicated as the leading candidate for taste receptor signaling in humans. Like other receptors there are several SNP variations reported. The impact of these genetic variations and the overall level of CD36 production in obesity are currently under intense investigation. The level of CD36 produced in the taste receptor cells changes over time. During the nighttime when food intake is typically low, CD36 levels decrease. Ever feel like avoiding fatty food after eating fat? A high-fat diet also seems to decrease the amount of CD36 produced in taste buds of humans suggesting that a chronic high-fat intake decreases our appetite for fatty foods. Variants that do not change the production of fatty acid receptors could be responsible for a poor satiation of fat intake. Like sweet taste receptors, hormones that impact appetite regulate CD36. Initial studies using hormones that provide a "full feeling" (satiation) decreased CD36 and fat intake in mice. However this had not been determined in humans and care must be taken not to extend the results too quickly. How this knowledge can impact cooking is yet to be truly investigated but has many interesting possibilities.

2.7.2 Flavor Intensity

There is a significant diversity to how we perceive, distinguish between, and recognize the intensity of tastants. The answer to how we achieve this lies in the manner in which each tastant binds and activates its taste receptor(s) on the taste receptor cell. How the tastant fits (lock and key fashion) into the receptor and small amino acid changes to the receptor both can influence the ability of how the tastant binds and sets the signal to the brain. The better the fit, the more tastant molecules will bind to the receptor for longer times and the greater the signal of that flavor is sent to the brain. The number of tastant molecules binding to receptors is one of the ways we distinguish intensity. More tastant molecules lead to a larger number of receptors binding to tastants and the greater the signal that will be sent through the taste buds to the neocortex. However there is a threshold or a minimal concentration of tastant that will trigger a response. The big question that remains unsolved is how we distinguish between types, levels, and mixtures of tastants. How can we tell the difference between different peaches or chocolates? Each taste receptor has a different story providing for its unique ability to respond to the thousands of different tastant molecules we recognize.

2.8 OLFACTION, THE OTHER WAY TO TASTE: BASICS OF SIGNAL TRANSDUCTION

Most of us remember a time when we had a serious head cold or sinus infection. The familiar taste of everything from our coffee to our favorite comfort food was wrong or simply missing. This is not because colds and sinus infections alter our taste buds; this is due to the fact that much of what we consider our sense of taste actually comes from olfaction.

A human's ability to smell chemicals is better developed than our ability to taste chemicals. It is estimated that humans can smell more than 400,000 different substances with the great majority being unpleasant smells (80%). We smell with receptor cells located in the olfactory epithelium located in the superior portion of the nasal cavity with projections through the cribriform plate to the olfactory bulb (Fig. 2.2). The olfactory receptor cells both structurally and functionally are different from taste receptor cells. Olfactory receptors are modified neurons, not modified epithelial cells. The key difference here is that olfactory receptors have their own axons that send electrical signals to the brain. They do not have to release neurotransmitter to stimulate the nerves to the brain.

The olfactory epithelium consists of olfactory receptor cells that have cilia, which is the location of the receptor proteins for smell. These cilia protrude into a mucus layer covering the epithelium. Support cells are also present that help maintain an appropriate environment for the receptor cells including the production of mucus. Finally, there are basal cells that are responsible for the production of new receptor cells. Basal cells are needed because the olfactory epithelial cells have a life span of 4–8 weeks and thus need to be replaced regularly.

The surface area of the human olfactory epithelium is approximately $10\,cm^2$ and, while not very large, it is sufficient to allow us to detect odorants at concentrations of only a few parts per trillion. This is equivalent of being able to find one specific drop of water in an Olympic size swimming pool. It is the size of the olfactory epithelium and the number of receptor cells that determine our olfactory acuity. For example, we all have heard of tracking dogs that can detect the scent of someone who had passed by a location hours earlier. The surface area of the olfactory epithelium of most dogs is greater than $170\,cm^2$, more than 17 times larger than that of a human. Additionally, dogs have approximately 100 times more receptor cells per cm^2 than humans. For dogs with the highest sense of olfactory acuity like the bloodhound, this means a 10,000,000-fold greater olfactory acuity than an average human.

With the human's ability to smell over 400,000 chemical compounds, you might think that signal transduction in the olfactory system would be quite complicated. It is not. The variation in olfactory sensory transduction comes in the number of odorant receptors and not the intracellular signaling process used. Humans have over 350 genes that code for odorant receptor proteins. The extracellular binding domains of these odorant receptor proteins have odorant binding sites with each receptor type being subtly different from the others. Each odorant receptor cell expresses a single type of odorant receptor protein, and each type of receptor protein binds to a specific type of odorant. This means that the activation of each type of receptor cell corresponds to a specific odor.

Odorant receptor proteins function as GPCRs similar to those involved with sweet, umami, and bitter tastes. When you inhale, odorant molecules dissolve in the mucus of the olfactory epithelium. The odorant diffuses through the mucus to the surface of the olfactory receptor cilia and find an odorant receptor. The sequence of events from that point on is the same for all odorants (Fig. 2.18).

The steps involved in olfactory sensory transduction are:

1. The odorant binds to a specific olfactory receptor protein in the plasma membrane of the cilia of an olfactory epithelial cell.

2. The receptor that is a GPCR stimulates the activation of G_{Olf}, the heterotrimeric G protein that functions in the olfactory system.

3. The G_{Olf} α subunit releases GDP and binds to GTP in the activation process.

4. α-GTP dissociates from the $\beta\gamma$ subunits.

5. α-GTP binds to and activates the enzyme adenylate cyclase.

6. Adenylate cyclase converts ATP to the second messenger cAMP.

7. cAMP binds to the cAMP-dependent cation channel causing the channel to open.

8. The cAMP-dependent cation channel is a nonspecific channel that changes membrane permeability to Na$^+$ and Ca^{2+}.

9. The net current flow of Na$^+$ and Ca^{2+} entering the cell leads to membrane depolarization and an increase in Ca^{2+} concentration.

10. The increase in Ca^{2+} opens a Ca^{2+}-dependent Cl$^-$ channel. This allows Cl$^-$ to leave the cell causing an even greater depolarization.

11. The olfactory receptor cells are modified neurons. Thus when the membrane depolarization reaches threshold, the receptor cells fire an action potential that runs into the olfactory bulb then to the brain.

12. Olfactory transduction ends when the odorant diffuses away from the protein receptor and is broken down by enzymes in the mucus. The cAMP production in the cell stops, and the existing cAMP is broken down.

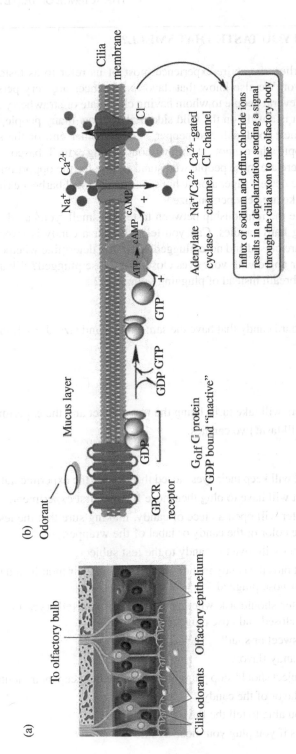

FIGURE 2.18 Sensory signaling in olfactory epithelium. (a) Odorant compounds bind to receptors in the nasal passage activating neural signals to the olfactory bulb. (b) Odorants bind to specific classes of GPCRs initiating a signaling cascade leading to a depolarization of membrane potential.

In the figure:

(a)
To olfactory bulb
Olfactory epithelium
Cilia odorants

(b)
Odorant
Mucus layer
GPCR receptor
GDP
Golf G protein
- GDP bound "inactive"
GDP GTP
GTP
ATP → cAMP cAMP
+
Adenylate cyclase
Na⁺/Ca²⁺ channel
Ca²⁺-gated Cl⁻ channel
Na⁺ Ca²⁺
Cl⁻
Cilia membrane

Influx of sodium and efflux chloride ions results in a depolarization sending a signal through the cilia axon to the olfactory body

BOX 2.3   CAN YOU TASTE THAT SMELL?

As presented earlier, the entire experience most of us refer to as taste is better described as flavor. We also know that flavor preferences are very personal and diverse. I know several people to whom having chocolate or strawberry ice cream instead of vanilla is a walk on the wild side. I also know many people who "use both spices on their food, salt and pepper." On the other end of the spectrum, I know many people who cannot eat chili without adding extra Tabasco sauce first; and of course, there are those people who stand in line for the opportunity to sign a waiver to try the newest super atomic hot sauce as the local barbecue rib festival. The flavors we like are very personalized.

To demonstrate the relationship between taste and smell, you can do a simple experiment using hard candies. Can you tell whether a candy is sweet or sour when your eyes are closed and nose plugged? Can you determine whether a candy is lemon, lime, or apple with your eyes closed and nose plugged? What happens if you hold your breath instead of plugging your nose?

Materials
- Five pieces of hard candy that have the same shape and size. Jolly Ranchers are one example.
- A partner.

Process
- The two partners will take turns being the test subject and the experimenter.
- Each partner will taste two candies.

Experiment 1
- The test subject will keep their eyes closed throughout the experimental process.
- The test subject will have to plug their nose for this first experiment.
- The experimenter will open a piece of candy, making sure that the test subject does not see the color of the candy or label of the wrapper.
- The experimenter will give the candy to the test subject.
- The test subject puts the candy in their mouth, closes their mouth, and tastes the candy with their nose plugged.
- The experimenter should ask whether the subject can tell: (yes this happens with your eyes closed and nose plugged).
  - Is the candy sweet or sour?
  - What is the candy flavor?
- Next the test subject should stop plugging their nose but keep their mouth closed.
  - What is the flavor of the candy?
  - How were you able to tell the flavor?
  - What happens if you plug your nose again?

Experiment 2
- Once again the test subject will keep their eyes closed throughout the experimental process.
- Instead of plugging their nose, this time the test subject will hold their breath while tasting the candy.
- Repeat the same process as before.
- The experimenter should ask whether the subject can tell: (yes this happens with your eyes closed and nose plugged).
  - Is the candy sweet or sour?
  - What is the candy flavor?
- Next the test subject should stop holding their breath but keep their mouth closed.
  - What is the flavor of the candy?
  - How were you able to tell the flavor?
  - What is the difference between the two experiments?
  - Record your observations and repeat the process with the other partner being the test subject.

2.9 TEXTURE, TEMPERATURE, AND PAIN

In addition to taste and smell, the texture of our food, the temperature of our food, and activation of pain receptors also contribute to our eating and drinking experience. Touch and pressure receptors exit in our mouth and tongue just as they do in other surfaces lining our body. We use these to recognize the crispness of a fresh garden salad or the smooth texture of whipped cream. We recognize the change in texture to foods as they are prepared or eaten in different ways. For example, the texture of a crisp, tart apple compared with the chewier, sweeter apples in our favorite apple pie. Both are pleasing to eat but have different texture qualities. Likewise, temperature receptors exist throughout our bodies including on the tongue and in the mouth. We have learned to enjoy a steaming bowl of soup on a cold, rainy day, and most of us enjoy the treat of some ice cream during the heat of summer. Throughout our lives we learn how certain foods are prepared, and we identify a proper temperature for serving and eating these foods. For example, most people in our parents' generation only drank coffee hot. They related the odor and taste of coffee to a hot beverage. For them the idea of iced coffee sounds foolish, and many of them will never develop a taste for drinking cold coffee. Inversely, many of us have picked up our bottle of soda after it has sat in a car on a warm day, and as soon as we take a drink, we realize that we have made a mistake. The beverage just isn't as refreshing warm as it is cold. During our lives we learn to have preferred temperatures for specific food items, and for many of us it is difficult to change those preferences.

The sensation of temperature contributes to foods in ways other than the food being truly hot or cold. There are specific food and beverage additives that give a false sense of hot or cold. For example, people will comment on a cool or fresh sensation when consuming food or beverages containing spearmint or menthol. This happens because the cold receptor cells contain a cation channel, TRPM8, which is activated by these compounds. Conversely, we have all eaten spicy food that we identify as "hot" whether the temperature of the food was hot or cold. The spiciness in hot peppers, for example, is caused by the chemical capsaicin. The capsaicin gives us the sensation of heat by directly activating the TRPV1 ion channel that is present in pain receptors that are normally activated by temperatures that could cause damage to our bodies, temperatures we identify as too hot.

2.10 THE ABSENCE OF TASTE AND SMELL

The loss of the ability to taste and smell can be devastating. Losing your sense of smell can decrease your ability to identify foods by as much as 90%. For many people this takes away the pleasure of the food they eat. As we have seen our ability to taste and smell varies across the population. The National Institutes of Health indicates that 25% of Americans are non- or minimal tasters, 50% are medium tasters, and 25% are supertasters. In general, women are more accurate than men in identifying odors. The sense of smell is most acute in humans between 30 and 60 years of age and then it begins to decline. One of the challenges as many people grow older is the loss of sense of smell that leads them to lose interest in their favorite foods and in some cases eating in general. At this point, food supplements are typically needed. Those people identified as nontasters typically have receptor deficits for both taste and smell, and they describe the sensation of eating very differently from others.

2.11 CONCLUSION

As we look at the science of food and cooking you will read many descriptions of food items. The descriptions will talk about the appearance of certain foods and the textures of other food items or ingredients. The most important descriptions will be based upon how the food tastes and smells. As you do different experiments and try different foods, you will gain a greater understanding of these descriptions. You will also learn that flavor preferences of foods vary greatly from one person to another. Your best friend will make their favorite new drink sound absolutely amazing and when you taste it your response may be blah! No problem. Everyone's tastes are different. The joy of food and cooking comes from learning from others and discovering exactly what your favorite tastes, smells, and flavors are so you can describe them to your friends and family.

REFERENCES

[1]   Marijke, M.F., Barbro, N.M. and Paul de Vos. (2010) A brief review on how pregnancy and sex hormones interfere with taste and food intake. *Chemosens. Percept.* 3: 51–56.

[2]   Yang, Q. (2010) Gain weight by "going diet?" Artificial sweeteners and the neurobiology of sugar cravings. *Yale J. Biol. Med.* 83(2): 101–108.

[3]   Maguder, T. (2006) Quinine was a lifesaver during the civil war. Available at: http://fredericksburg.com/News/FLS/2006/082006/08122006/212900 accessed on November 2, 2015.

[4]   Oka, Y. Butnaru. M., von Buchholz, L., Ryba, N.J. and Zuker, C.S. (2013) High salt recruits aversive taste pathways. *Nature.* 494(7438): 472–475.

3

MILK AND ICE CREAM

Guided Inquiry Activities (Web): 7, Carbohydrates; 8, pH; 9, Higher-Order Protein Structure; 10, Fat Intermolecular Forces, Solids, and Oils; 12, Emulsion and Emulsifiers; 16, Milk

3.1 INTRODUCTION

In this chapter we will use our understanding of science to investigate the composition of milk (lactose, fats, and proteins) and the impact of these macromolecules on taste and cooking and the process of pasteurization and homogenization. The ability of different people to drink milk will be investigated based on genetic variation and diversity for lactose intolerance. We will include the nature of protein and lipid cages of foams and whipping cream, acid-producing microbes, ice cream, and freezing point depression. Finally the information in this chapter includes the physical properties and chemical changes of milk that take place during preparation of butter and cream.

Milk is a basic nutritional component for all mammals providing both nutrients and immune protection from the mother to the newborn for over 4000 species. Animals such as goats, buffalo, sheep, and cow have been used to provide milk and dairy products for humans. Earliest evidence points out that as long as 8000–10,000 years ago, sheep and goats were farmed for their milk in ancient Iran and Iraq. Special cheeses, creams, and other products are still made from goat and buffalo milk, while cow's milk produces drinking milk for many.

So just what exactly is milk? The official US Code of Federal Regulations (CFR) define milk as "the lacteal secretion obtained from one or more healthy milk-producing animals, e.g. cows, goats, sheep, and water buffalo, including, but not limited to, the

The Science of Cooking: Understanding the Biology and Chemistry Behind Food and Cooking,
First Edition. Joseph J. Provost, Keri L. Colabroy, Brenda S. Kelly, and Mark A. Wallert.
© 2016 John Wiley & Sons, Inc. Published 2016 by John Wiley & Sons, Inc.
Companion website: www.wiley.com/go/provost/science_of_cooking

following: lowfat milk, skim milk, cream, half and half, dry milk, nonfat dry milk, dry cream, condensed or concentrated milk products, cultured or acidified milk or milk products" [1] (Fig. 3.1). While technical, it is obvious from this description that milk is a complex mixture of components. Simply put, milk is a liquid secreted by the mammary glands after the birth of young. The milk produced for the first few days is called colostrum, which is pale yellow to clear and has a high concentration of antibodies from the mother to protect the newborn against disease. After a few days, the mature white milk that we are more familiar with is produced and harvested by humans for drinking and cooking.

The basic components of milk are water, fat, casein, whey, and lactose. The relative concentration of each component of milk can change depending on the animal (Table 3.1). The many components of milk can be divided into two nonhomogeneous phases: the slightly acidic, aqueous (i.e., water) phase and the oil or fat phase (Fig. 3.2). The water or aqueous phase is often called the serum. As you may recall, water and oil/fat do not mix—they are immiscible. In milk, the fat phase is dispersed in many tiny fat droplets, also called globules, which are coated with proteins and two phospholipid membranes with a cytoplasmic layer between (Fig. 3.3). If left to stand, the fat phase of raw milk will eventually separate from the watery serum. The aqueous serum also contains dissolved proteins and sugars. The bulk of the milk proteins is found in the aqueous phase of milk and is classified as casein or whey proteins.

FIGURE 3.1 FDA definition of milk. The definition of food and drink is provided and regulated by the US Food and Drug Administration (FDA). iStock 22018405.

TABLE 3.1   Milk Components across Species.

Species	Water	Fat	Casein	Whey	Lactose
Human	87.1	4.6	0.4	0.7	6.8
Cow	87.3	4.4	2.8	0.6	4.6
Buffalo	82.2	7.8	3.2	0.6	4.9
Goat	86.7	4.5	2.6	0.6	4.4
Sheep	82.0	7.6	3.9	0.7	4.8
Horse	88.8	1.6	1.3	1.2	6.2
Rat	79.0	10.3	6.4	2.0	2.6
Ass	88.3	1.5	1.0	1.0	7.4
Reindeer	66.7	18.0	8.6	1.5	2.8
Camel	86.5	4.0	2.7	0.9	5.4

http://evolution.berkeley.edu/evolibrary/news/070401_lactose;
http://www.sciencedaily.com/releases/2005/06/050602012109.htm

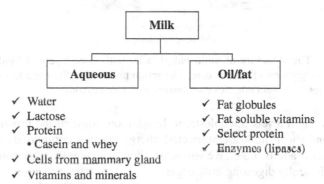

FIGURE 3.2   Composition of milk. Milk is made of two phases, aqueous (mostly water) and oil.

Because there is considerable nutritional value in milk and dairy products, milk use throughout the world remains high. Russia, Finland, and Sweden consume the most milk (130–180l per person per year), while the United States consumes 83l of milk per person per year. A glass of milk provides much needed nutrients for humans. In milk, lactose (a disaccharide) serves as an important source of energy for the human body. Those in the United States typically underconsume calcium, vitamin D, and potassium. Milk is an easy way to make up the deficiency in these important nutrients. Milk also provides two of the "essential" fats that are required for healthy living but not made by humans. Some of the proteins found in milk (whey proteins) are rich in branched-chain amino acids that are often sold as a dietary supplement for their potential role in supporting muscle recovery and preventing mental fatigue. In addition protein from milk provides amino acids used as building blocks for a diverse range of biological molecules including fats, sugars, new proteins, and nucleic acids, the building blocks of DNA. Unfortunately, there is a downside to consumption of milk. While milk contains important fats needed for energy, whole milk and dairy

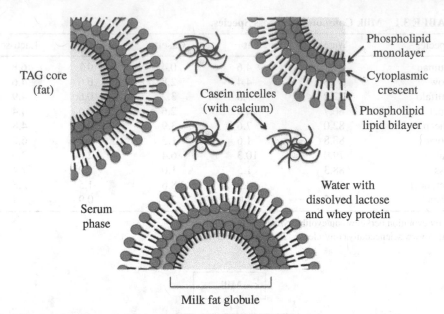

FIGURE 3.3   **The structure of milk.** Fat globules with three layers of lipid membranes encase the triacylglycerol (TAG) fat core. The serum phase with dissolved lactose and whey proteins contains the casein micelles coordinated with calcium ions.

products have high levels of cholesterol. In addition, some of the proteins found in cow, goat, and sheep milk are suspected allergy-causing compounds for 2.5% of infants and some adults. Later, we will learn about how many people throughout the world have difficulty digesting milk sugar.

3.2   BIOLOGY AND CHEMISTRY OF MILK: SUGAR, PROTEIN, AND FATS

3.2.1   Milk Sugar

Carbohydrates are organic molecules made up of carbon, oxygen, and hydrogen (Fig. 3.4). Simple sugars are carbon-based ring structures more formally known as monosaccharides, while disaccharides are made of two monosaccharides. One of the benefits of drinking milk is the energy available from carbohydrates. Nearly 5% of cow's milk is carbohydrate, with most of the sugar in the form of lactose. Lactose provides almost half of the calories of milk and gives milk its sweet flavor. Lactose is a disaccharide made of two simple sugars, glucose and galactose, linked together by a special covalent bond called a glycosidic bond (Fig. 3.5). Glucose is an important monosaccharide that makes up the primary source of potential energy for tissues and cells. The brain uses about 120 g of glucose (about 1/2 cup of solid sugar) each day! Fortunately, milk is a good source of glucose but is not available until lactose is digested into the two monosaccharides. Galactose is also an important nutritional

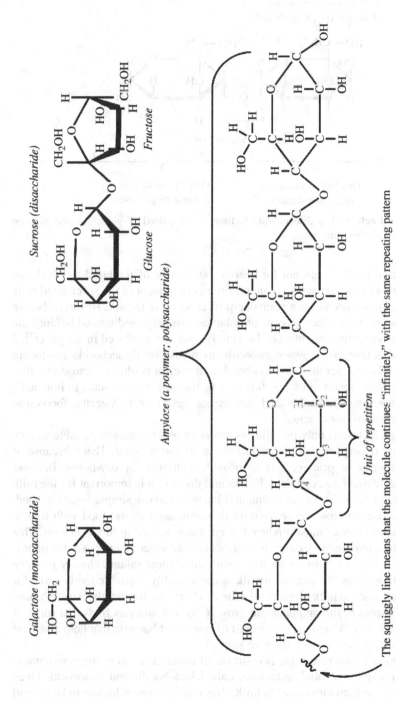

The squiggly line means that the molecule continues "infinitely" with the same repeating pattern

FIGURE 3.4 The forms of carbohydrates. Carbohydrates, primarily of hydroxylated carbons (hydrocarbons aka carbohydrates), are found in mono-, di-, and polysaccharide forms.

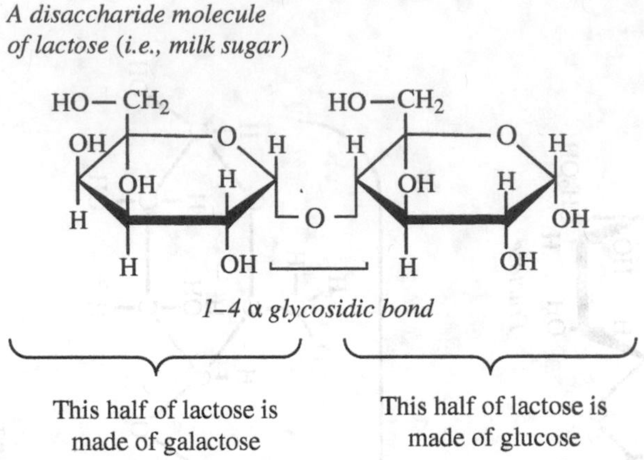

A *disaccharide molecule*
of lactose (i.e., milk sugar)

1–4 α glycosidic bond

This half of lactose is   This half of lactose is
made of galactose   made of glucose

FIGURE 3.5   Lactose. The disaccharide lactose is comprised of galactose and glucose joined by a special glycosidic bond.

component used by the brain not for energy, but as one of the building blocks of nerves and brain tissue. Galactose is an important component of the specialized cells that insulates the nerves and is a component of connective tissues. However, before lactose can be used for either glucose or galactose, the glycosidic bond holding the two sugar molecules together must be broken. An enzyme produced in the gut called lactase binds lactose and a water molecule to break the disaccharide producing glucose and galactose for the body to absorb and use. In adults, galactose (another disaccharide) also supports immune function as this sugar contributes to how antibodies function. Some commercial bakeries use galactose to sweeten foods and reduce the tartness of some acids.

Even though lactose is only one fifth as sweet as other sugars (e.g., table sugar), there is a good reason for milk to use lactose as its flavor agent. That's because of milk's evolutionary or primary role to provide nutrients for newborns. Because infants are particularly susceptible to illness and disease, it is important for the milk to remain safe to drink. Many environmental bacteria can use simple sugars in food, including glucose, to grow; these bacteria then contaminate the food with toxins. However, many microbes must produce the enzymes needed to digest lactose. This takes time. Once exposed to lactose, microbes can take several hours to a day to produce the needed enzymes to digest the sugar, while most infants already produce plenty of lactase. Thus, the lactose in milk is not a readily available food source for most microbes, and the milk remains safe for newborns to drink. Other sugars found in milk also impact health and bacterial growth. Recent analysis finds that some of the components in milk support the growth of important bacteria that help an infant grow and thrive. Nature at work for your safety!

One exception to this rule is the two strains of bacteria found in the environment primed and ready to metabolize lactose, called lactobacilli and lactococci. Once these bacteria have been introduced to milk, they readily convert lactose to lactic acid causing the milk to sour or curdle (the beginning of yogurt and cheese). Unlike many other bacteria, these bacteria do not produce significant toxins, and ingestion of these microbes is safe. In fact, as lactobacilli and lactococci make the milk more acidic by the production of lactic acid, the milk is less livable for many of the dangerous microbes.

3.2.2 Lactose Intolerance

So what happens if a person doesn't produce lactase and continues to drink milk and eat ice cream? If lactose is not broken down in the gut, water rushes into the intestines via osmosis creating a bloated feeling and watery stool. If this wasn't bad enough, the natural bacteria found in the human intestine are able to digest the lactose, producing carbon dioxide, methane, and hydrogen gases. Of course this all results in flatulence, cramps, bloating, and diarrhea. This syndrome is called lactose intolerance and occurs in most adults, and, in a few cases, lactose intolerance is found in young children.

The cause of lactose intolerance is due to a lack of production of the enzyme lactase that breaks down lactose to glucose and galactose in the gut (Fig. 3.6). Most children produce the enzyme; however, lactase production decreases with age. The reasoning for this is pretty straightforward. Milk is primarily a food for infants. As humans get older, their diet provides glucose and other carbohydrates, reducing the need to produce lactase. Over time, genes coding for lactase are switched off (or not activated) resulting in less and less enzyme being produced. Because making protein takes energy, avoiding protein production that is not needed is a way the body saves

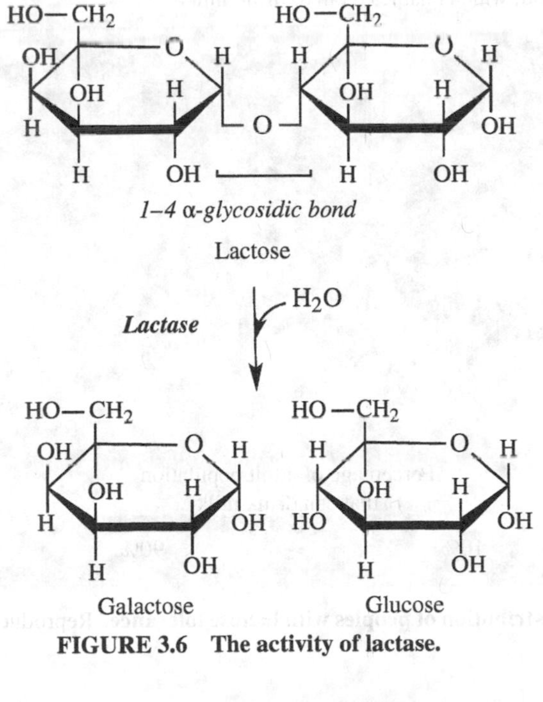

1–4 α-glycosidic bond

Lactose

Lactase

Galactose   Glucose

FIGURE 3.6 The activity of lactase.

energy in creating unnecessary protein. In fact, other than people from Northern Europe and parts of central Africa, few adults retain the ability to metabolize lactose. Globally, over four billion people have lactose intolerance; in fact only 39% of the world's population have been found to be lactose *tolerant* (Fig. 3.7). Like many other cultures, Northern Europeans domesticated animals to produce milk providing a good source of nutrition and energy. Yet, as late as 7000 years ago, a genetic mutation in ancient Europeans arose that allowed for the continued production of lactase. Because of the more temperate climate, dairy cattle could be maintained year-round, and milk and dairy products could be safely stored and consumed long after milking. This allowed those with the mutation to take advantage of the energy and nutrients stored in milk long after breast-feeding and provided a natural advantage to those carrying this mutation [2, 3]. Strangely, the trend of colder climates and persistent lactase production does not align with the observation of lactose tolerance in nomadic tribes from Africa and the Middle East. The current scientific thought is that their frequent migration allowed these tribes to raise and maintain dairy cattle by avoiding the extreme temperatures of the region.

So what to do if your stomach doesn't agree with ice cream, milk, or other lactose-rich foods? Avoid those foods or take Lactaid! The enzyme lactase is purified from human-friendly yeast such as *Aspergillus niger* and can be purchased in pill form or premixed in milk or other dairy products. Lactaid includes a small dose of

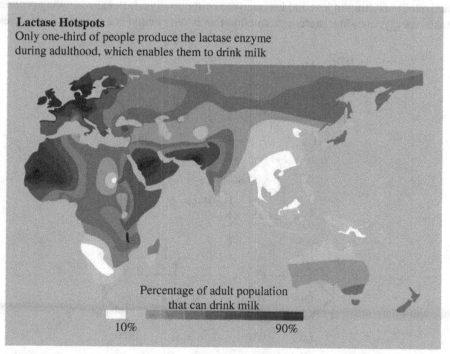

Lactase Hotspots
Only one-third of people produce the lactase enzyme during adulthood, which enables them to drink milk

Percentage of adult population
that can drink milk

10%   90%

FIGURE 3.7   Distribution of peoples with lactose tolerance. Reproduced with permission from Ref. [3].

the enzyme in each pill or already added to milk to predigest the lactose. The milk is then safe to drink and will not cause the lactose effect on those with lactose intolerance syndrome.

3.2.3   Lactose and Ice Cream

Lactose affects the flavor of milk and can cause metabolic problems, but lactose can also have an important effect upon the texture of milk-based food. Ever wonder why ice cream that has been partially melted and refrozen has a gritty feel? The sandy texture is due to lactose that has crystallized. Lactose can flip its —OH on the glucose portion of lactose. When dissolved, lactose converts between these two shapes, α-lactose and β-lactose. The versions of lactose are called anomers, and while they are chemically the same, there are a few differences. When the —OH is below the ring of glucose, lactose is considered to be in the α-anomer. When the —OH is above the ring, lactose is in the β-anomer. At room temperature, there is more β-lactose than α-lactose. At high cooking temperatures, the β-lactose is less soluble and falls out of solution leaving more α-lactose in the food. At low temperatures, the α-lactose anomer crystallizes into very hard solid grains, giving refrozen ice cream that gritty feel. Therefore, milk that has been cooked, then cooled, and later frozen will have the gritty texture of crystallized α-lactose (Fig. 3.8).

3.2.4   Milk Proteins

About 3.3% of milk content is protein and, like lactose, protein is found in the aqueous or liquid phase of milk. While there are thousands of milk proteins, most often milk proteins are easily thought of as divided into two basic classes of proteins, casein protein and serum or whey protein. If you were to add vinegar or another acid to milk, you would create white clumps of fat (i.e., the curds) surrounded by a yellowish white liquid (i.e., the whey)—in other words, you would make cottage cheese! Curds are the fats and proteins that become insoluble when milk is acidified. Whey is the mixture of water and proteins still in solution after the milk has been acidified and curdled. Milk proteins are divided into the acid soluble (those proteins that are stable at pH 4.6 or greater and remain in solution) and acid insoluble (those proteins, mostly caseins, that denature and precipitate as the milk turns acidic). You will learn more on curds in Chapter 5.

3.2.5   Proteins

Proteins are long polymers of hundreds of individual building blocks called amino acids. When bonded together, proteins fold into long tangles forming specific shapes giving the protein many of its valued properties. There are 20 common amino acids that make up protein chains. All but nine of these amino acids are produced by the human body, and the remaining nine amino acids (called the essential amino acids) must be included in the diet. Milk provides all 20 amino acids with the whey and other proteins (Fig. 3.9).

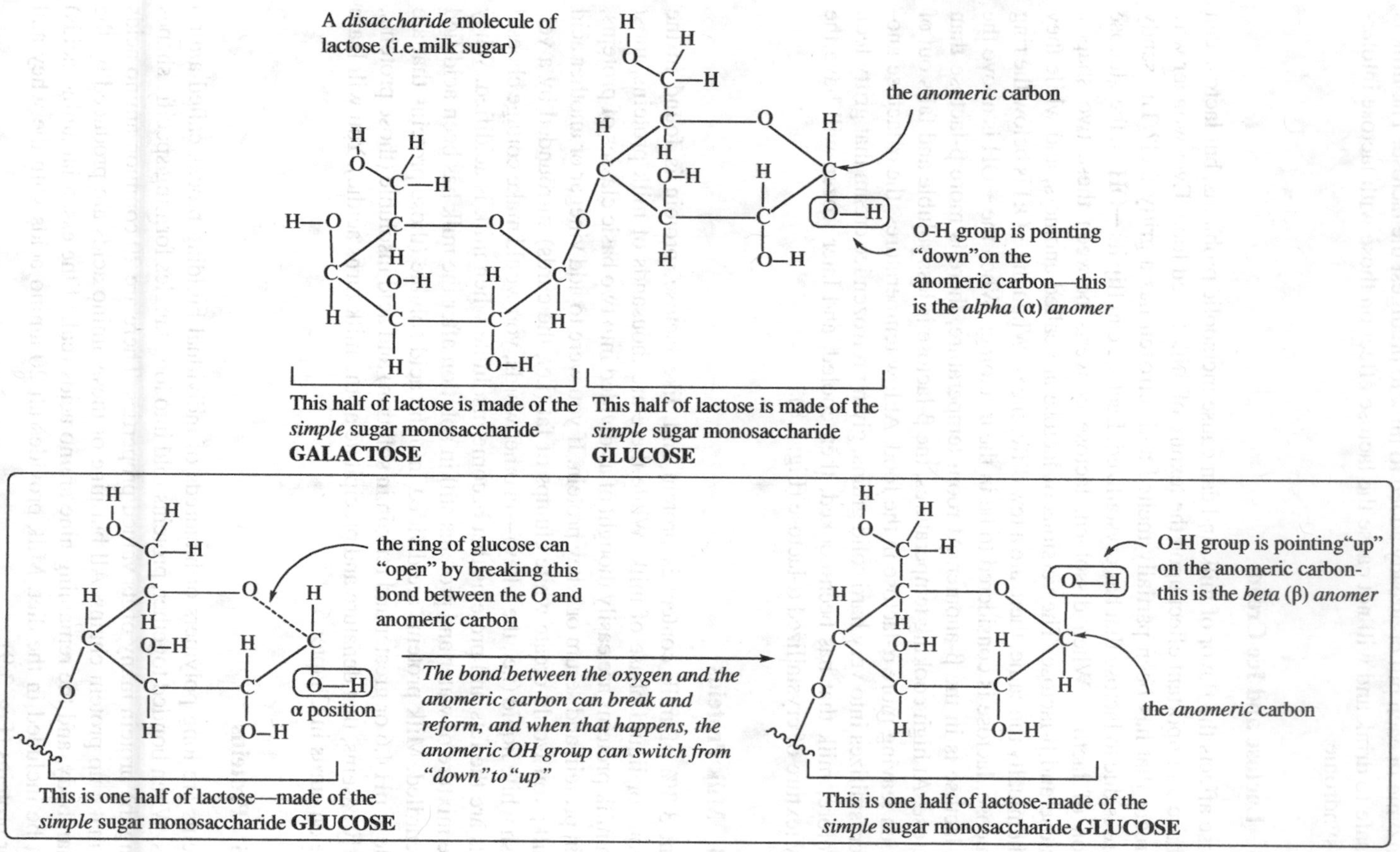
A *disaccharide* molecule of lactose (i.e.milk sugar)
the *anomeric* carbon
O-H group is pointing "down"on the anomeric carbon—this is the *alpha* (α) *anomer*
This half of lactose is made of the *simple* sugar monosaccharide **GALACTOSE**
This half of lactose is made of the *simple* sugar monosaccharide **GLUCOSE**
the ring of glucose can "open" by breaking this bond between the O and anomeric carbon
α position
The bond between the oxygen and the anomeric carbon can break and reform, and when that happens, the anomeric OH group can switch from "down" to "up"
O-H group is pointing"up" on the anomeric carbon-this is the *beta* (β) *anomer*
the *anomeric* carbon
This is one half of lactose—made of the *simple* sugar monosaccharide **GLUCOSE**
This is one half of lactose-made of the *simple* sugar monosaccharide **GLUCOSE**

FIGURE 3.8 Lactose-flipping anomers in ice cream.

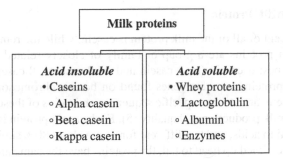

FIGURE 3.9   Organization of milk proteins.

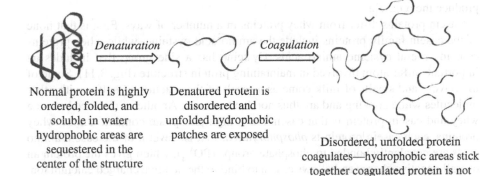

Normal protein is highly ordered, folded, and soluble in water hydrophobic areas are sequestered in the center of the structure

Denatured protein is disordered and unfolded hydrophobic patches are exposed

Disordered, unfolded protein coagulates—hydrophobic areas stick together coagulated protein is not soluble in water

FIGURE 3.10   Native and denatured protein. Heat, acidic, alkali, or other conditions will lead to the loss of protein structure called denaturation. Some proteins, when denatured (unraveled), can tangle together forming insoluble aggregates. Reproduced with permission from Ref. [4], figure 4.6.

When proteins are folded correctly, they are considered to be in their native state. When heat, acid, and chemical reactions alter the ability of the protein to hold their shape, the protein is unraveled or denatured. Depending on the protein and the conditions, denatured proteins can coagulate and form highly tangled webs of proteins that are insoluble precipitates (Fig. 3.10). For milk, this last precipitation is how curds (coagulated insoluble denatured proteins) are formed. Whey proteins remain in their folded, native state when milk is acidified, while those proteins that unravel and denature (i.e., the caseins) form clots of proteins we know as curds. However, both casein and whey proteins stay native and folded at high temperatures, giving milk the ability to withstand cooking and boiling without clumping. This high-temperature stability is why milk is used for creams and sauces and in hot drinks like coffee, tea, and hot chocolate.

3.2.6   Casein Milk Protein

Eighty-two percent of all of the milk protein is casein, while the remaining are whey proteins. Casein proteins are a group or family of closely related proteins, which include three types: α casein (α-s1 casein and α-s2 casein), β casein, and κ casein (kappa casein) protein. Separate genes found on bovine chromosome six code for each type of casein, and it is the specific sequences and ratios of these casein proteins that determine milk production and quality [5]. Each casein protein is a long polymer of about 209 amino acids, and the different forms are about 80% similar in sequence. That means, when lined up head to tail, the proteins have the same amino acids at the same positions 80% of the time. The extent of this amino acid similarity indicates that these proteins are highly *homologous*—that is, each of the individual casein proteins likely came from a common ancestral gene coding for a protein found in ancient milking cows. Cow's milk contains all four caseins, while humans do not produce the α casein.

Casein proteins differ from whey proteins in a number of ways. First is that none of the casein family proteins include the amino acid cysteine, while whey proteins contain several cysteine amino acids. Cysteine has a side group that includes an important sulfur atom involved in maintaining protein structure (Fig. 3.11). Some of the flavors and smells of milk come as a result of the reaction of sulfur with other molecules while cooking and are thus not due to casein. Another difference between whey and casein protein is that casein has a high phosphate content. Unlike whey proteins, each casein protein is *phosphorylated*, that is, several of the casein amino acids are covalently bonded to phosphate groups (PO_4^{3-}), which gives the protein an overall negative charge and allows casein to bind to the positive charged calcium ion. The casein–phosphate–calcium complex is critical for calcium delivery in humans. Noncomplexed calcium is poorly absorbed in the gut of humans and is eliminated from the body. For calcium to be absorbed (the ability of a molecule to be transported through the intestine and into the body is called bioavailability) through the gut and into the bloodstream, the calcium ion must be complexed with another molecule. Most frequently this is through interactions with casein phosphate or by either vitamin D or citrate (a molecule found in fruit).

Caseins are found in the aqueous (water) phase of milk where they form large clusters called micelles. A micelle is a globular complex of amphipathic molecules found in a water environment (Fig. 3.12). The water-fearing or hydrophobic part of the molecule faces the middle or interior of the micelle. The charged, water-loving (hydrophilic) portion of the molecule found on the outer portion of the micelle

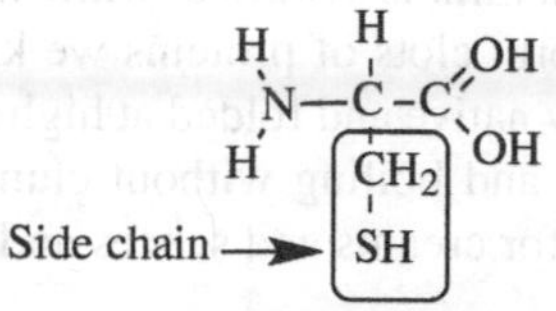

FIGURE 3.11   The amino acid cysteine.

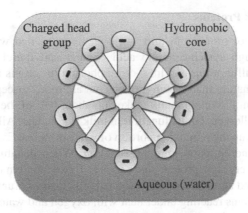

FIGURE 3.12   Model of a micelle.

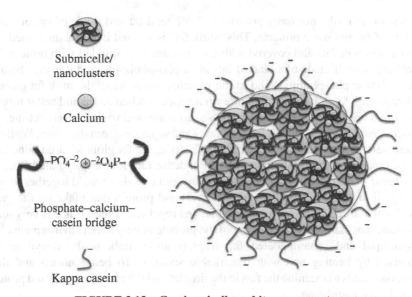

Submicelle/
nanoclusters

Calcium

$-PO_4^{-2} \oplus {}^{-2}O_4P-$

Phosphate–calcium–
casein bridge

Kappa casein

FIGURE 3.13   Casein micelle and its components.

interacts with the water and other molecules suspended in milk. Casein proteins form micelles or is more currently thought of as a mixture of nanoclusters of calcium phosphate/casein in the inner core of the micelle. The calcium acts as a bridge between the negative phosphate charges of adjacent casein proteins and acts as glue to hold the cluster together. On the extreme outer surface of these micelles are the κ casein proteins (Fig. 3.13). These proteins form what is called a hairy layer and play a critical role in keeping micelles from clumping together and forming curds. How caseins work to keep micelles from clumping together is an important feature of milk and will be further discussed in Chapter 5.

3.2.7 Whey Milk Proteins

An acid stable set of proteins in milk are collectively called whey protein. Whey proteins have a high cysteine amino acid content and therefore relatively high concentration of sulfur atoms and little or no phosphorus atoms and are much more difficult to acid-denature than casein proteins but are more susceptible to heat denaturation than casein. β-Lactoglobulin makes up about half of the whey proteins and is a major human allergen, but its function in milk is not clear. Albumins, antibodies, and enzymes make up the remaining portion of whey proteins. Each has an important role in nutrition and supports immune function. When denatured by heat, whey proteins can form very tight protein curds such as those found in ricotta cheeses. The strong rotten egg and ammonia smell of cooked milk is the sulfur and nitrogen atoms found in whey proteins reacting under heat with oxygen and water to form hydrogen sulfide gas and ammonia.

3.2.8 Milk Fat

The rest of milk (the nonwater portion) is 3.5% total fat and made of various sized droplets of fat and some proteins. This "milk fat" is formed into globules filled with different kinds of fats and covered with a membrane skin studded with protein. The membrane is itself made of a kind of fat called phospholipids, and it is the chemical nature of these phospholipids and of the proteins surrounding the milk fat globules that keeps milk fat in solution. Imagine what happens when an oil and water mixture (think of an oil and water Italian salad dressing) is allowed to sit for a short time. The oil coalesces (comes together into one phase) and separates from the water. While the fat and water will separate if raw milk is left alone, the fat globules are maintained, and the cream is an accumulation of the less dense fat globules separated from the more dense water layer of milk. The reason the globules don't meld together to form one big oil layer is because the phospholipids and protein that make up the membrane form a strong cover for the fat droplet and repel each other. It is this very tough membrane that maintains the integrity of the globule in very harsh environments. The phospholipid and protein-coated fat droplets allow milk to be dehydrated or condensed by heating and with remarkable stability. To better understand these forces, one needs to examine the fats in the droplet and the phospholipids and proteins that make the membrane.

Fats, lipids, and oils all describe molecules that do not mix well with water, are hydrophobic, and are primarily made of carbon structures with hydrogen atoms (Fig. 3.14). Lipids and fats are essentially a generic name for single carbon chains (fatty acids) or groups of fatty acids and other atoms, all bonded to a three-carbon molecule called glycerol. Three fatty acids bonded to a glycerol backbone are called a triglyceride. While the fat that makes up most of milk fat is comprised of fatty acids and triglycerides, hundreds of different kinds of fats have been identified in milk. Fatty acids are carbon chains with an acidic carboxyl group on one of the carbon chains. The length of the carbon chain and nature of bonding can vary greatly. Short-chain fatty acids are those with four and six carbons that are produced by the cow later in lactation and give milk its buttery character. Longer chains (8–16 carbons long) are either

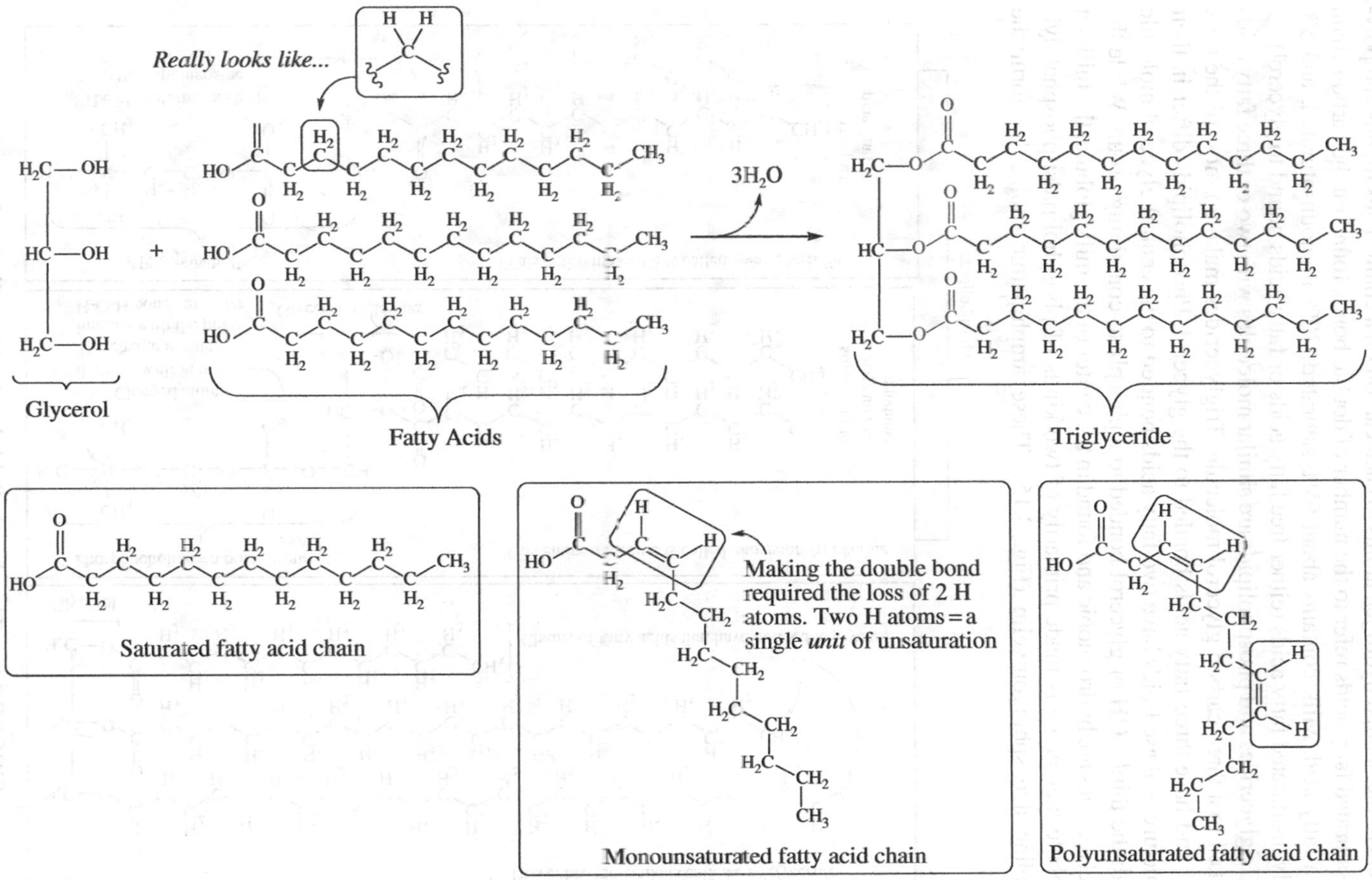

FIGURE 3.14 Structure of fatty acids and triglycerides.

produced by the lactating animal or come from the diet of the cow. The saturation refers to the bonded hydrogen atoms to each carbon on a chain. Thus mono- and poly-unsaturated fatty acids refer to the number of double bonds found on the carbon chain of a fatty acid. Milk contains about 65% saturated, 30% monounsaturated, and 5% polyunsaturated fatty acids (either free fatty acids or fatty acids bound to glycerol).

Triglycerides and phospholipids are similar molecules with two or three fatty acids linked to a three-carbon glycerol molecule. Triglycerides make up most of the milk fats and have three fatty acids bonded to the glycerol. Phospholipids differ in their structure as these lipids have two fatty acids bonded to the central glycerol molecule with the third —OH of glycerol bonded to a phosphate-containing group. While the triglycerides are hydrophobic and found in the center of the milk globule, the milk fat globule membrane is made primarily of two kinds of phospholipids: phosphatidyl-choline and sphingomyelin (Fig. 3.15). These amphipathic molecules form the

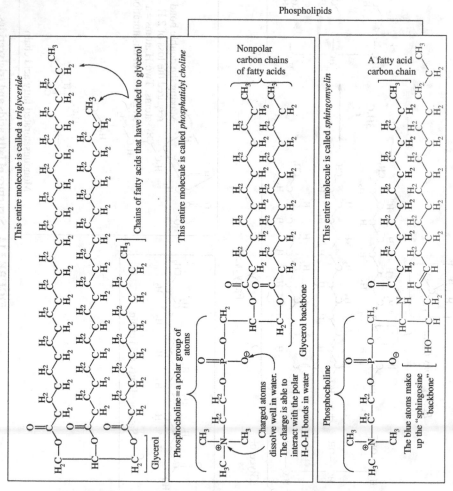

FIGURE 3.15   The phospholipids involved in milk membranes.

membrane by orienting their charged and polar groups toward the watery phase and their hydrophobic tail to the interior of the globule where the nonpolar triglycerides reside. The membrane provides a durable coating that slows down the coalescing of the fat into a single phase. This concept is important when cooking with milk or making creams and butters.

When raw milk is left alone, separation does occur. As the saying goes, "The cream rises to the top." While the fat globules remain intact, the density of fat is less than the density of water. This process of aggregating fat globules at the top of milk is called creaming, and the resulting fat globule layer is often separated to be used to make butter, whipped cream, or soft cheeses. If the rate wherein milk fat globules rise in raw milk is measured, you might be surprised to learn that it happens faster than should occur due to density alone. Some of the proteins coating the fat globules attract each other, causing the fat globules to aggregate and rise more quickly. This process of accelerated creaming is called cold agglutination and happens as the immune proteins in milk bind to each other and proteins on the fat globules bring about clustering and quicker separation than density differences of oil and water alone can account for.

3.2.9 Homogenization and Pasteurization

To avoid creaming and prevent contaminating microbe growth that will acidify the milk and potentially cause disease, raw milk is quickly shipped to local processing centers where raw milk is homogenized and pasteurized. These are two different processes that allow milk to be stored, shipped, and safely consumed long after milking the cow. Homogenization is a process that breaks the milk fat globules into much smaller and more uniformly sized fat globules. Fresh from the cow, milk fat globules range from 1 to 10 microns (μm); after homogenization the globules range from 0.2 to 2.0 μm in diameter. Milk is homogenized as raw milk is forced at pressure through very small diameter nozzles. Large milk globules are forced through a narrow opening, which causes the membranes to sheer, essentially breaking the tough globules coverings while mixing together and creating many smaller fat globules. The result of homogenization is the many smaller globules whose fat droplets are only partially covered by membranes. The exposed lipid droplets quickly become covered with casein proteins from the liquid phase of the milk. All of this results in smaller, casein-coated milk fat globules that are less likely to separate from the rest of the milk. The negative charge from the casein proteins ensures that the milk fat globules do not combine and form a solid layer of fat (Fig. 3.16).

Pasteurization is the process of quickly heating and then cooling the milk. Milk is heated to temperatures high enough to kill any contaminating bacteria or other microbes present in the raw milk but not enough heat to destroy its nutritive properties. The heat-stable proteins like casein and tough membranes of fat globules make milk very heat stable, and pasteurization is effective in sterilizing milk for long-term storage. The process of pasteurization was first used to process sake by Buddhist monks and formally used or invented for beer and wine by the French chemist Louis Pasteur in the 1860s. The concept of pasteurization is to heat the food or drink enough

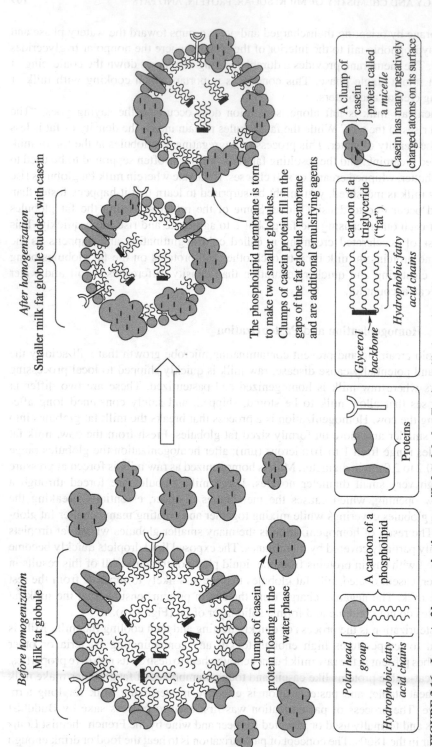

FIGURE 3.16 Impact of homogenization on milk fat globules. Forcing milk through small pores creates a sheering force that tears the micelle into smaller globules and encourages the integration of formerly independent casein micelles.

to kill microbes but not cook the food. As a form of sterilizing foodstuffs, pasteurization is about how high a temperature the food is exposed to and for how long. High temperatures are capable of killing most microbes after only a short period of time. Modern pasteurization heats milk to temperatures below the boiling point where nutrients degrade and proteins denature. Most milk is consumed regionally and doesn't need a long shelf life; therefore, it is pasteurized using high temperature for a short time. Raw milk is heated to 71.6°C for 15 s by pumping the milk through heated pipes and then chilled to the appropriate temperature. Treating milk with higher temperatures for slightly longer times (161°C for 20 s) followed by placing the milk into sterile containers creates sterile milk that has a very long shelf life and can be stored up to 6 months without refrigeration. Bacteria, spores, and other methods of disease transmission are eliminated. This method called ultrahigh-temperature processing (UHT) results in a partial loss of some of the vitamins (riboflavin, vitamin C, folic acid, and a few others), and the higher temperatures denature and cook some of the proteins (Maillard reaction—see Chapter 6) and alter some of the fats producing what some consider a slightly off-flavor. UHT milk is popular in regions without local dairy farms as it holds its usefulness longer. Many European countries use UHT milk for more than half of their milk needs due to space and reduced energy requirements, as the milk does not have to be refrigerated (Box 3.1).

BOX 3.1   COMMON MILKS FOUND ON THE MARKET

- Vitamin D or whole milk: Homogenized and pasteurized milk packaged with additional vitamin D added. None of the fat has been removed prior to pack aging (3.5% fat).
- Low-fat or skim milk: Milk in which some or nearly all of the milk fat has been removed. These milks range from 2% fat to less than 0.5% fat (nonfat or skim milk). Because the body of the milk is more watery without the fat, these milks are often supplemented with whey protein.
- Condensed milk: Sweetened or nonsweetened whole milk with much of the water boiled away. This milk was created to serve as a concentrated form of milk and to fight food poisoning during the US Civil War in 1865. Now this is commonly used for a range of different cooking and baking purposes. Originally, sweetened condensed milk has added table sugar to limit bacterial growth.
- Whipping and heavy creams: Cream is the fat globule layer from milk that has creamed. Differences between heavy (30%) and whipping cream (36–40% fat) are primarily in the concentration of fat. Both creams can be used to make whipped cream, although the more fat the better the resulting foam. Half and half is a mixture of milk with cream for a lower percent fat (10–18%).

3.2.10 Whipped Creams and Foams

Milk can be whipped into a foam, which is a type of colloid. To form a foam colloid, one substance (in this case air) is trapped in another. If low-fat skim milk is foamed (<2% fat), as in the foaming of milk for a cappuccino, it is the denaturing proteins that trap the air bubbles much like egg white foam. But if heavy cream (>35% fat) is foamed, as in the preparation of whipped cream, it is the fat that traps the air bubbles. The resulting creams and foams are unique for milk and different from other foams such as egg white foams. Both are types of colloids—where one substance (in this case air) is trapped in another (cages of fat—whipped cream or bubbles of protein and some phospholipids) (Fig. 3.17).

3.2.10.1 Whipped Creams   The directions for whipped cream are pretty simple. *In a large chilled bowl, add one cup of cold heavy cream and whip until peaks are stiff. Add sugar or vanilla just as the peaks are almost stiff.* Have you ever whipped cream that was warm or overwhipped the cream? You likely found cream whipped in this nature was oily or tasted kind of like butter. To better understand what is going on while making whipped cream, one has to understand what is happening to the fat globules of heavy cream during the whipping (Fig. 3.18). Mechanical agitation by the mixer beaters causes the membranes covering the globules to break and shear, forming smaller globules with sections of exposed fat (oil) droplet. Whipping also introduces small air bubbles to the cream, and the interaction of air with the globule also helps to disrupt or break apart the membrane. While whipping, the air bubbles become smaller and the sections of exposed fat within the globules clump or coalesce to form cages surrounding the small air bubbles. The trapping of air increases the volume of the now "whipped cream." As the fat globule membrane is disrupted, some of the casein protein will coat some of the exposed fat surface, which supports the fat–air interface. Stop whipping too early, and the cream remains unconnected milk fat globules from which air will easily disperse. Whip too long, and the globules

FIGURE 3.17   Whipped cream.

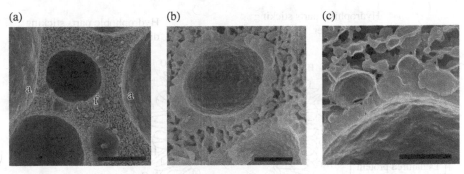

FIGURE 3.18   Electron microscopy of whipped cream. (a) Overview of (a) air and (f) fat globules. (b) Internal structure of air bubble highlights the partially coalesced fat. (c) Interaction of fat globules within the fat layer. Reproduced with permission from Douglas Goff, University of Guelph.

will break down into many smaller fat droplets with too little membrane to cover them. This will result in the beginning of butter, as the fat droplets are mostly naked oil and can readily coalesce into a solid mass of fat. The increase in volume of the whipped cream foam due to the trapped air is called *overrun*, a term often used when making ice cream. A high-fat cream will create a stiff foam capable of trapping a large volume of air bubbles; therefore, the higher the content of fat in the cream, the thicker the whipped cream and the greater the overrun or increase in volume. Thus, heavy cream or country creams with more than 36% total fat will allow for a stiffer, higher-volume cream, and a low-fat cream such as light cream will have a thinner, liquid-like whipped cream consistency. The UHT pasteurized creams are difficult to foam because of the chemical changes that occur in the milk fat globule membrane and proteins during pasteurization. Foaming a UHT pasteurized cream requires the addition of surface-active agents and stabilizers such as whey or gelatin proteins or complex carbohydrates such as gums to provide the needed additional support. So why do most whipped cream recipes direct you to use chilled cream and a cold bowl? Once the milk fat globule membrane is torn and the fat droplet is exposed, the exposed fat needs to remain solid to maintain the cage around the air bubbles. Like any other fat, once the material warms enough, the fat turns to liquid oil, the cage collapses, and the oil coalesces into a solid mass, giving the overwhipped or warmed foam a greasy, buttery feel to the tongue.

3.2.10.2   Foams   Imagine a latte or cappuccino and the milk foam sitting on top. The foam can make or break your drinking experience. In this context, the foam is made of denatured proteins that form a thin protein film holding small air bubbles in place. In the case of milk foams, the milk is "steamed" into a foam by forcefully bubbling hot steam into the milk. The whey proteins are denatured by the heat of steam and the physical agitation of mixing with the heated water vapor and air (Fig. 3.19). Under these conditions, the whey proteins lose their native shape and unravel such that the water-fearing, hydrophobic interior of the protein ends up facing the air and

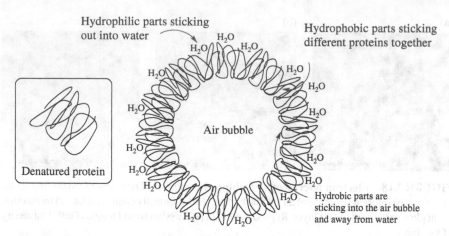

FIGURE 3.19 **Denatured protein and foam.** The layer of denatured protein forms a tenuous cage around the air pocket of milk foam.

the water-loving, hydrophilic portions of the proteins align themselves to face the water portion of the milk. The result is a thin film of protein surrounding small bubbles of air—a foam. The steam doesn't introduce the air as much as it mixes air with the milk and provides the thermal energy needed to denature the whey proteins. The delicate foams of lattes and cappuccinos not only require denatured protein but also an absence of fat. Fat itself is hydrophobic, and when mixed with denatured protein, the exposed hydrophobic regions of the denatured proteins will preferentially cluster around the fat, which will in turn inhibit the formation of the protein film around the air bubbles. Because of the higher content of protein and low fat content, low-fat and skim milk make some of the best milk foams.

3.2.11 Butter

Butter is the result of overwhipping a heavy cream. The milk fat globules, now mostly stripped of the protective membrane, aggregate and form into a solid mass of fat. Water and proteins left over from the cream are squeezed out of the solid fat (the resulting solid is now called butterfat) with small pockets of water distributed throughout the solid butter. The milk fat from which butter is produced is made of a range of different triglycerides composed of different fatty acid chain lengths ranging from 18 carbons with one double bond (oleic acid—32% of butter fat) to the short saturated chain butyric acid representing 3% of the butterfat (see Table 3.2). This complex mixture of fats gives butter its important characteristic of "spreadability." The physical change of melting requires that the interactions between fatty acids to be disrupted by the addition of energy in the form of heat. When the butter temperature is low, the carbon chains of the fatty acids stack tightly against one another, held in place by a type of weak attractive van der Waals interactions called London forces. Heat provides enough energy for the forces holding the fats together to be disrupted as the fat begins to vibrate, rotate, and move more freely with the increased added

TABLE 3.2   Fat Composition in Butter[a]

Fatty Acid	Structure	% Total Fatty Acid in Butter
Oleic acid	$CH_3(CH_2)_7CH=CH(CH_2)_7COOH$	31.9
Myristic acid	$CH_3(CH_2)_{12}COOH$	19.8
Palmitic acid	$CH_3(CH_2)_{14}COOH$	15.2
Stearic acid	$CH_3(CH_2)_{16}COOH$	14.9
Lauric acid	$CH_3(CH_2)_{10}COOH$	5.8
Butyric acid	$CH_3CH_2CH_2COOH$	2.9
Caproic acid	$CH_3(CH_2)_4COOH$	1.9
Capric acid	$CH_3(CH_2)_8COOH$	1.6
Caprylic acid	$CH_3(CH_2)_6COOH$	0.8
Linoleic acid	$CH_3(CH_2)_4CH=CHCH_2CH=CH(CH_2)_7COOH$	0.2
Linolenic acid	$CH_3CH_2CH=CHCH_2CH=CHCH_2CH=CH(CH_2)_7COOH$	0.1

From http://antoine.frostburg.edu/chem/senese/101/consumer/faq/butter-composition.shtml Good link
on evolution and lactose intolerance http://evolution.berkeley.edu/evolibrary/news/070401_lactose and
http://www.sciencedaily.com/releases/2005/06/050602012109.htm
[a] Reproduced with permission of 1997–2010 by Fred Senese.

thermal energy, and as more of the intermolecular forces are broken, the fat turns
from a solid into a liquid oil. The longer and straighter the fatty acid chain (saturated fatty acids are straight, while unsaturated are kinked or bent), the more
contacts each chain makes with the other fatty acid chains and the more heat
required to defeat the forces holding the fats in place. A fat that is made from one
type of triglyceride (a triglyceride is comprised of three fatty acids bonded to a glycerol
backbone) will have a sharp melting curve—there will be a definable temperature
at which the intermolecular forces holding the solid together give way and the solid
melts. In contrast, butter, which is made from triglycerides with a range of different
length fatty acids, will have a broad melting curve. The triglycerides made of
shorter-chain fatty acids will melt at the low-temperature end of the curve, while
the triglycerides made of longer-chain fatty acids are still solid. At higher temperatures the longer-chain triglycerides will have begun to melt but due to their large
size will slow down the faster-moving smaller fatty acids, creating a gentler slope
to that part of the curve. A fat consisting of one kind of triglyceride will quickly
make a transition from solid to liquid at the melting point, while butter with its
mixture of triglycerides and component fatty acid chains slowly changes from
solid to liquid over a much wider temperature range. Within this temperature range,
butter will be both a liquid and a solid and therefore the most spreadable. The
bottom line is that butter is spreadable because it doesn't have one type of fat
giving a sharp melting point from solid to liquid. See Table 3.2 for the different
types of fatty acids in milk fat.

3.2.11.1 Making Butter   Butter can be made from sweet cream or cultured cream.
Sweet cream is a term used to distinguish untreated cream from the cream separated
from milk after bacteria have been introduced and grown to acidify and partially

metabolize milk providing a sour or cultured taste. The cream is often concentrated by heating. To make cultured butter, the cream is inoculated with acid-producing bacteria, such as *Streptococcus cremoris, Lactococcus lactis,* and *Leuconostoc.* These microorganisms use the fats and sugars found in food to produce lactic acid and another flavorful compound, diacetyl, which give milk its slightly sour (lactic acid) and rich, buttery taste (diacetyl). Historically, the cream would be allowed to stand and sour from environmental bacteria. Now it is easy to introduce the correct living bacteria by adding cultured yogurt to fresh cream as cultured yogurt contains the proper living bacteria. Fermentation by these same bacteria also produces sauerkraut from cabbage and the sour in sourdough breads.

The cream is then cooled and aged to allow some of the fat to crystallize. Once churning begins, crystals seed the growth of additional crystals and help to tear the milk fat globule membranes. As the cream is churned, the milk fat globules are broken, and the fats coalesce first into granules and then into larger solid masses. Finally, the residual buttermilk liquid is separated, and the fat is kneaded to remove pockets of remaining buttermilk and to evenly distribute the fat. At this time (or sometimes during the churning process), salt is added as a 10% solution to end up with a 1–2% final concentration. In addition to enhancing the flavor, many harmful bacteria and other microorganisms that can use butterfat as a food source are not able to grow in salty conditions; in this way, the added salt is a preservative, increasing the shelf life of butter as it inhibits bacterial growth.

The buttermilk is the low-fat (0.5%), high-protein liquid remaining after the milk fat has solidified. Modern buttermilk is made by adding *Streptococcus lactis* bacteria to low-fat milk. As before, the bacteria will acidify the milk by producing lactic acid, which causes curdling of some of the proteins and produces the tangy flavor associated with buttermilk. Sometimes citric acid is added to aid in the process, and a second bacterium is added to create diacetyl to give the buttermilk a more buttery taste.

Butter adds a rich flavor to baked goods. Much of the flavor of butter comes from over 120 different compounds including the familiar fatty acids in triglycerides, lactones, methyl ketones, and diacetyl (Fig. 3.20). Short-chain fatty acids with four to six carbon chains contribute to flavor, while longer-chain fatty acids have little taste or aroma. While some of these fats are from the diet of the cow from which the milk came, many are produced when enzymes called lipases in the milk itself and from the bacteria used to culture the cream use water to cleave the free fatty acid from the glycerol backbone. One fatty acid, butyric acid, makes up only a small fraction (2.9%) of the total fat but is responsible for the rancid flavor and smell of rotten milk and butter. Depending on the time of year and what the cow was eating when milked, short-chain fatty acids can be more abundant in the triglycerides of the butter. Over time, the enzyme lipase can remove butyric acid from the glycerol backbone, causing the dairy product to smell and taste rancid.

While the triglycerides found in milk and therefore cream and butter are primarily comprised of fatty acids bonded to a glycerol backbone, a small percentage of triglycerides contain hydroxy acids (~0.3%) and keto acids (~0.85–1.3%) bonded to the glycerol. While the fraction of hydroxy acid and keto acid containing

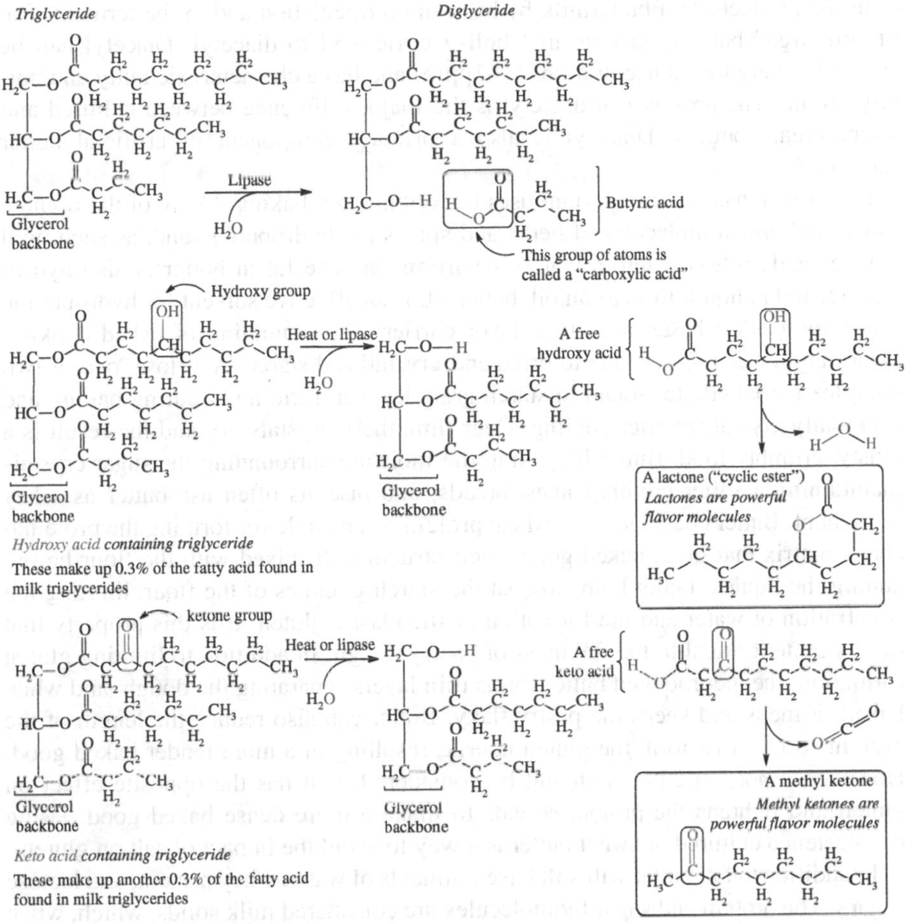

FIGURE 3.20 Generation of some of the aroma and flavor compounds of butter.

triglycerides is very small, these acids can produce powerful flavor molecules. When heat or a lipase cleaves a hydroxy acid from the triglyceride, the product hydroxy acid can lose a water molecule and cyclize into a lactone. When a keto acid is enzymatically or thermally freed from the glycerol backbone, the product keto acid loses carbon dioxide and forms a methyl ketone. Only a very small amount of these molecules is required to be noticed by the human senses. For example, the flavor threshold of small molecules containing methyl ketones is 0.02 ppm when in oil. When milk or butter is heated in making baked goods or ghee (clarified butter), lactones provide the nutty and fruity aromas and flavors of butter sauces. Lactones and methyl ketones are produced readily when butter is heated in cooking. The enzymes in bacteria used to culture cream can also catalyze these reactions and impart flavor to the butter. In the same way, blue cheeses owe much of their flavor and aroma to methyl ketones produced as fats are consumed by the mold *Penicillium roqueforti*.

In the production of buttermilk by bacterial fermentation and in the fermentation of "cultured" butters, bacteria metabolize citric acid to diacetyl. Diacetyl can be smelled at very low concentrations (1–2 ppm) and has a characteristic nutty and buttery aroma. The presence of diacetyl is the major difference between cultured and sweet cream butters. Diacetyl is also a primary component of artificial butter flavors.

Butter also has other important uses in cooking and baking. Many of the organic flavor and aroma molecules of herbs and spices are hydrophobic and, as such, will not be easily released into a watery environment. The fat in butter is also hydrophobic, and in liquid form as an oil, butterfat, is an effective solvent for hydrophobic flavor molecules; butter acts as a flavor carrier. The crumbling of baked cookies, bread, and other foods is due to dried and crystallized starch. Amylose, one of two complex carbohydrates found in starch granules, can form a gel during baking and eventually crystallize after cooling. Over time these crystals dry and the result is a crusty, crumbly food. Butter helps trap the moisture surrounding the sugar crystals maintaining a softer crumb. Cakes, breads, and biscuits often use butter as a key ingredient. Butter coats gluten, a wheat protein responsible for forming the proteinaceous matrix that gives baked goods their structure. If mixed with the flour before adding the liquids, butter helps to coat the starch granules of the flour, limiting the penetration of water and the formation of the elastic gluten. It is this property that makes butter critical to the flakiness of pastry dough. In addition to limiting gluten formation, the incorporated butter forms thin layers separating the dough, and when baked, it melts and keeps the pastry flaky. Butter can also reduce the length of the protein strands that form the gluten matrix, resulting in a more tender baked good. However, a note of caution should be considered. Salt has the opposite effect on gluten and tightens the protein strands to make a more dense baked good. Using unsweetened cultured or sweet butter is a way to avoid the impact of salt on gluten.

In addition to fat, butter will still have remnants of water, whey protein, and lactose sugars. The protein and sugar biomolecules are considered milk solids, which, when heated, fall out of solution and are easily burned. The sugar also provides a source of energy for microbes that shorten the shelf life of butter. One way around this problem is to remove the milk solids and water to produce clarified butter. The smoke point for butter is around 65°C (150°F), but removal of the sugars and proteins raises that smoke point to 450°F/230°C for clarified butter. Clarified butter is made by warming the butter gently to boil away the water, and the resulting protein floats to the top of the butter where it can be skimmed off, while the sugar and some of the proteins precipitate to the bottom of the melted butter, where they can be separated by pouring. The resulting yellow liquid is pure butterfat and can be used for dipping lobster, coating steak, and forming the base of a sauce (roux) and can take the high temperatures of sautéing. Without the sugars and proteins of butter, clarified butter is a more efficient storage of butterfat without refrigeration. Ghee, a clarified butter traditionally made from cow or buffalo curdled milk (yogurt), is widely used in Indian foods where it has been utilized for its storability.

I can't believe it's not… When margarine first came to market as a butter substitute, it was made with buttermilk and animal fat or tallow; modern margarine as a butter substitute is primarily produced from vegetable oil. Margarine is a water-in-fat emulsion of solidified vegetable fat, water, and in some cases skimmed milk for protein and cookability. The solidified vegetable fat begins as vegetable oil, produced from the seeds of corn, sunflower, flax, canola, and other plants. These oil-containing seeds are pressed for the oil and further processed by extraction with a volatile solvent such as hexane. The solvent is removed and oil refined by distillation (see Chapter 12 for more information on distillation). The resulting liquid is filled with mono- and polyunsaturated fats (Fig. 3.21).

These kinked fatty acids do not stack well and are easily melted at room temperature. In order to convert the liquid vegetable oil into a solid fat (margarine), the double bonds of the fatty acids have to be converted to saturated single-bonded carbon. The process of "partial hydrogenation" is the addition of hydrogen atoms to the double-bonded carbons. The conversion of liquid oil to solid fat is accomplished by bubbling hydrogen gas at high temperatures in the presence of a nickel catalyst. The result is an incomplete conversion of triglycerides containing unsaturated fatty acids that are liquid at room temperature to the solid forming saturated fatty acids (Fig. 3.22). This partial hydrogenation is an inexpensive way to produce a butter substitute. The partially hydrogenated oil is then mixed with water, salts, and emulsifiers for taste and cooking, in addition to gums that serve as thickening agents.

Some of the controversy in the consumption of margarine is due to a side product of the reaction. The double bond found in most plants and animals is arranged in what is called a *cis* configuration. This causes the carbon atoms on either side of the double bond to be on the same side (Figs. 3.22 and 3.23). The prefix *cis* is based on the Latin

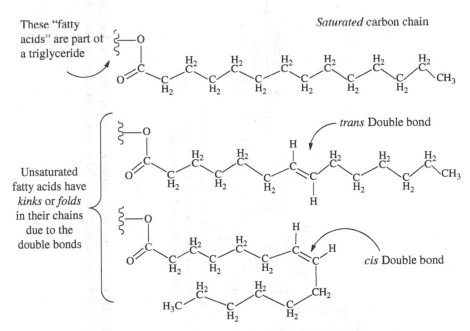

FIGURE 3.21   Unsaturated and saturated fatty acids.

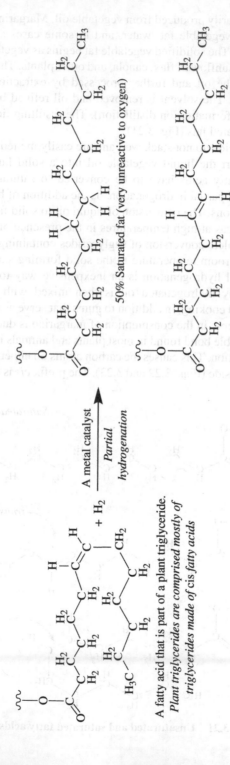

FIGURE 3.22   Partial hydrogenation. Heat and a metal catalyst help to add hydrogen atoms across the double bond of a plant saturated fatty acid. The result is a partial mixture of solid at room temperature of saturated fatty acid and a monounsaturated (*trans*) product.

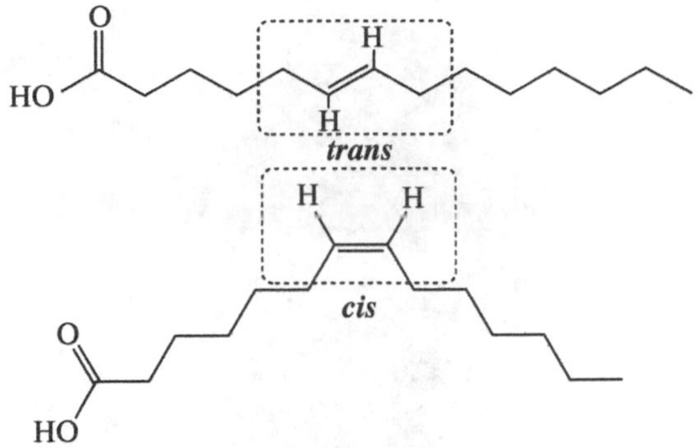

FIGURE 3.23 *Cis* and *trans* fatty acids. Notice the location of the hydrogen atoms on either side of the carbon double bond.

"cisalpine," meaning "on the near side of." This bond creates a bend or kink in the carbon chain and alters the physical character and melting point of the fat. The high temperatures used in hydrogenation create a handful of fats that are in the *trans* configuration. Here the carbons are still in a double bond, but instead of the bend, the *trans* double bond containing carbon chains is straighter—more like a saturated fatty acid. *Trans* fats are potent antimicrobials and, unlike *cis* fatty acids, they resist reaction with oxygen, which results in rancid fats and spoiled food. However, an increase in *trans* fatty acid consumption has been associated with a number of health risks. In particular, a diet high in *trans* fat increases the LDL or "bad cholesterol" responsible for forming plaques and heart disease. Modern hydrogenation processes use different pressures, times of reaction, and temperatures to give rise to fats free of *trans* fats.

Artificial or low-fat butter is another butter substitute that works well for spreading on food but not for cooking. These spreads are a mixture of vegetable oil emulsified with whey protein, buttermilk, and water. Depending on the product, many are very high in starches, gums, and milk proteins, all of which burn easily, making cooking with these butter substitutes difficult.

3.3 ICE CREAM

While the true origins of ice cream are somewhat unclear, there are a number of stories describing how ice cream was first created in ancient China via Marco Polo and other tales of the Roman Emperor Nero sending slaves to the mountains to collect snow for an ice cream-like treat. The rise in popularity and availability of ice cream certainly correlates to key scientific advances, particularly the concept of freezing point depression. The principle of freezing point depression states that adding a salt (solute) to an ice water mixture (solvent) reduces the temperature at which the mixture freezes (Fig. 3.24).

Ice cream is far from being a solid frozen bowl of cream. In fact, ice cream is a mixture of solids (ice and partially frozen milk fat), liquid (unfrozen cream and sugar

FIGURE 3.24   Ice cream. A careful control of chemistry (freezing point depression) and biology fatty globule composition can help make a better ice cream.

water), and pockets of air trapped in the freezing mixture by mixing. These three phases are scattered among each other forming a colloid. A colloid is a mixture with properties of homogeneous and heterogeneous mixtures and is formally defined as a microscopically dispersed mixture in which dispersed particles do not settle out. Ice cream can be described as an emulsion and a foam, both of which are examples of colloids. The formation of an emulsion (the solid phase of frozen fat globules and ice water distributed through the liquid phase of sugar water and cream) is typically unstable, but proteins and lipids coat the fats and stabilize the mixture from collapsing into two separate fat and water phases. These mediators between fat and water phases are called emulsifiers. The foam nature of ice cream is due to the trapped pockets of air created as the freezing cream is mixed. Overrun is the increase in volume of the ice cream before and after mixing due to this trapped air. Some ice cream can have an overrun of nearly twice the volume of the ingredients before freezing and mixing.

The ratio of each phase of the colloid is critical for the mouthfeel and creaminess of the ice cream. Too much fat and the ice cream will have the consistency of butter, too much sugar or milk solids creates a weak ice cream, while the emulsifiers limit the amount of crystals keeping the ice cream from becoming crunchy.

Federal standards (21 CFR § 135.110) require that ice cream contain a minimum of 10% milk fat and 20% milk solids—the solids refer to proteins and sugars like lactose or sucrose. Most ice creams include stabilizing emulsifiers to minimize the formation of ice and fat crystals that decrease the taste of ice cream. Fat is important for taste, providing both a creamy feel to the tongue and sweetness. The proteins and sugar add body or chewiness to the ice cream. A number of different stabilizers or emulsifiers can be found in ice cream. Added whey protein or gelatin protein from muscle tissue is used to coat the fat and provide the body. Custards use egg yolks, which have the phospholipid lecithin as an emulsifier. Another commonly used emulsifier is Polysorbate 80. This is a complex carbohydrate with a long unsaturated fatty acid bonded to it. As an emulsifier in ice cream, Polysorbate 80 can be found in fairly high concentration where it keeps the ice cream scoopable. The carbohydrate portion of the molecule interacts with water and protein, while the fatty acid tail of Polysorbate 80 hydrophobically interacts with the fat globules. This coating keeps the fat and water phases together. Stabilizers include complex carbohydrates (starches and gums) and are commonly found in the ingredient list of commercial ice cream. A common additive used to reduce the formation of ice crystals is alginate. Also a complex carbohydrate, alginate is isolated from the cell walls of algae. Alginate contains many —OH functional groups and readily binds water through hydrogen bonding. The extensive hydrogen bonding of alginate limits the flow of water and forms a gel that acts as thickener. The organization of the water–carbohydrate complex also defeats the formation of ice crystals. The cell wall carbohydrate from red algae (seaweed), carrageenan, is used in place of alginate in many foods (Fig. 3.25).

Ice cream can come in many confusing grades and styles. Superpremium and premium ice cream has low overrun and high fat content with the best quality ingredients. Standard ice cream has more overrun (air) than superpremium or premium ice cream and meets the minimum requirements of 21 CFR § 135.110. Fat-free ice cream has less fat than the CFR standard and must have less than 0.5% fat per serving. In contrast, light ice cream is a description of the amount of calories coming from fat. Light ice creams must have less than half of its total calories per serving from fat. Low and reduced fat ice creams fall somewhere between light and fat-free in their fat composition. Standard vanilla ice creams, also called Philadelphia-style ice cream, differ from French vanilla in that French-style ice creams, like custards, use egg yolks as an emulsifier, while standard or Philadelphia-style ice creams (also called New York) use no egg or just the egg whites. Gelato is a frozen ice cream-like dessert that has higher fat and almost no overrun. Sherbet stretches the ice cream-like properties with fruit juice and some milk fat, whereas sorbet is not an ice cream at all! Sorbet contains no milk or cream and is instead a frozen puree of fruit with added alcohol or wine to reduce freezing temperature. Soft serve ice cream is low fat (3–6%) with up to 60% air overrun.

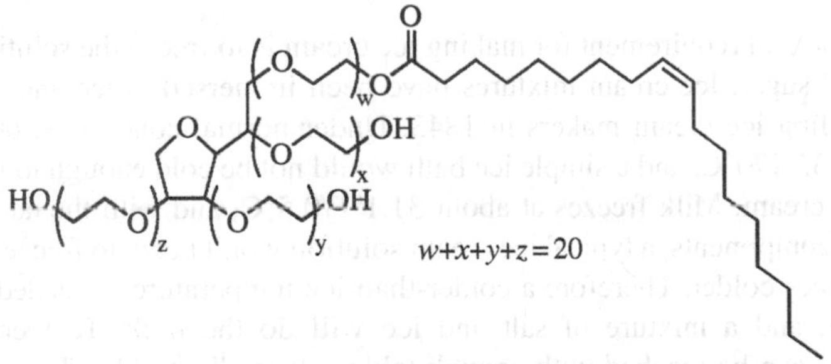

$$w+x+y+z=20$$

FIGURE 3.25 Polysorbate 80. An emulsifier often used in ice cream and other food and cosmetic compounds is made from fatty acids and other organic compounds.

Making ice cream is pretty straightforward and while an ice cream maker helps, it can be done without a machine. A simple base recipe is a combination of milk, heavy cream, sugar, and salt. From this base, flavorings including vanilla and chocolate can be added and are as diverse as there are ice cream creations. Richer custard or French-style ice creams include adding egg yolks as emulsifiers followed by heating and cooling the mixture. Cream and milk are added to a mixture of egg yolk and sugar, which is then cooled before freezing. With your ice cream mixture complete, you are ready to freeze it, but now comes the work! Air must be introduced, crystallization must be limited, and the fat and liquid phases must be kept together while freezing. This is all accomplished by mixing. Mixing can be accomplished by hand by placing the liquid ice cream into a larger container of ice, water, and salt. The salted ice bath will have a lower temperature than ice water alone (see Box 3.3) allowing the sugar and fat water "ice cream" to freeze. Ice cream makers maintain a constant mixing as the liquid ice cream mixture begins to freeze. Once frozen, the ice cream can be eaten or left in the freezer to "harden." At freezer temperature (−4°F/−20°C) about only 75% of the water is frozen, and the rest is a liquid sugar–water mixture. Rapid and deep freezing causes most of the liquid water to freeze without forming unwanted crystals. Partial thaw and refreeze cycles will increase the amount of the liquid phase, and larger crystals will form, giving the ice cream an off-taste and crunchy tooth feel (texture).

BOX 3.2   BAGGIE ICE CREAM

One fun way to make ice cream at home is to use two strong sealable baggies. The inner bag is mostly filled with the liquid ice cream, sealed with strong tape, and placed inside a larger one gallon-sized plastic bag filled with crushed ice and a cup of salt. Roll or shake the bags for 10 or 15 min, and remove the inner bag and enjoy.

BOX 3.3   COLLIGATIVE PROPERTIES: FREEZING POINT DEPRESSION. OR WHY THE ICE AND SALT BATH?

A very practical requirement for making ice cream is to freeze the solution of fat, water, and sugar. Ice cream mixtures have been immersed in ice and salt baths since the first ice cream makers in 1843. Under normal conditions, pure water freezes at 32°F/0°C, and a simple ice bath would not be cold enough to freeze the liquid ice cream. Milk freezes at about 31.1°F/0.5°C, and with the added sugar and other components, a typical ice cream solution won't begin to freeze until five or so degrees colder. Therefore a colder-than-ice temperature is needed to make ice cream, and a mixture of salt and ice will do the trick. Temperatures of −4°F/20°C can be reached with enough table salt (sodium chloride) and −40°C with calcium chloride hexahydrate salt.

Mixtures of ice and salt do not "melt" the salt, but instead the combination of the two lowers or depresses the melting point of the ice water. This is called a colligative property. These are characteristics or properties of a solution that depend on the number of particles of solute (in this case salt) in a solvent (water or ice). These properties are not impacted by the chemistry of the compounds dissolved in the water or the size, just the number of particles. For freezing, the more particles dissolved in the solvent, the greater the impact on freezing point depression. Salts are ionic molecules comprised of a cation and an anion attracted to one another by opposite charges. The ions will separate in water as each is solvated by water molecules. Sodium chloride will dissolve into the sodium cation (Na^+) and chloride anion (Cl^-). Thus, for each molecule of NaCl dissolved in water, both sodium and chloride particles are available to impact the freezing point depression. This can be measured using the following equation:

$$\Delta T_f = i K_f c_m$$

where ΔT_f is the change in freezing point, i is the number of ions present, K_f is a constant for the solvent, and c_m is the concentration of particles dissolved in the water.

From this equation, you can see that more particles mean a higher total concentration, which creates a greater change in freezing point. Sucrose, lactose, and other sugars do not dissociate into ions when dissolved in water, so one molecule of NaCl will have twice the impact on freezing point than will sucrose. This is how an ice–salt bath can achieve a temperature low enough to absorb heat from the ice cream.

REFERENCES

[1] Title 21, Vol. 8, Ch. 1, Pt 1240, subpart A, Section 1240.3(j), Release 13.

[2] Bloom, G. and Sherman, P.W. (2005) Dairying barriers and the distribution of lactose malabsorption. *Evol Hum Behav.* 26: 301–312.

[3] Andrew, C. (2013) Milk revolution. *Nature.* 500: 20–22.

[4] Boyer, R.F. (2005) *Concepts in Biochemistry.* 3rd edn, p. 110. Wiley, Hoboken.

[5] Meyer, K. (2009) Factor-analytic models for genotype × environment type problems and structured covariance matrices. *Genet Sel Evol.* 41: 24. doi:10.1186/1297-9686-41-24

4

METABOLISM OF FOOD: MICROORGANISMS AND BEYOND

Guided Inquiry Activities (Web): 7, *Carbohydrates;* 14, *Cells and Metabolism;* 15, *Metabolism, Enzyme, and Cofactors;* 18, *Starch*

4.1 INTRODUCTION

Metabolism is the collective processes that cells use to transform matter for energy, to create and store new molecules, and to respond to their environment. Understanding cell theory, the diversity of cells, and how different organisms use food from the environment to make new molecules that can impact our cooking and baking is a fascinating and important part of the food or cooking. Realizing how metabolism impacts plant and animal cells helps a cook or chef make informed choices on selecting food and preparing their dish. Focusing on the metabolism of microorganisms is crucial for preparing cheeses, wine, beer, bread, etc.—the list is endless.

If someone asked you to describe how microorganisms relate to food, what would you say? Most people immediately think about meats, fresh fruits, or vegetables that are contaminated with *Escherichia coli* or *Salmonella*, which causes foods to be pulled from grocery store shelves or people to get ill from eating a very rare hamburger. The relationship between microorganisms and food is typically thought of as detrimental, as in these situations the microorganism is unwanted, is harmful to human health, and grows in an uncontrolled manner.

Many foods, however, are dependent upon microorganisms for their preparation and production. Fungi (e.g., yeast) are important in the production of breads and

The Science of Cooking: Understanding the Biology and Chemistry Behind Food and Cooking,
First Edition. Joseph J. Provost, Keri L. Colabroy, Brenda S. Kelly, and Mark A.Wallert.
© 2016 John Wiley & Sons, Inc. Published 2016 by John Wiley & Sons, Inc.
Companion website: www.wiley.com/go/provost/science_of_cooking

---

BOX 4.1   NOT ALL MICROORGANISMS ARE GOOD FOR YOU

Not all cells and microorganisms are helpful to humankind. Many foodborne ill-nesses are caused by the ingestion of harmful microbes. According to the Center for Disease Control, one in six Americans (nearly 48 million people) will get sick from food poisoning in their lifetime, and 3000 will die of foodborne diseases. Although molds, viruses, and parasites all pose a potential risk of food contamination, some strains of bacteria are the most common culprits. One strain of bacteria, *Salmonella*, is responsible for a quarter of foodborne illnesses and impacts 1.2 million cases annually in the United States. Most of the transmission comes from food, water, or contact with infected poultry. Raw or unpasteurized milk and milk products pose a special risk of *Salmonella* illness. *Escherichia coli O157* is a strain of bacteria that lives in the intestines of some cattle, swine, and deer. Improperly cooked ground meat that is contaminated by *O157* causes hemorrhagic diarrhea and intense abdominal cramps, even resulting in death in extreme cases. Heating milk to 161°F/72°C for 20 s is enough to kill most harmful bacteria including *Salmonella* and *E. coli O157* often found in milk. Another common form of food poisoning is from the bacterium *Clostridium botulinum*. This bacterium, commonly found in soil, produces several neurotoxins that act to stop the nerves from signaling, caus-ing paralysis. Although botulism food poisoning is rare, the proteins that produce the neurotoxins are commonly used for cosmetic reasons to paralyze nerves (Botox injections). How can you reduce your chances of getting food poisoning? Wash fresh fruits and vegetables (even the outer parts of the food that you do not eat, like watermelon rinds and orange peels). Keep your food preparation and eating areas clean. Cook foods, as most cells and microbes do not survive in high cooking temperatures due to disruption of the cell membrane and protein denaturation.

---

alcoholic beverages. The presence of protists (e.g., molds) leads to a tangy, pungent, and flavorful blue cheese. Different species of bacteria are essential to the production of sourdough breads, soy sauce, yogurt, and pepperoni, to name a few. In contrast to the harmful bacteria that cause human illness, some microorganisms are purposefully added to and proliferate in a highly controlled environment, leading to a product that is safe and tasty to eat or drink. In this chapter, you will learn about various classes of multicellular organisms and microorganisms that are important to the chemistry and biology of food and how these organisms generate the energy resources required for them to survive and proliferate and some of the molecules that microorganisms produce in these processes that are important to food and cooking (Box 4.1).

4.2   THE BASICS OF THE CELL

The smallest unit able to sustain life is called a cell. A cell can be simply defined as a container of small and large molecules that are essential for the survival of an organism. Unicellular organisms, such as algae, bacteria, and protozoa, contain all the molecular machinery within a single cell that is necessary for their survival. When cells evolved to live as a collection or group, some organisms became, by definition, multicellular and develop cells that have distinct functions. Although the behavior and biology of the cells in uni- and multicellular organisms vary dramatically, several characteristics are common among cells. Cells are organized into specialized compartments called organelles. Every cell needs a barrier called a cell membrane or a cell wall that allows specific molecules into or out of the cell. A cell needs to import or make its own food for energy and the creation of new molecules. Common for both plant and animal cells is the thin outer membrane called the plasma membrane. Made of a combination of lipids (phospholipids and cholesterol) and proteins (associated at the surface and those that go all the way through the membrane—transmembrane proteins), the plasma membrane regulates the traffic of molecules entering and leaving the cell. Flavor molecules bind to transmembrane proteins found in specialized taste bud cells, signaling to other integral membrane proteins that a flavor molecule is in the food. The region of cells between the nucleus and plasma membrane is called the cytoplasm. Within the cytoplasm are various organelles and proteins. Enzymes, the cell's catalysts, are responsible for the breakdown and synthesis of molecules in a cell. A diverse and dynamic range of other molecules including carbohydrates, small molecules, nucleotides (DNA and RNA), and 20,000–30,000 different proteins also play a key role in the function of the cell. Importantly, residing within DNA is the genetic blueprint or hereditary information of the cell, which is passed from cell to cell during cell division.

Organisms are classified into two groups, prokaryotes and eukaryotes, based upon the characteristics of their cell or cells. The earliest forms of life are the unicellular prokaryote organisms. The distinguishing cellular feature of prokaryotes is the absence of a nucleus, an organized compartment/organelle that stores the cell's DNA. Bacteria are a species of prokaryotes. Plant, fungi, and animal cells contain a nucleus and are thus eukaryotes. Not only do eukaryotes possess a nucleus, but they also contain several types of specialized organelles or structures including mitochondria, chloroplasts, nucleus, and endoplasmic reticulum.

4.2.1 Plant Cells

Plants are multicellular organisms with specialized cells (Fig. 4.1). These have both mitochondria and chloroplasts to process energy for the plant cell. In addition to nuclei, endoplasmic reticulum, and other eukaryotic organelles, plant cells have a rigid structurally complex cell wall made of a number of large and small molecules. The cell wall is a particularly important part of food and cooking. The plant cell wall is a mixture of complicated carbohydrates and proteins giving plants rigid support. Specific to food and cooking are some very interesting components. The cellulose, hemicellulose, and pectin are complex carbohydrate polymers that add fiber to food. Cellulose and hemicellulose serve as thickeners and emulsifiers in some recipes. The pectin in fruit is used to solidify jellies under acidic conditions. About 10% of the plant cell wall is composed of polyphenolic compounds called lignins. Phenols are organic ring structures with the —OH (hydroxyl) functional group attached. Lignins are a complex set of diverse compounds found in plants and algae (Fig. 4.2). A special

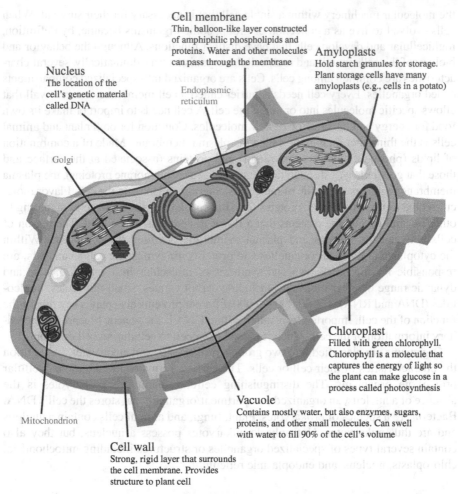

Cell membrane
Thin, balloon-like layer constructed of amphiphilic phospholipids and proteins. Water and other molecules can pass through the membrane

Amyloplasts
Hold starch granules for storage. Plant storage cells have many amyloplasts (e.g., cells in a potato)

Nucleus
The location of the cell's genetic material called DNA

Endoplasmic reticulum

Golgi

Chloroplast
Filled with green chlorophyll. Chlorophyll is a molecule that captures the energy of light so the plant can make glucose in a process called photosynthesis

Vacuole
Contains mostly water, but also enzymes, sugars, proteins, and other small molecules. Can swell with water to fill 90% of the cell's volume

Mitochondrion

Cell wall
Strong, rigid layer that surrounds the cell membrane. Provides structure to plant cell

FIGURE 4.1   The plant cell.

subfamily of polyphenols called lignin (a mixture of polyphenols) gives food and drink an astringent flavor. When oxidized by plant cell enzymes some polyphenols become brown, taste bitter and help plants fight infection.

4.2.2   Animal Cells

Animal cells do not have a plant cell wall or chloroplasts and are highly diverse depending on the tissue each cell comes from. Most animal cells will have nucleus, mitochondrion, Golgi, and other components. For food and cooking we focus on meat. Muscle cells have a slightly different nomenclature that is described later in this book (Fig. 4.3).

Perhaps you are thinking, how do these cells come together to make a functioning organism or a plant or animal tissue that might be cooked or eaten? Cells do not naturally

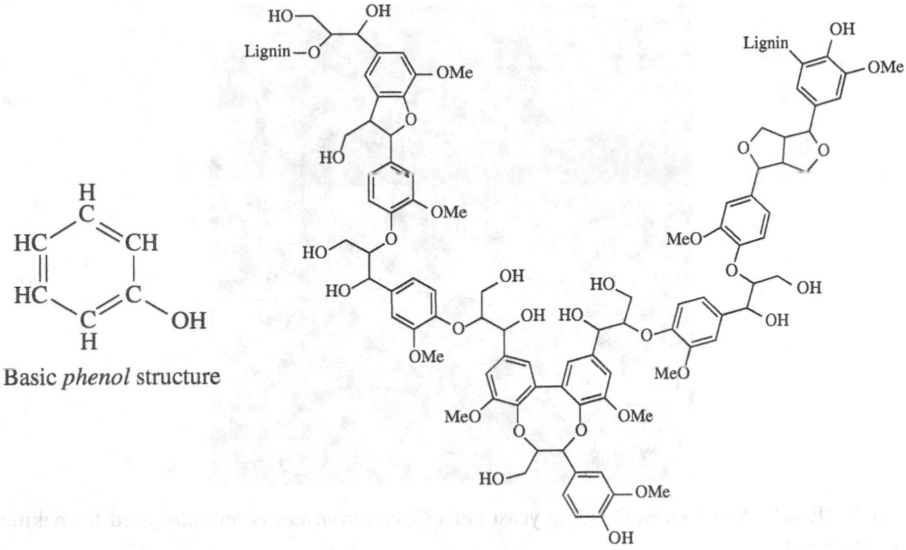

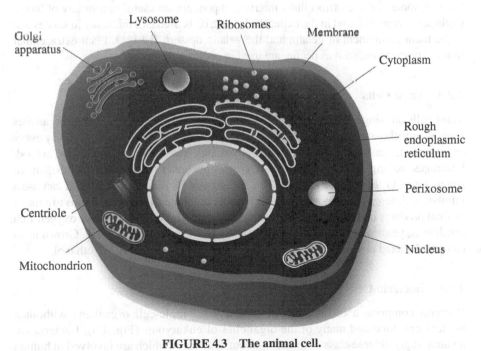

Basic *phenol* structure

General *lignin* structure

FIGURE 4.2 Polyphenols. Lignin is a highly diverse and modified polymer of phenol. Lignin is a major component of plant cell walls and in our food and drink.

FIGURE 4.3 The animal cell.

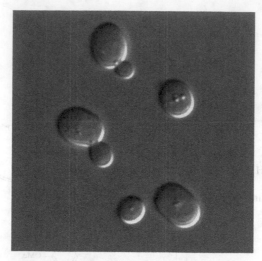

FIGURE 4.4 Yeast cells. Budding yeast cells (*Saccharomyces cerevisiae*) used for baking and brewing.

"stick" together. Groups of cells organized into plant and animal tissue are embedded in a jelly-like substance that acts as "cellular glue." The material surrounding the cells is a mixture of carbohydrates and proteins that are collectively called the extracellular matrix. This matrix is the mortar that holds the cells together into a cohesive mass. As in a tissue, some of these extracellular matrix components are useful in a variety of foods. Collagen, a protein found in the extracellular matrix, is used as a thickener in sauces and is the main component of gelatin and the gelatin dessert, JELL-O. Plant extracellular matrix components, such as pectin, are used in cooking fruit pies.

4.2.3 Yeast Cells

Yeast cells are single-cell eukaryote cells in the fungi kingdom containing organelles such as mitochondria and an enclosed nucleus, but no cell wall. Found in nearly every environment, yeast is very diverse with many strains. Yeast is terribly important in food, beverages, baking, and cooking (Fig. 4.4). Baking bread requires the metabolism of sugars into CO_2 gas to give bread its rise. Because of its mitochondria, yeast can use a number of different food sources and convert carbohydrates, fats, and proteins to a range of final products including carbon dioxide, ethanol, and acetic acid. Under oxygen-rich conditions, yeast will produce a mixture mostly of CO_2 and some ethanol. Grown in an oxygen-depleted environment, yeast will switch to producing primarily ethanol.

4.2.4 Bacteria Cells

Bacteria comprise a broad class of prokaryotic single-cell organisms without a nuclear envelope and many of the organelles of eukaryotes (Fig. 4.5). Bacteria are an amazingly diverse class of microorganisms, some of which are involved in human

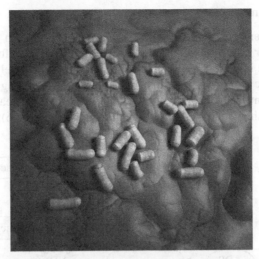

FIGURE 4.5 Illustration of bacteria cells. *Lactobacillus* bacteria are the lactic acid-producing bacteria used for cheese and other fermented products.

health and in disease. Throughout this chapter we discuss bacteria as part of the preparation of food. Bacteria that produce and can live in acidic conditions are used to make cheese, and some of the same bacteria are encouraged to grow on meat as it cures and ages. In this case, the bacterium helps produce acid that inhibits unhealthy bacterial growth. Lactic acid-producing bacteria are used to produce yogurt, sauerkraut, and fermented meats and sausages. Some winemakers will use bacteria to encourage new flavors including a green apple malonate flavor (malolactic fermentation) to wines. Yet others will encourage bacterial growth in wine to produce wine vinegar.

Having a better understanding of the nature of a cell and how a cell is organized will help you better understand ingredients and steps in cooking and baking.

4.3 INTRODUCTION TO BASIC METABOLISM

In order for a cell to survive, grow, and divide, it must have a source of energy. The mechanism by which any cell or organism breaks down and builds the molecules necessary for its function and proliferation is called metabolism. A cell's or organism's metabolism comprises all the chemical reactions that are essential to its survival; thus it is a vast, complicated series of reactions that are specific to a particular organism or cell type. These metabolic processes that are critical to the life of the organism are also critical to food and cooking. In order to understand this connection, we need to discuss the details of some metabolic pathways and why metabolism is necessary for a cell.

4.3.1 Metabolism

Cellular metabolism consists of many specific pathways, each of which accomplishes a particular task. Some of these pathways are responsible for synthesizing the molecules that are important to the function or structure of the cell. These pathways are called anabolic pathways, from the Greek prefix *ana* meaning "up." Other pathways are important in the breakdown of biomolecules; these reactions are termed catabolism, from the Greek word *kata* meaning "down" (Fig. 4.6).

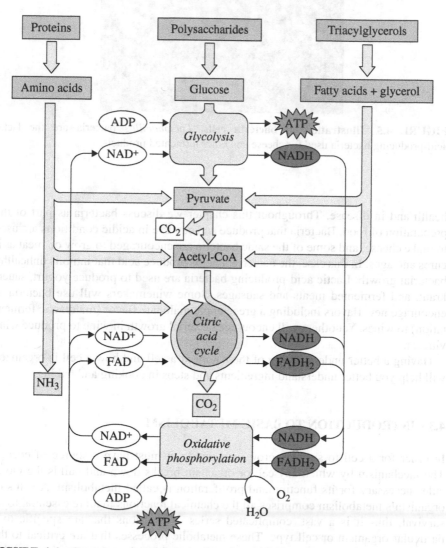

FIGURE 4.6 Overview of catabolism. Several metabolic pathways including glycolysis are involved in using and generating molecules necessary to meet the cell's needs. Voet Voet and Prat, Figure 14-3. Reproduced with permission of Voet Voet and Prat 4th Edition.

This type of covalent bond is special. It is called a phosphodiester bond, and breaking it during a chemical reaction releases energy!

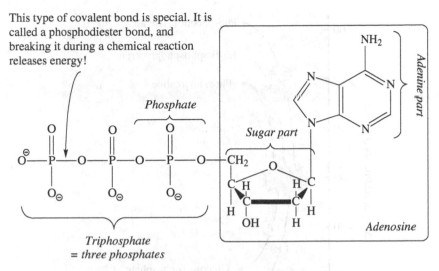

FIGURE 4.7 **Adenosine triphosphate (ATP).**

The role of the anabolic pathways or anabolism might be obvious; a cell needs to generate biological macromolecules, such as proteins and nucleic acids, that are necessary for its function. Just as it requires energy (in the form of effort and time) to make a cake, building molecules also requires energy. When a chemical reaction requires an input of energy to occur, you have an endergonic reaction. Thus, anabolic reactions are endergonic. Catabolic pathways play two roles within a cell. In catabolism, large molecules are broken down into smaller molecules called metabolites, and energy is released to carry out the work of the cell. Here, you might think about eating the cake as a catabolic process. You have to make a minimal effort, moving a bite of cake to your mouth, to yield a significant product, a delicious bite of light, fluffy cake in your mouth. Thus, anabolic and catabolic processes are closely tied together; the energy and metabolites generated through catabolism are used to make new molecules (anabolism) that are essential for survival of the organism. The energy source that connects anabolism and catabolism is a molecule called adenosine triphosphate (ATP) (Fig. 4.7).

As the name implies, ATP consists of an adenosine covalently linked to a triphosphate. The adenosine portion of the molecule is made up of an adenine base and ribose sugar. The triphosphate consists of three phosphate groups (PO_4^{3-}) that are bonded to one another via covalent bonds called phosphodiester bonds. ATP is the energy "currency" of the cell. It is generated in catabolic reactions and it is used up in anabolic reactions. How does this work and where does the energy in ATP come from? ATP's energy comes from the energy contained within the phosphodiester bonds. When the phosphodiester bond of ATP is broken in a reaction with water, called a hydrolysis reaction, about 30 kilojoules/mole (kJ/mol) worth of energy is given off under standard biological conditions. When this reaction is

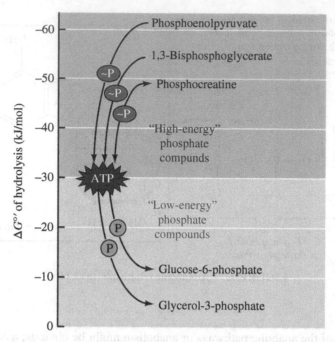

FIGURE 4.8   Energy of ATP. ATP has an inherent energy called Gibbs free energy (ΔG). While ATP is considered the energy currency of the cell, there are higher and lower energy compounds in the cell. Voet Voet and Prat, Figure 14-3. Reproduced with permission of Voet Voet and Prat 4th Ed.

linked/coupled to another reaction that requires energy to occur (like an anabolic reaction), that 30 kJ/mol worth of energy can be used to drive the unfavorable reaction forward (Fig. 4.8).

4.4   CATABOLISM OF GLUCOSE (GLYCOLYSIS OR FERMENTATION): GLUCOSE TO PYRUVATE

To launch our discussion of metabolism, we will first concentrate on the breakdown of the molecule glucose (Fig. 4.9). Not only is glucose a main energy source in most organisms and cells, but the breakdown of most other energy-rich substances (like fats, complex carbohydrates, and proteins) begins with their conversion into one of the intermediates in the pathway for glucose catabolism. Glucose catabolism can occur under aerobic or anaerobic conditions. In aerobic respiration, the complete breakdown of glucose to carbon dioxide and water occurs in the presence of oxygen. Aerobic respiration is a complex, multipathway process that utilizes oxygen as the final electron acceptor for the electrons that are lost from the glucose molecule (Fig. 4.6). Because the carbons in glucose molecule are fully oxidized to carbon dioxide, aerobic respiration of glucose yields the greatest amount of energy for the organism.

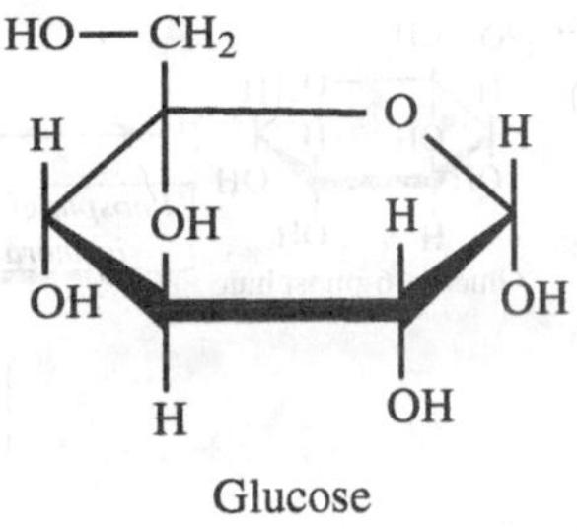

Glucose

FIGURE 4.9   Glucose.

Most organisms are also able to carry out anaerobic respiration or fermentation, which doesn't require the presence of oxygen. Anaerobic respiration generates less energy for the organism, as there is no net oxidation of the glucose molecule; however it is an essential process for organisms to survive in situations where there is an absence of or limited amounts of oxygen. Some organisms even prefer to carry out fermentation! Although there are major differences in the energy generated and oxidation processes for an organism under aerobic versus anaerobic conditions, both processes initially break down glucose using the same pathway called glycolysis.

4.4.1   Glycolysis

Glycolysis, derived from the Greek roots *glykos*, meaning sweet, and *lysis*, meaning "loosening" or "splitting," involves a sequence of 10 enzyme-catalyzed reactions that converts one six-carbon molecule of glucose into two three-carbon molecules of pyruvate. The first five steps of the pathway are called the "investment" or the "preparatory" phase of glycolysis because energy is consumed (see the use of an ATP molecule in step one) to break the six-carbon glucose molecule into two three-carbon units. The next few steps are referred to as the payoff phase of glycolysis, as they include an oxidation reaction that generates NADH and that generate ATP through direct addition of a high-energy phosphate from PEP to ATP in a process called substrate-level phosphorylation (Fig. 4.10). The final product of glycolysis is pyruvate; from a single molecule of glucose, two molecules of pyruvate are generated.

Any metabolic pathway, including glycolysis, is often viewed as a complicated series of enzymes, substrates, and products. However, all metabolic pathways do have some common themes. The name of the enzyme often describes the features of the reaction it catalyzes. For example, step one of glycolysis is catalyzed by hexokinase. *Hex* is the prefix that is associated with six (in this case the six-carbon glucose), and *kinase* is the enzyme name for a reaction that involves addition of a phosphate group. The oxidation/reduction reaction of step six is catalyzed by glyceraldehyde 3-phosphate dehydrogenase, providing a hint as to the substrate of the reaction (i.e., glyceraldehyde 3-phosphate) and the type of reaction that occurs (i.e., a dehydrogenation or oxidation/reduction reaction).

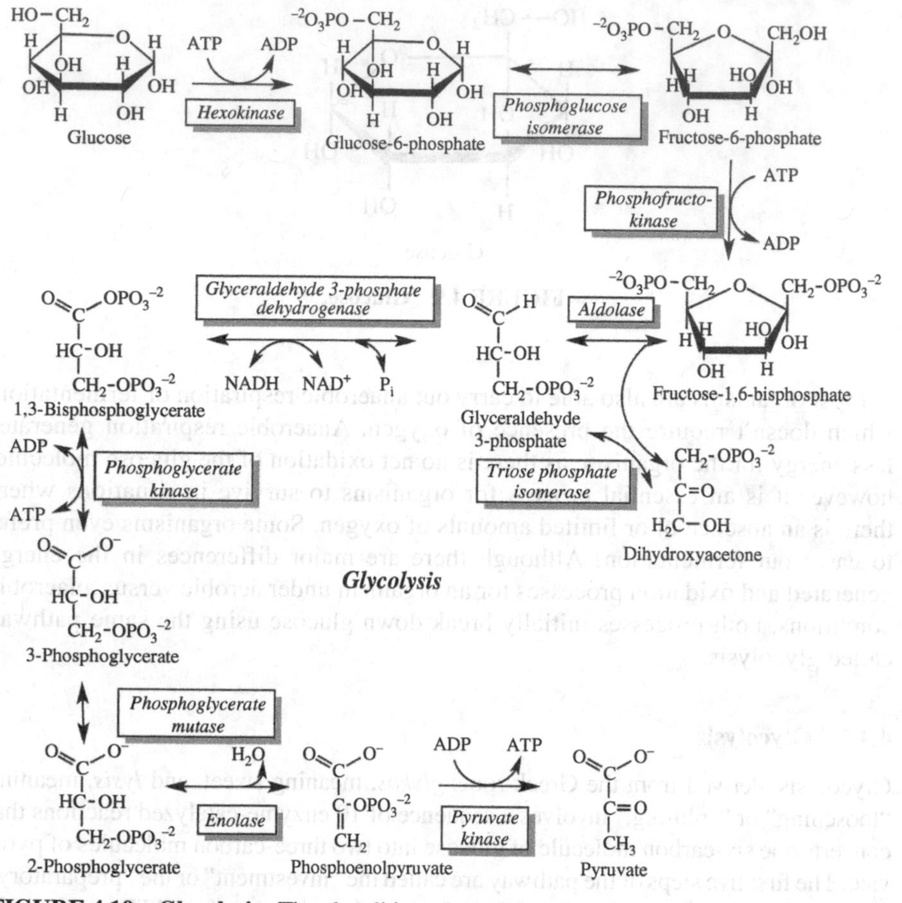

FIGURE 4.10 Glycolysis. The glycolitic pathway involved in fermentation to produce ATP.

4.5 FATES OF PYRUVATE: NOW WHAT?

As mentioned earlier, glycolysis is used by almost all organisms under aerobic and anaerobic conditions. One of the most important features of glycolysis is the final product of the pathway, pyruvate. Nearly every cell type from simple single-cell organisms to higher-order multicellular eukaryotes has the capacity to generate ATP from glucose via glycolysis. What happens to the pyruvate produced by glycolysis depends on the type of cell or organism, the nutritional state of the environment, and the access to oxygen (Fig. 4.11). Cells with a functional Krebs cycle and in oxygen-rich environments (aerobic) will continue to metabolize the pyruvate to fats through the Krebs cycle making new compounds from the carbohydrate (short- and long-chain fats, amino acids, and acids are just a few examples), CO_2 and ATP. Cells without mitochondria or those with mitochondria but exposed to low-oxygen environment (anaerobic) must regenerate the nicotinamide adenine

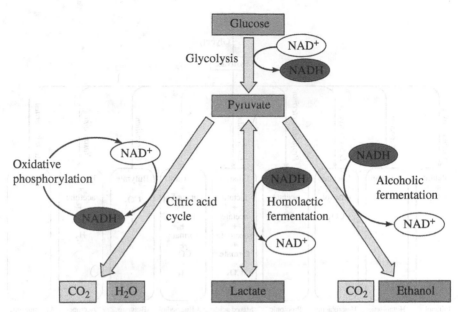

FIGURE 4.11 **The fates of pyruvate.** Depending on organism, cell type, and availability of oxygen, pyruvate can be metabolized to carbon dioxide and water, lactate, or carbon dioxide and ethanol. Voet Voet & Pratt, 4th Edition.

dinucleotide (NAD^+) to continue glucose metabolism. The reason for this is pretty simple. Each cell has a limited number of NAD^+/NADH molecules, but glucose and the environment only limit other foodstuffs feeding into glycolysis. Thus NAD^+ required for several steps of glycolysis is limiting, and when all or most of the NAD^+ is converted to NADH, glycolysis shuts down and the cell (and organism) can no longer produce the necessary ATP to survive. Cells with mitochondria can therefore shift their metabolism from aerobic to anaerobic metabolism to allow for continued ATP production. However bacteria and other cells without mitochondria use a different pathway (Fig. 4.12).

4.5.1 Bacteria

Continued metabolism of pyruvate in bacteria is highly varied depending on the cell type. Remember bacteria do not have mitochondria and some bacteria strains do not have the enzymes of the Krebs cycle and thus must convert NAH back to NAD^+ to continue to utilize sugars for ATP production. *E. coli* is the common bacteria found in the gut of humans and other than a few strains are harmless. There are several enzymes in *E. coli* that will react with pyruvate. Some produce lactate, and others ethanol and acetate, among others. The bacteria used to make Swiss cheese, *Propionibacterium*, produces propionate, acetate, and CO_2 responsible for the taste and gas holes of the cheese. *Lactococcus* bacterium produces primarily lactate in anaerobic conditions using the enzyme lactate dehydrogenase. This is how the dairy industry produces

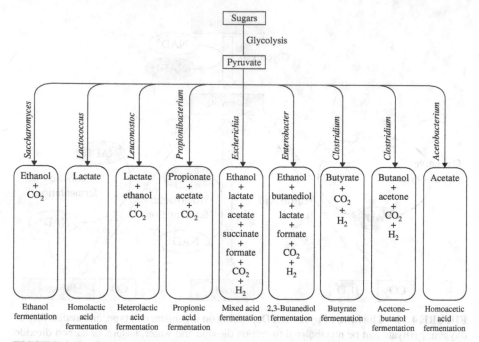

FIGURE 4.12   Major pathways for fermentation of sugars including organisms involved and end products formed. Taken with permission from Ref. [1]. http://www.els.net.

acidic conditions in fermented milk produces and cheese. The major fermentation endpoints of yeast and bacteria are detailed later in this chapter [1].

4.5.2   Yeast

Yeast metabolism of pyruvate has two metabolic fates. The yeast cell can use pyruvate to generate CO_2 and ethanol producing two total ATPs (Fig. 4.11). A second pathway is found in the mitochondria where pyruvate is further oxidized to CO_2 water and much more ATP. While both pathways can metabolize pyruvate, oxygen depletion controls the switch from respiration (CO_2 and ATP in the mitochondria) to fermentation (production of CO_2 and ethanol). Depending on the conditions, yeast will slowly convert ethanol to short-chain aldehydes and ketones, acetaldehyde, acetate, and acetyl-CoA.

4.5.3   Mammalian

Animal mammalian cells have two options. Like yeast, the presence of oxygen allows the cells with mitochondria to metabolize pyruvate to CO_2, producing the maximum amount of ATP. However if the oxygen levels fall due to exercise or other factors, anaerobic metabolism of pyruvate shifts to produce lactate. This has an important impact in meat before harvest.

BOX 4.2   FERMENTATION VERSUS GLYCOLYSIS

Let's consider the definition of glycolysis and fermentation. Glycolysis, as you've learned, is the oxidation of glucose to pyruvate. Fermentation is a broader term that includes glycolysis but includes whatever pathway is used to oxidize NAD^+ back to NADH in anaerobic conditions. Thus fermentation will include glucose to pyruvate and then the continued metabolism to lactate, ethanol, acetate, carbon dioxide, or other compounds depending on the organism and conditions. Fermentation has a more loose historical definition that involves ethanol in wine and beer. Therefore fermentation is a combination of different metabolisms to produce ATP from glucose and regenerate NAD^+ from NADH.

To review, in the presence of oxygen, pyruvate undergoes further oxidation to the molecule acetyl-CoA, which can be completely oxidized to CO_2. This process comprises aerobic respiration, which generates a large amount of ATP, H_2O, and CO_2. Under anaerobic conditions or conditions where fermentation is favored, the energy needs of the organism are met by the modest ATP yield of glycolysis as no additional ATP is generated in the further breakdown of pyruvate. In these cases, however, an organism still requires a means to regenerate NAD^+ so that glycolysis can continue as the organism's primary means to make ATP. The metabolic processes that occur to regenerate NAD^+ that do not require oxygen are called fermentation processes. As we have begun to learn, fermentation is very important in cooking and foods such as yogurt, cheese, beer, wine, soy sauce, and pepperoni (Box 4.2).

4.6   AEROBIC RESPIRATION: THE TRICARBOXYLIC ACID CYCLE AND OXIDATIVE PHOSPHORYLATION

In the presence of oxygen, the pyruvate generated in glycolysis is fully oxidized to CO_2, the energy of which is used to make ATP. Three different pathways are necessary to carry out this complete oxidation. The pyruvate is first processed by a large enzyme complex called the pyruvate dehydrogenase (PDH) complex. The PDH complex, as its name implies, oxidizes and decarboxylates the pyruvate into two molecules: a two-carbon molecule called acetyl-CoA and CO_2 (Fig. 4.13). The acetyl-CoA can then enter the second pathway, called the tricarboxylic acid (TCA) cycle, which allows for further oxidation of the molecule (Fig. 4.14).

The TCA or Krebs cycle (named after Hans Krebs who first elucidated the pathway) conjugates or attaches acetyl-CoA to oxaloacetate to make a molecule called citrate. Citrate is a TCA, hence the name of the pathway.

Although the TCA cycle is complicated, there are a few features of the pathway that are important to our discussion of food and cooking. The first relates to the pathway's namesake, citrate. Have you ever heard of citrate or citric acid? Interestingly, citrate or citric acid is a molecule that gives many drinks and foods a sour taste. It is

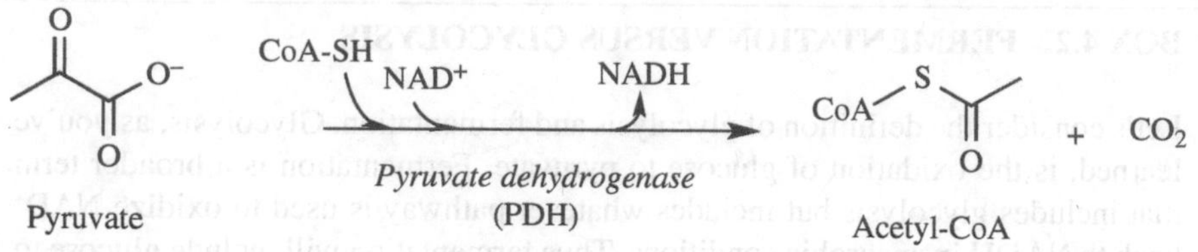

FIGURE 4.13 Pyruvate dehydrogenase. The enzyme important for the entry of pyruvate into the Krebs cycle.

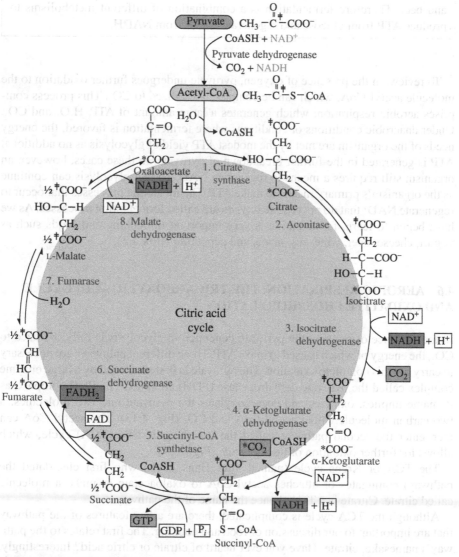

FIGURE 4.14 The Krebs cycle. Voet Voet and Prat. Reproduced with permission of Voet Voet and Prat.

naturally present at a relatively high concentration in limes, lemons, and other citrus fruits and is added to enhance the tangy, sour flavor of citrus-flavored soft drinks. The citric acid also helps to give the drink an acidic pH (between 2.5 and 4), which prevents microbial growth and emulsifies some of the added flavor molecules to keep them in solution. It is interesting that the citric acid used in the food industry is made by a mold called *Aspergillus niger*. The mold is allowed to metabolize various sugars (see Section 4.4.1), and the citrate produced is purified and utilized within the food and other industries.

The second important feature of the pathway is this: for every acetyl-CoA that enters the pathway, two additional CO_2 molecules are generated. Thus, upon completion of glycolysis, the pre-TCA cycle, and the TCA cycle by a six-carbon glucose molecule, six carbons have been fully oxidized to CO_2. Why is CO_2 important to food? This is the same CO_2 that generates the pockets of air that are present when yeast breads rise.

However, it is still not immediately obvious how this oxidation yields energy (and ATP) for the organism. Remember that an oxidation must be paired with a reduction, so when carrying out an oxidation reaction a molecule needs to be present to accept elections. From the six oxidation steps present in glycolysis, the pre-TCA cycle, and the TCA cycle, 12 electrons are transferred to and held by the coenzymes NADH (which we have already talked about) and $FADH_2$, another important electron carrier. If NADH and $FADH_2$ do not get rid of these electrons (become oxidized themselves), then there are two problems. The first is that the cell will soon run out of NAD^+ and FAD, so these metabolic processes will cease and the organism will die. Although we would like disease-causing microorganisms to die, we would not want the yeast in our bread dough to die prematurely. The second is that we haven't yet yielded a large amount of ATP from these oxidative steps. In the last pathway associated with aerobic respiration, called the electron transport chain or oxidative phosphorylation, NADH and $FADH_2$ are oxidized back to NAD^+ and FAD, and the energy associated with the oxidation of these molecules is used to drive the synthesis of ATP.

4.7   THE ELECTRON TRANSPORT CHAIN

The electron transport chain is the last pathway associated with aerobic respiration of a cell. It is in this pathway that we finally observe the oxygen dependence of aerobic respiration. How does this happen? The electron transport process requires molecular oxygen as the final acceptor of the electrons from NADH and $FADH_2$. The summarized reactions associated with the electron transfer are shown below:

$$NADH + H^+ + \tfrac{1}{2} O_2 \rightarrow NAD^+ + H_2O$$

$$FADH_2 + \tfrac{1}{2} O_2 \rightarrow FAD + H_2O$$

As you can see, in the process of transferring the electrons to oxygen to generate water, the coenzymes (NADH and $FADH_2$) are reoxidized, which is essential for the

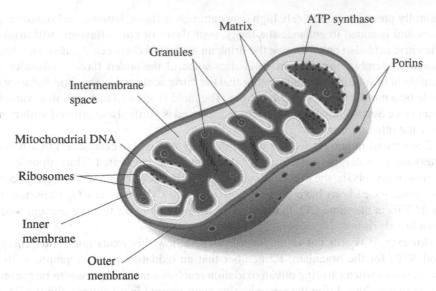

Matrix
ATP synthase
Granules
Porins
Intermembrane
space
Mitochondrial DNA
Ribosomes
Inner
membrane
Outer
membrane

FIGURE 4.15   Mitochondria.

survival of the organism or cell. These reactions make the electron transport process seem very simple, but they do not show how these electron transfers allow for the generation of ATP.

The electron transport chain comprises a multistep process that involves a series of electron carriers that are embedded within the mitochondrial membrane of the cell. As the electrons are transferred from the NADH or $FADH_2$ through a series of electron carriers that are present within the membrane (who will eventually transfer the electrons to the ultimate acceptor, molecular oxygen), protons (positively charged hydrogen ions) are pumped from the mitochondrial membrane into the intermembrane space of the mitochondrion (Fig. 4.15).

As you learned in Chapter 1, when you increase H+ within a solution, the pH decreases and the solution becomes more acidic. A similar phenomenon occurs within the mitochondrion. As H+ is pumped into the intermembrane space, the pH of that space decreases relative to the rest of the mitochondrion, creating a "proton gradient" since there is a difference in the number of H+ ions in the intermembrane space relative to the rest of the mitochondrion. The system wants to equilibrate or "even out" the concentration of H+ ions on the inside and outside of the membrane, similar to what happens when you put a turkey into a salt brine. The difference in water concentration between the turkey and the brine draws moisture from the interior of the bird to the surface, where it combines with the salt and other seasonings, eventually seasoning the entire bird. In the mitochondrion, the protons cannot pass through the hydrophobic mitochondrial membrane without the assistance of a protein. The protein that recognizes and allows the protons to flow back into the mitochondrial matrix is called ATP synthase. The energy associated with the difference in proton concentration and separation of charge across the inner

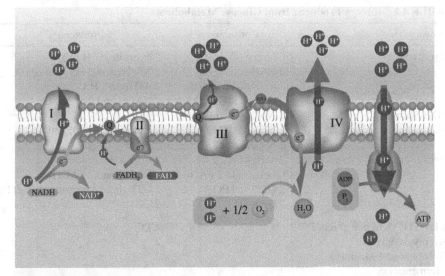

FIGURE 4.16 Electron transport system. Illustration of inner mitochondrial membrane reactions. From the NADH dehydrogenase complex, electrons are transported along a chain of proton pumping cytochrome electron carrier protein complexes. Reactions involve NAD and flavin adenine dinucleotide (FAD). Oxygen is involved in the step leading to the ATP synthase complex. The final stage involves an ATP/ADP transporter that channels the energy-carrying molecules ATP and ADP.

mitochondrial membrane drives the synthesis of ATP as proteins flow back into the matrix and the concentration and charge difference is reduced (Fig. 4.16).

Due to this process, for every NADH or $FADH_2$ that enters the electron transport chain, approximately 2.5 and 1.5 ATP molecules are generated, respectively. Thus, a single glucose molecule that is completely oxidized via aerobic respiration generates 30–32 molecules of ATP (Table 4.1).

Aerobic respiration gives an organism or cell a large energy yield so that it can carry out other processes required for its survival and reoxidizes coenzymes so that aerobic respiration can continue. Thus, this is the most efficient way for an organism to generate the energy necessary for survival.

4.7.1 Additional Metabolic Fates of Pyruvate: Fermentation

The role of a living organism in fermentation was discovered by Louis Pasteur in the late 1850s due to bacterial contamination within fermentation vats within an alcohol distillery. However, long before Pasteur's discovery, humans and other animals enjoyed food and beverage products of the metabolic, fermentative activity of a microorganism. Remember, in both fermentation and aerobic respiration, glucose is first broken down to pyruvate via glycolysis. However, the two pathways then split. In the presence of oxygen, the pyruvate continues to be oxidized through the TCA cycle and the electron transport chain, yielding high amounts of ATP for the

TABLE 4.1   Energy Produced from Glucose Metabolism.

Pathway	Consumed	Produced
Glycolysis	Glucose ($C_{14}H_{12}O_{14}$)	2 pyruvate ($C_3H_3O_3$)
	2 NAD$^+$	2 NADH and 2 H$^+$
	2 ADP + 2 phosphates	2 ATP and 2 H$_2$O
Pyruvate oxidation	2 pyruvate ($C_3H_3O_3$)	2 NADH and 2 acetyl-CoA
(PDH complex)	2 NAD$^+$ + 2CoA-SH	
Ethanol fermentation	2 pyruvate ($C_3H_3O_3$)	2 ethanol ($C_2H_{14}O$) and 2 CO$_2$
	2 NADH and 2 H$^+$	2 NAD$^+$
Krebs (citric acid) cycle	2 acetyl-CoA	2 CoA-SH
	6 NAD$^+$ and 2 FAD	6 NADH and 2 FADH$_2$
	2 GDP and PO$_4^{3-}$	2 GTP (later converted to 2 ATP)
	2 CoA-SH	2 CO$_2$
Total ATP yield per glucose	**ATP**	
Glycolysis only	2	
Glycolysis and ethanol fermentation	2	
Glycolysis and Krebs (with PDH and electron transport)	**30–32**	

The table is calculated based on *one* glucose metabolized. Remember that each NADH can produce approximately 2.5 ATP if mitochondria and oxygen are present and FADH$_2$ can produce approximately 1.5 ATP.

organism and products, like citric acid, that are important in foods. In the absence of oxygen or due to the characteristics of the microorganism, pyruvate isn't fully oxidized to CO$_2$. Rather, pyruvate is converted to another molecule that allows for the regeneration of NAD$^+$ and organism survival. The two major fermentation pathways are distinguished by the end product of the pathway. In some animals and many bacteria, the final product is lactate; thus the process of anaerobic glucose catabolism is called lactate fermentation. In most plant cells and yeast, the process is termed alcohol fermentation because the final product is ethanol. Both lactate fermentation and alcohol fermentation are important in food production and cooking; the specific pathway utilized by the organism depends upon the enzymes and conditions available to the organism.

4.7.2   Lactate Fermentation

Lactate fermentation is the anaerobic process of glucose catabolism that results in the production of lactate or lactic acid (Fig. 4.17). The pyruvate generated from glycolysis is reduced (it gains electrons, you know this because it gains hydrogens), and NADH is oxidized to NAD$^+$ (it loses hydrogens) in a single reaction step catalyzed by an enzyme called lactate dehydrogenase. That's it! The process of lactate fermentation is much simpler than the TCA cycle.

When you consider the reaction in the context of glycolysis, you can write the overall reaction and conversion of glucose to lactate as shown in Figure 4.17b. Again, lactate fermentation is not particularly efficient in terms of ATP production as only

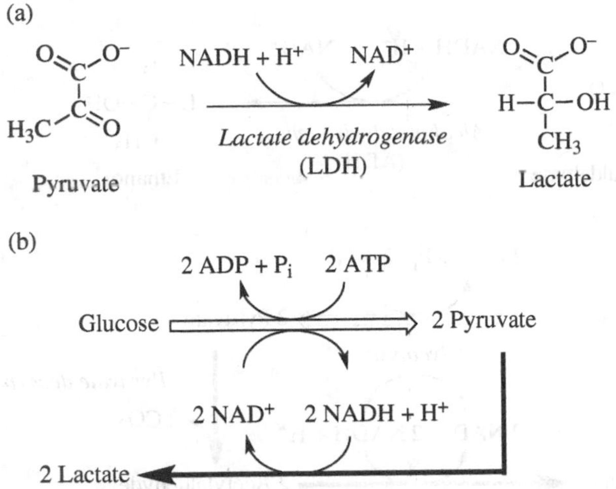

(a)

Pyruvate   Lactate dehydrogenase   Lactate
                (LDH)

NADH + H⁺   NAD⁺

(b)

2 ADP + P$_i$ 2 ATP

Glucose ⟶ 2 Pyruvate

2 NAD⁺   2 NADH + H⁺

2 Lactate ⟵

FIGURE 4.17 Lactate fermentation. To resupply the consumed NAD⁺, under anaerobic conditions, cells can utilize lactate dehydrogenase (LDH) enzyme. (a) LDH-catalyzed reaction. (b) Overall metabolism of glycolysis and LDH.

two ATPs are generated relative to the 30–32 ATPs obtained from aerobic respiration; however, it allows for regeneration of NAD⁺, which is critical for the organism such that it can continue to carry out glycolysis to obtain some energy from glucose and its intermediates. For people interested in the chemistry and biology of food and cooking, however, lactate fermentation does more than allow the organism to survive. As its name implies, lactic acid is an acid! Its production changes the pH of the environment in which it exists, and when pH changes, the behavior of molecules changes too. Lactic acid that is produced through the fermentation of bacteria causes milk to curdle, which leads to various dairy products such as yogurt, sour cream and cheese, and preserves and the flavorful fermented meats such as pepperoni and salami.

4.7.3 Alcohol Fermentation

Alcohol fermentation is the anaerobic process of glucose catabolism that results in the production of ethanol. In this two-step process, the pyruvate generated from glycolysis is decarboxylated (i.e., it loses a carbon atom as carbon dioxide) and is converted to acetaldehyde. Acetaldehyde is reduced (it gains electrons and hydrogen) by NADH to yield ethanol (Fig. 4.18).

The enzyme that catalyzes the conversion of pyruvate to acetaldehyde is called pyruvate decarboxylase because a carbon is removed from pyruvate in the form of carbon dioxide gas. The second enzyme is called alcohol dehydrogenase. When you consider the breakdown of glucose via glycolysis and alcohol fermentation, the full reaction is this:

$$Glucose + 2\ ADP + 2\ Pi \rightarrow 2\ CO_2 + 2\ ATP + 2\ ethanol$$

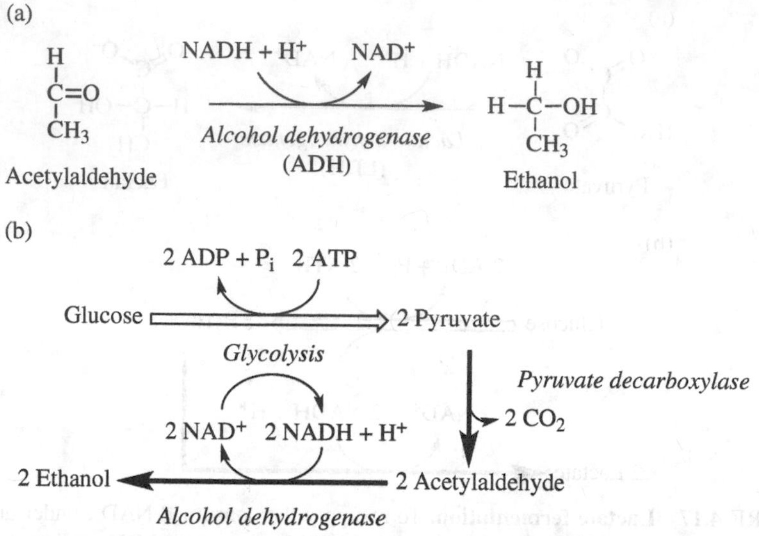

(a)

Acetylaldehyde

$$\text{Acetylaldehyde} \xrightarrow[\text{(ADH)}]{\substack{\text{NADH} + \text{H}^+ \quad \text{NAD}^+ \\ \text{Alcohol dehydrogenase}}} \text{Ethanol}$$

(b)

FIGURE 4.18 **Alcohol fermentation.** (a) The reaction catalyzed by alcohol dehydrogenase and (b) integration of ADH supporting anaerobic fermentation in yeast and a few novel bacterial strains.

As with lactate fermentation, this process is not a high energy yield, but it allows for the regeneration of NAD^+ (which is critical for the organism) and production of alcohol and carbon dioxide, both of which are important to the chemistry and biology of food and cooking. Alcohol fermentation by yeast is a key process in the production of beer, wine, and various baked goods. In yeast breads, the carbon dioxide produced through fermentation is what causes the dough to rise. The ethanol becomes part of the bread's aroma when it vaporizes during baking. For the brewer, ethanol makes the beverage alcoholic, and the carbon dioxide contributes to the carbonation (i.e., gas bubbles).

4.7.4 Other Products of Fermentation

Although lactate and ethanol are the most common fermentation products for organisms in general, many other molecules can be produced through fermentation pathways that are microorganism specific. In the production of Swiss cheese, the hole-making bacterium *Propionibacterium shermanii* converts the lactic acid product of fermentation to propionic and acetic acids and carbon dioxide. The carbon dioxide gives the cheese its distinctive holes, and the propionic and acetic acids provide the distinct aroma and nutty, sweet flavor of Swiss. Other fermentation processes yield acetone, isopropyl alcohol, and butyrate, which are components of the smells of rotten food (Fig. 4.19).

4.8 METABOLISM OF OTHER SUGARS

So far, we have assumed glucose to be the starting point for metabolism and the most important source of energy for an organism. Glucose is a very important energy source for aerobic and anaerobic respiration; however it is not the only available

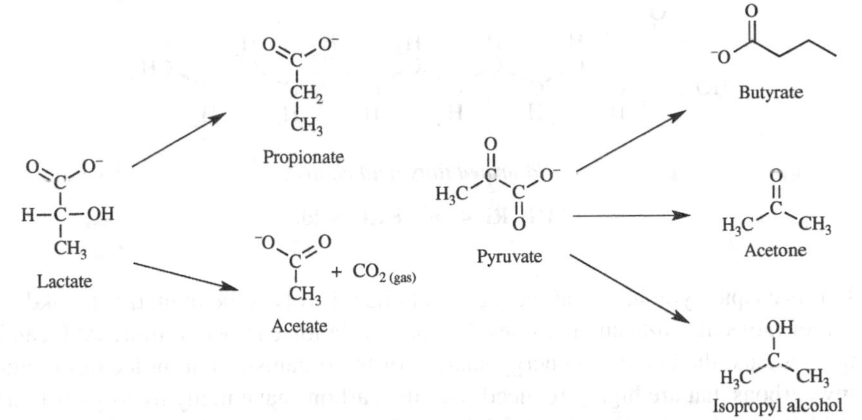

FIGURE 4.19 Additional fermentation products. Several alternative routes of fermentation will produce unique and interesting flavor and aroma compounds in fermented foods and drinks.

energy source for these processes. In fact, in many organisms, cells, and food-related processes, glucose is not the starting point. How do these other sources provide us with energy? In the same way as glucose! Most monosaccharides or simple sugars are structurally similar. When these nonglucose monosaccharides encounter the right enzyme, they are converted to a glycolysis intermediate. For example, the fructose in the honey on your toast can enter the glycolytic pathway at the product of step 2, fructose-6-phosphate. Complex carbohydrates such as lactose, a disaccharide, and starch, an oligosaccharide, are first broken down into monosaccharide units and then enter glycolysis.

Why do organisms utilize the same primary pathway for the breakdown of so many of their carbohydrates? By having essentially all carbohydrates feed into the same pathway reduces the need to have completely independent pathways for the breakdown of all the different carbohydrates that may be available to an organism. Thus, it is biologically efficient for an organism to have a common metabolic pathway for the catabolism of carbohydrates. In food production, this metabolic efficiency is also advantageous, as it allows for diversity in the types of carbohydrate source that the microorganism uses in the food or cooking process. In yeast-based bread doughs, honey (fructose), table sugar (sucrose), or the starch found in the flour can be used as a carbohydrate source for the fermenting yeast.

4.9 METABOLISM AND DEGRADATION OF FATS

Sugars are not the only source of energy to cells and the only molecules whose breakdown is important to food and cooking. The metabolism and degradation of fats are also very important, as fats are an important energy source for organisms and fat metabolism and degradation yield many molecules that contribute to the taste and aroma of foods. In Chapter 1, you learned about the hydrocarbon molecular structure of fat molecules (Fig. 4.20).

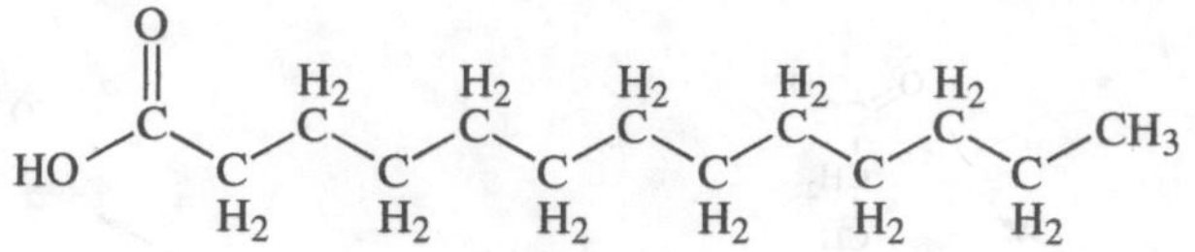

Saturated fatty acid chain

FIGURE 4.20   Fatty acid.

In this chapter, you have learned that catabolism involves the oxidative breakdown of larger molecules into smaller ones. If more oxidation can occur, more ATP can be generated, and the better the energy source for the organism. Fat molecules contain many carbons that are highly reduced (i.e., the carbons have many hydrogens bound to them); thus they have great potential to undergo oxidation. This "oxidative potential" makes a fat the most important long-term energy storage molecule for many organisms. However, when you look at the molecular structure of a fatty acid, it looks nothing like a glucose or any other sugar molecule! How are these molecules metabolized? How does the breakdown of fat impact food and cooking?

Most fats are stored as triacylglycerides (Fig. 4.21), containing a glycerol and three fatty acids.

In order to break down the triacylglycerol, the fatty acids are first cleaved away from the glycerol in a hydrolysis reaction to yield a molecule of glycerol and three "free" fatty acids. The glycerol, which is structurally similar to intermediates within the glycolysis pathway, can enter glycolysis following a two-step conversion to dihydroxyacetone phosphate.

In contrast, fatty acids are structurally distinct from carbohydrates; they hardly contain any oxygen at all! Thus, many more oxidation reactions need to be carried out to convert all the carbons in a fatty acid to CO_2. Fatty acids are broken down in a stepwise process called beta-oxidation that involves oxidation and removal of successive two-carbon units. This process generates acetyl-CoA, NADH, and $FADH_2$, which can enter the TCA cycle and electron transport chain to yield energy in the form of ATP. The longer the fatty acid chain, the greater the amount of energy yielded for the organism (Fig. 4.22).

The breakdown of fats and fatty acids is also key to the flavor, aroma, and properties of many foods, in positive and sometimes not so positive ways. The molds found in blue cheese such as *Penicillium roqueforti* transform fatty acids into methyl ketone molecules, such as 2-pentanone, 2-heptanone, and 2-nonanone, that create the cheese's characteristic and pungent aroma.

However, oxidation of fatty acids by oxygen in the air leads to some undesirable products. Over time, the oxygen reacts with fats, breaking them down into smaller, volatile fragments, which have an odor that we often characterize as rancid. An expired box of crackers or bag of tortilla chips may have this rancid scent. A triglyceride of butyric acid, found in butter, has a pleasant taste and aroma. However, as butter ages, the butyric acid is cleaved from the glycerol, leading to an unpleasant odor that some would characterize as vomit-like (Fig. 4.19). When you smell an old

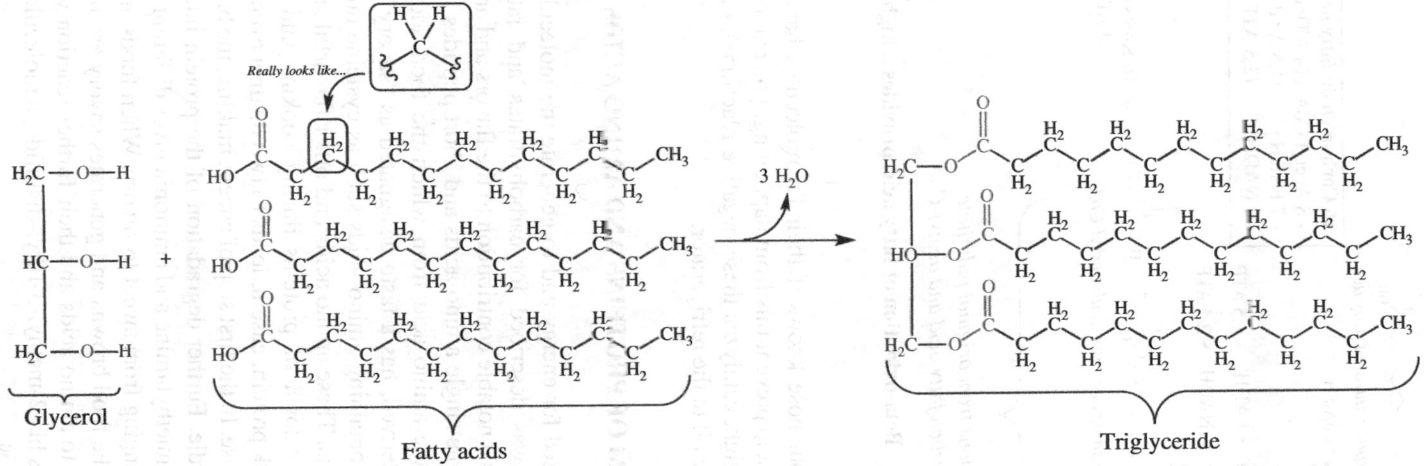

FIGURE 4.21 Formation of triacylglycerol.

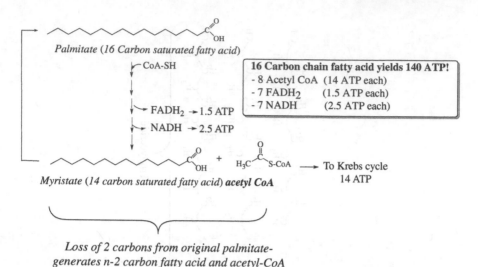

16 Carbon chain fatty acid yields 140 ATP!
- 8 Acetyl CoA (14 ATP each)
- 7 FADH$_2$ (1.5 ATP each)
- 7 NADH   (2.5 ATP each)

Palmitate (16 Carbon saturated fatty acid)

CoA-SH

FADH$_2$ → 1.5 ATP

NADH → 2.5 ATP

Myristate (14 carbon saturated fatty acid) **acetyl CoA**

To Krebs cycle
14 ATP

*Loss of 2 carbons from original palmitate-
generates n-2 carbon fatty acid and acetyl-CoA*

FIGURE 4.22   Beta-oxidation of fatty acid provides a high ATP yield.

bottle of olive oil, your nose knows if their fat molecules have been oxidized in detrimental ways. To help prevent this from happening, you can keep oils in the dark (the energy from the light catalyzes these negative reactions), or you can slow the reaction by placing the oil in the refrigerator.

4.10   METABOLISM OF PROTEINS AND AMINO ACIDS

Proteins are metabolized for energy and to generate the molecular building blocks for an organism, as was described for carbohydrates and fats. However, their breakdown also makes a notable contribution to the flavors and aromas of uncooked and cooked foods. Some single amino acids and short peptides have tastes of their own. Glutamic acid, the amino acid from which the food additive monosodium glutamate (MSG) is derived, has a taste designated as "savory" or "umami." The breakdown of sulfur-containing amino acids such as cysteine during cooking may lead to an "eggy" aroma. These amino acids and other flavorful small molecules are derived from the breakdown of proteins during cooking and food preparation. Degradation of the milk protein, casein, leads to two distinct events important in the process of making cheese. In the first step of cheese making, the breakdown of casein causes the milk to curdle. Further degradation of the protein into amino acids and small molecules like trimethylamine and ammonia gives different varieties of cheese their distinct flavors ranging from sweet to savory. When foods are seared or grilled at high temperatures, the food browns and generates savory aromas and flavors due to protein degradation to amino acids and their further reaction with sugars to produce molecules such as bis-2-methyl-3-furyl-disulfide, a molecule associated with a "meaty" odor (Fig. 4.23).

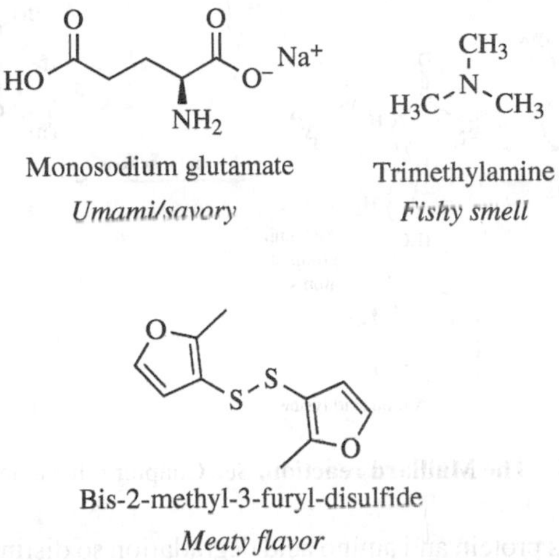

Monosodium glutamate

Umami/savory

Trimethylamine

Fishy smell

Bis-2-methyl-3-furyl-disulfide

Meaty flavor

FIGURE 4.23 Amino acid products that have odor/flavors.

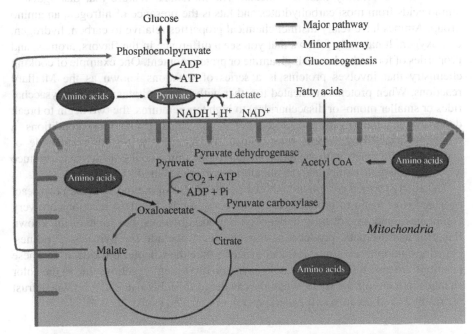

FIGURE 4.24 Amino acid metabolism. Amino acids feed into several pathways.

Earlier in this chapter, you learned that both fatty acid and carbohydrate metabolisms converge on the central catabolic pathways of glycolysis and the citric acid cycle. You see a similar phenomenon in the metabolism of protein and amino acids; following cleavage of a protein into its amino acid components, the carbons of the amino acids find their way into the citric acid cycle where energy is generated in the form of ATP (Fig. 4.24).

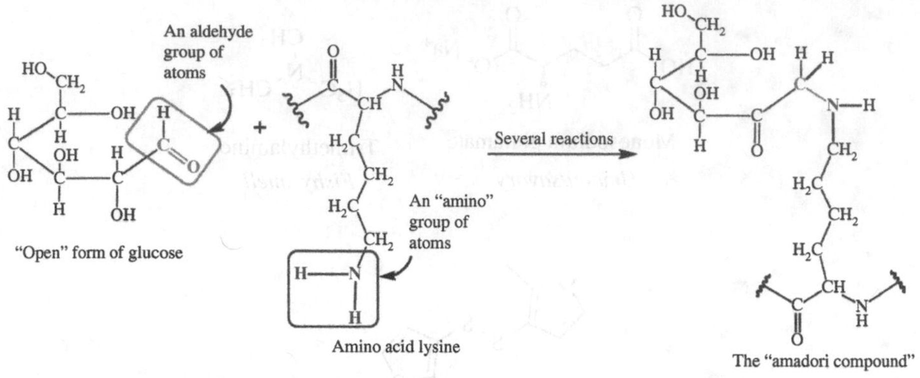

FIGURE 4.25   The Maillard reaction. See Chapter 6 for a detailed reaction.

Then, what makes protein and amino acid degradation so distinctive? Why do high-protein foods brown when you cook them at high temperatures? And why do cooked meats smell differently than raw meats? The molecular feature that distinguishes amino acids from most carbohydrates and fats is the presence of nitrogen, an amino group. Amines have really distinct chemical properties relative to carbon, hydrogen, and oxygen. It naturally follows that you see a difference in the flavors, aromas, and properties of foods that have a high amine or protein content. One example of cooking chemistry that involves proteins is a series of reactions known as the Maillard reactions. When proteins are heated together with carbohydrates (either polysaccharides or smaller mono- or disaccharides) to high temperatures, the two begin to break down into smaller sugars and proteins or amino acids. Under these conditions, a reducing sugar "ring" opens into a linear form, and the resulting aldehyde or carboxylic acid reacts with the amine group (of an amino acid or protein) to produce a new intermediate compound, an amadori compound (Fig. 4.25).

The newly formed compound (shown as a D-glucosamine in the figure) can react over and over again to produce additional compounds. Maillard reactions are very complicated and are not fully understood by food chemists, but some of the known flavorful and aromatic products of this chemistry include the pyrroles, pyridines, pyrazines, thiophenes, thiazoles, and oxazoles (see the Chapter 6 for details). These reactions are also referred to as "nonenzymatic browning" reactions due to the color changes in many of the foods when they undergo this chemistry: the browned crust of freshly baked bread, a soft baked pretzel, or a grilled steak.

4.11   METABOLISM AND DIET

Human cells carry out glycolysis, fermentation, the Krebs cycle, fat metabolism, and amino acid degradation processes in order to generate the energy and nutrients that are necessary for our survival. Individuals with a balanced diet utilize sugars, proteins, and fats as an energy source depending upon the needs of the body. Sugars and carbohydrates tend to provide a more immediate source of energy, while proteins and fats are utilized as a more long-term energy source. Do you wonder if there is a metabolic advantage to using only one food source? What would happen if you just ate protein or fat or carbohydrates? You don't have to wonder, as this is what some fad diets encourage you to do. High-protein diets, which recommend that a dieter receive approximately 30–50% of their caloric intake from protein, are widely promoted as an effective method for losing weight. High-protein diets often lead to rapid weight loss and happy, satisfied dieters. However, the weight loss is short lived because the weight loss is due to water loss. By eating only protein, the body transitions into starvation mode because the brain and most other tissues prefer to utilize carbohydrates as their energy source. Thus, the body uses stored blood sugar from the liver and muscles for energy; this results in muscle breakdown. Since muscle is mostly water, a dieter loses weight very rapidly, within the first few days of the diet. If carbohydrate restriction continues, the brain will eventually use fat stores as fuel, leading to a state called ketosis. Ketosis does bring not only weight loss but also a host of other problems, including heart issues, irritability, headaches, and kidney function impairment. High-protein diets also lack key nutrients that are present in plant-based foods. The healthiest and most effective methods for weight loss are a balanced diet, reduced caloric intake, and exercise.

4.12   IMPORTANT REACTIONS IN METABOLISM: OXIDATION AND HYDROLYSIS

Enzymes catalyze a wide range of chemical reactions. Two very important reactions are found throughout metabolite pathways and are crucial to truly understand some interesting changes that occur when cooking and baking. One set of reactions involves the transfer of electrons and hydrogens (oxidation/reduction), and the other uses water to break a bond or produces water to make a bond (hydrolysis and dehydration). These are only two very common chemical changes we see in food and cooking. Let's start with oxidation and reduction.

4.12.1   Oxidation and Reduction

As we learned earlier, ATP is an important product of metabolism and can be produced through reactions catalyzed by glycolysis. In addition to the direct phosphorylation of ADP seen in glycolysis, ATP can also be generated indirectly through an oxidation reaction. What is an oxidation reaction? Historically, if an element reacted with oxygen to produce an oxide, you would classify the reaction as an oxidation reaction. You know this type of reaction well, as it describes the conversion of elemental iron (Fe) to iron oxide (Fe_2O_3), also known as rust:

$$2\,Fe\,(s) + 3\,O_2\,(g) \rightarrow 2\,Fe_2O_3\,(s)$$

Now, an oxidation reaction has a much broader meaning. Oxidation reactions occur when an element, compound, or ion loses electrons (Fig. 4.26).

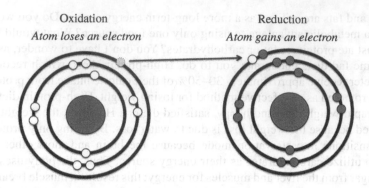

Oxidation
Atom loses an electron

Reduction
Atom gains an electron

FIGURE 4.26   Oxidation and reduction.

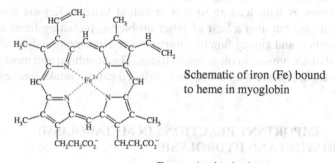

Schematic of iron (Fe) bound
to heme in myoglobin

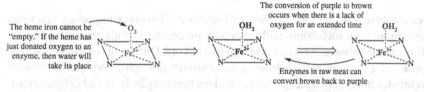

The heme iron cannot be
"empty." If the heme has
just donated oxygen to an
enzyme, then water will
take its place

The conversion of purple to brown
occurs when there is a lack of
oxygen for an extended time

Enzymes in raw meat can
convert brown back to purple

Myoglobin heme binding oxygen = red
(*this myoglobin is ready to donate oxygen*)

Myoglobin heme has changed to iron 3$^+$
and can no longer bind oxygen = brown

Myoglobin heme binding water = purple
(*this heme has just donated oxygen*)

FIGURE 4.27   Myoglobin and iron. The oxidation state of iron held in place by the heme
of myoglobin is shown here.

How do you know when an oxidation reaction takes place? You assign an oxidation
number to each atom in a compound, which indicates whether the atom is electron
rich, neutral, or electron poor. If the oxidation number for a particular atom changes
over the course of a reaction, you have an oxidation reaction.

Oxidation reactions occur all around us and are key in food preparation and cooking
processes. For example, some of the color differences that you see in meat are due to
numerous oxidation reactions that occur in the iron (Fe) that is found in the muscle
protein, myoglobin. As the iron changes its oxidation state (between +2 and +3) and
its bonding partner (O_2 or H_2O), it changes color, resulting in a color change in the
meat (Fig. 4.27).

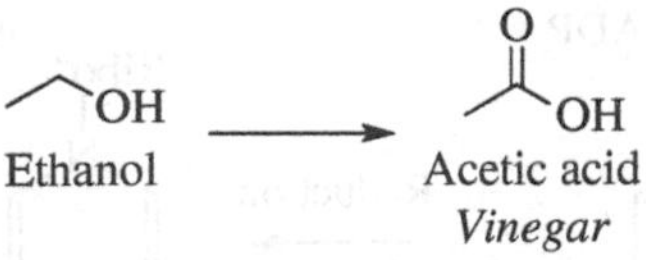

FIGURE 4.28   Ethanol conversion to acetic acid.

In raw fresh meat, iron has an oxidation state of +2; this means that the iron has two fewer electrons than protons, since protons are positively charged and electrons are negatively charged. Iron in an oxidation state of +2 in myoglobin typically has a red or pink color associated with it. As meat cooks, the iron is oxidized and loses one electron, so its oxidation state increases by one (there are now three more positively charged protons than negatively charged electrons). In other words, the iron is oxidized to Fe^{3+}. Iron in myoglobin with a +3 oxidation state is typically brownish in color, as is observed in well-done beef.

A different type of oxidation reaction occurs when you watch a cut apple turn brown. A browning apple actually undergoes two different oxidation reactions. In the first reaction, the oxidation reaction is simply identifiable by the addition of oxygen to a molecule, similar to what you see in the elemental iron to rust reaction. Oxygen is added to the monophenol and electrons are lost (which is not obvious in this example) to yield catechol.

In the second oxidation reaction (catechol to o-quinone), no additional oxygen is added, but the catechol undergoes a dehydrogenation reaction; it loses hydrogens. This is another way to identify and define oxidation reactions, particularly those in which oxidation numbers are not obvious. Following o-quinone production, a variety of enzyme- and nonenzyme-catalyzed reactions take place to yield colored compounds like melanin, which is the color that we observe in our browning apple.

Have you ever noticed that a partially used bottle of wine eventually develops a sour taste? Upon exposure to air (for a long time), the ethanol in the wine oxidizes to acetic acid (Fig. 4.28). This process doesn't happen very readily on its own, but occurs in the presence of enzymes that are produced by a bacterium called *Acetobacter aceti*. When vinegar is made, this type of bacteria is purposefully added to wine or other fermented alcoholic beverages to produce the vinegar that you might use in making salad dressings or dyeing Easter eggs.

Using the myoglobin, apple browning, and vinegar reactions as examples, you now have the expertise to identify oxidation reactions. If there is a loss of electrons (myoglobin), a gain in oxygen (browning and vinegar), or a loss of hydrogen (browning and vinegar), you have an oxidation reaction. To be clear, in all of these reactions, a compound, element, or ion loses electrons. In the case of the browning apple and vinegar, it is harder for a novice to identify the electron loss.

Where do the electrons go? In these examples, it appears that the electrons just get lost to some unknown place in the surroundings. However, this is not the case. When oxidation reactions occur, the electrons lost by one molecule must be transferred to or gained by another molecule. The molecule that gains the electrons is reduced. If a molecule loses electrons, another molecule must be available to accept the electrons.

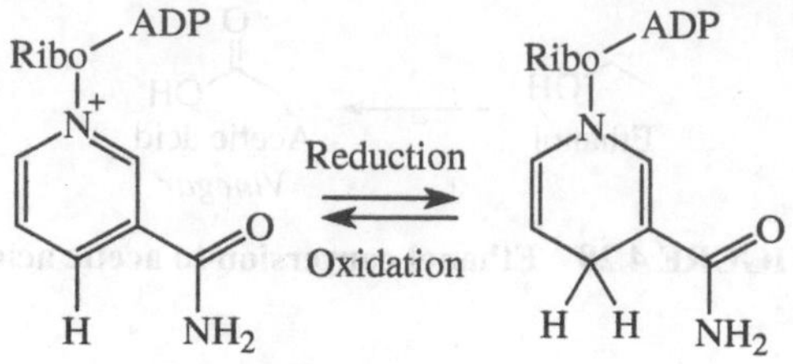

FIGURE 4.29 Redox of NAD⁺/NADH.

In other words, whenever an oxidation occurs, a reduction must also occur. If you think about oxidation/reduction reactions in terms of hydrogen and oxygen, the molecule that is oxidized gains oxygen or loses hydrogen, while the reduced molecule gains hydrogen or loses oxygen (Fig. 4.26).

Within a cell or microorganism, the ultimate acceptor of the electrons lost in an oxidation reaction is oxygen. In the process, the oxygen also gains two hydrogen ions and is converted to H_2O. However, oxygen does not directly and immediately accept the electrons in every cellular oxidation reaction that occurs. Rather, cells and microorganisms use an "intermediate" electron acceptor, called a coenzyme, that functions as a temporary electron carrier. A common coenzyme involved in metabolic processes is nicotinamide adenine dinucleotide (NAD⁺).

NAD⁺ serves as an electron carrier by accepting two electrons and two hydrogen cations (protons) from the molecule that is being oxidized, thereby generating NADH plus a proton. In the example shown, the L-malate is oxidized, while the NAD⁺ is reduced to NADH (Fig. 4.29). NAD⁺ is largely derived from niacin, a B vitamin that is essential for the survival of cells and organisms. Good sources of niacin include animal products (i.e., meat, eggs, seeds, and legumes). However, only small amounts of the NAD⁺ are present in a cell. In order for metabolic processes and oxidation reactions to occur within a cell or microorganism, recycling of NADH to NAD⁺ is a key part of metabolism and how microorganisms meet their energy needs. If you enjoy eating a crusty piece of bread with cheese or drinking a beer, you have benefited from the recycling of NADH to NAD⁺ by yeast and bacteria.

REFERENCE

[1] Volker, M. (2008) *Bacterial Fermentation.* Goethe Universität, Frankfurt/Main.

5

CHEESE, YOGURT, AND SOUR CREAM

Guided Inquiry Activities (Web): 6, *Higher-Order Protein Structure and Denaturation;* 7, *Carbohydrates;* 8, *pH;* 9, *Fats, Structure, and Properties;* 10, *Fats Intermolecular Forces;* 14, *Cells and Metabolism;* 16, *Milk*

5.1 INTRODUCTION

The components of dairy that are critical to the production of cheese, yogurt, and sour cream will be emphasized: curds and whey, lactose as a bacterial energy source, and fats. The science of making cheese, yogurt, and sour cream; the involvement of different types of bacteria (starters and finishing), acid, or temperature; and the role of proteins and fats in flavoring/aromas will be discussed in detail. Processed cheeses (components and science of melting) will be covered in the context of the use of different cheeses for cooking purposes (melting, gratins, and soups).

5.1.1 Cheese

Its mention may conjure up thoughts of melted, stringy mozzarella on pizza; sharp, hard Cheddar on a cracker; or pungent, crumbly gorgonzola in a lettuce salad. Few other single foods have such diversity in texture, flavor, aroma, and appearance. The diversity, which is derived from variation in the components, the details of the cheesemaking process, and the creativity of the cheesemaker, results in hundreds of different varieties of cheese that are produced for both economic and culinary values.

However, early ventures in cheesemaking were not about economics or taste at all. It is hypothesized that approximately 5000 years ago, the peoples of the

The Science of Cooking: Understanding the Biology and Chemistry Behind Food and Cooking,
First Edition. Joseph J. Provost, Keri L. Colabroy, Brenda S. Kelly, and Mark A. Wallert.
© 2016 John Wiley & Sons, Inc. Published 2016 by John Wiley & Sons, Inc.
Companion website: www.wiley.com/go/provost/science_of_cooking

FIGURE 5.1   Cheese. A durable form of milk and nutrition and a tasty way to eat with crackers.

Middle East and Central Asia discovered that upon exposure to the hot sun and warm temperatures, milk soured and curdled. This state of milk is not particularly palatable to most (think about pouring "chunky" milk on your breakfast cereal); however, when this sour milk was drained and salted, a tastier and, most importantly in the ancient Middle East, less perishable food that resembled yogurt resulted (Fig. 5.1). It was also discovered that the milk curd became more firm and cohesive when the curdling took place in the presence of a piece of animal stomach! It may sound disgusting, but 5000 years ago, you had to use what was available for your cereal bowl.

During the Roman Empire when longer-distance travel was more common, cheese making and the cheese trade industry began to flourish; the number of types of cheese expanded significantly due to differences in local climate, animals, and feed, even within the same country. The result was a remarkable diversity of traditional European cheeses, which number from 20 to 50 in most countries and several hundred in France alone. Although the European Union still leads the world in cheese production, export, and consumption [1], cheese is produced and enjoyed around the world (Fig. 5.2).

In this chapter, we will study the science behind the process of cheesemaking: the molecular basis for milk **curding**, what is **drained** from the milk curd when it is **pressed**, the flavorful aromatic molecules that are formed during salting and **ripening** and the time in which a cheese might sit for days, weeks, months, or years before it is eaten. In addition, we will study the differences in the cheesemaking process and components that result in various cheeses. By the end of this chapter, you will understand why Velveeta™ melts beautifully in your macaroni and cheese, while stringy mozzarella is the best cheese for the top of your pizza (Fig. 5.3).

FIGURE 5.2   Cheese as an ancient food. Pictorial evidence of the cheesemaking process from a fourteenth-century artwork of an eleventh-century Arab medical health book showing the preparation of cheese. Unknown master—book scan, https://commons.wikimedia.org/wiki/File:9-alimenti,_formaggi,Taccuino_Sanitatis,_Casanatense_4182.jpg. Used under Public Domain via Wikimedia Commons https://commons.wikimedia.org/wiki/Commons:Licensing#Material_in_the_public_domain.

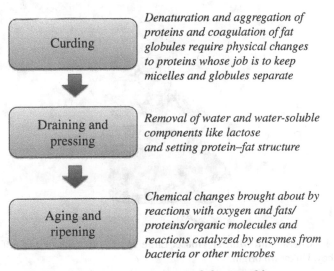

FIGURE 5.3   The process of cheesemaking.

5.2   MILK CURDLING AND COAGULATION

Have you ever thought about what Little Miss Muffet was eating? What exactly are curds and whey? Miss Muffet's curds and whey could have looked like a bowl of long-outdated, sour, chunky milk from your refrigerator, which sounds disgusting (Fig. 5.4). However, you might enjoy the products of controlled coagulation every day, like cheese, yogurt, and sour cream. You might ask, what exactly comprises a milk curd, that soft, coagulated (chunky) part of soured milk, and why does it form? To start, it is important to remember that milk is made of two phases or components. The milk globules with fats and proteins are one part of milk. Milk globules are suspended in an aqueous or watery phase filled with different proteins. Milk stays in this form until something happens to the proteins on the globules that keep the fat from coalescing into curds. Simply put, curds form and separate from the rest of the liquid milk because of the properties, characteristics, and interactions of the milk protein, casein.

5.2.1   The Milk Proteins: Casein and Whey

As you recall from Chapter 1, proteins are polymers of amino acids that are connected to one another via dipeptide bonds. Proteins, in their active native and functional form, have a distinct structure that typically maximizes intermolecular interactions with molecules that have similar properties (e.g., polar interacts with polar) and minimizes interactions of unlike components (e.g., polar with nonpolar).

Milk contains two types of protein, casein (mixed in and on the fat globules) and whey (in the aqueous phase). Both proteins are very distinct from one another in terms of structure and their impact on cheesemaking and the final cheese product. Generally speaking, casein provides the basis of a cheese's solid structure and texture

FIGURE 5.4   Cheese and Mother Goose. Little Miss Muffet sat on her tuffet eating her curds and whey. Unknown origin.

(a)   -Lys-Ala-His-Gly-Lys-Lys-Val-Leu-Gly-Ala-
      Primary structure (amino acid sequence in a polypeptide chain)

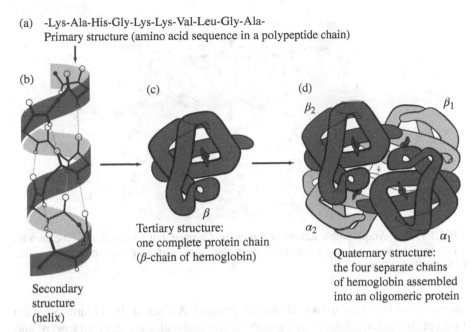

FIGURE 5.5   **Level of protein structure.** (a) Primary structure, (b) secondary structure, (c) tertiary structure, and (d) quaternary structure. Voet, Voet and Pratt, figure 6.1—pp. 128. Reproduced with permission of Voet, Voet and Pratt 4th Edition.

and governs the retention of fat and moisture in the cheese, while the whey proteins remain soluble in the milk liquid and are largely removed from the final cheese product during draining. You might ask, what is so special about casein? What makes it so different from whey (Fig. 5.5)?

5.3   CASEIN

As we learned in the milk chapter, casein, the most abundant protein in milk, is actually a family of related proteins that is organized into four different subtypes: α(s1)-casein, α(s2)-casein, β-casein, and κ-casein. The casein proteins play a key role in development as they help to transport calcium (in the form of calcium phosphate) to the bones. The mechanism that the caseins use for transporting calcium is actually related to cheese production.

Casein proteins have amphipathic character. This means that some parts of the structure of casein are hydrophilic, while other portions of the molecule are hydrophobic. Amphipathic molecules, like fatty acids, often aggregate or clump into a spherical shape and form a large molecular complex called a micelle (Fig. 5.6) in order to maximize interactions between "like" chemical groups and minimize

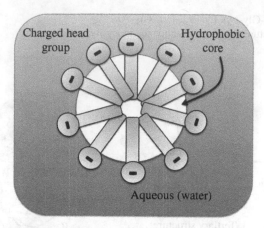

FIGURE 5.6 Amphipathic nature of a micelle. Charged, polar head groups (shown with a (−)) face the water solvent, while the hydrophobic portion of the compound aggregate due to hydrophobic forces.

interactions between unlike chemical groups. As noted in Figure 5.6, the hydrophilic, polar, charged head groups of the molecules are directed toward and interact favorably with the polar water, while the hydrophobic nonpolar groups are contained on the inside of the micelle sphere, where there is little to no inter-action with water (or other polar groups). Protein micelles are a bit more compli-cated than fatty acid micelles, as the individual proteins are larger and more structurally complex, and the polar and nonpolar components of a protein are interspersed throughout the protein sequence and structure. However, a micelle protein structure can still form with a largely nonpolar interior and a polar/charged exterior.

What holds the casein proteins in a micelle structure in milk, and why is it impor-tant? Calcium. Yes, calcium in milk does more than just help to make your bones strong. It actually holds the four types of casein proteins tightly in the micelle com-plex. Moreover, this protein–calcium complex helps to keep milk fat globules from coalescing into curds (Fig. 5.7). How does this work at the molecular level? One of the casein proteins, κ-casein, is a negatively charged protein that sits on the outside of the micelle. The negatively charged micelle exterior limits the size of the protein complex, as the negative charges repel each other, preventing too many negatively charged proteins from joining the complex (Fig. 5.8). Because the micelles remain relatively small, there are many of them in the solution, so they interfere with/prevent the milk fat globules from getting too close to one another. All of this is due to the deprotonated carboxyl groups of κ-casein (COO⁻). However, any disturbance in pH or other chemical changes that remove the negative charges from the κ-casein proteins will result in the aggregation of milk fat and protein, leading to curdling. To produce cheese, a cheesemaker disturbs the casein micelles to encourage the aggregation of the casein proteins into a solid-like mass that forms the basis of the cheese curd.

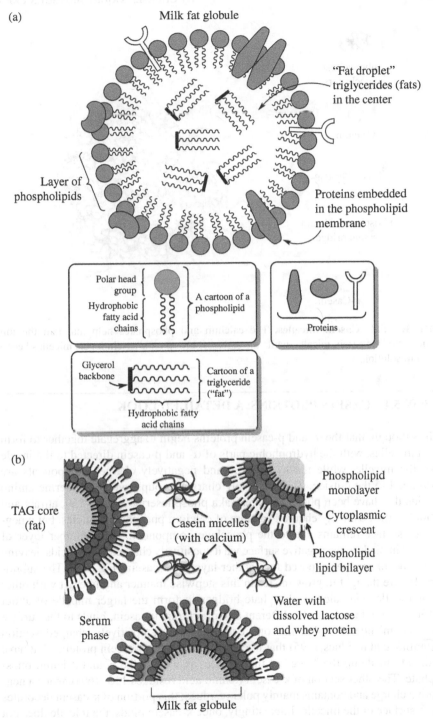

FIGURE 5.7 Fat globules, whey, and casein. (a) Milk fat globule. (b) The composition of milk is shown as the three-phospholipid membrane of the globules and the water-soluble casein micelles.

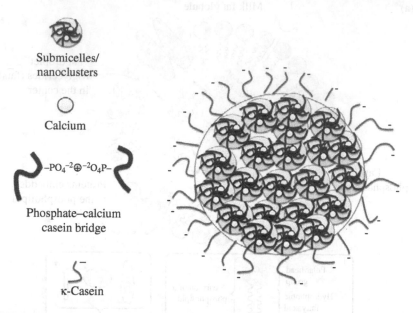

Submicelles/
nanoclusters

Calcium

$-PO_4^{-2}\oplus^{-2}O_4P-$

Phosphate–calcium
casein bridge

κ-Casein

FIGURE 5.8   Casein micelles. The calcium and phosphates help maintain the inner structure of the casein micelle, while the carboxyl group on κ-casein repels micelles keeping them in solution.

BOX 5.1   CASEIN PROTEINS: A DETAILED LOOK

It is thought that the α- and β-casein proteins begin to aggregate together to form submicelles, with the hydrophobic parts of α- and β-casein directed to the inside of the micelle, while the hydrophilic and negatively charged components are directed to the outside. The negatively charged groups, which are serine amino acids that have been phosphorylated (aka phosphoserines) (Fig. 5.9), attract and bind to the positively charged calcium of calcium phosphate particles, but a negative surface remains due to the phosphate component. Thus, another layer of α-casein binds to the negative surface via its positively charged amino acids, leaving the phosphoserines exposed for another layer of α-casein (Fig. 5.8). The submicelles are thought to grow in size in this stepwise manner and interact with other submicelles via calcium phosphate bridges to form the larger micelle of about 150–300 nm in size until a different casein protein, κ-casein, binds to the surface of the micelle. κ-casein has one hydrophobic yet positively charged section (amino acid residues 1–95) that interacts with the other casein proteins that have already made up the "core" of the micelle, as well as the bound calcium phosphate. The other section of κ-casein (amino acid residues 113–169) carries a negative charge and contains mainly polar residues; this portion of κ-casein decorates the surface of the micelle. Interestingly, once κ-casein binds, the micelle does not get any larger. How does κ-casein prevent the casein micelle from growing?

Evidence suggests that it has something to do with κ-casein's inability to bind calcium phosphate and perhaps a structural change that occurs when it binds to the growing micelle to contain more β-sheet character. In addition, the negatively charged amino acids of κ-casein interact favorably with the water, allowing the micelle to be soluble (negative charges and polar molecules interact well). However, since negative charges repel one another, κ-casein effectively prevents two micelles from interacting with one another and keeps the micelles small enough to remain soluble. Limiting the size of the micelle is important because when a single molecule or aggregate of molecules becomes too large, then it may no longer be soluble in a solution and it precipitates or "falls out" of solution. When you make cheese, this is exactly what occurs!

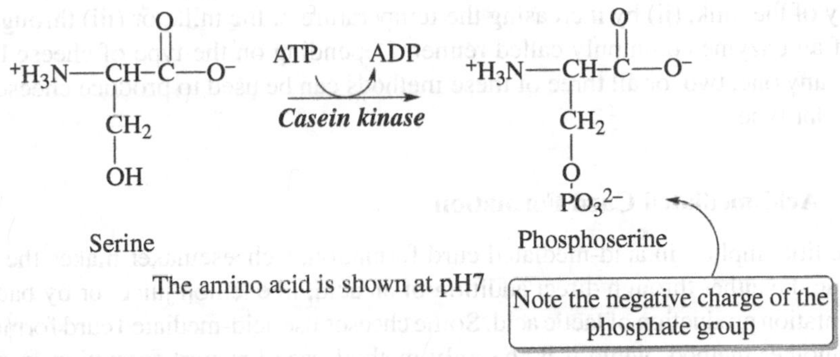

Serine   Phosphoserine

The amino acid is shown at pH7

Note the negative charge of the phosphate group

FIGURE 5.9 Serine and phosphoserine. The enzyme, casein kinase, transfers a phosphate group from ATP to the amino acid serine in casein to create a negatively charged phosphoserine.

5.4 WHEY

Whey comprises all of the noncasein protein components of the milk, about 20% of the total milk proteins. Approximately 20–40 different proteins make up the family of proteins called "whey," which have numerous different roles within an organism ranging from defensive proteins, transport proteins, to enzymes. The most abundant whey proteins are β-lactoglobulin (unknown function), α-lactalbumin (functions in lactose synthesis), serum albumin (a transport protein), and immunoglobulins (defensive proteins). However, what distinguishes the whey proteins is that they do not interact with other proteins to form large complexes or micelles in milk but fold into compact, globular structures that are soluble in the milk water under most conditions (Fig. 5.9).

They do not interact with the caseins within the casein micelle structure, nor do they aggregate together during cheese production. Given that the whey does not form part of the cheese curd, it is typically considered a by-product of cheese production, and approximately 630,000 tons of whey protein is produced annually from the worldwide manufacture of cheese. However, in a few cheeses and dairy products like ricotta and yogurt, the whey proteins play an important role in production, texture, and the final product, in addition to providing nutritional benefits.

5.5   MORE MILK CURDLING

Now you know that the main components of the cheese curd are the casein proteins and that, somehow, the casein proteins that are soluble in milk become insoluble during cheese production. Your next questions might be: how is curd formation induced? Is it the same or different in different types of cheeses? Does it matter how the curd is induced?

Curd formation is commonly induced in one of three ways: (i) by increasing the acidity of the milk, (ii) by increasing the temperature of the milk, or (iii) through the use of an enzyme commonly called rennet. Depending on the type of cheese being made, any one, two, or all three of these methods can be used to produce cheese of a particular type.

5.5.1   Acid-mediated Curd Formation

As the title implies, in acid-mediated curd formation, a cheesemaker makes the milk more acidic either through direct addition of an acid, like lemon juice, or by bacteria fermentation production of lactic acid. Some cheeses use acid-mediated curd formation with another method, while it is the only method used for curd formation in many fresh cheeses, such as Chévre, cream cheese, goat cheeses, and ricotta. First, we will learn about how acid induces the formation of the cheese curd at a molecular level, and then we will go into some of the details about the different types of acid-mediated curd formation processes and the resulting cheeses produced from each type.

5.5.2   The Molecular Basis for Acid-mediated Curd Formation:
Why does Acid cause Milk to Curdle?

In order to induce a milk curd to form, you have to disrupt the casein micelles in some way. One way to cause this molecular disruption is by changing the pH of the milk. As you recall from Chapter 1, a protein's structure and its characteristics are usually very sensitive to pH; when you change pH, you change the protonation state of acidic and basic amino acids within the protein. Generally speaking, a decrease in pH (a more acidic solution) will cause some amino acids to become more protonated, while an increase in pH (a more basic solution) will cause some amino acids to become more deprotonated (Fig. 5.10). Continue to raise the pH and the interaction between acidic and basic amino acids is lost as the amino groups lose their proton at pH greater than 9.

The native structure of a protein is often stabilized by a number of intermolecular interactions that occur between its component amino acid side chains. If an interaction depends upon an amino acid having a particular charge (positive or negative), as soon as you alter the pH of the solution, you will change the relative amount of the amino

(a)   (b)

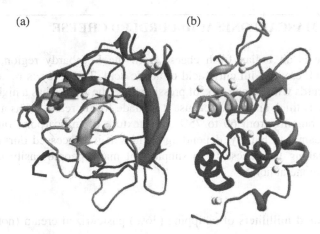

FIGURE 5.10   Lactalbumin. Two types of whey proteins are (a) β-lactalbumin and (b) α-lactalbumin.

acid that will have that charge, thereby potentially destabilizing the protein's structure. For example, if there is an interaction between a negatively charged amino acid (i.e., glutamate) and a positively charged amino acid (i.e., lysine) within a protein structure at a particular pH (Fig. 5.10), as soon as you decrease the pH of the solution, the interaction might be less likely to occur or be weaker. Thus, some protein molecules might denature/unfold in the solution.

If we come back to relate this concept to the formation of cheese curds, milk has a pH of approximately 6.5. At pH 6.5, the casein and whey proteins are soluble in the milk. The caseins are largely folded and incorporated into the micelle structure described earlier with the negatively charged domain of κ-casein at the surface, while the whey proteins have their compact, globular individual protein structures. The negatively charged domain of κ-casein is due to the presence of a high number of deprotonated glutamic acid (8) and aspartic acid (2) amino acids within the surface domain at pH 6.5.

As the milk pH decreases to a value of 5.5, the surface κ-casein protein and other casein proteins begin to lose some of their negative charges due to protonation of negatively charged amino acids like aspartate and glutamate, which makes them neutral. This "neutralization" impacts the casein micelle in two ways. With loss of negative charges, the interaction between calcium and casein is reduced; thus some calcium phosphate is lost from the casein micelle structure. With less calcium, the submicelles that comprise the larger micelle complex are not as attracted to one another and dissociate from one another, and the micelle begins to fall apart. As the milk pH continues to drop, the fragmented micelles become even less negatively charged and repulsion between one micelle fragment and another is reduced even further. The fragmented casein micelles, other free non-micelle protein in the milk, and milk fats begin to aggregate to prevent exposure of the hydrophobic protein and fat components to the polar milk water. These large networks of protein and fat precipitate out of the milk liquid as a soft curd, fragile gel.

BOX 5.2   MASCARPONE: ACID-CURDLED CHEESE

Mascarpone is an Italian fresh cheese from the Lombardy region, made by
curdling milk cream with citric acid or acetic acid. The whey is removed from
the casein curds without any type of pressing; thus the cheese has a high moisture
content and is thick and soft. Because it is made from cream, it has a very high
fat content ranging from 60 to 75%. The texture of mascarpone ranges from
smooth, creamy, to buttery, depending on how it is processed during cheesemaking. Making the cheese is so simple that many people easily make their
own mascarpone at home.

Recipe:
- Five hundred milliliters of whipping (36%) pasteurized cream (not ultrapas-
  teurized)
- One tablespoon of fresh lemon juice

  Bring 1 inch of water to a boil in a wide skillet. Reduce the heat to medium-
low so the water is barely simmering. Pour the cream into a medium heat-
resistant bowl, and then place the bowl into the skillet. Heat the cream, stirring
often, to 190°F/89°C. It will take about 15 min of delicate heating. Add the
lemon juice and continue heating the mixture, stirring gently, until the cream
curdles. All that the whipping cream will do is become thicker, like sour
cream. The back of your wooden spoon will have a thick layer of the cream,
and you will see just a few clear whey streaks when you stir. Remove the bowl
from the water and let it cool for about 20 min. Meanwhile, line a sieve with
four layers of dampened cheesecloth and set it over a bowl. Transfer the mixture into the lined sieve. Do not squeeze the cheese in the cheesecloth or press
on its surface (be patient; it will firm up after refrigeration time). Once cooled
completely, cover with plastic wrap and refrigerate (in the sieve) overnight
or up to 24 h.

5.5.3   How is Milk made more Acidic?

There are two mechanisms whereby milk can be made more acidic. The easiest way
that you can make milk more acidic is simply by adding an acid! There are many
acids that are used in cooking and are effective enough to instigate the curdling process, such as lemon juice, citric acid, or vinegar. If you have ever eaten ricotta (often
used in lasagnas), mascarpone (an Italian soft cheese used in the dessert Tiramisu),
or pannier (a cheese common in South Asian cuisine), you have eaten cheeses that
are made by the direct addition of an acid to milk. What is really neat about these
cheeses is that you can easily prepare them in your own kitchen and that they don't
really melt! They simply get drier and stiffer when you heat them up. The chemical
explanation behind this phenomenon is the aggregation of the casein proteins in the

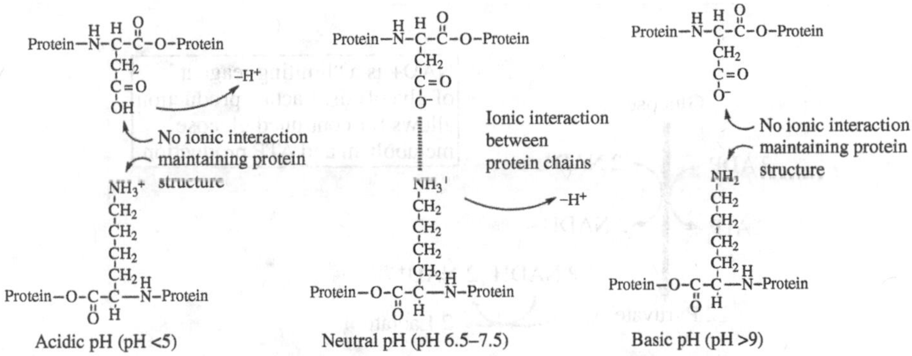

FIGURE 5.11   The impact of pH on ionic interactions and protein structure.

presence of acid. When the acid curd is heated, the casein aggregates have strong interactions; the first thing that is disrupted upon exposure to heat is the milk water. The water evaporates and the casein proteins become more concentrated and dried out, rather than stringy and more liquid-like (which requires the presence of water). Because of this trait, acid-curdled cheeses retain shape upon exposure to heat and can be fried.

The second mechanism used to make milk more acidic is more indirect and utilizes the process of bacteria fermentation that you first learned about in Chapter 3. Since you may not think about bacterial fermentation on a regular basis, we will highlight a few of the key points here, as it relates to milk coagulation in the cheesemaking process. In fermentation, bacteria utilize the enzyme lactate dehydrogenase to convert the glycolysis products of pyruvate and NADH to lactic acid and NAD^+ (Fig. 5.11).

Fermentation is critical for survival of the organism, as it regenerates the NAD^+ necessary to produce ATP through glycolysis. However, the lactic acid that is produced also makes the environment in which the bacteria are growing more acidic! Thus, if you allow bacteria to grow in milk that is being used to produce cheese, you have an internal source of acid, the lactic acid that is produced by the growing bacteria. Moreover, the lactic acid produced doesn't just affect the formation of the milk curd, but the presence and amount of lactic acid generated also impact the taste and aroma of a cheese. Depending on the cheese being made, two strains of bacteria may be used: a starter bacteria (such as lactobacteria) to begin the acid production and a second strain of bacteria, commonly called a nonstarter, or ripening bacteria (Fig. 5.12). The finishing bacteria will vary depending on the type of cheese being made. Some will produce gas to make Swiss cheese, and others will produce other flavors as the bacteria eat protein, fat, and sugars producing new flavorants. The common characteristic of finishing bacteria is they can thrive in a lower, more acidic pH produced by the starting cultures. Many of these bacterial strains can handle the higher heats used to complete the cheese production.

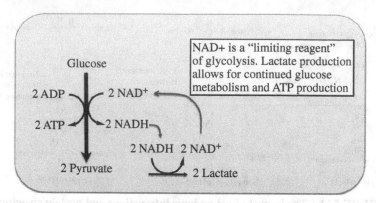

FIGURE 5.12   Glycolysis and NAD⁺/NADH. NAD^+ is quickly used during respiration/fermentation and is replaced in bacteria and some animal cells through the production of lactate.

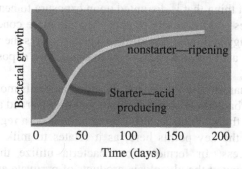

FIGURE 5.13   Bacterial culture and cheese.

5.6   LACTOBACTERIA AND FERMENTATION

In order for lactic acid fermentation to occur, you need a source of sugar and an organism that can utilize the sugar for energy. As you learned in Chapters 1 and 4, the primary sugar found in milk is lactose. Lactose is composed of a glucose molecule and a galactose molecule, linked together via an *o*-glycosidic bond (Fig. 5.13).

Given that both glucose and galactose are sugar sources for glycolysis, you might think that lactose is an ideal sugar source for every organism—two sugar molecules for the price of one disaccharide! However, lactose cannot enter glycolysis in its disaccharide form. Lactose must be cleaved at its glycosidic bond in order to produce the more "useful" glucose and galactose; there are only a limited number of bacteria that contain the necessary enzymes to carry out this process. The type of bacteria that are utilized in the curdling phase of cheese production is called (broadly) the lactobacteria. Lactobacteria are naturally present and thrive in raw milk, given that they have a nearly endless supply of sugar (~5 g lactose/100 g of milk). As the bacteria naturally present in the milk ferment (i.e., *use the lactose as an energy source,*

A disaccharide *molecule*
of lactose (i.e., milk sugar)

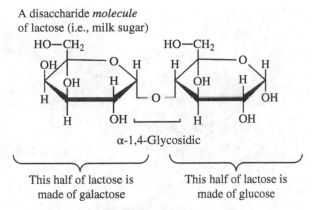

This half of lactose is
made of galactose

This half of lactose is
made of glucose

FIGURE 5.14   Lactose.

making lactic acid), the milk becomes increasingly acidic and begins to curdle. You might be thinking, but isn't all milk pasteurized? Doesn't pasteurization kill all of the bacteria, lactobacteria, and other potentially harmful types? Pasteurization does kill harmful, disease-causing bacteria like *Salmonella*, *Listeria*, and *Escherichia coli*, which are associated with many foodborne illnesses. It also destroys most of the lactobacteria that are native to the milk and critical for curd formation. Thus, for cheeses that are produced from the lactic acid generated through bacterial fermentation, one or more "starter bacteria" are added to the pasteurized milk during the cheesemaking process to ferment lactose and reduce the milk pH to induce curdling.

Broadly speaking, there are two groups of lactobacteria that are used to "start" or induce curdling: moderate-temperature (i.e., mesophilic) lactococci and the heat-loving (i.e., thermophilic) lactobacilli (Fig. 5.14).

These lactobacteria are both homofermenters, as they only produce lactic acid as a product of fermentation and not ethanol, so that is beneficial from the perspective of cheese production. Moreover, some lactobacilli are beneficial to human health, as you have heard advertised for dairy products with "probiotic" bacteria. The *Lactococcus* family is relatively small with only seven identified species, while the *Lactobacillus* species has approximately 50 members. What makes a cheese-maker choose a particular bacterial strain or type? The type of bacteria that is used depends upon the specific steps involved cheesemaking process for a particular cheese type (like temperature), as well as other characteristics of the bacteria that allow for the production of other molecules and by-product that are distinct for a particular cheese type. If the milk is subjected to temperatures of up to approximately 70°C during curd formation, then lactococci will effectively ferment and reduce the pH of the milk adequately to induce the formation of the curd. Most cheeses are acidified by the lactococci. The mesophilic bacteria can survive and thrive under these conditions, fermenting the lactose to lactic acid. However, for the cheeses that undergo a high-temperature cooking step, including mozzarella and the Italian hard cheeses, the thermophilic lactobacilli bacteria must be utilized since these bacteria

FIGURE 5.15   Dry and fresh yeast.

thrive and survive under high-temperature conditions. To almost all cheeses that are not prepared from milk that is curdled via direct addition of an acid (i.e., citric acid, lemon juice, etc.), a type or types of starter bacteria are added during the curdling phase of the cheesemaking process to acidify the milk, to induce the formation of the curd, and to contribute to the cheese ripening and aging process. However, also in the preparation of almost all cheeses, another agent is used to make the curd more robust, elastic, and strong—this agent is an enzyme that is called rennet.

5.6.1   Enzyme-mediated Curd Formation

A second method used to induce curd formation and coagulation of casein proteins involves an enzyme called rennet. Rennet is a general name for an enzyme or enzyme mixture that contains protease enzymes. Proteases cleave or degrade proteins into smaller pieces by breaking dipeptide bonds (Fig. 5.15).

Proteases are very important in biological system efficiency and play a key role in the degradation of proteins that are no longer needed and in our ability to recycle amino acids. However, protease activity is not only important for proper function of a biological system but also relevant in many aspects of food chemistry and is the basis for the warning label on your gelatin box about the addition of fresh pineapple, kiwi, or papaya.

Traditional rennet is made from the fourth stomach of a milk-fed calf. The effects of rennet on cheese production and curd formation are thought to have been discovered when ancient peoples stored their milk in pouches made from an animal stomach and discovered that the milk soured and coagulated more rapidly than stored in a different type of container. Through genetic engineering, a pure version of the calf enzyme that is critical to curd formation, called chymosin or rennin, is produced in a bacterium, mold, and a yeast and is purified from these model organisms. Most cheese in the United States is made with these engineered rennets. Traditional

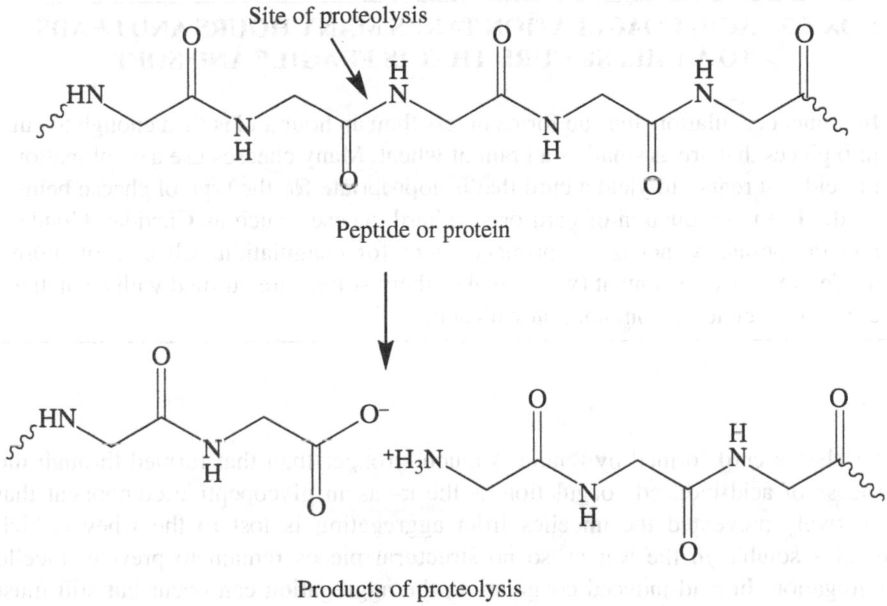

Site of proteolysis

Peptide or protein

Products of proteolysis

FIGURE 5.16 Hydrolysis of the protein backbone. Enzymes (proteases) or strong acids can break the protein backbone in a process generally called proteolysis.

rennet from a calf stomach is often required for traditional, artisanal European cheese production. An alternative enzyme is often used to avoid the animal origin of rennet. Vegetable rennet is actually closely related to rennet chymosin and isolated from thistle.

Some proteases are not very specific. They will break dipeptide bond links at any amino acid or at amino acids that have a certain property (i.e., *positively charged, aromatic*, etc.). These enzymes are very useful if you want to break a protein into its individual amino acid components during the process of biological protein degradation. Other proteases fulfill a specific role and are very specific for the types of amino acids that they will recognize and proteins that they will cleave. The role of a protease called chymosin, one of the components of rennin, is critical in curd formation during the process of cheese production. Chymosin is a protease that catalyzes the cleavage of a single dipeptide bond between Phe-105-Met106 (phenylalanine and methionine amino acids at positions 105 and 106 in κ-casein; Fig. 5.16).

As you recall from our discussion about κ-casein earlier in this chapter, the negatively charged tail of κ-casein (i.e., amino acid residues 116–169) assists in keeping the casein micelle soluble and small. The proteolysis cleavage event that is carried out by chymosin effectively removes the negatively charged tail of κ-casein from the casein micelle (called the κ-casein glycopeptide), essentially leading to a similar molecular outcome that was discussed in acid-mediated curd formation. Without the negative charge, the casein micelles fragment, calcium is lost from the micelle, the individual casein proteins aggregate, and a rubbery, strong cheese curd is formed.

BOX 5.3   ACID COAGULATION TAKES MANY HOURS AND LEADS TO A CHEESE CURD THAT IS FRAGILE AND SOFT

In rennet coagulation, the cud forms in less than an hour and is firm enough to cut into pieces that are as small as a grain of wheat. Many cheeses use a combination of acid and rennet to yield a curd that is appropriate for the type of cheese being made. In the production of hard or semihard cheeses, such as Cheddar, Gouda, and Parmesan, rennet is the primary agent for coagulation. Cheeses of more moderate moisture content (which makes them softer) are curdled with a smaller amount of rennet in combination with acid.

The cheese curd formed by rennin is much stronger than that formed through the process of acid-induced coagulation as the κ-casein glycopeptide component that effectively prevented the micelles from aggregating is lost to the whey (which remains soluble in the water), so no structural pieces remain to prevent micelle aggregation. In acid-induced coagulation, the aggregation can occur but still must overcome the structural presence of the κ-casein glycopeptide.

As stated earlier, the rennet used in cheese production typically contains a mixture of proteolysis enzymes including chymosin, pepsin (an enzyme important in protein breakdown in the digestive system that is most effective at cleaving peptide bonds that involve a hydrophobic, aromatic amino acid), and lipase (an enzyme that catalyzes the hydrolysis of fats). The nonchymosin proteases cause degradation of proteins and other molecules during curd formation to a lesser and less specific degree than chymosin. In addition to its role in curd formation, this degradation is also important to the texture and flavor of the finished cheese product.

5.6.2   Temperature-mediated Curd Formation

The final mechanism used to assist with curd formation in cheese production is temperature. By cooking the coagulating milk at a temperature in which the starter bacteria thrive, the microbes metabolize lactose and undergo fermentation more readily. Thus, at higher temperatures, more lactic acid is produced and the pH of the milk is decreased more rapidly. Proteins also denature at higher temperatures, causing them to unfold and form nonnative structures. Denatured casein proteins will disperse from a micelle and begin to aggregate, causing them to precipitate and curdle out of solution. Thus, in addition to the supplementation with starter bacteria and rennet, increasing the temperature of pasteurized milk to between 78 and 104°F (26 and 40°C) promotes the coagulation of the milk and fermentation of the starter bacteria. For most cheeses, the optimal temperature lies between 86 and 95°F/30–35°C; however some goat cheeses are made from goat milk that coagulates at 68–77°F/20–25°C, while Pecorino cheeses are coagulated from milk held at 95–104°F (35–40°C).

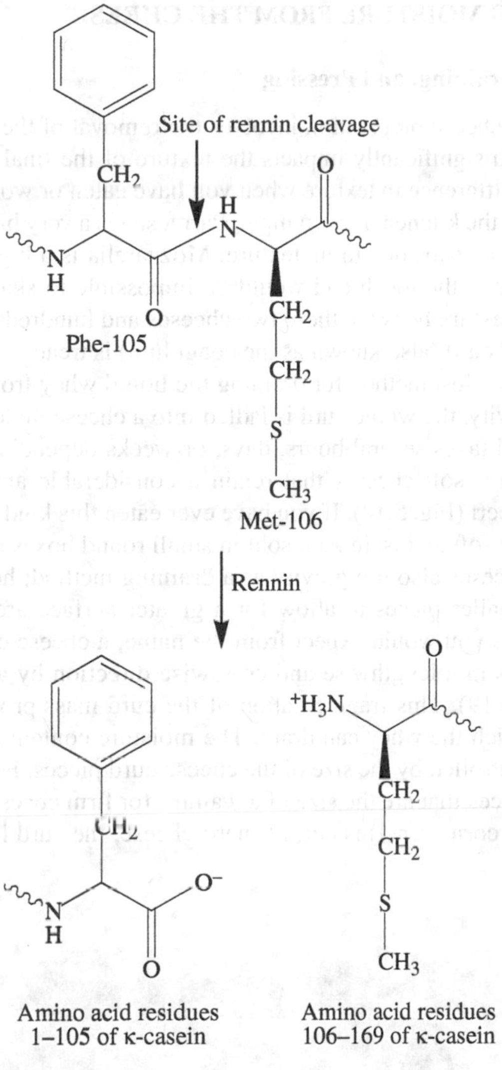

FIGURE 5.17   Actions of rennin on protein. The proteolytic action by rennin on casein results in its digestion to peptides.

During the process of acid-mediated, enzyme-mediated, or milk sitting on your kitchen counter on a hot summer day-mediated curdling, the casein proteins clump together, forming a solid curd mass, while the whey and other solution proteins and peptides remain in solution. However, if the process stopped here, then most cheeses would resemble yogurt or cottage cheese: a curdled mass of protein sitting in a sea of whey, water, and other nonaggregated molecules (Fig. 5.17). In order to prepare a more solid and drier cheese mass, the solid curd must be separated from the liquid through the process of cutting, draining, and pressing.

5.7   REMOVING MOISTURE FROM THE CHEESE

5.7.1   Cutting, Draining, and Pressing

The next phase of cheese preparation involves the removal of the moisture from the cheese curd, which significantly impacts the texture of the final product. You may have noticed this difference in texture when you have eaten or worked with different types of cheeses in the kitchen. For example, Parmesan is a very hard, dry cheese that is easily shaved into strips due to its texture. Mozzarella is a high-moisture cheese that is much softer to the touch and would be impossible to shave into thin strips. The difference in texture between these two cheeses and hundreds of others is based on how the formed curd (also known as the coagulum) is treated.

Gravity is the simplest method for draining the liquid whey from the cheese curd. In draining by gravity, the whole curd is ladled into a cheese mold and is allowed to drain. This method takes several hours, days, or weeks depending upon the cheese type and is used for soft cheeses that retain a considerable amount of moisture, including Camembert (Fig. 5.18). If you have ever eaten this kind of cheese, you can attest that it is very soft and is, in fact, sold in small round boxes for protection.

Some firmer cheeses also use gravity as a draining method; however the curd is first "cut" into smaller pieces to allow for a greater surface area from which the whey can drain. As you would expect from the name, a cheese curd is literally cut into smaller pieces in a lengthwise and crosswise direction by a knife, saber, or a Swiss harp (Fig. 5.19). This fragmentation of the curd mass provides a larger surface area from which the whey can drain. The moisture content of the final cheese product can be controlled by the size of the cheese curd pieces. For soft cheeses, the curd is cut into pieces that are the size of a walnut; for firm cheese, a medium-sized curd (the size of a corn kernel) is cut; for hard cheese, the curd is cut to the size of a grain of rice.

FIGURE 5.18   Cheese curds. Worker with cheese curds mixed in the watery whey.

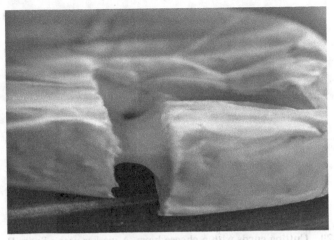

FIGURE 5.19 Camembert cheese.

Although cutting the curd does lead to a firmer cheese due to the increase in moisture lost, gravity alone does not adequately remove the whey-containing liquid necessary for a hard or even a semihard cheese. Thus, once cut, the curds are often stirred and warmed to expel additional whey. What does cooking do to the curd? Through heating the cheese curd, additional liquid-containing whey is released from the curd and the curd shrinks in size; thus it becomes more solid or firm. The increase in temperature also helps denature curd proteins. In the native state, the surface amino acids of most proteins will bind water molecules via hydrogen bonds. Once a protein is denatured, the protein unfolds disrupting the water-loving amino acids on the surface and exposes the hydrophobic amino acids normally located in the interior of proteins. A denatured protein will have less binding capacity, and water will be released while the protein molecules aggregate with one another.

This temperature increase also activates the finishing bacteria and increases the activity of enzymes released from bacteria into the curds. Activating the bacteria and enzymes in the cheese has two effects: the acidity of the curd is increased and the bacteria produce aromatic and flavorful molecules that will contribute to the flavor of the final cheese product (we will talk more about the chemistry of these molecules later). As you might expect, different cheeses are warmed to different temperatures; a higher warming temperature leads to greater shrinkage of the curd and a more firm cheese. For example, a Cheddar curd is warmed to a temperature of 100°F/38°C, the firmer Gruyere is warmed to 120°F/48°C, and the hard Parmesan and Romano cheeses are warmed to 130°F/54°C during cheese cutting.

After cooking the curd, it needs to be separated from the whey. The separation might be carried out by lifting the curds out of the cooking kettle with a piece of cheesecloth, or allowing the curd/whey mixture to flow into molds (where the two separate), or placing the mixture into a larger rectangular vat to drain (Fig. 5.20).

You might think that the cheese is now adequately firm with enough whey removed to move on the next stages of cheesemaking, but that is still not the

FIGURE 5.20   Cutting curds with a cheese harp. A worker uses a Swiss Harp to cut the curd into smaller squares.

FIGURE 5.21   Separating the curds from the whey. The use of a cheese cloth to drain the curd from the whey for Parmesan cheese.

case for most cheeses. After this initial separation step, the curd is placed into a mold (if it is not already molded) and is pressed to expel more whey and draw the curd into a cohesive mass.

Simple or gentle pressing, which involves stacking one layer of curds on top of another layer of curds, is the only pressing method used for semihard cheeses with a high moisture content like manchego. However, if more whey needs to be removed (i.e., hard cheeses) or to encourage rind formation, the cheese is pressed mechanically (Fig. 5.21).

During the process of molding and pressing, the bacteria continue to acidify the cheese, which contributes to whey expulsion, and make those flavorful molecules.

FIGURE 5.22   Cheese press. Mechanically pressing cheese before aging.

5.7.2   Cheddaring

Cheddar cheese undergoes a different method, called cheddaring, during the draining and pressing stage of cheese preparation. In the preparation of Cheddar cheese, the curd is left to sit (it is not separated from the whey) and is allowed to spread out laterally. During this time, it draws together into a mass called a curd cake. The curd cake is cut into pieces, which are stacked on top of one another for a period of time, and then they are relayered until the curds have a stringy texture, like a chicken breast. This method expels whey, as the curd continues to be acidified by the active bacteria. The curd mass is then milled into small pieces, salted, transferred to molds, and pressed.

5.7.3   Salting

Perhaps you have looked at the nutritional values for a variety of cheeses and noticed the sodium content in a 1 oz serving: blue, 396 mg; Cheddar, 176 mg; part-skim mozzarella, 150 mg; and Parmesan, 433 mg. Where does this sodium come from and what does it do? Cheeses contain sodium because the cheesemaking process involves the addition of salt either during or at the end of the production process. Although salting contributes to the taste on many cheeses, it is more than a taste additive that raises your blood pressure. Salting draws more moisture out of the curds, inhibits the growth of bacteria (both the helpful lactobacteria and harmful microbes), and alters the activity of enzymes that are important to the ripening or finishing stages of cheese production.

First, let us discuss the moisture removal. Through osmosis, salt pulls moisture from the cheese. As a general example, if you have two objects that contact one another and one object is saltier than the other, the saltier object will pull moisture from the less salty object, and salt will move from the saltier object to the less salty one (Fig. 5.22). More specifically, osmosis is the net movement of solutes (dissolved particles) from a region of high concentration to a lower concentration across a semipermeable membrane. In this case, proteins coagulated in cheese act as the semipermeable membrane.

You may have taken advantage of the power of osmosis if you have ever placed a piece of bread into a canister that contains hardened brown sugar. Initially, the bread is soft (i.e., high moisture content) and the brown sugar is hard. After some period of time, the moisture from the bread is pulled into the sugar, and when you open your canister, the bread is hard and the brown sugar is soft. We take advantage of osmosis in so many ways in the kitchen! But let's now think about osmosis in the context of salting the cheese. Salt is put on the surface of freshly formed cheeses or cheese curds. The cheese/curds are less salty and have a higher moisture content than the salt. As such, the salt on the surface migrates into the cheese (very slowly), and some of the moisture in the cheese is pulled out. After some period of time, the salt grains on the cheese's surface "melt" as they intermingle with the whey-containing liquid, and the cheese becomes less moist and saltier.

Cheddar and Colby are good examples of cheeses that are salted in the curd stage of cheese preparation. Because the salt is mixed into the curd (which can be mixed), a homogeneous balance of moisture and "saltiness" is attained within the mixture. In other cheeses, such as Pecorino and German blue, the molded cheese is rubbed or sprinkled with salt (often multiple times). In these cases, the salt is concentrated on the outer layer of the cheese and only gradually penetrates to the interior. Some amount of time is required to achieve a salt and moisture balance between the interior and the exterior of the cheese and leads to the formation of the rind. The rind not only holds a cheese together but also keeps the cheese from drying out. For Camembert, the salting process is complete after 4–6 days, and the formed rind holds the soft interior of the cheese in place.

Many types of cheeses are salted by using a brine, a liquid solution with a salt concentration of 15–20%. It is a useful method, as it enables the cheese to be evenly salted (the cheese literally sits in this liquid, salty brine). The cheese might remain in the brine bath for anywhere from 30 min to several days. For example, a 10 kg piece of Cheddar will sit in brine for 40 h.

As stated earlier, salt also reduces bacterial growth and enzyme activity. You may be thinking, don't I want the bacteria and enzymes to be flourishing? After all, the lactobacteria are our friends in the cheesemaking process. This is true; however other bacteria, including disease-causing ones, also flourish in a moist, warm environment. Salting will reduce the growth of unwanted and potentially harmful bacteria and molds, as well as slow the growth and enzyme activity associated with the beneficial starter and ripening bacteria. This reduction in the function of the bacteria is likely an important step in the cheesemaking process in terms of the degree of acidity of the cheese and the molecules produced by the bacteria.

5.8 RIPENING OR AFFINAGE

At this point, your cheese might look like a salty, crumbly, or rubbery curd, which might not be the most appetizing visually, aromatically, or in terms of taste. In the final step of cheesemaking, the finishing phase, the cheese is allowed time to ripen.

During ripening, the starter bacteria (still present) and other added bacteria (called the finishing or ripening bacteria) and their corresponding enzymes transform this unappetizing mass into a delicious cheese over a period of 2 weeks (mozzarella) to two or more years (Cheddars and Parmesans). Thus, temperature and humidity play a significant role in governing how and which bacteria grow and thrive, which impacts flavor, texture, and aroma of the cheese. The French term for ripening is *affinage*, which means "end" or "ultimate point." As such, at times this stage of cheese making is carried out by an *affineur*, a cheese tenderer or finisher. The *affineur* takes care of the cheeses in the cheese-ripening cellars until the cheese has ripened adequately for packing and sale. As has been a theme within this entire chapter, the ripening process varies considerably for different cheese types. Thus, first, we will discuss some of the basic techniques and chemistry involved in the ripening phase and then go into detail about how ripening is carried out in the production of certain types of cheeses. However, not all cheeses undergo the ripening or aging process. In fresh cheeses that are prepared by acid or acid/heat coagulation methods, the cheeses are ready to eat as soon as the curds are processed.

5.8.1 Work in the Ripening Cellar

Many cheeses are ripened within a ripening cellar or special storage room that is specific for ripening cheese. A ripening cellar imitates the conditions of a cave, which was historically used for this particular purpose (have you ever heard of the caves of Roquefort?). The cellar is controlled for humidity and temperature, depending upon the type of cheese that is being ripened. The majority of cheeses are ripened between 46 and 60°F (8–10°C), with a relative humidity of 85–95%. The climate of the cellar is controlled by the ambient temperature and humidity, as well as the movement of air through the space.

The care that a particular cheese receives in the ripening cellar is dependent upon cheese type. Work in the ripening cellar by an *affineur* involves tending the surfaces of the cheeses and controlling temperature and humidity. The *affineur* regularly brushes, rubs, or washes the surfaces of the ripening cheeses with salt brine and ripening bacteria or coating cheese surfaces with salt. All cheeses must be turned from time to time to ensure even development of the bacteria and to prevent deformities in the cheese shape. The *affineur* also determines when the cheese has acquired its proper appearance, aroma, taste, and texture, so that it can be packaged for sale.

5.8.2 The Chemistry of Ripening: How does it Happen?

The chemistry of the ripening process is dependent upon the presence and activity of bacteria and the corresponding activity of their various protease and lipase (these enzymes break down fats) enzymes that break down the proteins, carbohydrates, and fats found in the milk to smaller molecules. The bacteria, yeast, or molds that carry out this work, some of which have been present throughout the entire cheese production process and others that are added during this final phase of production, contain enzymes that are essential to the ripening process. If ripening occurs due to the presence of starter lactobacteria, then the cheese will ripen in a homogeneous fashion, since the bacteria are present in the entire body of the cheese. This type of ripening is common in hard and semihard cheeses like Parmesan, Cheddar, and Gouda. However, the soft cheese Camembert is ripened from the outside inward due to a coating of the *Penicillium camemberti* mold. In blue cheeses, the blue-green *Penicillium roqueforti* mold is added to the cheese by piercing the cheese mass and adding the mold to the middle. The cheese is then pierced further to allow oxygen access to the bacteria. The piercing speeds up ripening and causes the cheese to be ripened from the interior to the exterior.

5.8.3 The Chemistry of Ripening

What exactly happens during the ripening stage? What are these flavorful and aromatic molecules that are produced during ripening, and where do they come from? The aromatic, flavorful, and distinct molecules that we smell in a particular type of cheese are due to the degradation of three types of molecules in milk: protein (casein), lipids/fats, and lactose. The specific molecules that are formed depend upon the particular microbes that are present during the ripening phase of cheese production, as well as the type of milk used.

Casein proteins can be degraded to their individual amino acid components by proteases present in the ripening bacteria. The amino acids themselves have flavorful properties that range from sweet to savory. For example, alanine has a sweet taste, tryptophan has a bitter taste, and the sulfur-containing amino acids, cysteine and methionine impart a "meaty" or "eggy" taste to dishes. The amino acid glutamate (also known as MSG) is often added to savory dishes to enhance their flavor. When casein is degraded, all of these amino acids are produced and affect the flavor of the cheese (Fig. 5.23).

Some of these amino acids will remain in that molecular state; however many will be broken down further into smaller amines, including trimethylamine (which has a fishy taste), putrescine (smells of spoiling meat), or ammonia. Although in this context these flavors sound disgusting, small amounts present within a particular cheese bring about a complex, rich, and unique flavor.

Fats also have an important role in the flavor and texture of cheese (Fig. 5.23). In fact, the same ripening enzymes that degrade proteins will also metabolize and chemically change fats to produce a symphony of aromatic and flavorful compounds. Triacylglycerol fats from milk fat are broken down by lipases (enzymes that degrade lipids) into free fatty acids in the curd. The fatty acids then undergo additional oxidation/reduction chemistry. In beta-oxidation, the fatty acid is oxidized at the carbon that is "beta" to the carboxylic acid. This allows for a decarboxylation reaction, which reduces the fatty acid chain by two-carbon units, yielding a ketone or alcohol. In delta-oxidation, an oxygen atom is added to the carbon four units away from the carboxylic acid to yield a δ-hydroxyacid. These molecules can cyclize to a lactone. The fatty acid can also be converted to an ester. The relevance of this chemistry is that some short fatty acids have a peppery, sheep-like, or goat-like taste or aroma.

Semipermeable membrane

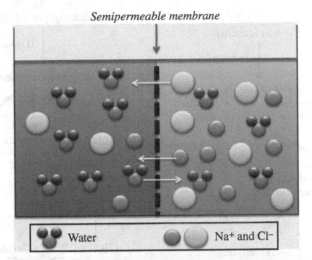

FIGURE 5.23 **Osmosis.** Water, sodium, and chloride ions (salt) will spontaneously move from a high to a low concentration. For cheese, a high salt solution will result in a lower water concentration, and the water molecules in cheese will flow from the cheese to the brine.

The breakdown of lactic acid and citric acid, both of which are products of bacteria fermentation, produces important molecules like diacetyl (a "buttery" taste), ethanal (also known as acetaldehyde), and ethanol (Fig. 5.23).

In order to bring this ripening chemistry to life, let's discuss some of the molecular details of the ripening of several cheeses that have very different properties: blue, Swiss, Camembert, Cheddar, and Muster, and Limburger.

5.9   BLUE CHEESES, MOLDS, AND CHEMISTRY

Molds are microbes that require oxygen to grow, can tolerate drier conditions than bacteria, and produce a significant number of protease and lipase enzymes that impart a distinct texture and flavor to certain cheeses. The standard ripening molds used in cheese production come from the genus *Penicillium*, from which the antibiotic penicillin is derived. The blue *Penicillium* are unique molds in their ability to grow in the low oxygen conditions in small cavities and veins within cheese. *Penicillium roqueforti* gives Roquefort cheese its veins of blue, its name, and its flavor. *Penicillium roqueforti* and its close relative *P. glaucum* also color the interior and flavor of Stilton and gorgonzola. Where does the flavor of a blue cheese come from?

The typical blue cheese flavor comes from the *Penicillium* mold's metabolism of milk fat. *P. roqueforti* degrades 10–25% of the triacylglycerol fat in its cheeses, first liberating short-chain fatty acids from the triglyceride (Fig. 5.24). These fatty acids give a peppery taste but are transformed further by the *P. roqueforti* to δ-hydroxy-acids and β-ketoacids that give a peppery taste and then further transforms the fatty

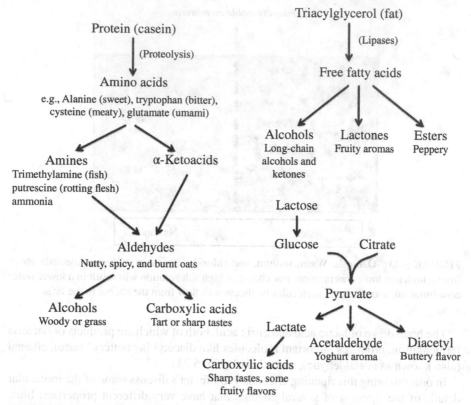

FIGURE 5.24 Aging cheese—proteins, fats, and lactose. Microbial cultures or the remnants of bacteria cells continue to metabolize biological molecules into new products, many of which have unique aromas and flavors.

acids to methyl ketones and alcohols that give the characteristic aroma of a blue cheese. Heptan-2-one and nonan-2-one are most often linked with blue cheese smell; pentan-2-one is usually described as fruity. Other compounds, including alcohols and esters, round out the flavors; thus in gorgonzola, heptan-2-one and nonan-2-one are key impact molecules, but 1-octen-3-ol, 2-heptanol, ethyl hexanoate, and methylanisole are also important odorants (Fig. 5.25).

5.9.1 Camembert Cheese, More Molds, and Chemistry

Penicillium camemberti is the mold microorganism that is smeared on the surface of Camembert and Brie cheese to form the characteristic white crust, buttery taste, ammonia-like aroma, and soft texture. Let's think about how a microorganism on the exterior of a cheese can do all of this chemistry!

The smeared *P. camemberti* breaks down lactic acid (the product of the metabolism of lactose) to carbon dioxide and water. Because of the loss of the acid, the pH of the cheese's surface increases to a value of approximately 7.5. At this high pH

FIGURE 5.25 Blue cheese. Roquefort cheese finished with *Penicillium roqueforti* mold.

(for a cheese), the calcium phosphate present within the casein curds becomes insoluble; thus it precipitates on the surface of the cheese. This precipitation creates a calcium concentration gradient between the interior and exterior portions of the cheese, where the interior is at a much higher concentration than the exterior. Just like in osmosis, the calcium phosphate moves away from the interior, essentially breaking apart the casein micelles on the interior of the cheese (remember that the calcium phosphate helps to bind the micelles together). Hence, as a Camembert cheese ripens, its interior softens (Fig. 5.18).

What about the taste and aroma of Camembert? The aromatic and flavorful molecules produced primarily come from extensive lipolysis that is carried out by the *P. camemberti* molds and fatty acid catabolism and proteolysis. The distinctive odor of Camembert is due to the generation of several alcohols, aldehydes, ketones, and sulfur compounds as products of lipolysis, fatty acid metabolism, and proteolysis. 1-Octen-3-ol and 1-octen-3-one produced by the activity of the *P. camemberti* on the fatty acid, linoleic acid, give a mushroom taste to the cheese. Butane-2,3-dione produced upon the metabolism of citrate gives the buttery taste (Fig. 5.25).

However, this is the first time that we have really mentioned the sulfur compounds in relation to the taste of a cheese. Thanks to the ripening bacteria, *Geotrichum candidum* and *P. camemberti*, proteolysis of casein and further chemistry of the amino acid products lead to several molecules that are strong contributors to the taste and aroma of Camembert. The amino acid methionine is broken down to several sulfur compounds that impart a flavor of garlic and cabbage. Methanethiol is an important odorant, and although by itself methional's boiled potato smell is not particularly pleasant, in combination with other volatiles, it contributes to the cheese aroma of Camembert. Although it is rare that one molecule carries the smell of a particular cheese, but S-methylthiopropionate, $CH_3CH_2C{=}O(SCH_3)$, does indeed smell like Camembert.

5.9.2 Swiss

An important bacterium in Swiss cheese cultures is *Propionibacterium shermanii*, the hole-making bacteria. The propionibacteria consume the cheese's lactic acid during ripening, converting it to a combination of propionic and acetic acids and carbon dioxide gas. The carbon dioxide gas creates the bubbles or the characteristic eyes in the cheese due to the collection of the gas at the weak points within the curd. The copper cauldrons used to prepare Swiss cheese damage some milk fats, and the liberated fatty acids are modified to esters and lactones, which have aromas of pineapple and coconut.

5.9.3 Cheddar

Lactic acid bacteria that initially acidify the milk and initiate curdling persist in the drained curd and generate many of the flavor molecule products of ripening, including the semihard and hard cheeses, including Cheddar, Gouda, and Parmesan. Although the number of bacteria often drops dramatically during cheesemaking, the enzymes survive and continue to degrade proteins into savory amino acids and aromatic by-products. Thus, proteolysis and amino acid catabolism are responsible for the aromas associated with Cheddar cheeses. The lactic acid bacteria, along with added nonstarter bacteria, contribute to the formation of the aromatics that we associate with Cheddar. Amino acids are converted to keto and hydroxyl acids by *Lactobacillus*, while *Lactococcus* strains further convert them to carboxylic acids. In the case of the *Lactococcus*, the first step in the degradation of amino acids is transamination, leading to the formation of α-ketoacids. For example, the amino acid phenylalanine is converted to phenyl pyruvic acid and is further degraded to the flavor compounds, phenyl lactate and phenyl acetate. Methanethiol, as a product of the degradation of methionine, is associated with Cheddar-type sulfur notes in good-quality Cheddar but does not produce Cheddar flavor. The recognizable Cheddar aroma is produced by a mixture of 2,3-butanedione, methional, and butyric acid. Methional is considered to have a boiled-potato-like aroma, while methanethiol adds garlic tones to the flavor of Cheddar. Butyric acid derived from lipolysis has a cheesy and sweaty odor, which is considered an important component of Cheddar flavor.

5.10 THE SMELLY CHEESES: MUSTER AND LIMBURGER

Smelly Limburger cheese originated in northern Europe and for centuries was a favorite in Germany and in blue-collar eighteenth-century America. The bacteria that give Muster, Limburger, and other cheeses a pronounced stench that can clear a room (and contribute more subtly to the flavor of many other cheeses) are *Brevibacterium linens*. Brevibacteria are another smear bacteria that appear to be natives of salty environments and grow only in warmer conditions (20–20°C) at salt concentrations of up to 15%; these salty conditions inhibit the growth of most other microbes. However, unlike typical finishing bacteria, brevibacteria do not tolerate acidic conditions well and require oxygen to grow. Thus, these bacteria are added during the ripening process by wiping a cheese with a salt brine that causes the characteristic sticky, orange-red smear of brevibacteria to flourish. Due to the presence of the bacteria, extensive lipolysis and proteolysis occur at the surface; hence, it is considered to be a surface-ripened cheese. This causes significant formation of several carboxylic acids including the volatile butanoic, 3-methylbutanoic, and hexanoic acids, which have aromas that have been likened to "sweaty feet." Other aroma molecules that contribute to the smell of Limburger include methionine-derived methanethiol and methyl thio-acetate, $CH_3C(=O)SCH_3$. The reason these cheeses have the sweaty feet smell is because another habitat of these bacteria is the human skin and quite likely the original source of the culture. Yes, someone, somewhere, likely transferred bacteria from their skin, grew the cheese in a salty, warm, and oxygenated condition, and ate it. But then again, that is how a lot of the food we eat started.

5.11   COOKING WITH CHEESE

When used as an ingredient in cooking, cheese can add flavor, texture, and appearance to a food in a great diversity of ways. In the case of a great grilled cheese sandwich, you want a cheese to be stretchy and stringy upon melting. In fondues, the cheese should melt into a smooth, cohesive, and saucy texture. In some cases, you don't want the cheese to change texture at all, such as a pasta dish made with goat cheese. In no case do you want a cheese blob that is an oily, lumpy mess spread over the top of your steamed broccoli. What causes different cheeses to be cooked and used in different ways in the kitchen? As we have seen before, cooking properties of cheeses are founded in intermolecular interactions and acid–base chemistry

5.11.1   Melting Cheese

Melted cheese is a staple ingredient in a number of popular culinary dishes that cross ethnic boundaries. Think about the importance of a good cheese melt to your lasagna, fondue, or quesadilla. However, even among these three dishes, the type of melt desired ranges from stringy to smooth and flowing. First, let's talk about what happens to the molecules during the melting process, and then we will go into why some cheeses melt in a stringy way, while other cheeses provide a better melt for sauces or soups (Fig. 5.26).

As you know, during the melting process, a cheese makes a transition from a solid to a more liquid-like state in the presence of heat. Simply put, the cheese is able to "flow." With any state of matter change (i.e., *solid to liquid, liquid to gas*), interactions between molecules change. In the case of cheese, the interactions and molecules that are relevant to melting are the milk fats, proteins, water, and calcium. As you might expect, interactions change depending on the melting temperature. The fats are the first molecules within the cheese to melt; this occurs at 90–100°F (32–38°C). During the melting of fat, the intermolecular interactions between individual fatty acids become less robust or tight. This fat melting softens the cheese and beads of melted fat come to

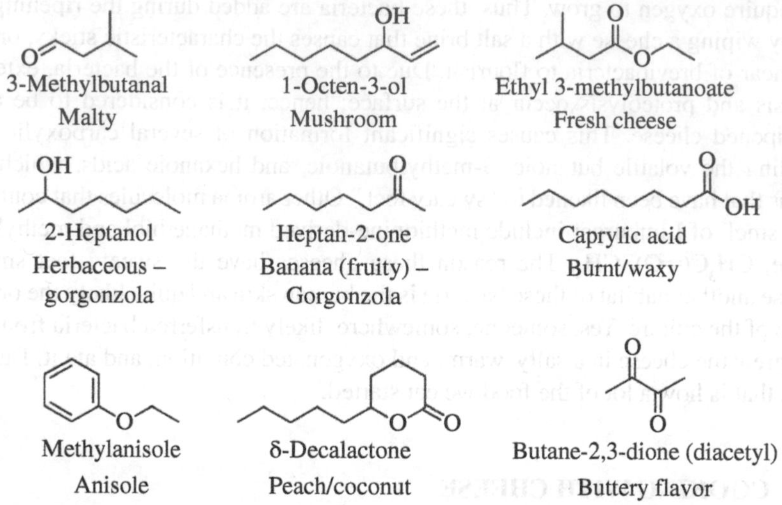

3-Methylbutanal
Malty

1-Octen-3-ol
Mushroom

Ethyl 3-methylbutanoate
Fresh cheese

2-Heptanol
Herbaceous –
gorgonzola

Heptan-2-one
Banana (fruity) –
Gorgonzola

Caprylic acid
Burnt/waxy

Methylanisole
Anisole

δ-Decalactone
Peach/coconut

Butane-2,3-dione (diacetyl)
Buttery flavor

FIGURE 5.26   Some of the aroma and flavor compounds of cheese.

the surface of the cheese. As the temperature continues to increase up to 130°F/54°C (for soft cheeses), 150°F/68°C for Cheddar/Swiss hard cheeses, and 180°F/82°C for Parmesan, the intermolecular interactions between the individual casein proteins break (remember: this is what holds the casein curd together as a solid mass) and the protein mass begins to flow as a thick, viscous liquid.

The melting behavior and properties of a particular cheese are also determined by the amount of water present within the cheese matrix. In a low-moisture cheese like Parmesan, the interactions between individual protein molecules are very strong because there is very little water present to interfere with the interactions. Thus, a higher temperature is required for melting; moreover, when melted, there is not a lot of "flow" due to the lack of water. Thus, individual Parmesan shavings don't combine; they remain separate within a baked pasta dish. In contrast, the shreds of a high-moisture mozzarella melt together and readily flow.

Have you ever tried to melt Cheddar cheese in the microwave and ended up with an oily cheese blob? As mentioned earlier, most cheeses expel some oil/fat during the melting process when the intermolecular interactions between fat molecules are reduced during the early stages of heating a cheese. As the protein matrix begins to break down due to the loss of protein–protein interactions at the higher melting temperatures, more fat is leaked to the cheese's surface. The higher the fat content of the cheese, the more likely this event is to occur.

If you have ever tried to make a fondue with mozzarella cheese, you know that stringy cheeses don't work well, unless you want to string your fondue-dipped bread across an entire room. Melted cheese is stringy when calcium is present to link one intact casein molecule to another one into a long, stretchy fiber. This phenomenon requires a few conditions. First, the casein molecules must be largely intact and not extensively degraded by protease enzymes. Thus, cheeses that are not aged for an extended period of time are more stringy than those that are well aged. Second, the cheese must be of low to moderate acidity, as with greater acidity comes a greater loss of calcium from the protein matrix. Third, the cheese must be of moderate moisture and salt content. High moisture and salt will also interfere with the ability of the proteins to be cross-linked by calcium. Cheeses are the stringiest at their melting point or the point at which the hot dish has cooled enough to eat. They get stringier as the dish is stirred and stretched. The most common stringy cheeses are mozzarella, Emmental, and Cheddar.

In contrast, cheese sauces and soups melt nicely into a smooth, even, and often creamy texture, without any note of stringiness at all. How is this achieved? First, by not using a cheese that is prone to being stringy! In short, use a moist cheese like Colby or Jack or a well-aged grating cheese like Parmesan or Pecorino. In addition, grate the cheese finely when adding it to the soup/sauce, heat the dish as little as possible after addition of the cheese, and minimize stirring.

Another way to cook with cheese (or create a cheese sauce) is to stop the melting fat from congealing together by melting the cheese in a blocker of some sort. Cooking cheese with a roux is a common method to use stringy cheese in your dishes. A roux is a classic French culinary technique that uses flour and butter to create a base or a roux for cheese and other sauces or gravies. Flour and fat (often butter) are cooked to allow the starches to coat with fat and water from fat or butter to either bind to the starches of flour or boil off. The proteins and expanded starch–fat complex then will interfere with the melting fat, preventing it from forming large interacting strings. The result is a smooth flowing cheese sauce. You may recognize the consistency of a cheese roux from your favorite mac and cheese dish.

5.12 PROCESSED CHEESES

Anyone who has used Velveeta to make nachos or macaroni cheese knows that it has great melting properties (Fig. 5.27). Velveeta and many other cheeses that you purchase, receive in gift boxes, and consume are known as processed cheeses. Why did the concept of a processed cheese come about? Processed cheese was a practical solution to a problem with cheese deterioration that occurred during the shipment of cheeses to distant countries. As a solution to this problem, in 1911, Emmental was mixed with citric acid salts to produce a cheese that was much less perishable. The same result occurred in the United States in 1916 when Cheddar was treated with phosphoric acid salts. The results of these processes were cheeses that deteriorated less readily and melted without becoming rubbery with no fat separation. In the 1920s, the process spread rapidly to other countries (Fig. 5.28).

Today, processed cheese is a cheese that uses surplus, scrap, and unripened cheese products as its basis. Manufacturers combine a mixture of sodium citrate, sodium phosphates, and sodium polyphosphates with a mixture of new, partly ripened, and fully ripened cheeses. The polyphosphates carry water into the cheese, remove calcium from the casein matrix, and bind to the casein. This loosens the protein matrix, as protein–protein interactions between individual casein molecules are

FIGURE 5.27 **Melting cheese.** The meltability of cheese depends on the type of cheese, protein, fat, and mineral content.

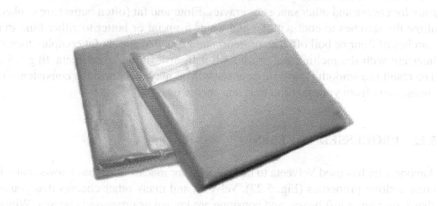

FIGURE 5.28 **American processed cheese.**

reduced and the calcium bridges that cross-link caseins are reduced. The reductions in protein–protein interactions, along with the high moisture content, generate a cheese that melts into an even, smooth, creamy blend.

REFERENCE

[1] Dairy: World Markets and Trade Circular, July 2012, USDA—Foreign Agricultural
     Service, Office of Global Analysis, Washington, DC.

6

BROWNING

Guided Inquiry Activities (Web): 5, *Amino Acids and Proteins*; 7, *Carbohydrate*; 17, *Browning*; 23, *Meat Cooking*

6.1 INTRODUCTION

We will describe chemistry/molecular changes that take place in cheese browning, chemistry of Maillard browning of proteins, chemistry of caramelization browning of sugars, and enzymatic browning including some basic chemistry and biochemistry of enzymes.

Brown food can be both incredibly appetizing and unattractive. Grilling produces a delicious flavor that cannot be matched by poaching, while a browned apple slice is something we try to avoid eating (Fig. 6.1)! The chemical reactions that bring us the goodness and the off-flavors of browning are everywhere. Chocolate cocoa is bitter and astringent until reactions leading to brown color and flavor are created. Coffee and beer are both better for the browning that occurs during their making. Plant enzymes catalyze a different set of browning reactions, in this case to provide first aid for an injured plant and to detract animals from eating and further damaging the plant. There are thousands of flavors and aromas produced by these reactions; a good scientist and cook will have an appreciation and understanding of the process by which these compounds are formed.

This process that creates an amazing diversity of new molecules is often called browning. However, this term is not quite accurate as many of the chemical products are not brown but yellow or even colorless. Moreover, browning reactions can be organized in a number of different ways: caramelization, Maillard reactions, ascorbic

The Science of Cooking: Understanding the Biology and Chemistry Behind Food and Cooking,
First Edition. Joseph J. Provost, Keri L. Colabroy, Brenda S. Kelly, and Mark A. Wallert.
© 2016 John Wiley & Sons, Inc. Published 2016 by John Wiley & Sons, Inc.
Companion website: www.wiley.com/go/provost/science_of_cooking

FIGURE 6.1   Browning is better. The impact of browning reactions on the taste and attractiveness of our food. Both are cooked to a safe temperature but one is much more tasty.

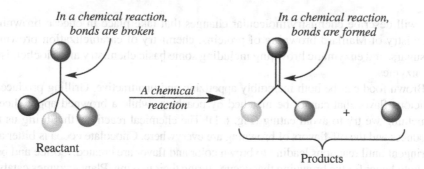

*In a chemical reaction,
bonds are broken*

*A chemical
reaction*

*In a chemical reaction,
bonds are formed*

Reactant

Products

+

FIGURE 6.2   The anatomy of a chemical reaction. A chemical reaction (chemical change) takes place when a chemical bond is broken or made.

acid browning, and fruit browning. A closer look reveals that each type of browning is a chemical process—bonds are broken and formed to make new molecules. In some instances, the browning occurs with the aid of enzymes and other browning reactions occur without a catalyst, also called nonenzymatic browning. To better understand the difference between an enzyme- and nonenzyme-catalyzed reaction, it will help to consider how a chemical reaction proceeds. A chemical reaction is the process by which one or more molecules (chemical substances) will change by the breaking and/or making of new bonds to form one or more new molecules (Fig. 6.2). The chemical equation for a reaction shows one or more starting substances called reactants forming new molecules called products. How fast the reaction happens is measured as the rate of the reaction.

6.2   CHEMICAL REACTION KINETICS

When a chemical reaction involves the collision of two reactants, the reacting atoms or molecules must collide with enough force and correct geometry to change the bonds. When a chemical reaction involves the breakdown of one molecule into two or more, the reactant must vibrate with enough thermal energy to break the bonds. How fast the reactants collide or vibrate to form products depends on a number of parameters. Heating increases the energy of the molecules involved in a reaction. Increasing the energy of a system makes molecules vibrate more and move around more rapidly. When a reaction involves the collision of two molecules, increasing the temperature increases the number of collisions over time. Compared with reactions that take place at room temperature, every 50°F/10°C increase in temperature nearly doubles the rate of a reaction. Increasing the concentration of reactants also helps to increase the number of collisions of reacting molecules over time.

One class of browning reactions, the Maillard reaction, can take place at room temperature, albeit very slowly. Heating food increases the number and successful collisions between sugar and protein molecules that make up the Maillard reaction and thus browning takes place more effectively at higher temperatures. To better understand this process, it is helpful to investigate a concept called activation energy (Fig. 6.3). Activation energy is the amount of energy needed to start and continue a spontaneous reaction, that is a reaction where there is less total energy in the products than the starting reactants.

To start a reaction the reacting molecules absorb energy from the surroundings and collide. Appropriately oriented collisions between reaction molecules may not result in a chemical reaction unless there is enough energy in the collision. The minimum energy of the collision is called the activation energy of a reaction. This is often shown by the energy diagram in Figure 6.3. The inherent energy of a substance (for our purpose a molecule from food) is shown on the vertical or Y-axis, while the horizontal or X-axis describes the process of a reaction, also called a reaction coordinate.

Let's consider the reaction between sugar and oxygen producing CO_2 and H_2O. The inherent energy of the starting reactants is higher than the energy of the products. Thus overall, energy is given off, and the reaction is considered exothermic or spontaneous. However, the bag of sugar in your pantry has yet to react with oxygen in the air! That is a good thing; there is a lot of potential chemical energy in a bag of sucrose if all of the molecules were to react at one time! What stops the reaction? From the energy diagram you can see there is an intermediate or activated complex of the reaction. To produce the activated complex requires additional energy—called the energy barrier, an amount of energy that is given off as the reaction proceeds. The activated complex (also called a transition state) is short lived, and the overall energy of activation is released as the activated complex forms products. The energy barrier requirements prevent sugar and oxygen from reacting without the addition of heat. That is why to get the combustion of sugar, one has to add heat to overcome the energy barrier and produce a nice brown crust on a crème brûlée (Fig. 6.4).

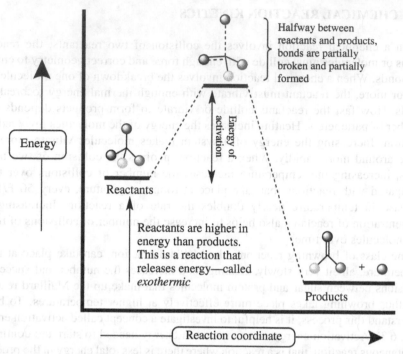

FIGURE 6.3 **Energy diagram for a chemical reaction.** The progress of a chemical reaction is shown on the horizontal axis (reaction coordinate) and the free energy of the substances involved in the reaction is shown on the vertical axis. The energy required for the reaction to reach the intermediate (also called a transition state) of the reaction is the energy of activation. A spontaneous reaction will end with less free energy than the starting compounds possessed.

Heating food helps to bring about reactions by increasing the energy of the reacting molecules in vibrations and collisions that overcome the energy barrier to products. Heating food increases the number of molecules that have the required energy to react. At room temperature, only a few molecules have enough energy for sugars and proteins to collide and form brown compounds. Thus, browning happens at a low temperature but very slowly. Increasing the temperature increases the number of molecules that have productive collisions and react. This is one of the reasons why we heat our food, to create more brown molecules!

Enzymes are protein catalysts. Catalysts reduce the activation energy for a reaction (Fig. 6.5). Thus, in a catalyzed reaction, it takes less energy to overcome the activation complex/energy barrier and more product molecules can be formed at a lower temperature. However, it should be noted that a catalyst does not change the total energy of the reaction. Heat can also speed up an enzyme-catalyzed reaction. The additional energy added to a system when cooking, as in a noncatalyzed reaction, results in more molecules with enough energy to overcome the activation energy barrier. However, remember that enzymes are proteins. Heat can also impact the structure of

FIGURE 6.4 Heat increasing the rate of a reaction—crème brûlée. Heat from the flame provides the energy needed for oxygen to react with sucrose sprinkled on the top of a desert.

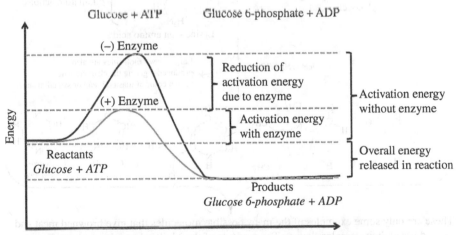

FIGURE 6.5 Effect of an enzyme catalyst on energy of a reaction.

the protein. Many proteins will denature around 45–65°C (113–149°F). Therefore, heating an enzyme-catalyzed reaction will increase the rate until the protein is heated above its denaturation point, at which time the reaction will slow down with increased temperature.

So we catalogue browning reactions in two categories: enzyme-catalyzed reactions (vegetable and fruit browning) and nonenzyme-catalyzed reactions. These reactions include caramelization, the Maillard reactions, lipid oxidation, and ascorbic acid browning.

6.3 THE MAILLARD REACTION

The Maillard reaction is a complex series of reactions between some sugars and the amino groups of amino acids and proteins (Fig. 6.6). This reaction creates the pleasant odor and flavor of baked bread, cooked meat, and buttered popcorn flavor, among hundreds of others. The reaction is also responsible for the brown color of beer, roasted meat, coffee, and chocolate. The reaction was first described and eventually named after the French physician/scientist Louis Camille Maillard. In modern days, he would be considered a biochemist. At the age of 16, Maillard entered the University of Nancy in Lorraine, France, where he earned a master's degree in science at the age of 19, became a physician, and began research on metabolism in 1903. There he studied metabolism of biological molecules, particularly the

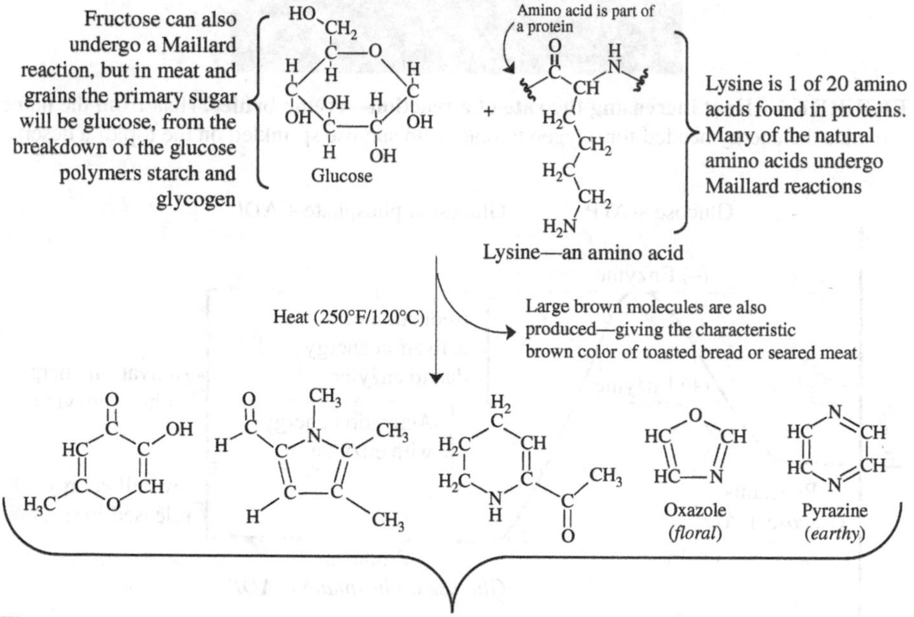

These are only some examples of the many possible molecules that give browned meat and toasted bread their complex flavors. The products of the Maillard reactions with different amino acids yield flavors ranging from floral and leafy to earthy and meaty. Aroma molecules made from the Maillard reactions include nitrogen atoms and sulfur atoms (not shown) from the amino acids.

FIGURE 6.6   The Maillard browning reaction. Maillard browning is a reaction between select sugars (reducing) and nitrogen atoms of many amino acids.

synthesis and breakdown of proteins. At the age of 34, he published the first article describing reactions between sugars and amino acids, starting with glycine and glucose [1]. Maillard established the formation of a class of dark-colored compounds he named "melanoidins." Eventually he studied and found the order of reactivity for the kinds of sugars that react with different amino acids, defining the reactivity for each. While he eventually published 14 papers on reactions between sugars and amino acids, specifics on how the reactions were taking place remained elusive until the early 1950s when chemist John Hodge from the US Department of Agriculture (USDA) defined the chemical steps in the early stages of the browning process. Interestingly, Maillard referred to the reactions that create brown flavorful molecules as "my reaction." However, the convention of calling these complex reactions the "Maillard reaction" had appeared in the literature (published in a review by General Foods and several chemical publications) by the early 1950s (Box 6.1). Thus, the large set of nonenzyme-catalyzed reactions between reducing sugars and amino groups are now called the Maillard reaction.

The initial steps of the Maillard reaction are the simplest and, therefore, the easiest place to begin. For those interested in cooking, this is where the magic happens. Heat, sugars, and proteins or amino acids result in the brown tasty and aroma-filled goodness of cooked food. A closer look at the chemistry shows that the reaction takes place at the $C=O$ bond (a carbonyl group) of a simple sugar and the nitrogen of an amino group found on amino acids and proteins. Under higher temperatures (to surmount the activation energy barrier!) around, 100–140°C/38–60°F, we begin to observe the formation of significant amounts of browning products (Fig. 6.6).

Sugar molecules can take on two different structures or shapes: a straight or open chain and ring or closed form of the molecule. In sugars such as glucose, the ring-opening process produces a $C=O$ or carbonyl at the *anomeric carbon*. See Figure 6.7 for examples of several reducing sugars and the anomeric carbon. When dissolved in the water of our cells and our food, these sugars can be found in both the ring-opened and closed conformations. The carbonyl group formed upon ring opening is what participates in the Maillard reaction, and sugars that can "ring open" to form a $C=O$ group are called "reducing sugars." However, in other sugars—such as fructose or

---

BOX 6.1 ADVANCED CARBOHYDRATE CHEMISTRY

Not all sugars work well in the Maillard reaction with amino acids. The carbonyl carbon of the aldehyde group is required for the reaction. Sugars that contain an accessible aldehyde are called "reducing sugars." A reducing sugar can act as a "reducing agent" in several chemical reactions including the Maillard reaction. They are given this name because of the reaction between the $C=O$ and another compound. Under the right conditions, the $C=O$ will react and cause the second compound to be "reduced" in its electron state. Thus, the sugar causes the other compound to be "reduced" and is called a "reducing sugar." This carbon is also called the anomeric carbon for reasons we will not worry about here.

sucrose—ring opening is not possible, and these sugars do not participate in the Maillard reaction. Examples of monosaccharides that can open and close to form carbonyl groups are glucose, ribose, galactose, and ribulose (Fig. 6.7). Disaccharides can be reducing sugars if at least one of the anomeric carbons is able to convert to the open chain format. The disaccharide found in milk, lactose, is composed of galactose bonded to glucose and has two anomeric carbons. The anomeric carbon of galactose is tied up in two C—O bonds and unable to convert to reducing sugar form. But the anomeric carbon of the glucose portion of lactose is able to open and close

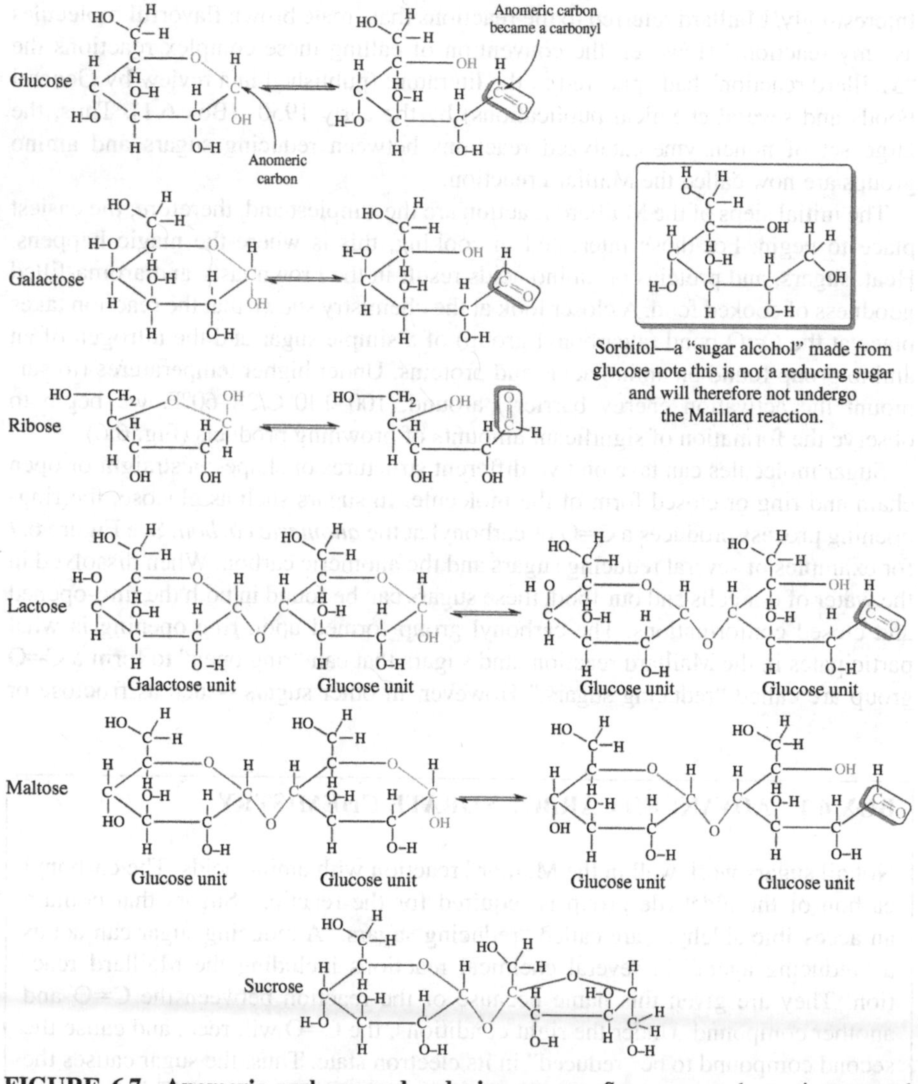

Sorbitol—a "sugar alcohol" made from glucose note this is not a reducing sugar and will therefore not undergo the Maillard reaction.

FIGURE 6.7 Anomeric carbons and reducing sugars. Some sugars have ring-opened forms in which the anomeric carbon becomes a carbonyl group (C=O), while in other sugars the anomeric carbon is trapped by two C—O bonds and cannot ring open.

and thus can react to chemically reduce other compounds. Maltose (formed from the breakdown of starch found in sweet potato and cereals and used to make candies, syrups, and ketchup) is another example of a disaccharide that is a reducing sugar. Foods with milk and potatoes brown for this reason. However, don't try to use table sugar (sucrose) to brown your food. Sucrose is a disaccharide of glucose and fructose has both of the anomeric carbons bonded to each other and is not a reducing sugar (Fig. 6.7).

Proteins and amino acids found in food are the second reactants in the Maillard reaction. The atom involved is the nitrogen of amino groups within amino acids, small proteins (peptides), or the amino groups of proteins themselves. In the reaction, the amino nitrogen bonds to the anomeric carbon of a reducing sugar. The combining of the reducing sugar (at the carbonyl carbon) and the nitrogen of amino groups is called a dehydration reaction and takes place with the loss of a water molecule. Very quickly the dehydration product rearranges to form a new compound of both the sugar and amino acid called an Amadori compound (Fig. 6.8).

At this point, you can begin to see the diversity of new molecules that can be formed by the Maillard browning reaction. There are many reducing sugars found in meat, vegetables, and fruits. Each of the 20 amino acids has an amino group, and one amino acid, lysine, has a side chain that contains an additional amino group. Some food scientists report that arginine can also form the Amadori compound of a Maillard reaction. Considering that 18 of the amino acids can react once and two of the amino acids can react with three different outcomes (the amino group, side chain, or both), the number of Amadori compounds possible with cooking is significant! If we limit the reducing sugars to eight of the more common reducing sugars, we can predict the possible number of initial products of the Maillard reaction. Looking at just the 18 amino acids that can only react with one of the eight sugars, there are $18 \times 8 = 144$. If two of the amino acids have three different combinations of sugar reactions, there are an additional $2 \times 3 \times 8 = 48$ possible products. Therefore in our simple calculation that ignores the many other sugars, peptides, and proteins that also react to form flavor molecules, there are 192 Amadori compounds that can be made from these reducing sugars and amino acids. When you then consider that there is a minimum of three different paths that each Amadori product can take to create the final compound, there are an additional $192 \times 3 = 576$ different compounds from eight different sugars and 20 amino acids. That is a lot of flavor, color, and aroma compounds! When you factor in the many additional reactions that follow the initial Amadori step, one can begin to comprehend the thousands of possible new products and the complexity of food color, flavor, and aroma.

There is an order of reactivity for both sugars and amino acids for the Maillard reaction. The most reactive reducing sugars are the smallest sugars. Five-carbon sugars such as ribose react faster than six-carbon sugars such as glucose, galactose, and fructose. Disaccharides are much slower to react than monosaccharides. Sorbitol and other sugar alcohols used as low-calorie substitutes do not react and if used in baking or cooking will not brown. Free amino acids react faster than peptides or amino acid residues within proteins. Lysine is the most reactive amino acid and browns very well with foods rich in ribose. Foods made with wheat such as bread are

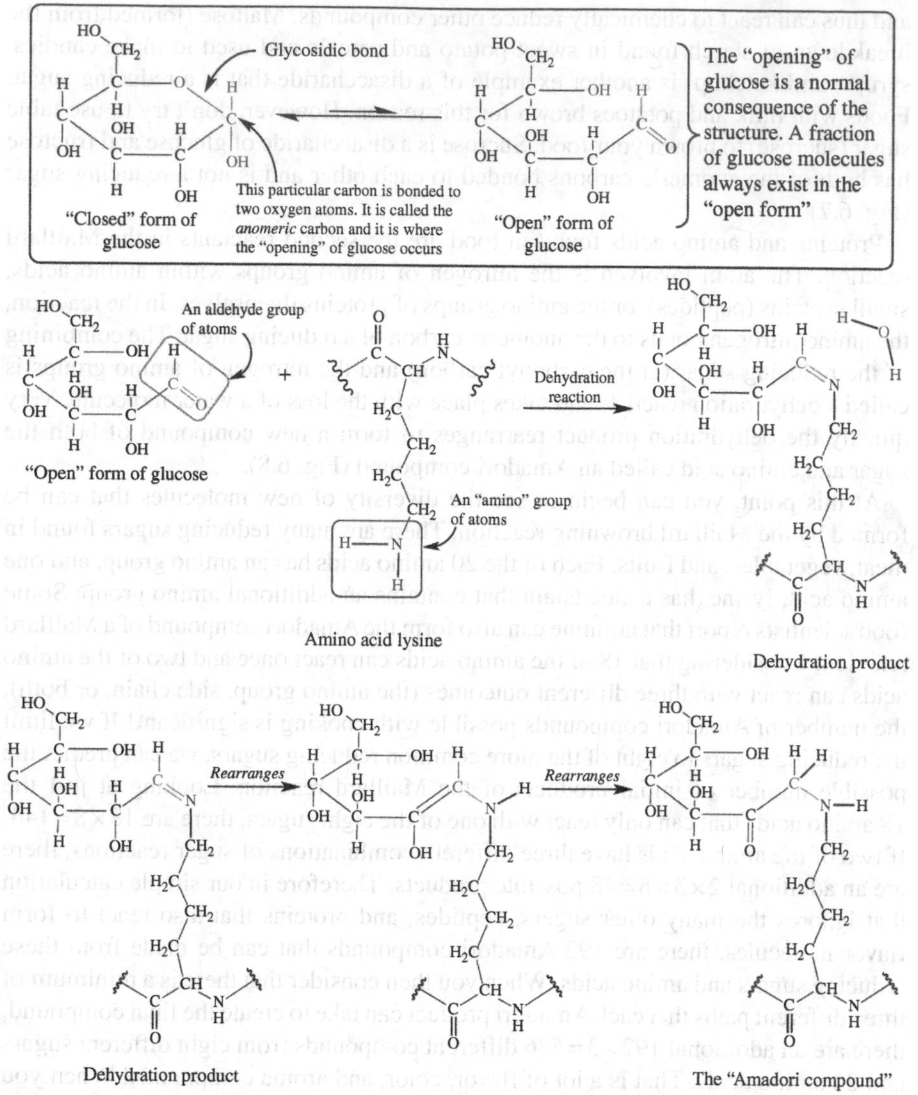

FIGURE 6.8   **The Maillard reaction.** The Maillard browning reaction requires a reducing "open" form of a sugar and the amino group of an amino acid. After the dehydration (also called a condensation) reaction, the compound will rearraign to form the intermediate "Amadori compound."

rich in lysine-containing proteins and produce a rich flavor and dark color when toasted. The whey protein in milk is also rich in lysine and is often used as a browning agent to increase browning in foods. Cheese and aged meat whose proteins have been degraded to peptides and amino acids and have a high concentration of sugars are particularly quick to brown. Ribose in meat is created by the breakdown of ATP and nucleotides making up DNA and RNA molecules. In baking, egg washes contribute to the flavor and browned color of baked good by providing amino acids, including lysine. The amino acid cysteine reacts well with ribose, creating a strong meaty flavor and aroma. In fact, through the Maillard reactions, over 200 different volatile aroma compounds from the reaction between cysteine and reducing sugars have been found in cooked meat. See Figure 6.9 for an example of potatoes cooked in different sugar–amino acid compounds (Table 6.1).

(a)   (b)   (c)   (d)

FIGURE 6.9 Browning potatoes. Potatoes dipped in a dilute solution of sugar (ribose) and an amino acid was lightly fried for 3 min and 45 s. (a) Water control—no sugar or amino acid, (b) ribose–leucine, (c) ribose–lysine, and (d) ribose–glycine.

TABLE 6.1 Intensity of Maillard Browning of Common Amino Acids and Sugars.

| Amino Compound | Absorbance at 420 nm | | | |
	+D-Glucose	+D-Fructose	+D-Ribose	+α-Lactose
Lys	0.947	1.04	1.22	1.23
Gly	0.942	1.07	1.34	1.49
Trp	0.826	0.853	0.972	1.32
Tyr	0.809	0.857	0.951	1.06
Pro	0.770	0.783	0.792	0.876
Leu	0.764	0.747	0.895	1.11
Ile	0.746	0.797	0.870	0.986
Ala	0.739	0.792	0.945	1 .06
Phe	0.703	0.751	0.800	0.941
Met	0.668	0.669	0.828	0.888
Val	0.663	0.800	0.772	0.900
Gln	0.602	0.644	0.633	0.639
Ser	0.600	0.646	0.679	0.751
Asn	0.560	0.578	0.565	0.560
His	0.535	0.573	0.529	0.609
Thr	0.509	0.601	0.590	0.600
Asp	0.353	0.426	0.378	0.336
Arg	0.335	0.331	0.370	0.312
Glu	0.294	0.338	0.341	0.320
Cys	0.144	0.202	0.150	0.273

Amino acids and sugars were mixed at a 1 : 1 ratio with a final concentration of 5 mM dissolved in 40 mM carbonate buffer pH 9.0. Each solution was autoclaved at 121°C for 10 min. The absorbance after each sample was treated was taken at 420 nm [2].

6.4 FACTORS THAT IMPACT MAILLARD REACTION BROWNING: pH, TEMPERATURE, AND TIME

There are a number of interesting factors that impact the rate and completeness of the Maillard browning reaction, the most impressive of which is pH. The acid level of a food can dramatically alter the rate and ability of a food to brown. To create their characteristic dark brown color, old-fashioned pretzels were dipped in lye and then cooked. Lye is a strongly basic/alkaline solution of sodium or potassium hydroxide (NaOH or KOH) and is used in a number of dishes including lutefisk and some types of noodles. In the case of the pretzels, the lye created a basic/alkaline pH on the outside of the pretzel dough, which increases the rate and quantity of Maillard browning reactions. Rates of browning reactions can be increased severalfold when the pH is above 7. Conversely, lower, more acidic pHs inhibit the Maillard reaction (Fig. 6.10). The explanation for this becomes clear if one considers the chemistry for the first stage of Maillard browning. The open/closed formation of reducing sugars is dependent on pH. In acidic solutions, reducing sugars are mostly in the ring conformation and unable to react. In more alkaline conditions, more sugar molecules are in the ring-opened form, which allows the anomeric carbon to react with the amino group. In addition, when the pH is low, the amino group becomes a protonated ammonium ion, which is very unreactive toward the reducing sugar. When the pH is raised, the proton is removed from the amino group and the reaction can occur. This explains why most marinades, which include an acid to break down proteins in tougher cuts of meat, take longer times and higher temperatures to brown. Figure 6.11 shows the acid/base chemistry impacting both the sugar and amino acid in the Maillard reaction.

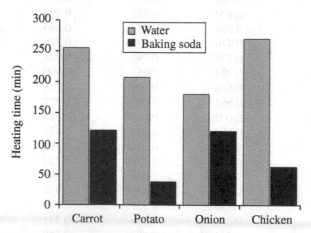

FIGURE 6.10 Alkali conditions speed browning time. 2.5 cm cubed sections of foods were dipped in water or a dilute solution of baking soda (sodium bicarbonate) and lightly fried for the indicated time. Note that decreasing the pH with baking soda decreased the time needed to brown each food item.

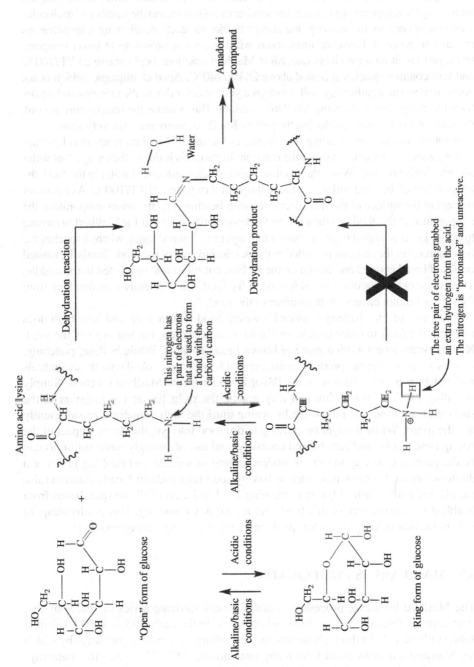

FIGURE 6.11 Acid/base chemistry of the initial phase of the Maillard reaction.

Temperature also impacts the rate of the reaction. As presented earlier in this chapter, increasing the temperature increases the number of collisions and the number of molecules with enough energy to overcome the energy barrier to reaction, allowing a spontaneous reaction to proceed. However, most foods will have some browning at lower temperatures, just not at an appreciable rate. Most Maillard reactions begin above 212°F/100°C and will continue quickly if heated above 284°F/140°C. Aged champagne, which is not allowed to warm significantly, still develops a yellow color due to the proteins and sugars found in the grapes undergoing Maillard reactions. But because the temperature is cool, the rates of the reactions producing the yellow Maillard compounds are very slow.

Another way in which heating food increases the rate of Maillard reaction is by evaporating water. Water content slows the reaction. In part, this is due to the energy that water absorbs during heating. While there remains a significant amount of water in the food, the temperature of the food will stay at the boiling point of water, 210°F/100°C. As the water content at the surface of the food decreases with heating (i.e., the water evaporates), the temperature of the food can then rise to the temperatures needed for Maillard browning to occur. Fried or grilled meat can have a thin layer of browned tissue where it touches the heat source, but the reaction is limited to this semidried area of the food. Similarly, toasted or baked bread is dried and brown on the surface but moist and unreacted in the middle. This is one of the reasons why oil is used to fry food. Oil efficiently transfers heat from the pan or heating element to the surface of the food.

Thus, one of the challenges when browning food is how long and how high does one heat the food in order to achieve the Maillard reaction, but not dry out the food. No one wants to eat a dried piece of brown bread or meat. While boiling, poaching, or sous vide cooking can produce moist, cooked food, food cooked with these methods will not brown, nor will it have the flavor bouquet of the Maillard reaction. Simply dropping the food onto a hot pan may not be the right fix, as the moisture at the surface of the food will inhibit the browning until the high temperatures sufficiently dry the tissue. While it might be enticing to dip food into lye, the flavor impact of the strong base can be problematic and unsavory, and use of strongly basic solutions can be dangerous. One way to use our understanding of science and cooking is to use a dilute solution of a weak base such as baking soda (aka sodium bicarbonate) to raise the pH and add a wash of lysine-containing food and a bit of ribose (purchased from health stores) to the surface of a food item for a quick browning. This is why tempura batters include baking soda—for quick and tasty browning during cooking!

6.5   MAILLARD IS COMPLICATED

The Maillard browning process is actually a very intricate series of reactions that start with the formation of the Amadori complex. In the early 1950s, USDA scientist John Hodge created a three-phase scheme describing the possible pathways by which the Maillard reactions could form many final products [2]. Three possible pathways combined with many different sugars and amino acids available to react in most foods give us an insight to the many possible products and flavors of browning (Fig. 6.12). See Table 6.2 for examples of some of these products [3].

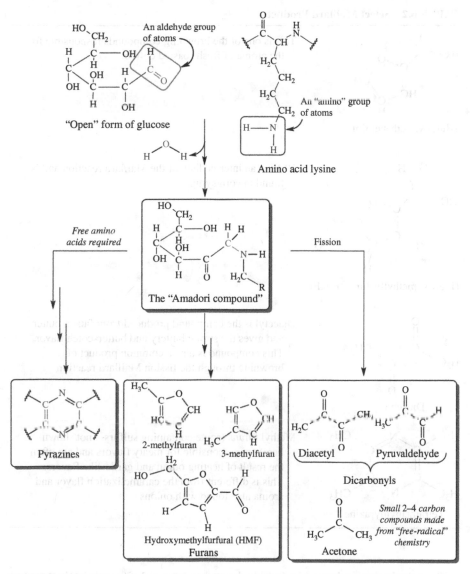

FIGURE 6.12 **The pathways of the Maillard reaction.** A simplified version of the reactions described by Hodge and others [2].

Shorter chain products are formed by the fission pathway. This pathway involves single electron compounds called free radicals. While complex in nature, this pathway can result in shorter 2- and 3-carbon compounds. These compounds can react with other sugars or amino acids, giving rise to interesting compounds including diacetyl (the smell of butter) and acetone (a fruity smell).

TABLE 6.2   Select Maillard Products.

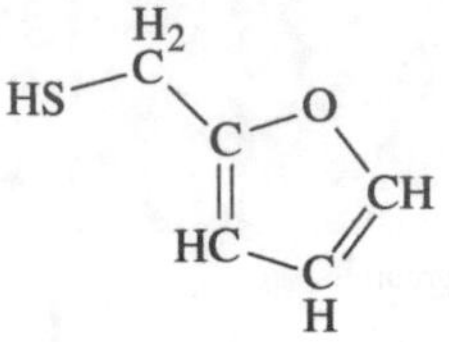

2-furanylmethanethiol

This is one of the browning compounds responsible for the aroma of fresh roasted coffee

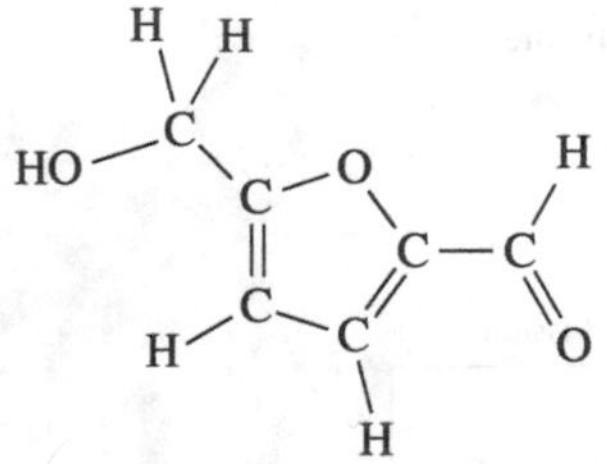

Hydroxymethylfurfural (HMF)

HMF is an intermediate of the Maillard reaction and is found in corn syrup

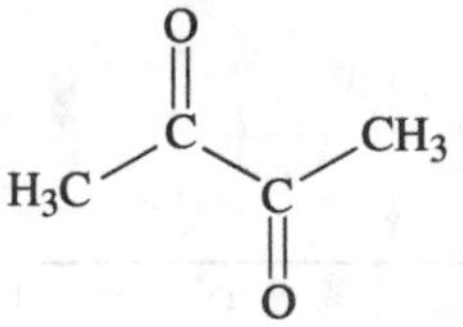

Diacetyl

Diacetyl is the compound produced from fats in butter and gives the strong buttery and butterscotch flavor. This compound is also a common product of browning through the fission Maillard reaction

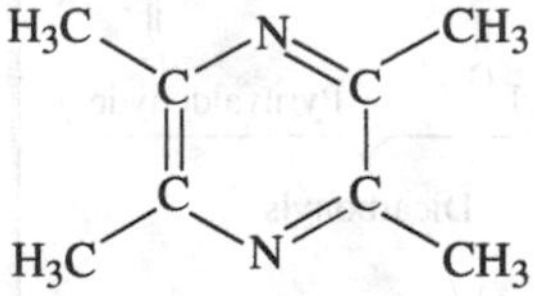

Tetramethylpyrazine

Methyl furans (some containing sulfurs—not shown here) are responsible for meaty flavors and are often the result of heating onion and garlic-like flavors. This is different from the caramelization flavor and aroma also found with onions

The dehydration pathway involves the loss of a water and nitrogen (amine) group. This pathway can produce ring structures including the furan-based molecules, which give roasted coffee smells, flavors of meat, and if sulfur is present a burnt bread odor and flavor.

A third pathway providing a large number of aromatic and brown-colored compounds especially in roasting is the Strecker degradation. This reaction can occur among the products of the Maillard reaction but can also occur apart from Maillard chemistry. In this reaction, dicarbonyl species (two C=O bonds) produced from breakdown of the Amadori product or elsewhere react with free amino acids to create aminoketones. These aminoketones condense and after oxidation (typically from

---

BOX 6.2   MAILLARD AND CANCER

Excuse me, but there is a strange chemical in my French fries! Acrylamide is a compound that when polymerized can be used by chemists and biologists to analyze protein size and make plastics. In 2002, much to the surprise of those who use acrylamide, scientists researching the Maillard reaction found small amounts of acrylamide in foods cooked at high temperatures, but higher levels of the compound were discovered in French fries and potato chips. Not a "natural" compound, acrylamide is the product of the Maillard reaction between asparagine—one of the 20 amino acids, found in nearly every food—and several common reducing sugars. Acrylamide is produced when foods are heated above 120°C/284°F by frying, broiling, or baking. At high concentrations, acrylamide is a known neurotoxin. Studies in rodents using concentrations of acrylamide that were a thousandfold higher than those found in food demonstrated no increase in cancer risk; however, it is not clear how well these studies translate into humans. The impact of acrylamide on cancer risk is still under investigation, but there is some hint of an increased risk for some types of cancer. The full impact of acrylamide on the human diet is, unfortunately, still unknown. While one could imagine avoiding intake of acrylamide, the initial study found the compound in 40% of the American diet. Cooking methods now used by industry and the restaurant industry are working on ways to reduce the heat that generates the Maillard-driven production of acrylamide.

---

oxygen in the air), pyrazines are formed. Pyrazines have strong flavors that add to the complexity of the Maillard bouquet.

In addition to their fabulous aromas and flavors, many of the final products from any one of these pathways can further react in polymerization reactions to create brown, high molecular weight pigments called "melanoidins." These pigments provide the dark color for beer and coffee as well as the brown color associated with browned meats and breads (Box 6.2).

6.6   CARAMELIZATION: BROWNING BEYOND THE MAILLARD

Caramelization involves another complex set of nonenzymatic browning reactions. While both Maillard reactions and caramelization create brown flavorful compounds, caramelization differs from the Maillard reaction in several important ways. Firstly, the reaction takes place between sugar molecules; no proteins or amino acids are involved in this reaction. Secondly, the required temperature needed to induce the reaction is higher for caramelization (starting at 320–356°F/160–180°C) than the Maillard reaction (typically 12–284°F/100–140°C). Finally, the reaction for caramelization is an oxidation reaction leading to long polymers of sugars with some shorter volatile compounds.

With the exception of most candies, which are primarily sugar and fats, most cooked dishes have Maillard and caramelization reactions happening concurrently, and both contribute to flavor. Braised beef, beer, and chocolate are just a few foods and drinks that owe much of their taste to a combination of Maillard reactions and caramelization. Caramels made solely from sugars are used in puddings or desserts including nougats, caramel brittles, and custards, while the process of caramelizing sugars in protein-containing foods contributes to the flavors one associates with browned onions, carrots, and coffee/cocoa beans. Browning describes both flavoring processes, but it is important for a careful cook to know the difference. Once both processes are understood, there is a world of opportunity to create interesting flavors and foods.

Caramelization is the process of degrading mono- and disaccharide sugars to form large polymers and volatile flavor molecules. These reactions happen when sugar is heated at or above its melting point. Like the Maillard reaction, caramelization requires heat to drive the complex, multistep reaction. The first part of the reaction is to melt the solid sugar crystals into liquid form. Then the sugars are further heated where they begin to lose OH groups and C atoms as smaller, often volatile compounds. Further heating will induce the collisions necessary to join many of the degraded sugars into large, viscous, dark-colored polymers. It is possible to heat the sugar too much; extended heating will create a burned, carbon (charcoal-like) residue. Complete oxidation (burning) of sugars will convert all the carbon, hydrogen, and oxygen atoms to CO_2 and H_2O—leaving nothing behind.

For solid, crystalline sugars—in contrast to liquid sugars like honey or syrups—there are two common techniques to make caramel, the wet and the dry methods. The dry method is simple and begins with a slow heating of solid sugar, such as sucrose (i.e., table sugar), until the solid is turned to a liquid. As a solid is heated, eventually the noncovalent intermolecular interactions holding the molecules in the crystal lattice are defeated, and the sugar molecules melt and begin to move more freely. This melting occurs as the sugar molecules acquire enough energy to overcome the intermolecular forces binding them into an orderly crystalline lattice. Melting does not occur instantaneously, because molecules bound to each other in a lattice must absorb the energy and then physically break the intermolecular forces. Typically, the outside of a crystal will melt faster than the inside, as it takes time for the heat to penetrate.

For many years, the conventional thinking was that addition of heat first melted sucrose into a liquid, and then the liquefied sugar broke down via chemical changes to a caramel product. However, Dr. Shelly Schmidt, a food chemist at the University of Illinois, published interesting research findings in a series of papers [4–7]. This work demonstrates that when heated slowly, some of the sugar molecules begin to chemically change into caramel products while still in a solid, crystal lattice form. This chemical change seems to take place before the molecules separate from solid into liquid form and not after melting takes place, as is the case for most solids. Dr. Schmidt showed that at lower temperatures and with a slow rate of heating, the sugar molecules began to chemically change and liquefy at 145°C/290°F while still solid, instead of at the expected temperature (190°C/380°F) and with higher rates of heating used to make the liquid sugar. The caramelized sugars made at this lower temperature did not contain the undesirable compounds made by caramelization at higher temperatures.

Some caramelization products are flavorful and have a pleasing odor, while others impart a bitter and burnt taste to food. Anyone who has caramelized sucrose too quickly at high temperatures understands the poor tasting, burned result. While still being investigated this is a shift in the traditional understanding of how solids make the transition from a solid to melted form, although how this all happens (what scientists call "mechanism") has yet to be determine. Dr. Schmidt's work impacts how food is processed to control the kinds of caramelization products formed. Applying this new understanding of how to make one set of compounds over the other by the careful application of heat and time yields more predictable results (Box 6.3).

The first step of caramelization chemistry requires reducing sugars. The reaction proceeds as reducing sugars lose a water molecule (dehydration) and rearrange into new compounds. This reaction requires the presence of the carbonyl (C=O) of a

---

BOX 6.3   CONTROVERSIES IN SCIENCE

Dr. Schmidt's work challenged the accepted ideas regarding the melting and decomposition of solids, and there is disagreement within the food science community over this work, highlighting the nature of science and critical peer review. In a commentary published in a scientific journal, several experienced food science experts published a series of critical questions and comments about the published work of Dr. Schmidt [8]. The reasoning for the work of Dr. Schmidt and the arguments by scientists led by Dr. Roos is that the published melting points for sucrose vary from source to source. Dr. Schmidt begins her work by explaining that these variations occur because of degradation/decomposition of the sucrose molecules before and during melting. The group opposing these conclusions believes the differences in melting point are due to impurities, differences in crystal sizes, and water mixed at the surface of the crystals. At the heart of the dispute is that many of the conclusions made by the Schmidt group are incorrect and it is further argued that such incorrect conclusions about melting versus decomposition devalue food science and mislead the popular media. The latter has taken place, as several food-related blogs discuss how sugars don't melt as they heat, but instead they caramelize while or instead of melting.

The commentary by Dr. Roos is filled with specific and very detailed scientific reasoning to question the decomposition phenomena. In the same journal, published alongside the initial questions, Dr. Schmidt rebutted each point to clarify her explanation of terminology and added new information to support her conclusions that "the loss of crystalline structure in sucrose is caused by the onset of thermal decomposition, not thermodynamic melting" [9]. So a nonscientist is left asking, "who is right"? The more appropriate question is what is actually happening? Scientists are more focused on interpreting and discussing experimental results to understand the nature of the world around us. This is a good example of how different scientists or groups of researchers present and peer-review publish opposing viewpoints and conclusions to complicated experiments. More work may be needed to understand the implications of both sets of scientists.

reducing sugar. Sucrose, as you may recall, is made of glucose and fructose and is not a reducing sugar. Therefore sucrose must first be converted to its individual components: glucose and fructose. This can happen in a few ways. In solution, sucrose will very slowly break down into glucose and fructose monosaccharides. This process is increased by heat (thus the melting of sucrose, or dissolving sucrose prior to heating) and accelerated by acids including tartaric and lemon juice. Two enzymes—invertase from plant cells and sucrase from animal cells—can also break the glycosidic bond of sucrose into glucose and fructose. The process of converting sucrose to its individual units is called inversion and will produce what is commonly called "inverted sugar." An examination of the temperatures required to start the caramelization reaction reflects the need to "invert" sucrose. Table 6.3 also explains why foods cooked and baked with fructose (including honey) brown faster and darker than those with sucrose or glucose.

Like the Maillard reaction, the diversity of products of the caramelization reaction is immense and depends on the sugars involved in the reaction and the chemical environment of the reaction. As we saw in Chapter 1, after the loss of water, individual sugar molecules can then react with the reducing C=O of another sugar to form a sugar anhydride. This is a condensation reaction, also called a dehydration reaction (Fig. 6.13), where two molecules are joined in a covalent bond by the loss of a water

TABLE 6.3   Temperature Required for Caramelization of Sugars.

Fructose	110°C
Glucose	160°C
Sucrose	160–180°C
Maltose	180°C

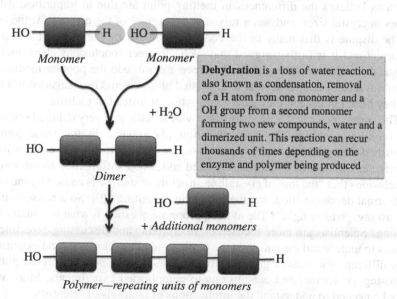

FIGURE 6.13   Polymerization reaction.

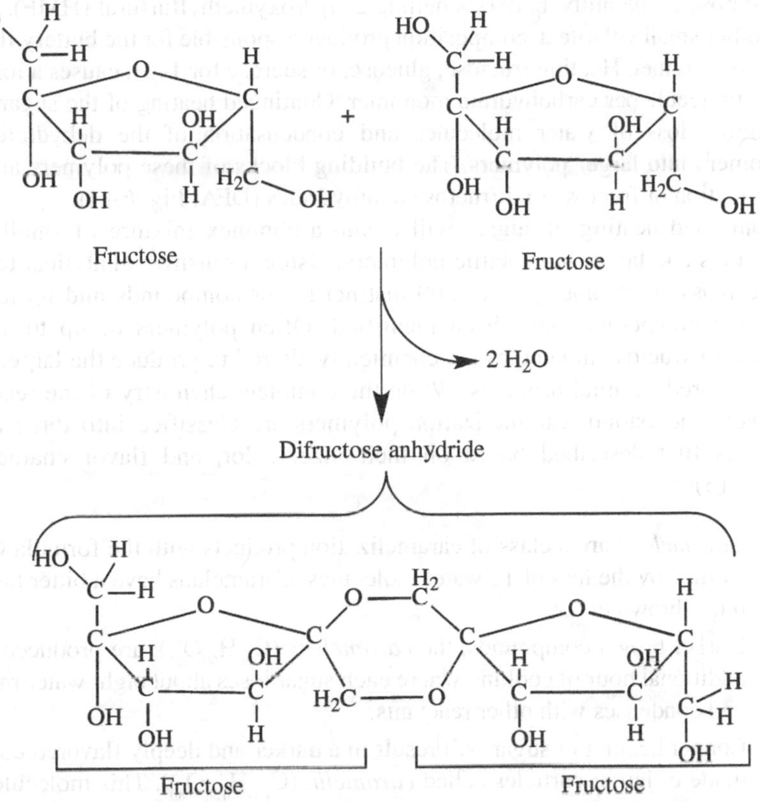

FIGURE 6.14 Dehydration reaction forming difructose anhydride.

molecule. The result is a larger molecule. Depending on the time, temperature, and process, this reaction will continue with some sections of the molecule breaking from the parent molecule (fragmentation) forming small aromatic molecules. Fructose will begin this reaction forming difructose anhydride (Fig. 6.14). As the sugar polymer continues to grow by dehydration condensation, there can be transformation of carbons, oxygen, and hydrogen molecules into different structures. Heating the sugars in acidic or basic conditions leads to a range of different products and colors where the protons and hydroxyl ions serve as catalysts for the chemical rearrangement of the intermediate caramelization compounds.

Several organizations including the US Food and Drug Administration classify the colors and flavors of the process based on the chemical solutions used during caramelization. When a sugar solution is heated under different conditions, the resulting caramel colors are grouped into four classes for different products and uses.

The caramelization reaction will continue to form both small volatile compounds and very large polymers of the sugar monomers. Initial heating will decompose and dehydrate the sugar monomers, and a population of the compounds will also give rise to smaller volatile molecules. Fructose and, to a lesser extent, glucose can decompose to the nutty, buttery smelling 2-hydroxymethylfurfural (HMF). Diacetyl is another small volatile decomposition product responsible for the buttery flavor and smell of caramel. Heating fructose, glucose, or sucrose for 1–2 h causes a loss of one water molecule per carbohydrate monomer. Continued heating of the sugar leads to additional loss of water molecules and condensation of the dehydrated sugar monomers into larger polymers. The building blocks of these polymers arise from modification of fructose to difructose dianhydrides (DFA; Fig. 6.14).

Continued heating of sugars will create a complex mixture of small volatile molecules and larger nonvolatile polymers. Using a sensitive analytical technique called mass spectrometry, over 300 distinct major compounds and up to several thousand compounds have been identified. Often polymers of up to six fused glucose or fructose molecules are chemically altered to produce the large, nonvolatile, colored caramel products. While the complete chemistry of the reactions is not well understood, caramelization polymers are classified into three complex mixtures first described based on their size, color, and flavor characteristics (Fig. 6.15):

1. *Caramelans* are a class of caramelization products with the formula $C_{12}H_{18}O_9$ formed by the loss of 12 water molecules. Caramelans have a bitter taste and a nutty-brown color.

2. Darker brown compounds, the *caramelens* ($C_{36}H_{50}O_{25}$), are produced after an additional hour of cooking where each sugar loses about eight water molecules and condenses with other reactants.

3. Longer heating of sugar will result in a darker and deeply flavored compound made of larger particles called *caramelin* ($C_{125}H_{188}O_{80}$). This molecule is very dark and does not dissolve well in water.

Caramel is one of the oldest and most widely used flavor compounds with over 50 metric tons of caramel produced and consumed in foods each year. As indicated in Table 6.4, the products of caramelization are used frequently in food and drink for many different properties including taste. For example, colas owe their color and some of their flavor to caramelization products. Some caramelization products can act as emulsifiers and help keep molecules in solution. Several of the caramels found in coffee and honey have antibacterial properties. One compound produced by both caramelization and via the Maillard reaction, 4-methylimidazole, or 4-MEI for short, is found in dark beers, coffee, and roasted meats and is part of the caramel coloring and flavor in some cola soft drinks. While a natural reaction during heating of these foods and drinks, some people have a concern about the dark flavored molecules. When mice were exposed to very high concentrations of 4-methylimidazole—levels that far exceed the average consumption of a human—a moderate but statistically significant increase in tumors was documented in the animals. This has led several to be concerned about drinking colas, coffee, or dark beer! While California now includes 4-MEI on its list of probable carcinogens, one would have to drink tens of thousands of cans of the soft drink a day to approach the level of concern for the compound!

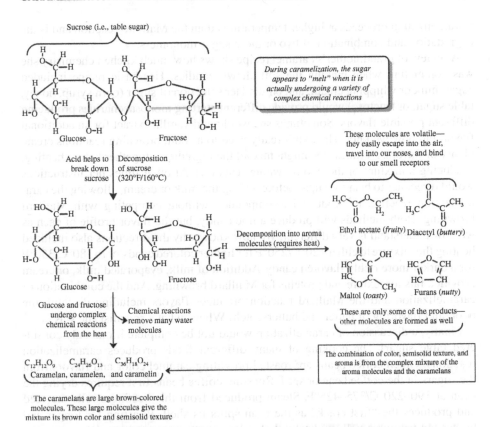

FIGURE 6.15   The many reactions of caramelization.

TABLE 6.4   Classification of Caramel Used in Food and Beverage.

Class	Classification	Preparation	Uses
I	Plain or spirit caramel	No ammonium or sulfur compounds can be used	Distilled high alcohol spirits such as whisky
II	Caustic sulfite caramel	High pH (NaOH) and sulfite (SO_3^{2-}) used	Beer, malt bread, sherry, malt vinegars
III	Ammonia caramel	No sulfites but ammonium compounds can be used	Beer, sugar candies, soy sauce
IV	Sulfite ammonia caramel	Both sulfite and ammonium can be used	Widely used for soft drink and in acidic solutions

Complicating our understanding of the process and products of caramelization is that these reactions take place side by side with the Maillard reaction. Many of the intermediates from both browning reactions can react to form new aromas, flavors, and colors. However, the distinction between the two processes remains simple. Maillard reactions take place at lower temperatures than caramelization, and the reaction is between a reducing sugar and amino group of an amino acid or protein.

Caramelization proceeds at higher temperatures than the Maillard reaction and is the degradation and combination of two or more sugar monomers.

A review of grandmother's caramel recipe shows how much kitchen chemistry she was conducting when making the soft chewy candies. Her basic recipe included sugar, milk or whipping cream, and butter. Her selection of sugar (corn syrup, honey, table sugar, or fructose) would provide different starting monosaccharides producing different possible flavors. Sometimes she would add vanilla extract for an additional flavor, but now we know she didn't really need to as both browning reactions create plenty of flavors. At times, she might mix all the ingredients together before heating, producing a mixture of the two browning reactions. At other times her instructions would direct her to heat the sugar before adding the milk or cream, allowing the caramelization reaction to produce its compounds without competing with Maillard browning. Both methods will produce a fine candy, but the flavor profile of each as we now understand is very different. We now know why the directions also included heating the mixture to either 120°C/250°F for lighter colored candy or to 180°C/350°F for a darker more richly flavored candy. Addition of milk, evaporated milk, or cream provided fats for texture and proteins for Maillard browning. And the combination of caramelization and the Maillard reaction produces flavors including diacetyl, the odor, and flavoring of butter and butterscotch! Whew, what a scientist she was!

Finally, a discussion of caramelization would not be complete if one only considered table sugar! The cooking of many different foods produces caramelization reactions. Green coffee beans are roasted providing caramel color and flavors from the sugars in the coffee bean or seed. Roasting coffee beans first requires drying the bean at 190–220°C/375–425°F. Steam produced from this heating swells the seed and produces the "first crack" as the bean splits its shell. Roasting for longer and higher temperatures will produce a darker bean with strong flavors. Of course, the flavors are a combination of browning reactions. Seeds contain many complex and simple sugars as well as proteins and amino acids to react and form colorful flavors and odors. Braising tough cuts of meat is a way to slow cook beef, poultry, or pork for a more tender result. Braised meat is typically browned on the stovetop by Maillard reactions before placing it into a tight fitting pot or kettle with a small amount of liquid. Addition of sugars (corn syrup or brown sugar) and vegetables high in sugar content such as carrots and onions are often used to add flavor as the meat is simmered at or near the temperature of caramelization.

Carrots, corn, potatoes, and onions have high sugar content and are great vegetables for caramelization. Adding a small amount of fructose to low sugar vegetables like broccoli and cauliflower allows them to brown during roasting. Baking potatoes doesn't get hot or dry enough to induce caramelization at 150°C/300°F. However, slices of potatoes brown nicely if warmed enough to release the steam (unpacking the complex starches) and then heated sufficiently to get the caramelization rolling at a higher temperature. Carrots and onions are particularly high in the reducing sugars fructose and glucose, which promotes browning reactions. Remember to get the reaction right; it takes time and high temperatures (up to 356°F/180°C) for the reactions to occur. Directions for onion and carrot caramelization stress the importance of heating the vegetable for nearly an hour. Of course, an impatient cook might use too much heat and burn the food instead of caramelizing it. Again, this is where you can use a little science of cooking to ease the process. Remember that caramelization takes place between reducing sugars that can form open or straight chains. The equilibrium between closed and open chain is shifted to the open form in basic or alkaline conditions. A sprinkle of baking soda helps to raise the pH of onions and carrots to the open chain, where the higher concentration of the open chain reducing sugar can dehydrate and polymerize into caramel goodness. Of course, the Maillard reaction is also taking place at a higher rate at alkaline pH, and more flavor is always better, isn't it?

6.7   ASCORBIC ACID BROWNING

Left alone, fresh-squeezed citrus fruit juices will brown in under 3 days. Sugars and amino acids both play a key role in this browning; however, there is another reaction involving vitamin C, also called ascorbic acid. Ascorbic acid is a water-soluble weak acid, also known as vitamin C, naturally found in citrus fruits and many vegetables. Citric fruit juice is especially rich in ascorbic acid and the breakdown of ascorbic acid generates brown pigments and off-flavors over time. This browning, like caramelization and the Maillard reaction, is a nonenzymatic browning process.

Now that we've studied the complex chemistry of amino acid and sugar browning reactions, it will surprise no one to learn there are also many possible products for the breakdown of ascorbic acid, and they depend upon the conditions of the juice during degradation. Up to 17 different ascorbic acid breakdown products have been identified so far, but two of these products are present. There are two key degradation products of ascorbic acid, furfural and dehydroascorbic acid. In the presence of oxygen (aerobic conditions), ascorbic acid is converted to dehydroascorbic acid and then further degraded to brown compounds. Dehydroascorbic acid produced in ascorbic acid degradation then reacts in the Maillard pathway taking part in the Strecker reaction producing brown pigments. In acidic and low (anaerobic) conditions (that found in concentrated, canned, and fresh citrus juice), furfural is one of the main components of ascorbic acid breakdown. This compound is then further degraded to form brown pigments after reacting with amino acids in the fruit juice (Fig. 6.16).

Ascorbic acid degrades to dehydroascorbic acid and furfural, both of which, through very different pathways, react with amino acids to create bitter off-flavor brown pigments, which decrease the quality of the juice. Several researchers in the food industry have found that treating the juice to remove amino can stop the final few steps of each browning pathway. While the browning reaction can take place with or without oxygen, it proceeds more slowly in the absence of oxygen. Reducing head space (the gap above the juice in the container) or exchanging the air for nitrogen gas can limit exposure of the juice to oxygen. One of history's oldest food additives, sulfites are used to compete with ascorbic degrading reactions, sparing the vitamin from breakdown. For example, sodium metabisulfite ($Na_2S_2O_4$) is used in wine, beer, and other foods and reduces browning severalfold even when juices are heated to over

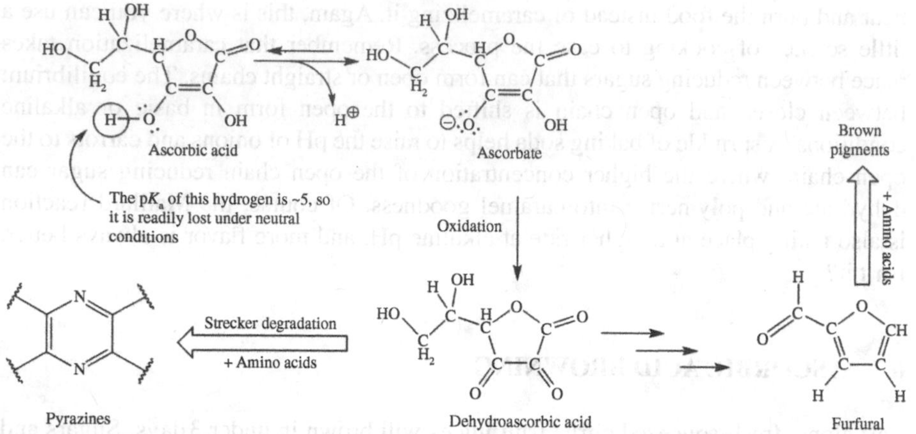

Ascorbic acid

The pK_a of this hydrogen is ~5, so it is readily lost under neutral conditions

Ascorbate

Brown pigments

+ Amino acids

Oxidation

Pyrazines

Strecker degradation

+ Amino acids

Dehydroascorbic acid

Furfural

FIGURE 6.16 Ascorbic acid and browning reactions common to fruit juices.

212°F/100°C. The sulfite inhibits the oxidation of ascorbic acid to dehydroascorbic acid, which prevents later reactions leading to the production of the brown pigments.

6.8 ENZYME-CATALYZED BROWNING

Cut into an apple or banana and leave it on the counter for an hour or so. The once enticing food is now brown and bitter tasting. Vegetables, fruits, and even some shellfish will all brown after being cut or damaged. Unlike caramelization or the Maillard reaction, this browning reaction is unpleasant and catalyzed by enzymes not heat. The browning occurs when plant compounds called phenols react with oxygen to form large, polyphenolic brown pigments.

Phenols are rings of carbon atoms with alternating single and double bonds and with at least one hydroxyl group (—OH) coming off the ring. For example, the amino acid tyrosine, found in almost every protein in nearly every organism, is a phenol. Polyphenols are compounds with several of these phenol rings bonded together (Fig. 6.17).

Phenols and polyphenols are found in nearly every living organism. They have interesting biological roles in a range of organisms including mold, bacteria, algae, and higher animals. There are over 8000 different phenolic compounds in plants with just as many diverse roles including the color and flavors of fruit, bark, and leaves. Tannins are a family of polyphenols found in plants; they provide color and function as pesticides for the plant, and some tannins are plant hormones regulating root development, fruit softening, and flowering. Tannins also have an astringent taste. Tannins are responsible for the dry pucker that accompanies a sip of strong black tea or red wine.

Common phenolic compounds found in food and browning reactions are grouped into four classes. The simple phenols include tyrosine and mono- and diphenols such as catechol, resorcinol, and dihydroquinone. The addition of a carboxylic acid group

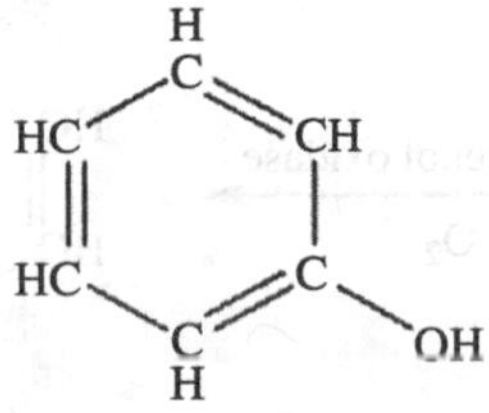

Basic *phenol* structure

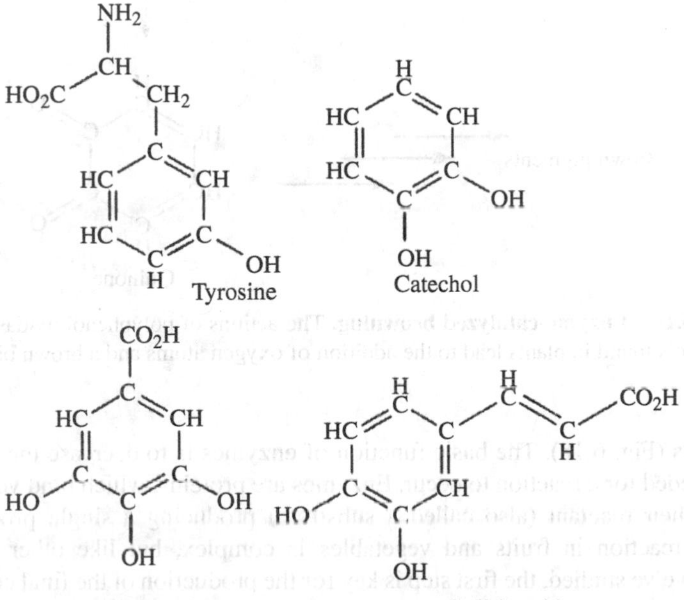

Tyrosine

Catechol

Gallic acid

Caffeic acid

Examples of *phenolics* found in plants

FIGURE 6.17   Phenols.

distinguishes another class called the phenolic acids. Gallic acid is found in tea, grapes, and apples and was used for dyes, possibly even as an invisible ink. One of the acid phenols is chlorogenic acid, a key substrate for polyphenol oxidase (PPO) in food browning especially potato, peach, and prunes. Chlorogenic acid is also present at relatively high concentration in coffee beans, and the metabolic products of the acid may be responsible for some of the heart health benefits of drinking coffee. Flavonoids are a diverse group of polyphenols providing color and the building blocks for plant hormones. Several of the polyphenols of the flavonoid family are targets for PPO and are responsible for the astringency of some fruits as well as color of dark teas.

Fruit and vegetable browning of phenol and polyphenol compounds is controlled by an enzyme that binds and starts the series of reactions that lead to bitter brown

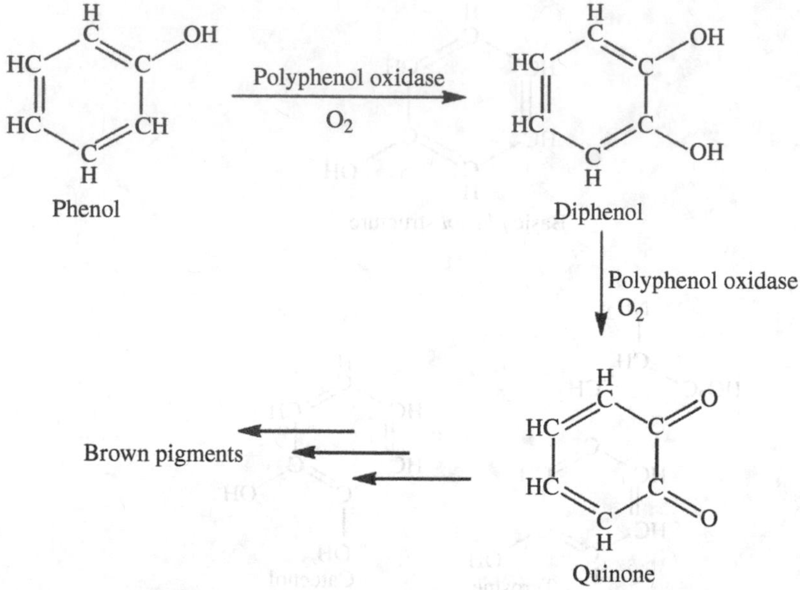

FIGURE 6.18   **Enzyme-catalyzed browning.** The actions of polyphenol oxidase on many of the phenols found in plants lead to the addition of oxygen atoms and a brown bitter tasting pigment.

compounds (Fig. 6.18). The basic function of enzymes is to decrease the activation energy needed for a reaction to occur. Enzymes are proteins, which bind very specifically to their reactant (also called a substrate) producing a single product. The browning reaction in fruits and vegetables is complex, but like other browning reactions we've studied, the first step is key for the production of the final compound.

The critical first step in enzyme-catalyzed browning starts with a phenolic compound, the enzyme PPO, and oxygen.

There are several common phenolic compounds found in food and browning reactions. The simple phenols include tyrosine and mono- and diphenols such as catechol, resorcinol, and dihydroquinone (Figs. 6.17 and 6.19). The addition of a carboxylic acid group to a phenol distinguishes the phenolic acids. Gallic acid is found in tea, grapes, and apples and was used for dyes, possibly even as an invisible ink. One of the acid phenols is chlorogenic acid, a key substrate for PPO in food browning especially potato, peach, and prunes. Chlorogenic acid is also present at relatively high concentration in coffee beans, and the metabolic products of the acid may be responsible for some of the heart health benefits of drinking coffee. Flavonoids are part of a diverse group of polyphenols providing color and the building blocks for plant hormones. Several of the polyphenols of the flavonoid family are targets for PPO and are responsible for the astringency of some fruits as well as color of dark teas.

PPOs are enzymes found in all plants and animals that catalyze the formation of quinones from mono- or diphenolic molecules. These enzymes are responsible for the

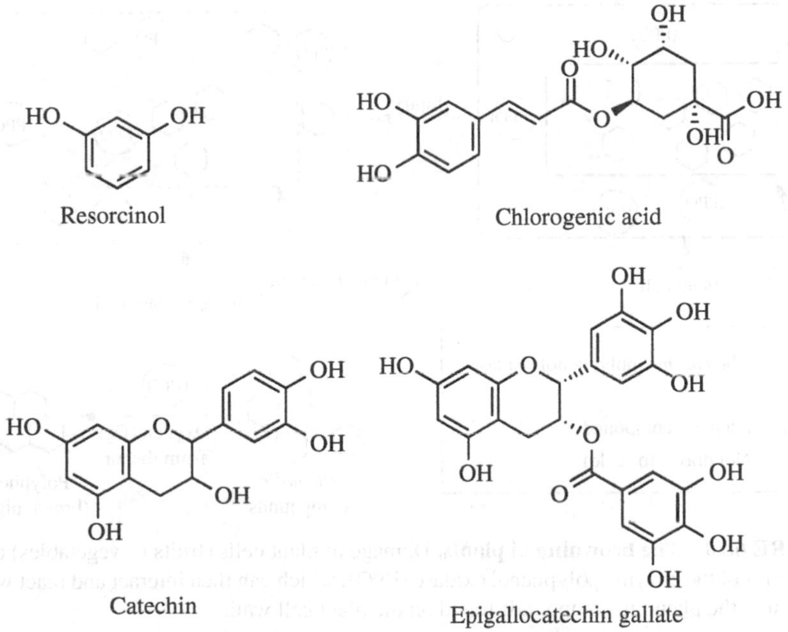

FIGURE 6.19 Examples of plant phenols.

first reactions in the browning process. There are three types of PPOs (tyrosinase, catechol oxidase, and laccase); each is a different protein made from different genes, but each protein binds and catalyzes the same basic reaction. Tyrosinase catalyzes the addition of an —OH group onto monophenols and then converts the product diphenol to quinones. Tyrosinase is also responsible for the synthesis of the neurochemical dopamine and melanin, the pigment in the skin, hair, and eyes. Laccase also converts diphenols to quinones. Catechol oxidase (aka diphenol oxidase) converts diphenols, and to a lesser extent monophenols, into quinone compounds. All three enzymes are called PPOs in the scientific literature and for the rest of this chapter. The bottom line is that these enzymes alter phenols to quinone compounds, which starts the browning reaction producing complex melanin products.

Plant cells are filled with a variety of phenols and polyphenols and two or more PPO enzymes. Yet, vegetables and fruits don't brown until cut or damaged. This is because the plant cell has done a brilliant job of separating the substrate from the enzyme, until a reaction is beneficial to the plant. PPOs are made as inactive longer proteins (called preproteins), and many of the phenols used in the browning reaction are sequestered in cellular compartments called vacuoles. Browning is slow to progress in intact vegetables and fruits, and only with aging does the reaction significantly occur. However, once cut or damaged, the PPOs are activated into shorter, mature enzymes, and the cellular destruction of the knife releases the phenols from the vacuoles allowing phenols to mix with enzyme and oxygen, beginning the process of browning (Fig. 6.20).

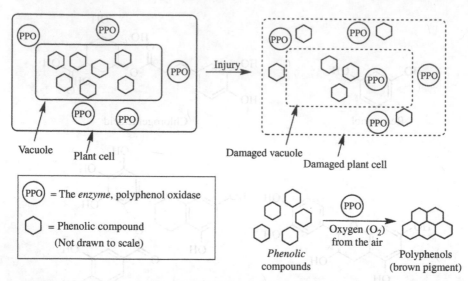

FIGURE 6.20   The browning of plants. Damage to plant cells (fruits or vegetables) causes the release of the enzyme polyphenol oxidase (PPO), which can then interact and react with its substrate—the phenolic compounds found on the plant cell wall.

Monophenols react with oxygen and PPOs to form diphenol compounds. The oxidase enzymes will then use a second molecule of oxygen to oxidize each alcohol of the diphenol to a carbonyl group (C=O) of the product quinone. Once formed, the quinones react with other compounds including amino acids to form large complex brown polymers. In animals, the brown melanin pigments found in skin and hair are formed by similar reactions that start with tyrosine.

The natural role for phenols and PPO enzymes depends on the plant in question. Some polyphenols are used by plants to generate plant hormones, and others are used to strengthen the cell walls, while other polyphenol products create color and UV light protection. Tea leaves, figs, and cocoa plants all utilize PPO-catalyzed synthesis of polyphenols to create a darker color and new flavors. There are several interesting scientific studies that suggest the brown compounds produced by PPO enzymes are beneficial to the plant. When insects bite or other physical damage occurs to the plant, the resulting wound provides an opportunity for microbial pathogens to infect the plant tissue. In this case, polyphenols are a defense mechanism, a kind of immune response for the plants. In tomato, tobacco, and other plants, wounding activates the PPO enzymes and the resulting enzyme-catalyzed browning produces polyphenol compounds that inhibit the parasitic growth of mold, fungus, and other microbes while promoting wound closure. The bitter taste of the brown compounds is thought to discourage further attacks on the plant by insects and herbivores.

In addition to being less attractive to insects, brown food is bitter and unattractive to humans. Nearly half of our vegetable and fruit crop each year is lost and wasted due to browning after the food is harvested and before it gets to our table. This represents a significant waste of farming resources and agriculture. One interesting approach to avoid such loss is the creation of genetically modified organisms (GMO)—in this case, a genetically modified plant that will not produce brown polyphenols. The Arctic Apple, a GMO product, is a nonbrowning apple where the eight genes for PPO enzymes have been silenced. The approach used to reduce the production of PPO in the apple uses a technique called RNA interference. This molecular therapy is being developed as a treatment for cancer, tissue transplantation, liver disease, as well as other human diseases.

Unless you can use RNA interference to stop browning food in your kitchen, the average person will have to use an understanding of enzymatic browning to block the reaction at home. To start, recall that inhibition of PPO will block the browning compounds from forming. Polyphenols, oxygen, and enzyme: stop any one of these three components and you've just created a way to keep food from browning. Limiting access to oxygen is easy but only mildly effective. Immersing vegetables and fruits in water prior to and during cutting reduces the availability of oxygen to bind to the enzyme. Potatoes, which quickly brown, are prepared this way for industrial production of French fries and potato chips. Some have experimented with packaging fruit in carbon dioxide gas to eliminate oxygen, but the results were limited in success. There are several compounds that bind and block the ability of PPO to react with polyphenols. As classic enzyme inhibitors, many of these compounds have nearly the same structure as the phenolic substrates of PPOs. These substrate-mimicking compounds compete with the natural enzyme substrate inhibiting the polyphenol reaction. Guaiacol, resorcinol, and phloroglucinol are all examples of such "competitive inhibitors."

Blanching vegetables and some fruits is an effective way to inhibit the enzyme and soften the cell wall. Blanching is the short submersion of a vegetable in boiling water for a short period of time. Some, but not all, PPO enzymes are denatured and rendered inactive after short periods of 212°F/100°C heat of boiling water. Longer exposures are needed to kill several forms of PPO and would result in cooking the vegetable (Box 6.4).

Some acids are also effective inhibitors of PPO. The pH optimum for PPO activity lies between 6.0 and 6.5. If the pH ranges greater than 1 unit above or below these points, almost no activity remains. Phosphate buffers are sometimes used as they can buffer the pH at either end of this range and are considered food safe. Citric acid,

---

BOX 6.4   PPO INHIBITORS

PPOs include tyrosinase, catechol oxidase, lactase, and others. There are very specific inhibitors for each enzyme, many of which are used in food preparation. Tyrosinase and catechol oxidase are inhibited by tropolone, cinnamic acid, and salicylhydroxamic acid but not laccase. Laccase is inhibited by azide, cyanide, thiocyanate, fluoride, and sulfhydryl compounds. New compounds with promise include kojic acid (found in fermented Japanese foods) and hexylresorcinol (also used in throat lozenges).

malic acid, and ascorbic acids are effective inhibitors of the enzyme; less for their acidic properties, but instead for other properties. Both citric acid and malic acid can bind and strip copper away from PPO enzymes. Copper is absolutely required by the enzyme. However, each acid will only partially inhibit the production of brown compounds in apples and other fruits or vegetables (Fig. 6.21). While ascorbic acid is involved with browning in citric juices, the acid is an effective inhibitor of PPO function and does not impart taste to food like malic and citric (sour and tart). Ascorbic acid will react with quinone compounds, reversing the PPO reaction. As long as ascorbic acid is present in excess, the quinone products cannot continue the reaction to form brown pigments. However, once the ascorbic acid is consumed in the reaction, browning can occur.

Calcium chloride is another modest inhibitor of PPO activity. When used alone, any one of these PPO inhibitors is moderately effective, but combinations of

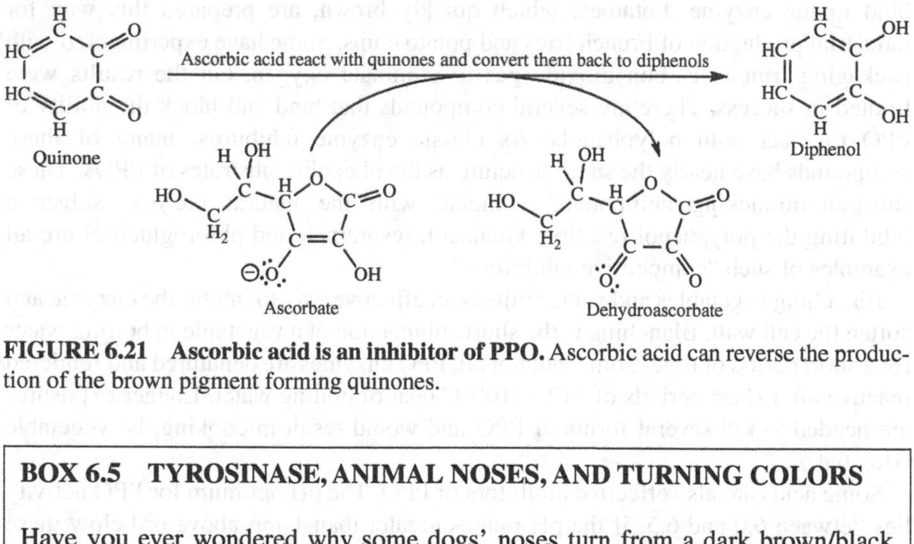

FIGURE 6.21   Ascorbic acid is an inhibitor of PPO. Ascorbic acid can reverse the production of the brown pigment forming quinones.

BOX 6.5   TYROSINASE, ANIMAL NOSES, AND TURNING COLORS

Have you ever wondered why some dogs' noses turn from a dark brown/black color to light tan or pink? The same basic phenomena take place with Arctic rabbits and Siamese cats. In each case, the PPO enzyme tyrosinase is involved. The enzyme variant in these animals is active in warmer temperatures. The mutation (two glycines are replaced by arginine or tryptophan amino acids) renders the enzyme inactive at lower temperatures. This "temperature-sensitive" PPO found in these animals serves as a camouflage of sorts. Rabbits and Burmese and Siamese cats both lose the pigment production of tyrosinase in cold temperatures and become white, blending in with the snowy background. In dogs, this effect is called "winter nose" and is more common in larger dogs such as Labrador Retrievers and Huskies. The biological significance of this phenomenon in dogs is unclear but troublesome for those living in northern climes. A pink or "Dudley Nose" in a dog is an unwanted characteristic that can disqualify a dog from a show (Fig. 6.22).

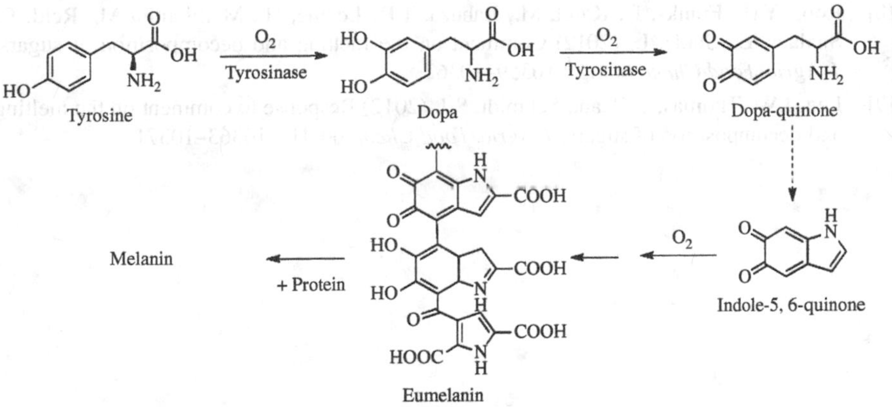

FIGURE 6.22 Tyrosinase starts the conversion of tyrosine to the pigment melanin. One of the metabolic fates of the amino acid tyrosine is to be converted to the dark pigment melanin. The activity of tyrosinase can be influenced by temperature.

inhibitors are even better at preventing browning. Several combinations of the acids mentioned previously and calcium ions have a strong effect on PPO activity. NatureSeal is a company producing a range of products to inhibit browning in packaged salads, fruit, and other produce. NatureSeal is a mixture of calcium and ascorbic acid. Used as a dip, the combination can inhibit browning in apples for up to 14 days. Using this technology, one fast-food restaurant sold nearly $0.5 billion in just sliced apples in 2010 (Box 6.5).

REFERENCES

[1] Hodge, J. (1953) Chemistry of browning reactions in model system. *J. Agric. Food Chem.*, 1: 928–943.

[2] Ashoor, S.H. and Zent, J.B. (1984) Maillard browning of common amino acids and sugars. *J. Food Sci.* 49: 1206–1207.

[3] Somoza, V. and Fogliano, V. (2013) 100 years of the Maillard reaction: why our food turns brown. *J. Agric. Food Chem.* 61: 10197–10197.

[4] Schmidt, S.J. (2012) Exploring the sucrose–water state diagram. *Manuf. Confect.* 92(1): 79–89.

[5] Lee, J.W., Thomas, L.C. and Schmidt, S.J. (2011) Investigation of the heating rate dependency associated with the loss of crystalline structure in sucrose, glucose, and fructose using a thermal analysis approach (Part I). *J. Agric. Food Chem.* 59: 684–701.

[6] Lee, J.W., Thomas, L.C., Schmidt, S.J. (2011) Investigation of thermal decomposition as the kinetic process that causes the loss of crystalline structure in sucrose using a chemical analysis approach (Part II). *J. Agric. Food Chem.* 59: 702–712.

[7] Lee, J.W., Thomas, L.C., Schmidt, S.J. (2011) Can the thermodynamic melting temperature of sucrose, glucose, and fructose be measured using rapid-scanning differential scanning calorimetry (DSC)? *J. Agric. Food Chem.* 59(7): 3306–3310.

[8]　Roos, Y.H., Franks, F., Karel, M., Labuza, T.P., Levine, H., Mathlouthi, M., Reid, D, Shalaev, E., Slade, L. (2012) Comment on the melting and decomposition of sugars. *J. Agric. Food Chem.* 60(41): 10359–10362.

[9]　Lee, J.W., Thomas, L.C. and Schmidt, S.J. (2012) Response to comment on the melting and decomposition of sugars. *J. Agric. Food Chem.* 60(41): 10363–10371.

7

FRUITS AND VEGETABLES

Guided Inquiry Activities (Web): 14, Cells and Metabolism; 18, Starch; 19, Plants; 20, Fiber and Cell Walls; 21, Plants and Color

7.1 INTRODUCTION

The world of edible plants not only includes the fruits that we know and love—apples, bananas, grapes, lemons, tomatoes and so on—but also roots, leaves, flowers, and seeds. The soft, juicy peach, the dense, opaque flesh of a potato, and the firm crunch of a celery stalk are evidence that plants can make a variety of molecules for many, many uses. Small sugar molecules and volatile aroma molecules make the peach appealing to eat, while the dense starch-filled cells of a potato store energy to last many months. The tough fibrous celery stalk is made of long polymers that hold the plant upright as it grows. Human beings have been plant eaters from the very beginning, and in fact animals depend on plants to construct complex molecules from the most basic raw materials: sunlight, minerals, air, and water. The energy of simple molecules and sunlight is stored in the more complex molecules that make up that plant material, and other organisms obtain that energy by consuming the plant. Furthermore, since plants cannot run away from predators or travel in search of the means to reproduce, they became remarkable chemists, making molecules that interact with our senses of taste, smell, and sight in delightful or unpleasant ways.

Think of when you've bitten into a fruit that has not yet ripened: a banana that is dense and not sweet or perhaps an apple that is tart or even bitter. Fruit undergoes significant chemical changes as it ripens. We must understand the different molecules that make up the structure of the plant and the chemical changes they undergo

The Science of Cooking: Understanding the Biology and Chemistry Behind Food and Cooking,
First Edition. Joseph J. Provost, Keri L. Colabroy, Brenda S. Kelly, and Mark A. Wallert.
© 2016 John Wiley & Sons, Inc. Published 2016 by John Wiley & Sons, Inc.
Companion website: www.wiley.com/go/provost/science_of_cooking

in order to understand how certain fruits like bananas or avocados will ripen after being picked, while others such as pineapples or grapes do not ripen after picking. In the same way, understanding the science of plant colors and the chemistry of plant cell wall structure will allow you to prepare tender yet vibrant green beans and avoid the limp and dull appearance of an overcooked green bean.

This chapter will begin with the chemistry and biology of plants and their cells. We will first study the molecules responsible for the structure of a plant and the molecules used by the plant to store energy. We will also examine the effects of cooking on these structural, energy storing, and colorful molecules during the preparation of delicious and appealing foods. Finally, we will consider the molecules responsible for the unique and appealing colors, tastes, and aromas of plants.

7.2 PLANT PARTS AND THEIR MOLECULES

Plants are living things that have solved the "problem of life" in a manner fundamentally different from animals (you and me!). Plants nourish themselves by building their own glucose out of water, air, and minerals and with the energy of sunlight. Animals cannot use sunlight for energy, and we cannot build our own glucose out of such simple materials. We need help in making the basic molecules of life, and thus we eat them—by eating plants or other animals.

7.2.1   Plant Cells

Since plants are living things, they are also built from *cells*. As you learned in Chapter 2, the cell of a plant shares some similarity to the cell of an animal: both contain a nucleus that holds the genetic material, and both contain a barrier or *membrane* that controls the flow of molecules in and out of the cell. In a plant cell however, the *cell membrane* is accompanied by a thick and rigid *cell wall* that provides the plant with the structural support to stand upright. See Figure 7.1 for an annotated illustration of the cell. The *primary cell wall* surrounds a plant cell as it grows and is responsible for structural integrity and mechanical strength; it is made of a flexible layer of proteins and complex carbohydrates including hemicelluloses, cellulose, and pectins. Sometimes, a stronger *secondary cell wall* is added to provide additional strength after the plant cell has stopped growing—for example, secondary cell walls are the major component of wood. Finally, a gluey layer of pectin—also called the *middle lamellae*—forms the exterior layer of the plant cell wall and helps adjacent cells adhere to one another. Cellulose, hemicelluloses, and pectins are all carbohydrate-based polymers that will be discussed in more detail in subsequent sections.

7.2.1.1 *Chloroplasts Contain Chlorophyll*   Plants are *autotrophs*. They can take energy from sunlight and store that energy in carbohydrate molecules. The green molecule *chlorophyll* not only makes plants green, but it is the molecule that captures the light energy and allows plant cells to photosynthesize. Animal cells contain

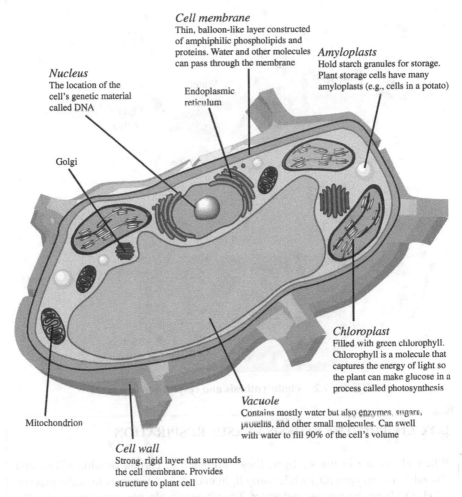

Cell membrane
Thin, balloon-like layer constructed
of amphiphilic phospholipids and
proteins. Water and other molecules
can pass through the membrane

Amyloplasts
Hold starch granules for storage.
Plant storage cells have many
amyloplasts (e.g., cells in a potato)

Nucleus
The location of the
cell's genetic material
called DNA

Endoplasmic
reticulum

Golgi

Chloroplast
Filled with green chlorophyll.
Chlorophyll is a molecule that
captures the energy of light so
the plant can make glucose in a
process called photosynthesis

Mitochondrion

Vacuole
Contains mostly water but also enzymes, sugars,
proteins, and other small molecules. Can swell
with water to fill 90% of the cell's volume

Cell wall
Strong, rigid layer that surrounds
the cell membrane. Provides
structure to plant cell

FIGURE 7.1   Anatomy of a plant cell.

mitochondria for conducting the chemistry that produces the molecules of metabolism, while in plants that role is played by the *chloroplasts* (Fig. 7.2).

7.2.1.2 Vacuoles   The plant *cell wall* is a firm but flexible container that holds mostly water. When water is abundant (Fig. 7.3), the water molecules travel into the cell's *vacuole* (through the wall and membrane), and the *vacuole* enlarges. Plants do not require water as frequently as animals due in large part to the water-filled *vacuole* that takes up most of the space inside the cell. Some vegetables contain so much water they could almost be considered water delivery devices. Lettuce contains 98% water by weight, and carrots, even though stiff and crunchy, are 88% water by weight. All this water is contained within the balloon-like vacuole. When water is abundant and the *vacuole* enlarges, all the other contents of the cell are squished against the

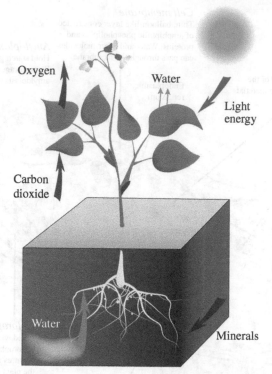

FIGURE 7.2   Photosynthesis and respiration in a plant.

BOX 7.1   PHOTOSYNTHESIS VERSUS RESPIRATION

When plants are in the sunlight, they "breathe" in carbon dioxide (CO_2) and "breathe" out oxygen (O_2) while using light energy and enzymes to make glucose ($C_6H_{12}O_6$) from the carbons and water. The glucose molecules are then used as the plant's own food—the glucose is essentially how the plant stores the energy of light (Table 7.1).

At the same time, the glucose molecules the plant makes by photosynthesis are *respired* in plant metabolism in the same way most organisms (including humans) consume glucose to run metabolism (Chapter 2). Plants take in O_2 and use it to oxidize the glucose via many enzymes—the end result is the release of CO_2, water, and energy for metabolism.

TABLE 7.1   Metabolism of Plants.

Photosynthesis	Respiration
A building-up process that stores energy	*A breaking-down process that releases energy*
$6CO_2 + 6H_2O + light \rightarrow C_6H_{12}O_6 + 6O_2$	$C_6H_{12}O_6 + 6O_2 \rightarrow 6CO_2 + 6H_2O + energy$
(carbon dioxide) (water) (glucose) (oxygen)	(glucose) (oxygen) (carbon dioxide) (water)

(a)   (b)   (c)

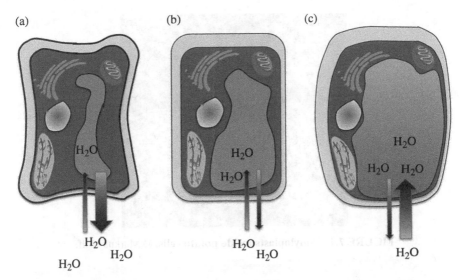

FIGURE 7.3   Plant cell, vacuole, and water. (a) The plant cell when water is low; water leaving the cell. Cell contents shrink away from the cell wall. Cell becomes limp and *flaccid*. Plant wilts. (b) Water in equilibria. (c) Plant cell when water is abundant. Water entering the cell enlarges the vacuole. Cell becomes firm or *turgid*. Plant material is firm and crisp.

cell wall, which bulges under the pressure. These water-filled, swollen cells make for a crisp, firm, *turgid* vegetable. If the cells are low on water, the *vacuole* shrinks, the pressure disappears, and cells become limp and *flaccid*. The water-rich vegetable is both crisp and juicy; the limp one is chewy and less juicy. If a vegetable has lost water, you can "reinflate" the water-depleted cells by soaking the vegetable in water for a few hours.

7.2.1.3 *Amyloplasts Store Starch*   Starches are long polymers of glucose. These long complex carbohydrates are how plants store glucose for later use, just like humans store energy in the form of fat molecules. Plants deposit *starch* molecules as a series of concentric layers in microscopic, solid *granules* that are collected into *amyloplasts* (Fig. 7.4). In plants, starch is found most often in one of two forms: *amylose* or *amylopectin*. The composition of plant starch is typically 20–25% amylose and 75–80% amylopectin, and the starch is packaged into granules of various shapes and sizes depending on the species of plant (Fig. 7.5). The starch content and granule size also impact the cooking qualities of the starch.

Putting the organelles of a plant together with the plant organ or type of vegetable material, Table 7.2 describes the type of plant, its organs with function, and the type of material responsible for the actions of the plant.

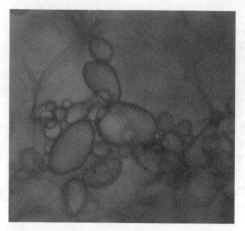

FIGURE 7.4   Amyloplasts inside potato cells. © Markus Nolf.

(a)   (b)

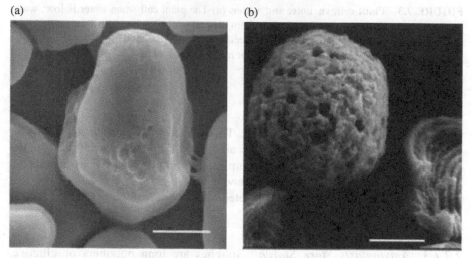

FIGURE 7.5   Microscopic view of starch granule. (a) A powerful microscope image of a starch granule. (b) A starch granule is exposed to the starch-degrading enzyme, alpha amylase. http://openi.nlm.nih.gov/faq.php#copyright. Used under Creative Commons Attribution License. J Insect Sci 2009; 9:43.

7.3   PLANTS ARE COMPRISED OF DIFFERENT TYPES OF COMPLEX CARBOHYDRATE

Carbohydrates are molecules made up of one or more sugars. The most common sugar we find in food is *glucose* (i.e., dextrose)—$C_6H_{12}O_6$ (Fig. 7.6) *Starch* is a *carbohydrate* and one of the polymers of glucose that you will find in plants (Fig. 7.7).

TABLE 7.2 Plant Organs by Function, Tissue, and Cell Type.

Organ		Function	Plant Material and Cell Type	Example
Vegetables	Roots	Anchor plant into ground; absorb nutrients	Tough fibrous material—cells have thick, cellulose-rich *cell walls*. Some roots swell up with storage cells full of *amyloplasts*	Inedible Carrots, parsnips, radishes, sweet potatoes
	Stems, stalks	Conduct nutrients to roots and leaves; give structural support	Fibrous material (stems, stalks)—cells have thick, cellulose-rich *cell walls*	Asparagus stems, celery stalks
	Tubers, rhizomes		Some stems swell up with storage tissue (tubers, rhizomes)—cells are full of *amyloplasts*	Potato, turnip, ginger
	Leaves	Produce sugar molecules by photosynthesis	Plant material is thin so gases can penetrate/escape. Almost no structural support—*cell walls* are thin and flexible. Cells have many *chloroplasts* and large air pockets between them for gases	Lettuce leaves, spinach leaves
	Flowers	Plant's reproductive organs	Brilliantly colored and scented. Cells are full of *chloroplasts* and *chromoplasts* (contain other colored molecules besides green), and the *vacuoles* contain molecules that give the flower a smell	Broccoli, cauliflower, artichoke
Fruits		Surround seed(s); sometimes the seed is a "pit"	Almost entirely storage tissue. Cells contain thinner *cell walls*. There are *amyloplasts* and storage *vacuoles* full of sugars and good smelling/tasting molecules	Apples, peaches, snow peas, and so on
Seeds		Contain plant embryo and its food supply	Tough outer layer—cells have thick cell walls with cellulose and lignin. Inner storage cells contain many *amyloplasts*, as well as proteins and oils	Wheat kernel, peas, corn, lima bean, nuts

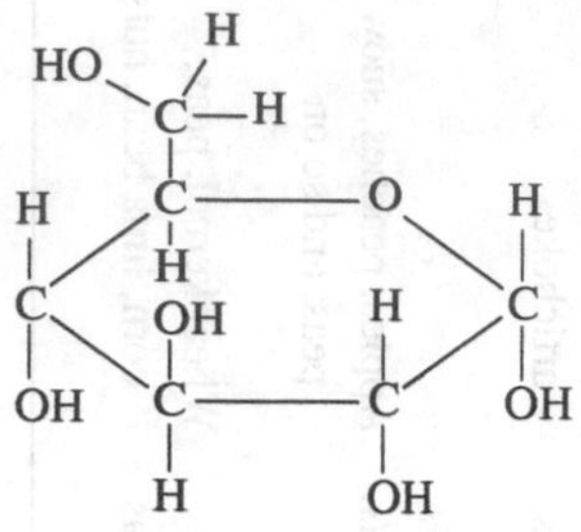

Glucose is a simple sugar—it contains only one *ring* and is rapidly metabolized in the body for energy

Starch is made of chains of glucose molecules. To utilize the starch as a source of energy, we must break the large starch molecule down into glucose molecules.

Glucose (a.k.a dextrose)

FIGURE 7.6   Glucose, a simple sugar.

7.3.1   Carbohydrates for Energy Storage

Living things need glucose for energy, and plants are no exception! Fruits and vegetables are parts of plants, and plants store their glucose in a chain form until it is needed for energy. Root and tuber crops are particularly good examples of starchcontaining plant organs. Beets, carrots, and potatoes store starch underground to survive the winter months.

If you examine the structure of *amylose* or *amylopectin* in Figures 7.7 and 7.8, you will notice that each is a combination of repeating units. The same molecular piece is repeated over and over again. *Amylose* and *amylopectin* are *polymers*—where *poly* = "many." The unit of repetition is called a *monomer*. In *amylose* and *amylopectin*, the *monomer* unit is a glucose. Glucose polymers are also called *polysaccharides*—where the root *saccharide* comes from the Latin *saccharum* meaning "sugar." Amylose and amylopectin are *both* polymers of the sugar glucose, but they differ in how the glucose monomer units are connected. In amylose, the glucose monomers are connected at the 1 and 4 positions or in a "head to tail" arrangement. In amylopectin, there are 1→4 or "head to tail" connections, but there are also 1→6 branches. As shown in Figure 7.3, a typical amylose polymer will be made of approximately 1000 glucose monomers, each connected 1 to 4 in a single, long chain. A typical amylopectin polymer is comprised of 5000–20,000 glucose monomers connected in many 1→4 chains that are linked through 1→6 branches. The enzyme *alpha amylase* is able to break starch molecules down into individual glucose units. In addition, the human body can break down *amylose* and *amylopectin* into individual glucose molecules using *enzymes* called *glycosidases*. The consequence is that an amylopectin polymer is very bushy. As we will discuss in the following, the different shapes of the amylose and amylopectin polymer impact the texture of the food when it is cooked.

7.3.2   Carbohydrates for Cell Walls and Cell Glue

The cell wall is responsible for the strength of the individual cells and ultimately the plant itself. The cell wall is made of complex carbohydrates (also known as polysaccharides) that form an armored shell around the cell's contents. When coupled with the cell glue—also made of complex carbohydrates—the cell walls of many cells

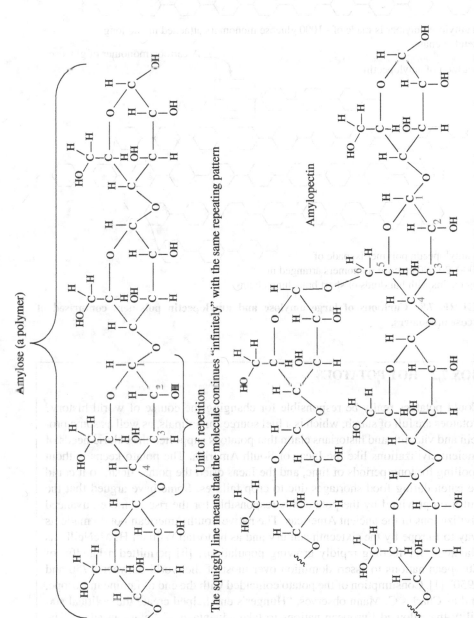

FIGURE 7.7 Amylose and amylopectin. Two types of starch. The numbers indicate how chemists count the atoms in each glucose monomer and distinguish how amylose and amylopectin are connected.

A cartoon of amylose

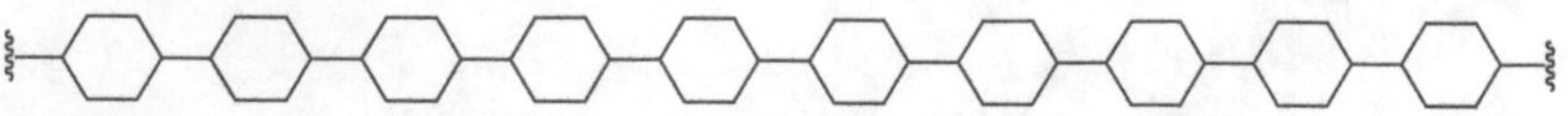

An amylose polymer is made of ~1000 glucose monomers attached in one long extended chain

A cartoon of amylopectin

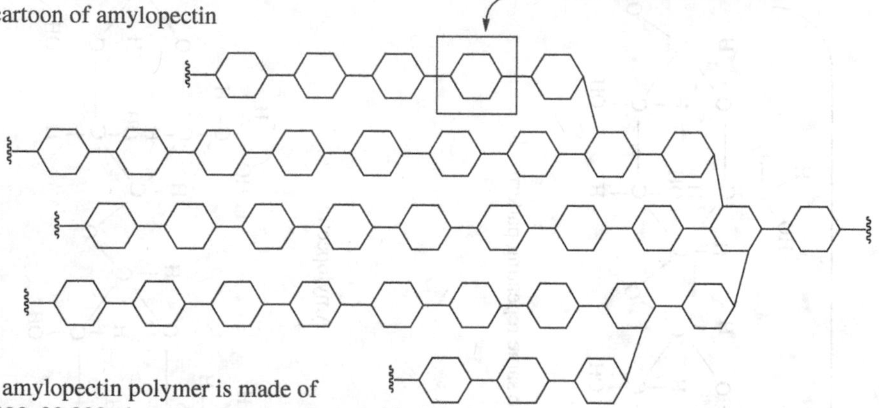

A cartoon monomer of glucose

An amylopectin polymer is made of ~5000–20,000 glucose monomers arranged in long chains with hundreds of short branching chains

FIGURE 7.8 **Cartoons of large amylose and amylopectin polymers comprised of glucose monomers.**

BOX 7.2   HOT POTATOES

Could potatoes really be responsible for changing the course of world history? Potatoes are full of starch, which is a fuel source for animals, as well as some protein and vitamin, and historians claim that potatoes supported the caloric needs of ancient civilizations like the Incas of South America. The potato keeps without spoiling for long periods of time, and the Incas dried the potato flesh, so it could be eaten during food shortages due to crop failures. Some have argued that the nutrition provided by the potato was responsible for the rise of more advanced civilizations in the ancient Americas. The native South American potato made its way to Europe by the sixteenth century and as historian William H. McNeill has claimed, "By feeding rapidly growing populations, [it] permitted a handful of European nations to assert dominion over most of the world between 1750 and 1950" [1]. Consumption of the potato coincided with the end of famine in Europe, and as Charles C. Mann observes, "Hunger's end helped create the political stability that allowed European nations to take advantage of American silver. The potato fueled the rise of the West."

Today, there are over 5000 varieties of potato, and potatoes are a major worldwide crop (fourth or fifth most important). Russet or red or Yukon or white, potatoes come in a variety of shapes, colors, and sizes. All potatoes are a type of plant organ called a tuber—or a swollen area of the stem that specializes in the storage of starch. The cells of a potato contain many *amyloplasts* full of starch granules. Picking the right potato for your cooking need depends on the type and quantity of starch within the tuber. Potatoes can be divided into two basic types: *boiling* or *waxy* potatoes that hold together well when heated versus *baking* or *floury* potatoes that fall apart when cooked. It is the quantity and type of starch that make the difference between waxy and baking potatoes. Russet potatoes are the classic example of a baking potato— they are ideal for making fries and for mashing. These potatoes are high in starch (up to 22%); the starch of the baking potato contains 26% of the straight amylose and 74% of the branched amylopectin. The cells of baking potatoes have weaker cell walls, and the cells are densely packed with starch granules that swell (gelatinize) easily when they come in contact with water. As baking potatoes are cooked, the cell walls and vacuole walls break open, leaking starch and water, and the cells separate from one another. The higher calcium content of a baking potato helps the starch to readily form a soft solid or gel [2]. The result is the fluffy or mealy texture of a baked potato. In contrast, waxy potatoes maintain their shape during and after cooking and are great additions to soups and salads. Since baking potatoes have a mixture of starch types and higher levels of total starch, they gel at lower temperatures (58°C/136°F) than boiling potatoes, which do not gel until 70°C (158°F). The low gelation temperature of the baking potato starch means the starch absorbs water and swells sooner in the cooking process, which puts pressure on the cell walls. The starch of a boiling potato has a different composition and a higher gelation temperature. A boiling potato also has more pectin and cellulose holding its cells together, so the cells do not burst open or separate as readily. These factors help the boiling potato maintain its shape during cooking. Some potatoes like the yellow or Yukon Gold have starch composition and cell wall strength that are inbetween classic baking or boiling potatoes, making these a versatile option for cooking. Now, when selecting a potato for baking, mashing, salads, or boiling, you can be sure to pick the right type!

connect; the strength of these many connected cell walls actually allows the plant to grow vertically *against* the forces of gravity!

All three layers of the plant cell wall are made of some or all of the following four components: cellulose, pectins, hemicelluloses—all carbohydrate polymers called *polysaccharides*—and a tough, woody polymer called lignin. The primary cell wall is typically a matrix of the three-polysaccharide polymers (i.e., cellulose, hemicelluloses, and pectins), which provides structure and mechanical strength to the cell. The construction materials for making a reinforced concrete wall are often used to describe the plant cell wall and matrix. When building a reinforced concrete wall, steel bars are embedded within a matrix of pourable concrete. In the same way, the plant cell walls have "bars" and "cement" [3]. The bars are the tough fibers of *cellulose* that act as a structural framework. In some plant cells that have a secondary cell wall, the very rigid *lignin* is also present, but most of the time we don't eat lignified plants (i.e., you don't chew tree bark). The tough lignin polymer is added to create the secondary cell wall only after the cell has stopped growing. Lignified cell walls are thick and rigid and can persist even after the cell has died. The "cement" is a semisolid, flexible mixture that fills the space inbetween the *cellulose* fibers and is made mostly of the soluble fiber components *hemicellulose* and *pectin*. When neighboring cell walls are close together (as in the tissue of a plant), a gluey cement material holds the cells together. The layer of cell glue or the middle lamellae is comprised of mostly pectin.

We encounter the cell walls of plants as the *fiber* in our diets. The fibers in celery stalks, the stone that surrounds a peach seed, and the seedpods of beans and peas are mostly cell wall material. Overall, fiber is defined as material in our plant foods that our digestive enzymes can't break down into absorbable nutrients. Nutritionally, we divide dietary fiber into *soluble* and *insoluble* types. Soluble fiber is dissolved in water, while insoluble fiber is not. Both forms do not break down in our stomach or intestines and instead pass through undigested. It may seem surprising that cell wall material is indigestible, when in fact, it is comprised of carbohydrate, just like starch! As we learned earlier, cellulose forms the "bars" of our reinforced concrete wall and therefore helps to form the structure that holds the plant up. Cellulose may be one of the *most abundant* glucose polymers on earth, but it is also *insoluble fiber*—indigestible carbohydrate. So why can't we digest cellulose?

Cellulose is another *polymer* of glucose, a *polysaccharide*—just like the starch polymers amylose and amylopectin. If you examine the structures of amylose and cellulose in Figure 7.9, you can see that—once again—the *type of connection* between the glucose monomers differs between amylose and cellulose. Both are linear, $1\rightarrow4$ connections of glucose, but the direction of the $1\rightarrow4$ connection differs for the two polymers. The direction of the connection can be assessed from the *glycosidic carbon*—the carbon that is bonded directly to two oxygen atoms. In Figure 7.9, the glycosidic carbons are in boldface type. In amylose, the oxygen atoms make the $1\rightarrow4$ connection point "down" from each *glycosidic carbon*—this is called an alpha ($1\rightarrow4$) or $\alpha(1\rightarrow4)$ linkage. In contrast, the oxygen atoms making each $1\rightarrow4$ connection in cellulose point "up"—this is called a beta ($1\rightarrow4$) or $\beta(1\rightarrow4)$ linkage. This seemingly insignificant difference between $\alpha(1\rightarrow4)$ and $\beta(1\rightarrow4)$ linkages is the reason that the carbohydrate polymer amylose is a digestible source of nutrition for humans, while cellulose is an indigestible component of *fiber*. The $\beta(1\rightarrow4)$ linkage of cellulose polymers is a bond that human enzymes *cannot* break, and, therefore, humans cannot digest cellulose. In fact, most animals cannot digest cellulose, but microorganisms like bacteria and fungi *can*, which is how plant material rots in nature.

Lignin is another component of *insoluble fiber*, but unlike the other components of fiber, lignin is not a polysaccharide. Lignin is a polymer of one of three types of *monolignols*. *Monolignols* are made from the amino acid phenylalanine, and so their structure is fundamentally different from the monomer glucose. The polymer lignin is very strong and tough; it gives mechanical strength to the cell wall, but due to its tough consistency, lignified plants are not typically eaten.

Hemicellulose and pectins are *soluble* fiber—this means they dissolve in water to make a clear, jelly-like semisolid. Hemicelluloses and pectins are also *polysaccharides*,

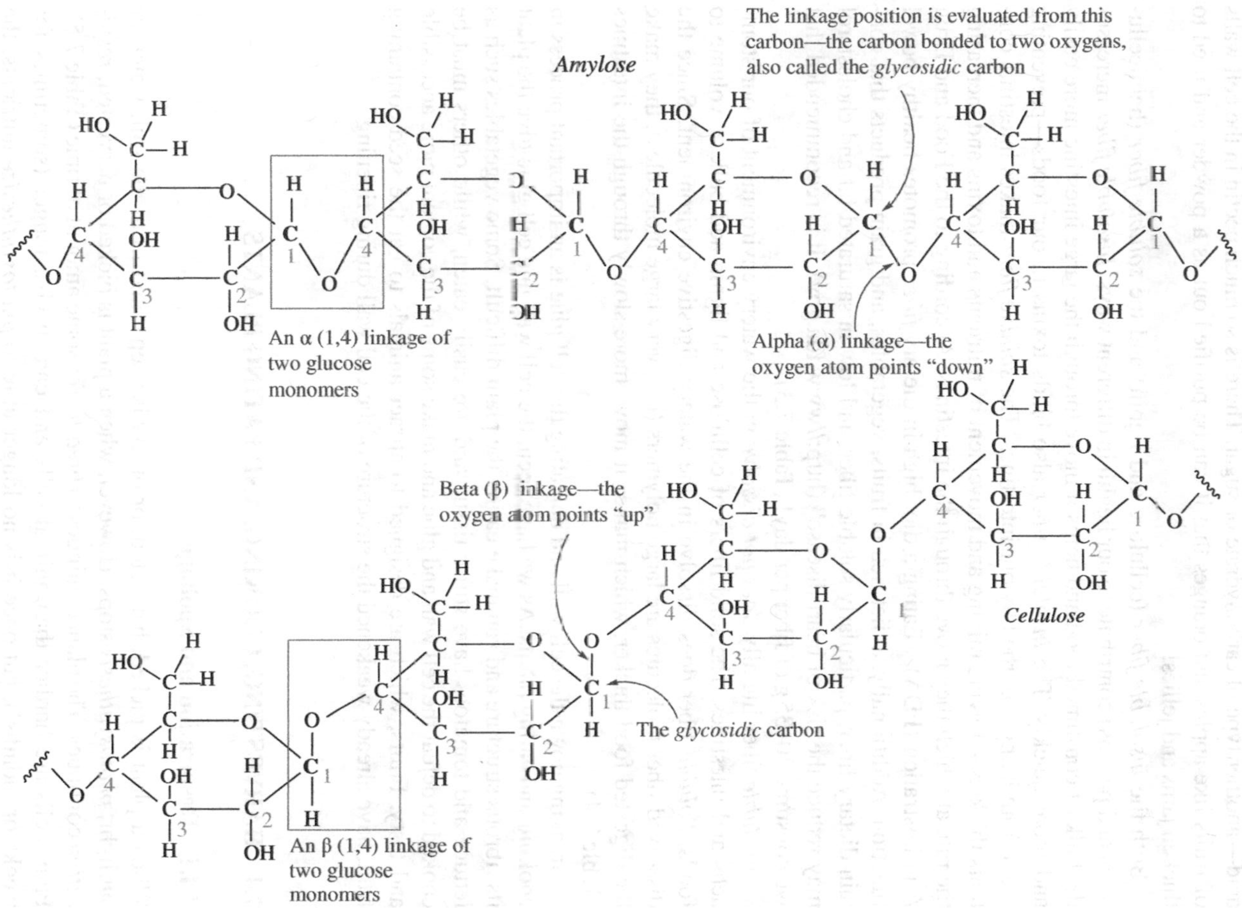

The linkage position is evaluated from this carbon—the carbon bonded to two oxygens, also called the *glycosidic* carbon

Alpha (α) linkage—the oxygen atom points "down"

Amylose

An α (1,4) linkage of two glucose monomers

Cellulose

The *glycosidic* carbon

Beta (β) linkage—the oxygen atom points "up"

An β (1,4) linkage of two glucose monomers

FIGURE 7.9 Glycosidic bonds of amylose versus cellulose.

but the monomer units of these polymers are not glucose. Hemicelluloses are also made of $\beta(1{\rightarrow}4)$ linkages of sugars such as xylose. Again, the $\beta(1{\rightarrow}4)$ linkages make hemicelluloses indigestible. Pectins are mostly polymers of a modified galacturonic acid—another type of carbohydrate or sugar. There is so much pectin in the cell walls of fruits like apples and oranges that it can be purified out as a powder and used to thicken jams and jellies!

Both the *insoluble fiber* (cellulose and lignin) and the *soluble fiber* (hemicelluloses and pectins) contribute to our health in different ways. *Insoluble fiber* increases the bulk of consumed food and helps it move through the large intestine more easily and more quickly. The *insoluble fiber* also binds toxins in our foods—preventing them from being absorbed by our intestines. The *soluble fiber* makes intestinal contents thicker—slowing mixing and movement of nutrients and toxins and increasing the rate at which they move through the intestines. According to the Food and Drug Administration (FDA), "Eating a diet high in dietary *fiber* promotes healthy bowel function. Additionally, a diet rich in fruits, vegetables, and grain products that contain dietary fiber, particularly soluble fiber, and low in saturated fat and cholesterol may reduce the risk of heart disease" (http://www.fda.gov). It is recommended that we consume 20–35 g of fiber per day! (Table 7.3).

Insoluble fiber literally *does not dissolve* in the watery environment of our stomachs and intestines. The solid fibers of cellulose and lignin add bulk or volume to foods. *Soluble* fiber does dissolve in the watery digestive environment. Since the dissolved fiber molecules are large polymers (i.e., very large molecules), they make the digested food thicker, which makes it move more slowly through the intestines (Table 7.4).

Breaking down the cell wall and defeating the cell glue is an important process in cooking and eating plants. As we have seen, the cell wall and cell glue give the plant its fibrous structure and can make eating the plant difficult. Some vegetables such as lettuce and tomatoes have softer tissue and are easily eaten, while others must be cooked to defeat the cell wall and glue and make some of the content more accessible and tasty. Fruits, which are designed to attract animals to eat the seed-containing fruit, have already weakened the structure of the cell wall during ripening.

7.4 HARVESTING, COOKING, AND EATING PLANTS

7.4.1 Respiration and Spoilage

When a plant is picked, harvested, or otherwise separated from the soil, water, or sunlight, *photosynthesis* stops. However, when a plant is picked or harvested, *respiration* continues. The plant continues to *breathe* for some amount of time (Table 7.5). Plant cells are hardier than animal cells and can survive longer (sometimes for weeks or months!), but once it is no longer able to *photosynthesize* nutrients, the plant will eventually *respire* away all its stored energy and accumulate waste products, and the cells will eventually die. We perceive this as the loss of flavor and changes in texture that occur with deterioration and spoiling of produce. Bacteria,

TABLE 7.3   The Components of Dietary Fiber: Indigestible Plant Material.

Component	Soluble/Insoluble	Polymer of Carbohydrate Monomers	Common Food Sources
Cellulose	Insoluble	Glucose	The stringy part of celery, wheat bran
Pectins	Soluble	Galacturonic acid (a carbohydrate)	Fruit (e.g., apples and citrus fruits)

Grayed out atoms are modifications of the galacturonic acid structure

Pink numbers indicate the 1,4 connection points in the polymer

(Continued)

TABLE 7.3 (Continued)

Component	Soluble/Insoluble	Polymer of Carbohydrate Monomers	Common Food Sources
Hemicellulose	Soluble	Pink numbers show the points of connection in the polymer 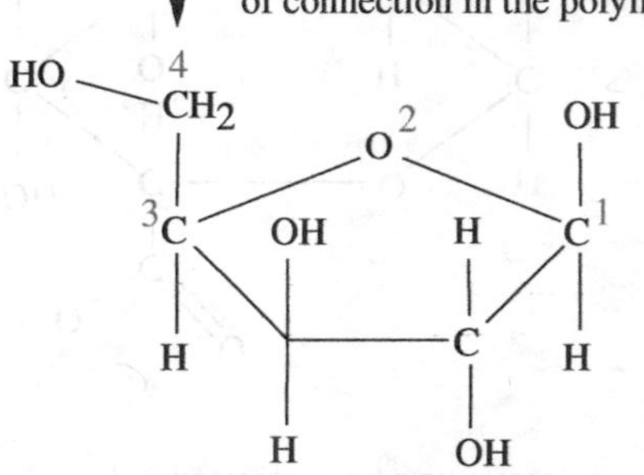	Many fruits, veggies, and grains
Lignin	Insoluble (major component of wood)	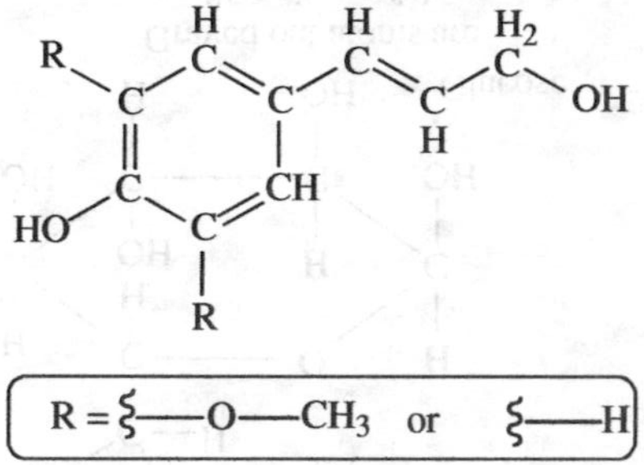 Xylose and other small carbohydrates	Woody asparagus and broccoli stems, the gritty parts of pear and guava flesh

Monolignols

$R = $ —O—CH_3 or —H

TABLE 7.4 Calories in Protein, Fat, and Carbohydrates.

Macronutrient	cal	kJ
One gram of protein	4	16.7
One gram of fat	9	37.7
One gram of carbohydrate	4	16.7

While digestible carbohydrate like amylose and amylopectin yields 4 cal per gram, the indigestible carbohydrate in *fiber* does not yield any calories—we cannot break it down for energy. So when considering a nutrition label, the dietary fiber should not count toward the total calories.

TABLE 7.5 Postharvest Respiration Rates in CO_2 ml/kg/h at 59°F/15°C.

Potato (cured)	17	Pear (Bartlett)	16 (after 1 day in storage)
Grape	16	Pear (d'Anjou)	9 (after 1 day in storage)
Lemon/orange	18–19	Lettuce	39 (head), 63 (leaf)
Apple	15 (fall), 25 (summer)	Brussels sprouts	200
Cabbage	28	Banana (ripe)	140
Carrot (topped with greens)	40	Asparagus	235
Peach	87 (at 20°C)	Strawberry	114

Data from agriculture handbook number 66 (HB-66) by the USDA and the agricultural research station.

yeasts, and molds can also accelerate spoilage by attacking the weakened plant cells and consuming their cell contents. How quickly a plant succumbs to spoilage is dependent on how fast it *respires* after harvesting. The more perishable a fruit or vegetable, the faster its *respiration rate will be*. Plants with high rates of *respiration* will consume all their available stored energy (the glucose and glucose polymers, starch) faster and die sooner.

The key to preserving the *lifetime* of the produce is to slow the rate of *respiration* (how fast the plant cells take up oxygen and release carbon dioxide). The most effective strategies for slowing *respiration* are to keep the plant cold and limit its oxygen supply. Plastic bags can limit oxygen but also trap moisture, which can encourage the growth of mold. Food suppliers understand the impact of respiration on ripening (as seen later in this section) and will store and transport food in low oxygen conditions (which limit the respiration) and cold temperatures to slow down respiration of fruits and vegetables. Unfortunately many tropical and subtropical fruits and vegetables cannot tolerate cold temperatures, resulting in bruised, brown fruit that may not ripen at all. However, there are many plants including bananas, berries, cabbage, avocados, pears, and onions that survive at very low oxygen levels (the gas is often replaced with carbon dioxide or an inert gas such as nitrogen) allowing for commercial worldwide shipping.

7.4.2 Ripening

Eating immature fruit is not a pleasurable experience. The fruit is dense, hard, and not sweet tasting. Ripening is an intentional, dramatic, and final phase of life for a fruit; it is the process that makes a fruit edible. We see this process occur as the tart, hard fruit softens and sweetens. As a fruit ripens, enzymes convert some of the starch (a tasteless glucose polymer) into free glucose molecules. Some of the glucose is further metabolized into fructose and other monosaccharides by other metabolic enzymes. The conversion of starch into these smaller sugars (i.e., monosaccharides) gives the mature fruit its characteristic sweet taste. Other enzymes break down the pectin in the primary cell wall, softening the rigid structure. When you bite into a ripe fruit—for example, a ripe peach—the flesh of the fruit is easily crushed, and "juice" squirts and drips everywhere. When a fruit is ripe, the weakened cell walls of the fruit give way to the pressure of your teeth, and the water- and sugar-filled vacuoles burst (i.e., the "juice"). The color changes of ripening include the degradation of green chlorophyll that reveals the orange and red carotenoid pigments that give fruits like bananas and tomatoes their characteristic color and the formation of anthocyanin pigments that impart pink, red, purple, and blue colors to fruits such as strawberries and blueberries. The pleasant aromas associated with ripening are the result of molecules made by the plant during ripening and include aldehydes, ketones, alcohols, esters, and terpenes. These volatile aroma molecules are made from a variety of proteins, carbohydrates, lipids, and vitamins present in the plant, and each species of plant has a characteristic blend of aroma molecules that give it a unique flavor (see Section 7.6). For a more comprehensive discussion of flavor, please see Chapter 2. The energy needed for these chemical reactions comes from the ATP produced in respiration. These changes are in marked contrast to vegetables that do not ripen in the same way that plants do. After harvest, vegetable respiration is slow and eventually fades to very low levels. Since vegetables do not soften and release flavor molecules on their own, we use cooking to soften the plant cell walls and break open the cells releasing flavor molecules.

The ripening of fruits occurs by one of two mechanisms. Those fruits for which respiration dramatically increases after becoming detached from the parent plant are considered *climacteric*. Bananas, apples, mangos, pears, peaches, avocados, and tomatoes are all examples of *climacteric ripeners*. Remember, a green banana will ripen to a yellow color while in your kitchen … far from the banana tree from which the fruit was harvested. Other fruits maintain the same rate of metabolic respiration after harvest, and these are called *non-climacteric*. *Non-climacteric* fruits include blueberry, cucumber, watermelon, peas, lemon, pineapple, strawberry, orange, grape, and grapefruit; these fruits will not ripen once they have been harvested.

The rise in production of CO_2 observed for *climacteric* fruits is due to dramatically increased respiration after harvest. This change in respiration is associated with the production of a simple gas called ethylene (Fig. 7.10). Not only was ethylene the first plant hormone ever discovered, it was also the first gas discovered to act as a hormone—essentially, ethylene is produced at one location in the plant, and then it travels through the plant to act on the fruit. Ethylene is produced within the plant

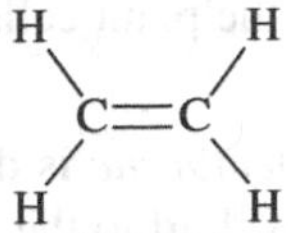

Ethylene
(C$_2$H$_4$)

FIGURE 7.10 Ethylene, the ripening plant hormone.

from the breakdown of the amino acid methionine, and it induces dramatic ripening. Once stimulated by ethylene, the fruit will increase its *respiration rate* several fold, and the color, texture, and taste will change rapidly. The coordinated ripening of large amounts of fruit at one time is thought to attract and encourage animals to ingest and spread the seed found within the fruit. Farmers and grocers have found that *climacteric fruits* can be harvested while still unripe ("green") and will ripen on their own by producing the *ethylene* gas. If the *ethylene* is trapped around the fruit, or the plant is intentionally exposed to *ethylene*, it will ripen much faster. For example, a common recommendation to accelerate fruit ripening is to place the fruit in a sealed, brown paper bag. This trick only works for climacteric fruits and depends on trapping the ethylene gas around the fruit to accelerate its ripening. In the same way, bananas are grown in Central and South America and shipped to North American grocery stores while still green. Exposure of these green bananas to ethylene gas produces rapid ripening before the bananas are placed on your grocery store shelves. Those mystery green bags that keep fruit fresh? They are reported to bind ethylene produced by the plants limiting the overripening. Unfortunately, these strategies will not work on *non-climacteric* fruits; they do not ripen after harvest.

In addition to ripening, ethylene controls other aspects of plant growth, development, wounding responses, and cell aging. The hormone signals to activate many genes in the plant responsible for the production of enzymes and other plant components. Many of the ethylene-sensitive genes code for proteins involved in metabolism and its regulation. Many of these genes alter the metabolism of fats and proteins as well as carbohydrates.

7.5 COOKING PLANTS

7.5.1 Changes to the Cell Wall during Cooking

During ripening, when a fruit or vegetable is cooked, the changes in texture that we observe are caused by changes in the *cement* molecules—the *hemicelluloses* and *pectin*. If the *cement* is weak, the cells will easily separate from one another when chewed. In the wild, animals have developed sharp teeth for tearing, along with flat teeth to grind the fibrous plant matter and strong jaw muscles to pulverize vegetable material. In the same way, humans have evolved modern strategies to break down plant material and make it easier to chew—heat and acid are two effective means of breaking down plant cell walls and turning flora into food.

When cooking a plant with heat, the plant cells break down in several ways:

- Above 140°F/60°C, the *cell membrane* is damaged and disintegrates, the *vacuole* ruptures and the cell loses fluid to the surroundings. The cells deflate and become *flaccid*, but the *cell walls* are still intact and will give the material a firm texture while chewing. This is why after sautéing onion, carrot, and celery for a few minutes in a little olive oil, water appears in the skillet. This is called "cooking out" the vegetables or cooking vegetables until they "release their juices." If the vegetable is boiled, the sugars and flavor molecules found in the vacuole are lost to the boiling water. Adding table sugar to potato or corn while boiling can help retain some of the sweet flavor that is lost in this way.
- As the plant material is heated hotter and longer, the *cell walls* begin to break down, and the cells begin to separate from one another.
- When the plant material contains *amyloplasts*, heating causes the starch granules within the *amyloplasts* to swell as the starch absorbs the water released by the disintegrating vacuole—such as the starch granules of a potato, for example. This is why cooking a potato results in a pasty or mealy texture—not juicy. Extended cooking can cause the starch to leak out of the damaged cells and/or gel into a soft solid.

When plant tissue is heated above >140°F/60°C, the cell walls begin to break down. The *cellulose* is resistant to heat and chemicals and remains largely unchanged. However, the polymers *hemicellulose* and *pectin* gradually break down into shorter molecules. These shorter pieces of *hemicellulose* and *pectin* dissolve in the cooking liquid, and the cell wall *cement* literally disintegrates. With no *cement*, the cells easily separate from one another, and the plant tissue becomes tender. If you keep boiling the plant material, eventually all the cell wall *cement* will be destroyed, and you'll be left with a *puree*.

The cell wall cement-dissolving part of fruit and vegetable tenderizing is dependent on the cooking environment:

Even though hemicellulose is a component of *soluble fiber*, it is not very *soluble* in acidic liquids. Therefore, in acidic conditions, hemicellulose resists breaking down into shorter molecules and dissolving away. When vegetables/fruits are steamed, very little additional fluid is added. The plant material is exposed mostly to the mildly acidic cell fluid itself and to the steam, which has a slightly acidic pH of 6. Consequently, steamed veggies retain their firmness. The reverse is true for alkaline conditions—which accelerate hemicellulose breakdown. Fruits/vegetables cooked in alkaline conditions quickly become mushy (Table 7.6).

The *calcium ion* is able to *cross-link* pectin molecules together—strengthening their interactions and making the cell wall cement stronger. Pectin chains are made of galacturonic acid monomers that are joined together in a polymer chain. As shown in Table 7.6 and Figure 7.11, the galacturonic acid monomer of pectin is sometimes modified by esterification, but within a pectin chain, some of the monomers are modified, and some are not. As we saw in Chapter 1, an ester is formed from the reaction

TABLE 7.6 The Effect of Cooking Liquids on Vegetable/Fruit Firmness.

Condition	Cooking Liquids	Effect on Vegetable/Fruit
Acidic conditions (pH<7)	Tomato sauce, fruit juices/ purees	Maintains firmness for longer
Alkaline conditions (pH>7)	Baking soda (NaHCO₃) added to water	Softens more readily
Neutral conditions (pH~7)	Plain water	Softens readily
High calcium	"Hard" water (i.e., water that contains CaCO₃)	Maintains firmness for longer
High sodium	Table salt (i.e., NaCl) added to water	Softens readily

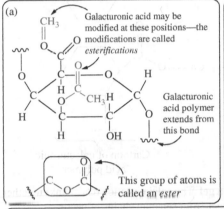

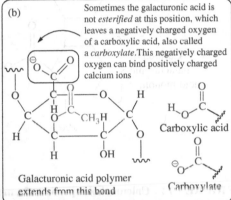

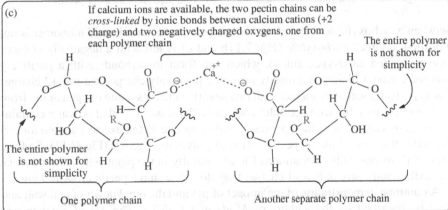

FIGURE 7.11 Modification of pectin. (a) Pectin is comprised of polymers of galacturonic acid, which are sometimes modified by esterification (gray atoms). (b) and (c) If the galacturonic acid is *not* modified by esterification, the free oxygen anions from carboxylic acid groups of two separate polymer chains can interact with positively charged calcium ions (Ca^{2+})—linking the two chains together.

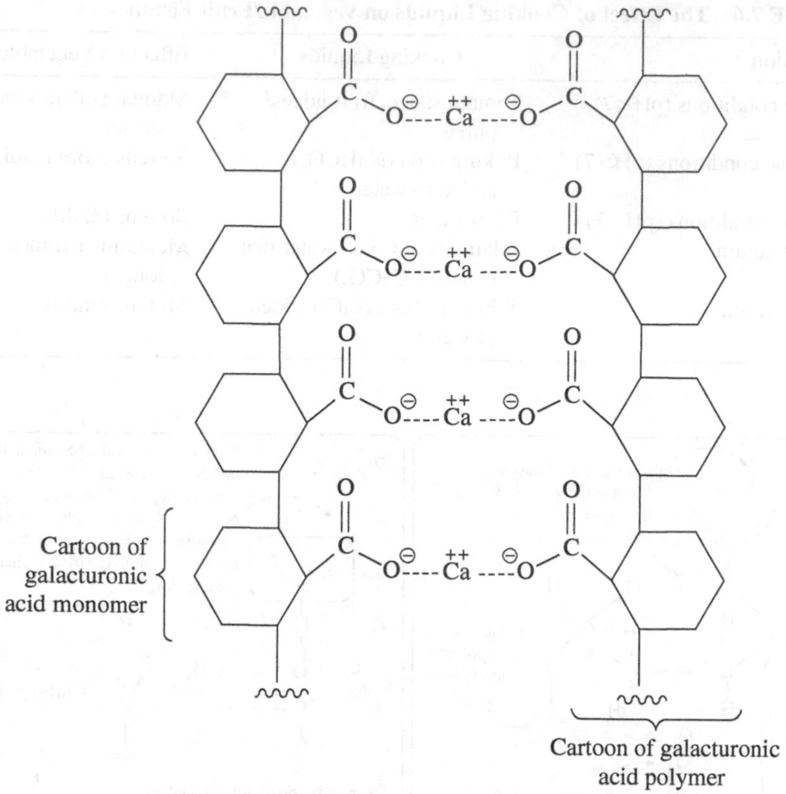

Cartoon of galacturonic acid monomer

Cartoon of galacturonic acid polymer

FIGURE 7.12   Calcium holds pectin to form a gel. The calcium cation (Ca^{2+}) links together two oxygen anions of a galacturonic acid monomer from separate chains of pectin.

between a carboxylic acid and an alcohol. When a galacturonic acid monomer is not modified, a "free" *carboxylate* (Fig. 7.11b and c) is left with a negatively charged oxygen (called an oxygen anion), which can form ionic bonds with a positively charged cation. Calcium is a readily available positively charged cation (+2 charge) that is perfectly suited to interact with *two* negative charges—one oxygen anion from each pectin chain. In fact, vegetables cooked in "hard water"—that is, water containing *calcium carbonate* ($CaCO_3$)—resist softening due to the excess *calcium ions* in the water that cross-link the pectin cell wall polymers together (Fig. 7.12). Sodium ions (Na^+) displace the calcium ions found naturally in the plant cell wall, and since the sodium ions have only a +1 charge, they do not promote pectin cross-linking.

A common demonstration of the impact of pH and the breakdown of cell wall and glue is described by the authors of *Modernist Cuisine: The Art of Science and Cooking*. In their book, the impact of pH on cooking beans is described by boiling raw beans under three different conditions: (i) water containing vinegar (acidic conditions), (ii) distilled water (slightly acidic to neutral), and (iii) water with baking soda (alkaline conditions). Under the alkaline conditions, the cell wall cement and cell glue molecules—pectins and hemicelluloses—dissolve readily, and the beans

BOX 7.3   SPHERIFICATION

Turning a flavorful liquid into a sphere that pops in your mouth is an invention of modern kitchen *alchemy* that uses the principles of calcium cross-linking of polysaccharides such as the calcium cross-linking of pectins discussed in this section. In the process of *direct spherification*, a flavorful liquid is blended with sodium alginate powder. The alginate is a polysaccharide (similar to pectin) derived from seaweed. In direct spherification, the liquid–sodium alginate mixture is then dripped into a bath of calcium chloride (Fig. 7.13a). The calcium ions displace the sodium and cross-link the chains of alginate to make a thin, flexible solid around the sphere of liquid. The result is a liquid encapsulated in a calcium cross-linked film of alginate. In *reverse spherification* a calcium-containing liquid is dripped into a bath of sodium alginate (Fig. 7.13b). Either method produces the liquid-filled spheres. In many ways, the spheres resemble the roe or fish eggs used to make caviar. The spheres are an attractive way to decorate dishes and add bursts of flavor in unexpected ways.

(a)   (c)

(b)

FIGURE 7.13   Modern chemistry in the kitchen. (a) Direct spherification. (b) Reverse or indirect spherification of liquid flavors. (c) Students preparing their own creation of a direct spherification.

quickly become soft and mushy. The beans cooked in acidic solution only slightly soften as the cell wall cement and glue are left largely intact. Finally, the beans cooked in the distilled water (pH ~7) soften at the best rate as the cell wall cement slowly disintegrates; a brief cooking time is all that is necessary for a tender vegetable. To minimize the impact of the slightly acidic cell contents, the beans should be cooked in a large pot of water. This dilutes the plant acids as the cells break down and

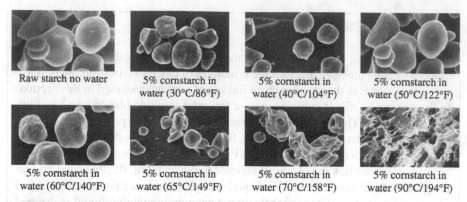

FIGURE 7.14   Starch in water at increasing temperatures. Dr. ZoeAnn Holmes, Professor Emeritus. Reproduced with permission of Oregon State University.

release their contents and thus reduces the time needed to soften the cell wall. As we will learn later in this chapter, acidic conditions will also degrade the green color of green beans, and cooking in a large pot of water can take care of both issues!

7.5.2   Changes to Starch during Cooking

When we eat fruits and vegetables—including starchy vegetables like potatoes or corn—we are consuming starch that our bodies later break down into glucose for energy. When we cook starchy vegetables like potatoes or corn and grains like rice, the methods we use make the starch polymers more pleasant to eat.

When starch granules are mixed with water at room temperature, not much happens to the starch. But if the starch in water mixture is heated, eventually the temperature gets high enough that the granule begins to swell up as water penetrates into the center (Fig. 7.14). Eventually, the granule absorbs enough water and swells to such a degree that it breaks apart into a network of starch molecules mixed with water. These networks of long starch molecules with water in between are called *gels*. The temperature at which the starch granule falls apart into this matrix of water and starch molecules is called the *gelation temperature* (Table 7.7).

The *gelation temperature* of starch is dependent on the kind of starch it is. The amount of amylose, the size of the amylose chains, and the size of the starch granule are all factors in the thickening power. Amylose forms stronger *gels* than amylopectin—this is because amylose has a linear (thread-like) structure, while amylopectin has the bushy, branched structure. The threads of amylose more readily form *gels*: three-dimensional networks of polymer that has trapped water molecules between the polymer chains. Weak, noncovalent attractions between the slightly negative oxygen atoms and the slightly positive hydrogen atoms of the O–H groups link the chains of amylose together. These weak, noncovalent attractions are called *hydrogen bonds*. Amylopectin molecules will form gels, but the gels are softer and less stable because the bushy amylopectin molecules cannot stack closely and form stable hydrogen bonds with each other as easily. When the starch granules are large, they

TABLE 7.7 Types of Common Thickening Starch and their Properties.

Type	Thickening Power	Gelation Temperature	% Amylose	Avg. Granule Size	Additional Factors that Affect Thickening Power
Wheat	+	126–185°F, 52–85°C	26	17 μm (50%), 7 μm (20%), other (30%)	High-protein content (~10–12%); variable granule size inhibits good amylose gels
Corn	++	144–180°F, 62–80°C	28	14 μm	Opaque gels
Potato	++++	136–150°F, 58–65°C	20	36 μm	Amylose polymers are unusually long; improves gels
Tapioca (cassava root)	++	126–150°F, 52–65°C	17	14 μm	Forms clear gels but stringy
Arrowroot	++	140–187°F, 60–86°C	21	23 μm	Forms clear gels but stringy

20% amylose means a 20:80 ratio of amylose to amylopectin.

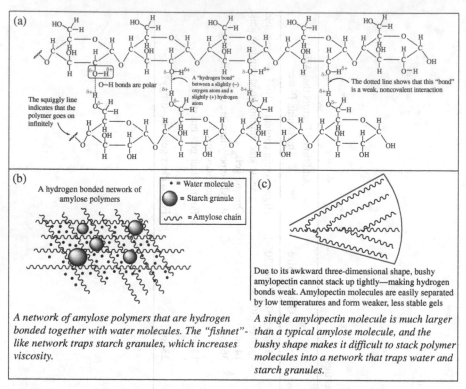

FIGURE 7.15 Hydrogen bonding, water, and starches. (a) Hydrogen bond formation between adjacent chains of amylose in the formation of a gel. (b) A cartoon depicting the gelation of amylose. (c) The bushy branched shape of amylopectin.

are easily trapped in the three-dimensional network of starch chains and water and therefore increase the viscosity of the liquid. The tightly ordered clusters of amylose molecules require higher temperatures, more water, and longer cooking times to *gel* (separate into a network filled with water molecules). Clusters of amylopectin molecules gel at lower temperatures.

When a cook adds starch to a liquid and then heats it to form a gel, the starch molecules leak out of the starch granules and the sauce thickens. When the sauce is thick enough, the cook will turn off the heat, and the mixture will cool. As the temperature falls, there is less energy for the molecules to move and groove; consequently they move less and form more stable hydrogen bonding networks with each other and with the water molecules interspersed among the polymer chains. If the temperature gets low enough, the starch molecules will *congeal* or solidify into a solid gel. Amylose chains reform their tightly packed hydrogen-bonded structures quickly because their linear shape makes tight packing easy and *hydrogen bonding strong* (Fig. 7.15). If there is little water, then the amylose chains will pack into a hard *crystalline* solid instead of a moist gel. Amylopectin molecules take longer to reassociate into hydrogen-bonded networks upon cooling because their bushy

TABLE 7.8   Rice Varieties and % Amylose.

Rice	% Amylose
Long grain	22
Medium grain	18
Short grain	12
Waxy ("sticky") rice	0–1

All rice is cooked by boiling in water, which heats the starch granules up, causing them to swell and leak starch. In the table, 22% amylose means a 22:78 ratio of amylose to amylopectin. So a 0% amylose starch has 100% amylopectin.

shape makes tight packing difficult, and they form weaker networks (softer gels) compared with amylose.

This process of heating starch to gelation temperatures and then cooling the gelled network into a solid is called *retrogradation*. The *retrograded* starch is *more compact* (i.e., harder) than the native starch. This is what happens when pie fillings, puddings, and other gel-like solids are made. It also explains why cooked rice turns hard in the refrigerator overnight.

Rice comes in three basic varieties shown later that vary in their amylose/amylopectin ratios (Table 7.8). Any starch is characterized by its "% amylose" content; therefore, the remaining starch is made of amylopectin. Long-grain rice is a high amylose starch (22%), and boiled long-grain rice will have a firm, springy texture when cooked and become inedibly hard when refrigerated overnight. Boiled "sticky rice" has a softer, stickier texture and hardens much less during overnight refrigeration; this is a result of the small percentage of amylose and the larger amount of amylopectin. The "bushy" amylopectin forms weaker, softer gels—so low % amylose (i.e., high % amylopectin) rice is "softer" rice.

In the same way, when potatoes are boiled, professional cooks recommend cooking the potato slices at approximately 320°F/160°C for approximately 20–30 min followed by *cooling them down* while standing (i.e., don't mash or otherwise disturb the potatoes) for 30 min. You can then finish the cooking process by briefly reheating the potatoes by steaming or simmering followed by gentle mashing. This process of heating → cooling → and briefly reheating, avoids the gluey texture that mashed potatoes can sometimes get by first gelling and retrograding the starch into a tightly packed network.

Other carbohydrate polymers are capable of forming gels. The cell walls of some fruits (e.g., citrus fruits and apples) are especially rich in *pectin*. As we saw earlier, pectin is a polymer of carbohydrate monomers (the monomer is the sugar galacturonic acid; see Table 7.3), and it forms the cell wall cement of the primary cell wall and the cell glue of the middle lamellae. When fruit is cut up and heated near to boiling point, the pectin chains break into smaller chains and dissolve into the surrounding fluid. Upon cooling, the pectin chains can *gel* in much the same way that the chains of amylose starch *gel*—a moist solid is created by the association of pectin chains into a three-dimensional network that traps water molecules. Boiling off water (to concentrate

TABLE 7.9 Plant Gums.

Plant Gums (Carbohydrate Polymers)	Uses
Agarose and alginates: cell wall polymers from seaweed Acacia gum (or gum arabic, from acacia trees) Guar gum: from seeds of *Cyamopsis tetragonobola* Xanthan gum: carbohydrate polymers produced by bacteria in industrial fermentation	These carbohydrate polymers can thicken and stabilize like starch and gel-like starch and pectin. They are added to jams, jellies, custards, sauces (salad dressings, ketchup, etc.), and used to create a smooth texture in ice cream

the diluted pectin chains) and adding sugar encourage better *gelation* of the pectins. The polar sugar molecules help to form the gels by using hydrogen bonds to connect neighboring pectin chains. Adding extra sugar is so effective at encouraging *pectin gelation* so that fruit cooked in a concentrated sugar solution will retain its firmness—the sugar helps the pectins gel within the structure of the fruit cell walls. Pectin can be purified from citrus fruits and apples, and the purified pectin powder can be added to crushed fruit (cooked or not) to give a firm gel. Food chemists have developed modified pectin that can gel with the aid of calcium ions that help cross-link the pectin chains together. The cook adds the calcium after the fruit and modified pectin have cooked together. This innovation allows food manufacturers to gel pectin without sugar and therefore produce "low-sugar" preserves and fruit jellies.

There are other plant carbohydrates that can gel and/or thicken like pectin and starch. These carbohydrates are usually called *gums*, and they come from a variety of plant sources (Table 7.9).

7.6 COLORFUL AND FLAVORFUL FRUITS AND VEGETABLES

What makes color? Why do we perceive most plants and plant parts to be green in color, while some are red, orange, yellow, purple … and so on? Molecules are the source of the colors we see.

When a ripening fruit loses its green color (think of green bananas…), it is the loss of the chlorophyll pigment that we see. Other plant pigments are also due to molecules.

All carbon-based molecules that have color share one structural feature in common—multiple *double bonds* (Figs. 7.14, 7.15, 7.16, and 7.17 and Table 7.10). The number and arrangement of those double bonds determine if the molecule will be able to absorb *visible light*. Molecules that can absorb *visible light* have color.

Colored molecules contain many *conjugated double bonds* (Figs. 7.17 and 7.18). It's not possible to easily correlate the structure of a molecule to its *exact* color. However, the presence of greater than five conjugated double bonds is a good indicator that the molecule will be colored. Most plants contain all of the types of colored molecules (chlorophylls, anthocyanins, and carotenoids) but in different concentrations.

(a)

Chlorophyll A = bright green

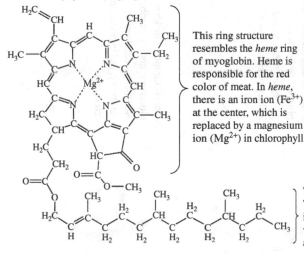

This ring structure resembles the *heme* ring of myoglobin. Heme is responsible for the red color of meat. In *heme*, there is an iron ion (Fe^{3+}) at the center, which is replaced by a magnesium ion (Mg^{2+}) in chlorophyll

(c)

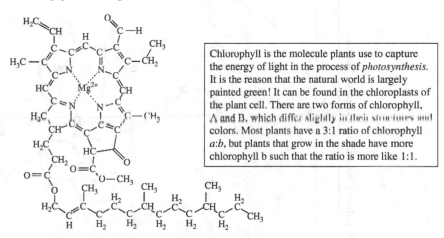

This "tail" is hydrophobic. These atoms become embedded in a hydrophobic membrane within the chloroplast and act as an anchor for the chlorophyll

(b)   Chlorophyll B = olive green

Chlorophyll is the molecule plants use to capture the energy of light in the process of *photosynthesis*. It is the reason that the natural world is largely painted green! It can be found in the chloroplasts of the plant cell. There are two forms of chlorophyll, A and B, which differ slightly in their structures and colors. Most plants have a 3:1 ratio of chlorophyll *a:b*, but plants that grow in the shade have more chlorophyll b such that the ratio is more like 1:1.

FIGURE 7.16   Why plants are green. The structures of (a) chlorophyll A and (b) B, the green pigments in the chloroplasts of plant tissue. (c) Chloroplasts, the plant organelle containing chlorophyll. (c) taken by Kristian Peters.

When a carbon based molecule has more than one double bond between carbons, those double bonds can be arranged in one of three ways:

Isolated double bonds   *Cumulated* double bonds   *Conjugated* double bonds

FIGURE 7.17   Double bonds. The arraignment of carbon double bonds has a significant impact on the light-absorbing ability of the compound.

TABLE 7.10 Plant Pigment Molecules, their Classes, and Common Sources.

Type	Molecule	Common Sources
Anthocyanins	Pelargonidin Cyanidin	(Reds) Ripe raspberries and strawberries, blueberries, plums, cranberries, pomegranates, and red kidney beans
	Delphinidin	(Purples) Grapes, blackberry, blueberry, cherry, cranberry, elderberry, acai berry, raspberry, red cabbage, and red onion

Carotenoids

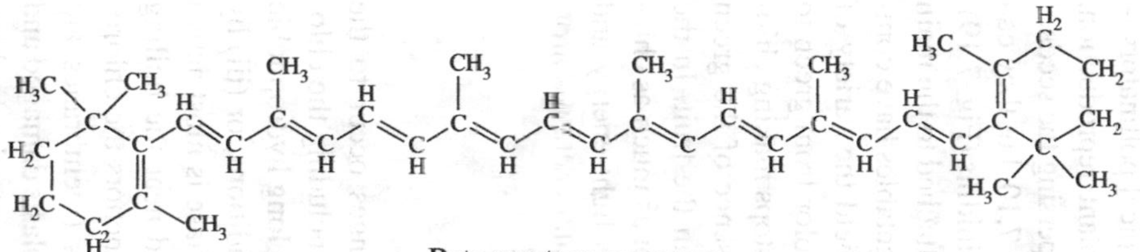

Carrots, sweet potato, butternut squash, and pumpkin

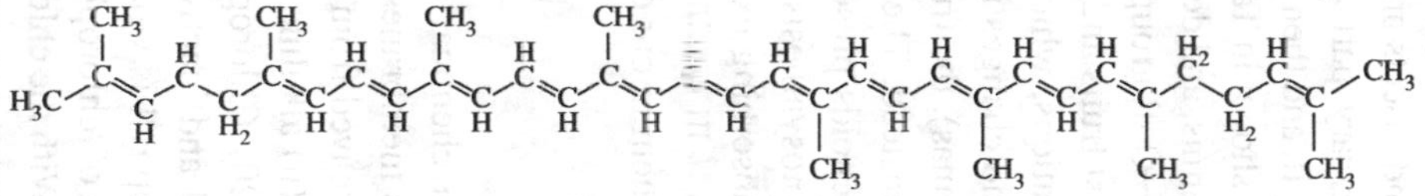

Tomatoes, red carrots, watermelons, and papayas

Beta carotene = orange

Lycopene = red-orange

While all colored molecules absorb *visible light*, not all colored molecules perform the same purpose within the plant (Fig. 7.18). The chlorophyll in chloroplasts is essential for the light-harvesting reactions of photosynthesis, while colored fruits are the plant's strategy for appealing to hungry animals that might eat the fruit and disperse the seeds, and still other colored molecules help defend the plant against intense light, harmful oxidation, or predators.

Anthocyanins are responsible for the red, pink, blue, and purple colors of flowers, leaves, fruits, and vegetables. Purple carrots, red cabbage, and red strawberries all owe their beautiful colors to anthocyanins. In fruits, the anthocyanins accumulate as the fruit ripens; this is obvious in the ripening of strawberries, blackberries, and blueberries. This is consistent with the primary role of the anthocyanin pigments within the plant: to attract birds, insects, and animals. Birds and insects are typical pollinators—they help the plant spread pollen around, a necessary part of plant reproduction. Animals help the plant by eating the fruit, digesting it, and then dispersing the seeds.

Anthocyanins all share the same basic structure shown in Table 7.10, but they can be broken into three main types: pelargonidins, cyanins, and delphinidins (Fig. 7.19).

The three types vary in the number of —O–H or —O–R groups attached to the main ring system, and have slightly different colors. Most fruits and vegetables have combinations of these pigments in different concentrations, which yield the variety of color that we see in nature. When a fruit ripens and changes in color from green to red or yellow (think of ripening tomatoes and bananas), the fruit stops making chlorophyll and begins making carotenoid pigments instead. The absence of the green chlorophyll also reveals the red and yellow carotenoids present in the fruit. In the plant, carotenoids also capture light energy for photosynthesis (not as much as chlorophyll, though), they help protect the plant by absorbing excess light energy and quenching harmful *reactive oxygen species*. As we will see in a later section, carotenoids can do the same for us when we consume them in our diet!

7.6.1 Cooking Colors

When green plant material is cooked, two main chemical changes occur to the chlorophyll. The heat breaks down the plant cell's membranes—including the chloroplasts. The chlorophyll pigment is chemically cleaved from the long hydrophobic tail by three different means: (i) acid conditions, (ii) alkaline conditions, or (iii) by the actions of the enzyme, chlorophyllase (Fig. 7.20). Chlorophyllase is most active at 150–170°F/66–77°C, and so is only denatured and inactivated near the boiling point of water. The hydrophobic tail of the chlorophyll is what anchors the chlorophyll molecule in the hydrophobic membranes of chloroplasts (membranes are mostly made of fat-like, hydrophobic molecules). With the chloroplasts damaged and the hydrophobic chlorophyll tail severed, the pigment can escape into the cooking liquid (if you are boiling) or into the slightly acidic plant cell fluid (if you are steaming or sautéing). One of the reasons we blanch (briefly plunging the vegetable or fruit into boiling water and then quickly submerging it in cold water) is to denature and inactivate chlorophyllase and thus maintain the green color of the food. *Acidic conditions* are the main reason that the pigment part of the chlorophyll can then lose

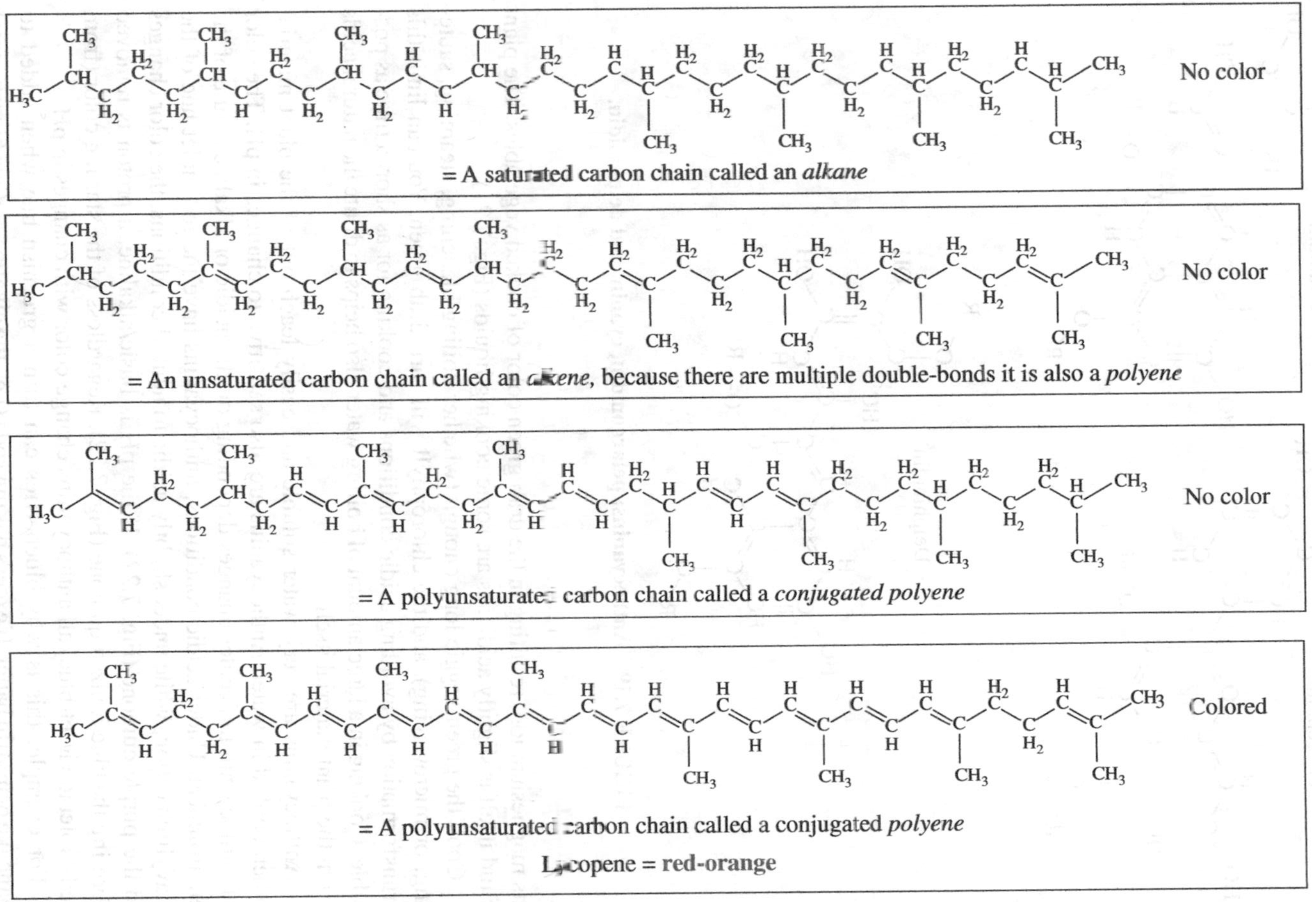

FIGURE 7.18 Impact of carbon double bonds on light absorption. Four different unsaturated carbon compounds are shown and only the conjugated compound lycopene has light-absorbing characteristics.

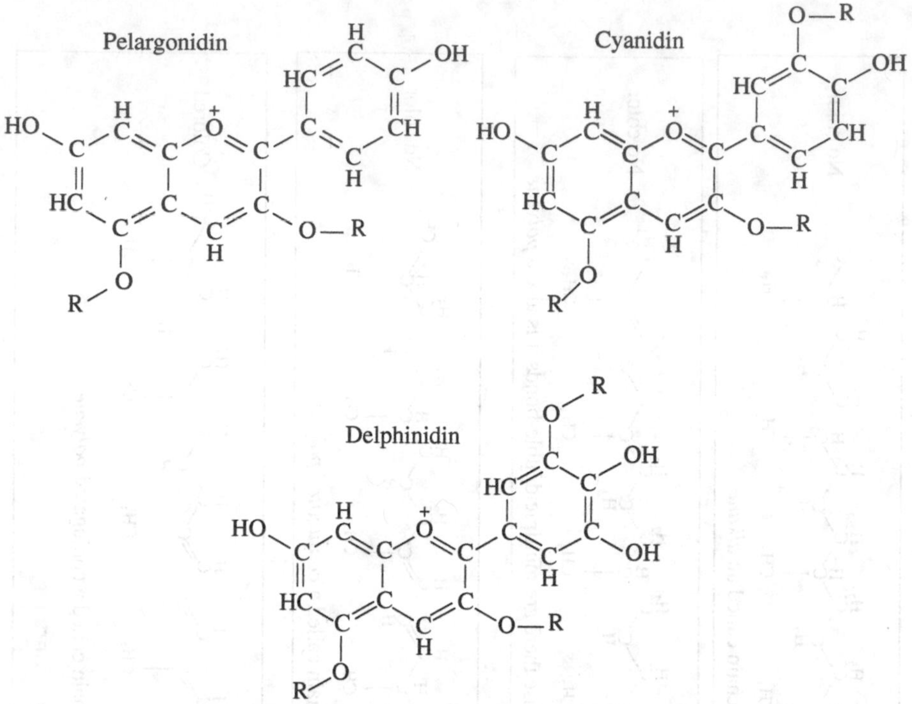

FIGURE 7.19 Anthocyanins: pelargonidin, cyanin, and delphinidin.

its magnesium ion, resulting in the drab green color of cooked vegetables. The plant fluid itself is slightly acidic, as are some cooking liquids (Fig. 7.21).

Cook the green veggie long enough (whether boiling/blanching, steaming, sautéing, or microwaving), and the chlorophyll will turn drab green. You can limit this transformation by avoiding acidic conditions and cooking for as short a time as possible. Cooking in a large amount of boiling water also helps to dilute the natural acids from the plant material itself.

Anthocyanins are very water soluble and easily leach out of the plant material when cooked in water. Anthocyanins are also sensitive to changes in pH. The color of anthocyanin molecules changes depending on the acidic or alkaline nature of the environment. Under acidic conditions, anthocyanins have the red-violet hue of the flavylium cation, while under slightly acidic to neutral conditions, the color changes to the purple quinone (Fig. 7.22). If the pH is basic/alkaline, a proton is removed forming the blue ionized quinone (Fig. 7.22). Regardless of the starting color (from red-violet to violet-blue) all anthocyanins change color with changes in pH.

For example, this is why blueberries can turn a greenish hue when added to pancake or muffin batter. If excess baking soda (i.e., the alkaline *sodium bicarbonate*) is used or incompletely mixed into the batter, the alkaline condition changes the color of the anthocyanin molecules in the blueberries. Similarly, mash raspberries

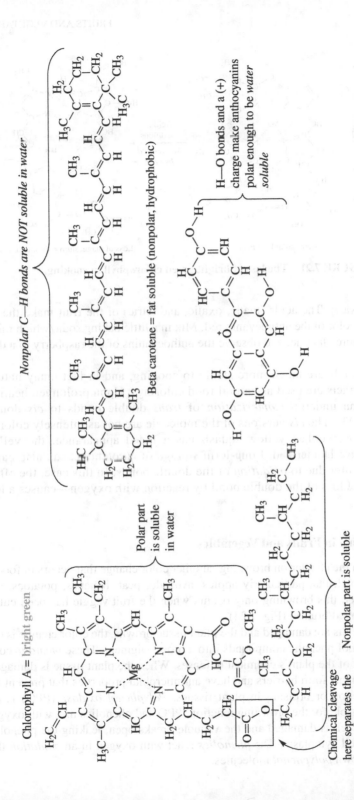

FIGURE 7.20 Plant pigments and their solubility.

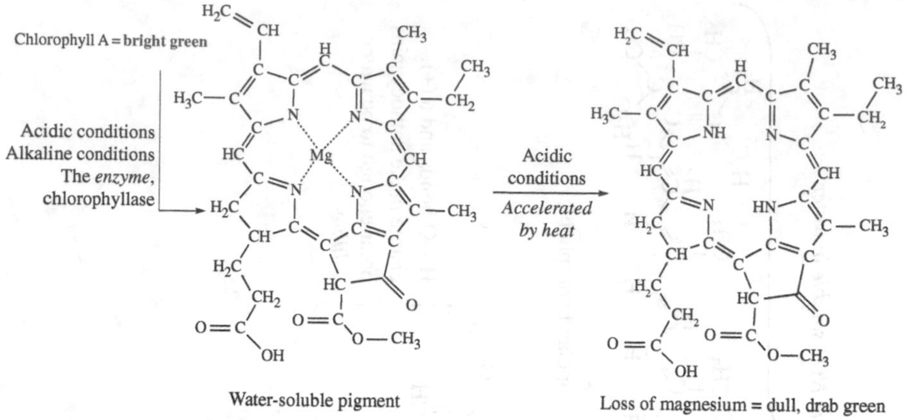

FIGURE 7.21 The loss of bright green chlorophyll in cooking.

with a bit of water. The acids (malic, oxalic, and citric) of the fruit make the pH acidic and the color of the anthocyanin red. Mix in a little baking soda, which raises the pH to be more alkaline, and observe the anthocyanins of the raspberry to a dark purple color.

The carotenoids are much more stable to cooking, and in fact many natural carotenoid extracts are used as natural food colorings. Upon prolonged heating, carotenoids can undergo *isomerization* of *trans* double bonds to *cis* double bonds (Fig. 7.23). The *cis* versions of the molecule are not as intensely colored. This is why overcooked yellow squash has a faded appearance; the yellow carotenoids have isomerized. Long-term storage of carotenoids can also cause fading of the color due to *oxidation* of the double bonds. In this case, the effect is the same; the loss of the double bond by reaction with oxygen—causes a loss of color.

7.6.2 Browning in Fruits and Vegetables

As discussed in the chapter on browning, another color change that occurs in food is the *browning* of fruits, particularly apples, avocado, pear, bananas, potatoes, and lettuce. However, this browning only occurs when the fruit/veggie has been cut or otherwise damaged/bruised (Fig. 7.24).

When plant cells are damaged and the plant tissue browns, the color change is due to the *oxidation* of phenolic compounds into a brown pigment. These *phenolic* compounds are part of the plant's chemical defenses. When the plant tissue is damaged, the brown pigments form barriers and have antimicrobial properties that prevent the spread of infection or bruising in plant tissues. *Polyphenol oxidase* (PPO) is the enzyme responsible for the browning reaction. PPO and phenolics mix with oxygen when the plant cell is damaged and the vacuole breaks open, leaking the phenolics into the plant cell cytoplasm. The *phenolics* react with oxygen in an *oxidation* that forms large, brown *polyphenol* molecules.

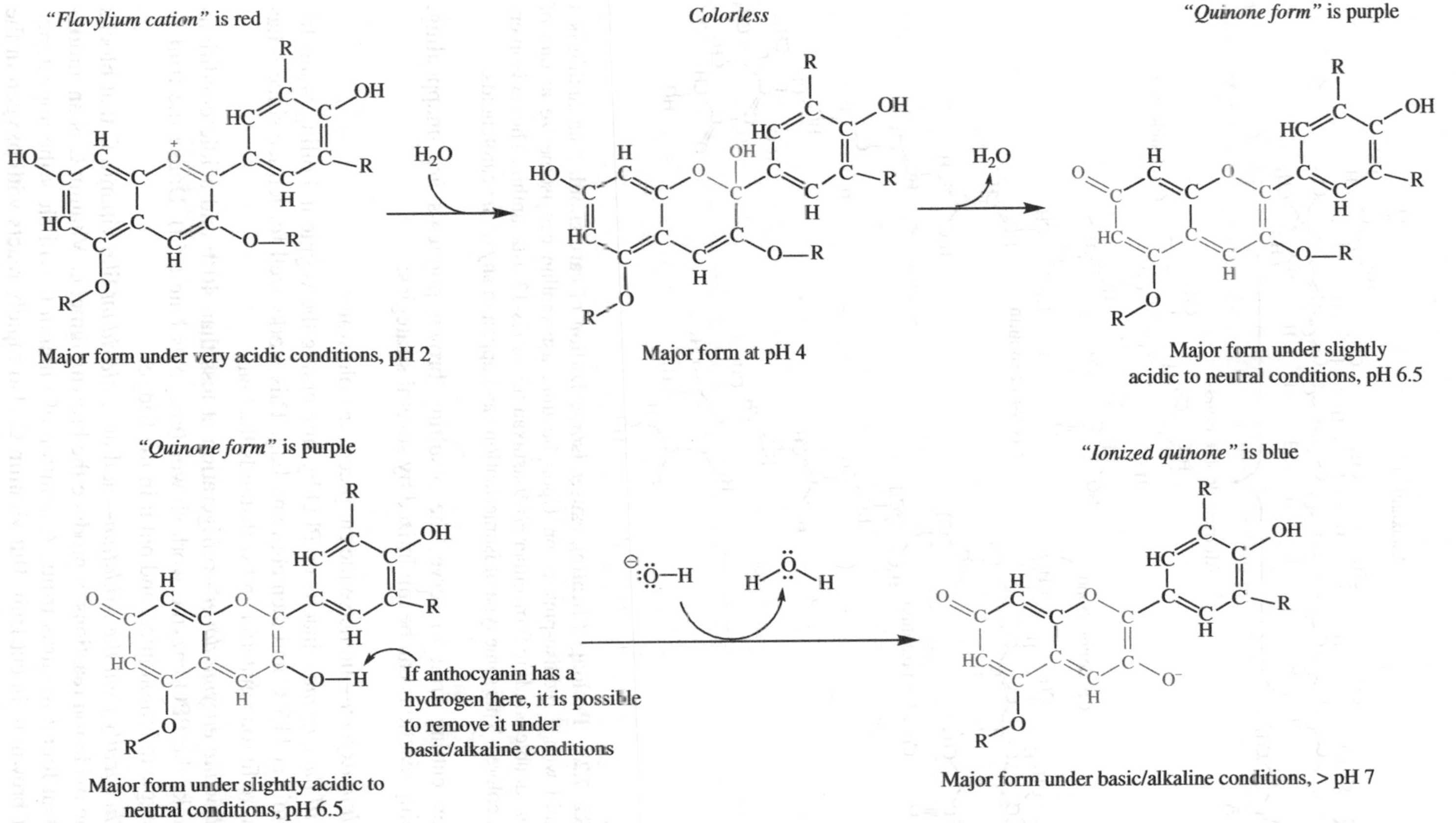

FIGURE 7.22 Changes to anthocyanins with pH. The grey flavylium cation form is present at low pH (*very acidic* conditions). As the pH increases, water reacts with the flavylium cation to produce the colorless carbinol. As the pH increases further, a purple quinone is formed, and water is released again. Finally, increasing the pH above 7 (*alkaline/basic pH*) will produce a blue ionized quinone.

"*Quinone form*" is purple

Major form under slightly acidic to neutral conditions, pH 6.5

Colorless

Major form at pH 4

"*Flavylium cation*" is red

Major form under very acidic conditions, pH 2

"*Quinone form*" is purple

Major form under slightly acidic to neutral conditions, pH 6.5

If anthocyanin has a hydrogen here, it is possible to remove it under basic/alkaline conditions

"*Ionized quinone*" is blue

Major form under basic/alkaline conditions, > pH 7

H_2O

H_2O

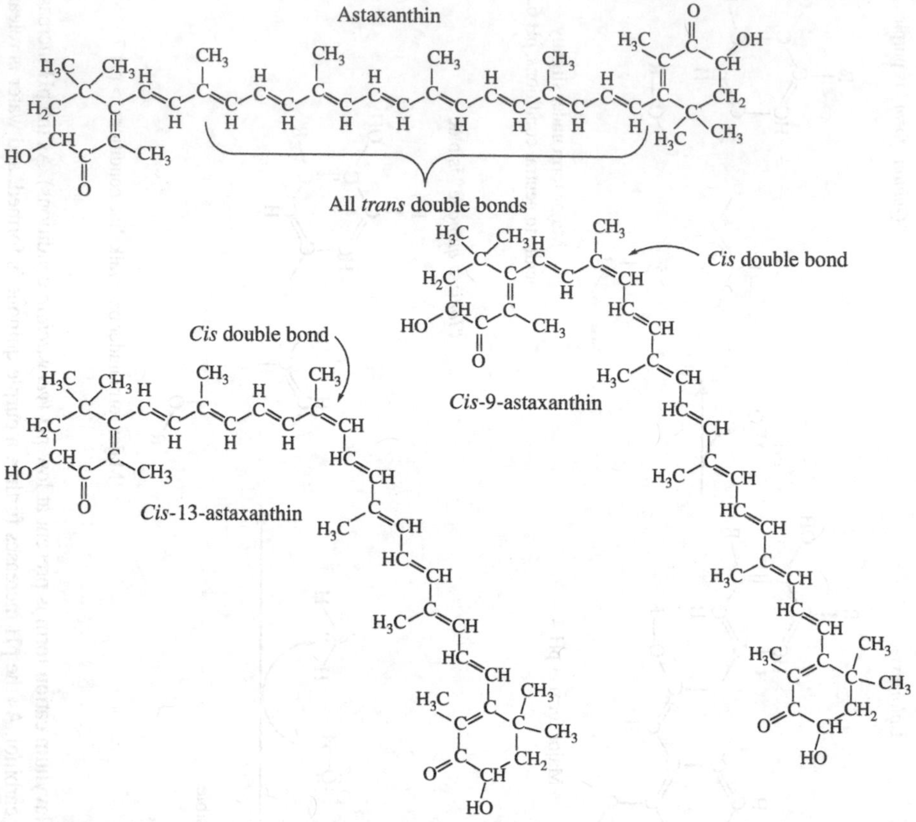

FIGURE 7.23 Prolonged heating causes isomerization of carotenoid. Astaxanthin is a carotenoid with a reddish-pink color. Upon heating, astaxanthin can isomerize at one of two *trans* double bonds to form either *cis*-9-astaxanthin or *cis*-13-astaxanthin. The *cis* isomers are less colored. This same type of isomerization can happen to any of the carotenoids.

When cutting fruit to serve, the resulting brown pigments are unappealing. Browning reactions can be minimized by several strategies:

- *Limit oxygen*—immerse the cut pieces in cold water.
- *Kill the enzyme*—inactivate PPO by immersing the veggie in boiling water for 3 min at 115°C and then chill and bag. This works well for lettuce since it has more fibrous material and withstands the heat.
- *Slow the enzyme down*—refrigeration at less than 40°F and acidic conditions make the PPO enzyme work slower (enzymes hate acid). Douse the fruit or veggie in lemon juice, and put it in the fridge.
- *Chemically fight the oxidation*—add an *antioxidant* like vitamin C that blocks the oxidation reactions to produce the brown pigments. Vitamin C is an antioxidant found in citrus fruits. A solution of vitamin C will slow the appearance of browning in cut fruit. But vitamin C also rapidly reacts with oxygen in the air and is destroyed by heat.

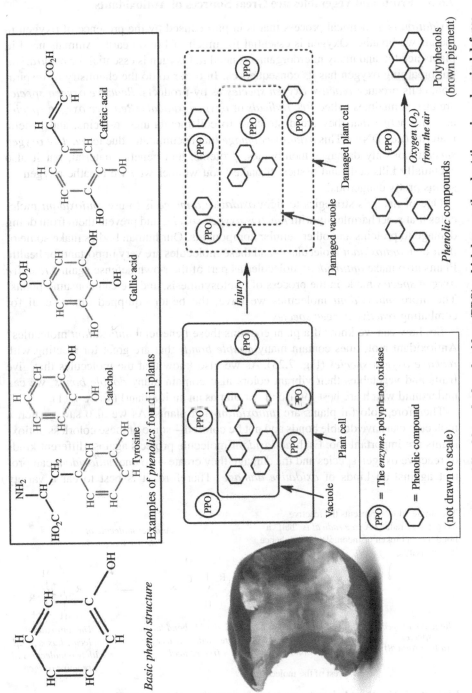

FIGURE 7.24 *Phenolic compounds are oxidized by polyphenol oxidase (PPO), which browns the plant tissue in a chemical defense mechanism.*

Basic phenol structure

Examples of *phenolics* found in plants

Tyrosine

Catechol

Gallic acid

Caffeic acid

Vacuole

Plant cell

PPO = The *enzyme,* polyphenol oxidase

= Phenolic compound

(not drawn to scale)

Injury

Damaged plant cell

Damaged vacuole

Phenolic compounds

Oxygen (O_2)
from the air

Polyphenols
(brown pigment)

7.6.3   Fruits and Vegetables are Great Sources of Antioxidants

Oxidation is a chemical process that is in part caused by the presence of oxygen in the air we breathe. Oxygen is essential for much of life on earth. Animals need it, plants need it, and many microorganisms need it. Oxygen is essential for *respiration*, but breathing oxygen has its consequences. In order to do the chemistry of respiration, cells produce *reactive oxygen species* as by-products. *Reactive oxygen species* are also sometimes called *free radicals* or *oxygen radicals*. *Reactive oxygen species* are so reactive that they are destructive to cell membranes, proteins, and genetic material (i.e., DNA). This *oxidative damage* that accumulates due to *reactive oxygen species* not only damages membranes, proteins, and genetic material, but it also eventually kills cells and tissues. It makes you wonder why we breathe oxygen ... seems pretty dangerous!

One of nature's strategies to fight *oxidative damage* is to use *antioxidant* molecules that react harmlessly with *reactive oxygen species* and prevent them from doing damage to proteins and other cellular components. Our human bodies make some of their own *antioxidant* molecules—and these molecules are very important for health. Plants also make *antioxidant* molecules as part of their own defense against *reactive oxygen species* made in the process of photosynthesis, and we can consume plants! The more *antioxidant* molecules we have, the better equipped our arsenal for combating *reactive oxygen species*.

So, how can we know if a plant contains these beneficial *antioxidant* molecules? Antioxidant molecules contain many *double bonds* that are great for reacting with *reactive oxygen species* (Fig. 7.25). As we also know that the molecules that give fruits and vegetables their vibrant colors also contain many *double bonds*, we can understand which are best for use in our diet as an antioxidant! (Table 7.11).

Therefore, colorful plants are *antioxidant*-rich plants. As we also saw earlier, a molecule can have double bonds and not be colored—some of these colorless antioxidants are important too. Each *antioxidant* molecule protects against different kinds of reactive oxygen species and the damage they create—no one *antioxidant* can protect against all kinds of *oxidative damage*. Therefore, it is best to eat a variety

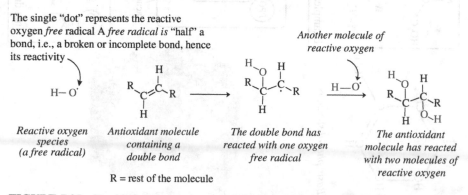

The single "dot" represents the reactive oxygen *free* radical A *free radical is* "half" a bond, i.e., a broken or incomplete bond, hence its reactivity

Another molecule of reactive oxygen

Reactive oxygen species (*a free radical*)

Antioxidant molecule containing a double bond

The double bond has reacted with one oxygen free radical

The antioxidant molecule has reacted with two molecules of reactive oxygen

R = rest of the molecule

FIGURE 7.25   **Double bonds are great for reacting with (neutralizing)** *reactive oxygen free radicals.*

TABLE 7.11 Antioxidant Molecules, their Colors, and Some Sources.

Carotenoids	Beta-carotene	Orange	Carrots, cantaloupe, apricots
	Lutein	Yellow	Corn, kiwi fruit, zucchini, pumpkin, spinach, kale, red grapes
	Zeaxanthin	Yellow	Orange pepper, corn
	Lycopene	Red orange	Tomatoes (fruit and juice), red delicious apples
Chlorophyll		Green	Spinach, green leafy lettuce (romaine)
Phenolics (tannins)		Colorless (brown)	Grapes, tea, cinnamon, clove, vanilla, basil
Vitamin C		Colorless	Citrus fruits, bell peppers, kiwi fruit, grapefruit
Vitamin E		Colorless	Sunflower seeds, almonds, spinach, salad greens

TABLE 7.12 Other Health Benefits of Plant Pigment Molecules: Not an Exhaustive List!

Molecules	Health Benefit	Sources
Anthocyanins	Slow development of heart disease	Grapes, berries, plums
Beta-carotene	Precursor to vitamin A, essential for vision	See Table 7.2
Lutein	Slows development of cataracts and macular degeneration	See Table 7.2
Phenolics, terpenes	Inhibit the growth of cancer cells	Grapes, rye, many fruits, and veggies

of colorful fruits and vegetables and therefore consume a variety of *antioxidant* molecules.

Grapes are very rich in antioxidant molecules, and the health benefits of moderate red wine consumption have been widely studied. Anthocyanins are responsible for the color of red grapes and wines. They "have antioxidant, antimicrobial and anticarcinogenic activity" and "exert protective effects on the human cardiovascular system." Anthocyanins have different biological functions in plant tissues, such as protection against damaging sun exposure, pathogen attacks, oxidative damage, and attack by *reactive oxygen species*. Polyphenols in the grape are generally responsible for *antioxidant* activity, but resveratrol—a particular phenolic present in grapes—is characterized by anticancer, antioxidant, anti-inflammatory, and cardioprotective activity. In the grapevine, resveratrol also defends the plant against attack by predators and pathogens (Table 7.12).

7.6.4 Flavor

Plants—especially fruits—can smell and taste wonderful. As we learned in Chapter 2, we experience flavor as taste—using our tongue—and aromas—using our nose. Both components contribute to our perception of flavor. Taste is experienced by our tongues in five major categories: sweet, salty, sour, savory, and bitter. As we saw earlier, the sweet taste of a ripe fruit is due to the glucose, fructose, and other sugar molecules that are liberated from the stores of starch by enzymes during the ripening process. The bitter taste of cruciferous and leafy greens like chard, chicory, and endive is actually a sign these vegetables are good for you! The molecules that make plants so healthy to consume are often called phytonutrients, and these include phenols, flavonoids, isoflavones, terpenes, and glucosinolates. These molecules promote human health, but they are almost always bitter, acrid, or astringent. Glutamic acid is an amino acid whose concentration is increased in mature tomatoes and mushrooms, giving these foods an umami or savory flavor. The sour oxalic, ascorbic, and quinic acids are altered in maturing pears and other fruits. Metabolism of glucose into the Krebs cycle and other pathways will produce succinic, malic, citric, and ketoglutaric acids. Immature fruit and some vegetables have high concentrations of these acids making the plant taste sour and tart. Loss of these organic acids accompanies the decrease of starch content during ripening.

Aroma molecules (or odorants) are experienced in the nose, not the mouth, so the molecules must evaporate and travel among the gas molecules in the air. Molecules that readily travel in air are called *volatile*. We can smell *volatile* molecules because they enter our noses with the air we breathe, and they travel up into our nasal passages when we crush plant material in our mouths. Using a technique called gas chromatography, chemists can actually measure the aroma molecules present in the air. This method was used to demonstrate that the floral and fruity aroma of bananas is due to over 40 different molecules! While isoamyl acetate (or "banana oil") is a significant contributor to the banana bouquet, it is not solely responsible for the complex aroma of banana. Each fruit and vegetable has its own palate of molecules responsible for its characteristic smell and taste. There are over 400 volatile molecules responsible for the unique aroma of tomatoes! During ripening, enzymes produce a massive number of small molecules that volatilize to the smell receptors in our nose and make the fruit appealing to eat.

Aroma molecules are typically *volatile* liquids that evaporate a little at a time—that is why we can smell them. Since aroma molecules evaporate into the air over time, fresh plant material will smell differently (and taste differently) than plant material that has been cooked or dried. This fact explains why fresh basil smells and tastes so different from dried basil, and it also explains why the flavor of bottled herbs and spices can change over time, limiting their useful shelf life. A bottled herb or spice is actually dried and pulverized plant material—leaves, seeds, flowers, and so on. The longer the plant material sits, the more volatile aroma molecules evaporate away.

Many aroma molecules are produced by the metabolism of amino acids and fats. For example, free fatty acids can give rise to aroma compounds. *cis*-3-Hexenal is a product of the free fatty acids linoleic and linolenic acid, and it is the molecule responsible for the intensely grassy green odor of freshly cut grass. *cis*-3-Hexenal also contributes to the aroma of ripe tomatoes. *cis*-3-Hexenal is somewhat unstable and readily isomerizes to *trans*-3-hexenal, which has a similar odor but is much more stable (Fig. 7.26).

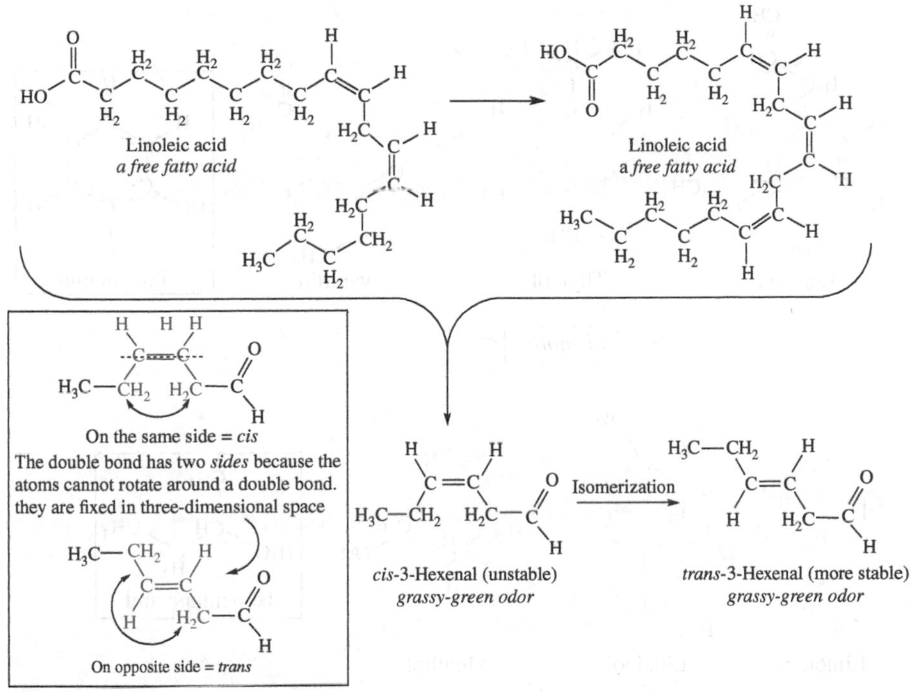

FIGURE 7.26 **The grassy green odor of cut grass and ripe tomatoes is caused by** *cis* **and** ***trans*-3-hexenal.** *cis*-3-Hexenal is made from free fatty acids (linoleic and linolenic acids) by enzymes in the plant, but it is relatively unstable. The *cis* form isomerizes to the more stable *trans* form.

TABLE 7.13 **Examples of Aromatic Herbs, Spices, and Fruits and Some of the Molecules that Contribute to Each Aroma.**

Herb/Spice/ Fruit	Terpenes	Phenolics	Herb/Spice/ Fruit	Terpenes	Phenolics
Basil	Cineole, linalool	Eugenol	Cloves		Eugenol
Peppermint	Pinene, menthol		Ginger	Cineole, citral, linalool	
Sage	Cineole, pinene		Lemon	Limonene, linalool, pinene	
Thyme	Pinene, linalool	Thymol	Cinnamon	Cineole, linalool	Cinnamaldehyde, eugenol
Vanilla	Linalool	Eugenol, vanillin			

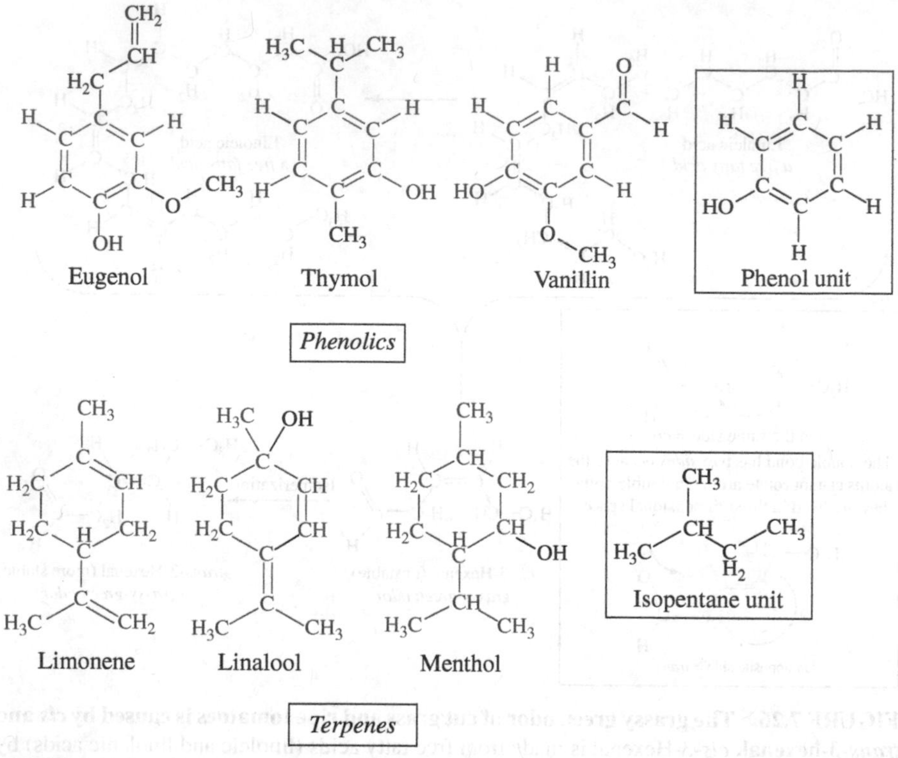

Eugenol   Thymol   Vanillin   Phenol unit

Phenolics

Limonene   Linalool   Menthol   Isopentane unit

Terpenes

FIGURE 7.27 Examples of terpene and phenolic aroma molecules found in plants.
Phenolics all share the basic structure of a phenol, while terpenes are some combination of
isopentane units.

Most aroma molecules fall into two structural categories—*terpenes* and *phenolics*. It's nearly impossible to predict exactly how a molecule will smell based on
its structure, but molecules of similar structure generally give related aromas
(Table 7.13). One plant will typically have many aroma molecules in different
concentrations:

- *Terpenes* provide citrusy, fresh, and floral aromas (Fig. 7.27). *Terpenes* are very
  volatile and therefore evaporate quickly when the plant material is heated. It's
  best to add the fresh herb, fruit juice, etc., to the dish after it's been cooked.
- *Phenolics* have more characteristic, distinct flavors that are generally warm or
  sweet (Fig. 7.27). They are less *volatile* and mostly *water soluble*—which
  means they persist longer in cooking—and they will *dissolve* in the water of the
  foods we cook and eat and in our saliva as we chew.

It is the particular combination of aroma molecules that give the food its unique
smell. Combine aroma molecules with other flavor molecules (taste, astringency, and
pungency) and you get the unique flavor of a particular fruit or vegetable.

REFERENCES

[1] Mann, C.C. *1493: Uncovering the New World Columbus Created*, Vintage Books, a division of Random House, New York, 2012.

[2] Koch, K. and Jane, J.-L. (2000) Morphological changes of granules of different starches by surface gelatinization with calcium chloride. *Cereal Chem.* 77: 22, 115.

[3] McGee, H. (2001) *On Food and Cooking*.

8

MEAT AND FISH

Guided Inquiry Activities (Web): 5, Amino Acids and Proteins; 14, Cells and Metabolism; 22, Meat Structure and Properties; 23, Meat Cooking

8.1 INTRODUCTION

This chapter will cover the important anatomy and physiology of muscle (cattle and fish), the role of myoglobin in biology and cooking, the science of how meat cooks, water movement, protein denaturation, the relationship between the cut of a steak and its connective tissues and age, and the biochemistry of rigor. Fish and shellfish will also be discussed along with browning reactions and the effects of salting, brining, and marinating.

The change in flavor, texture, and tenderness of raw meat after aging and cooking reflects the definition of meat as "muscle tissue that has undergone chemical and physical change." Some expand the meaning of meat to other animal tissues (liver, kidney, intestine, etc.), and the more formal definition of a meat food product by the USDA is "any tissue from the carcass capable of being used as human food." For cooking purposes, in this chapter, we will limit ourselves to skeletal muscles from mammals, bird, and fish and shellfish muscle.

Around two million years ago, early man developed a larger brain and smaller digestive track, and according to anthropologist Leslie Aiello, this is when our ancestors started eating meat. The switch in diet allowed for less energy devoted to digestion and more metabolic energy spent maintaining the larger, energy-demanding brains. Our ancestors were attracted to eating meat because of the high nutritional and caloric value owed to the protein and fat found in meat. However, some of these same

The Science of Cooking: Understanding the Biology and Chemistry Behind Food and Cooking,
First Edition. Joseph J. Provost, Keri L. Colabroy, Brenda S. Kelly, and Mark A. Wallert.
© 2016 John Wiley & Sons, Inc. Published 2016 by John Wiley & Sons, Inc.
Companion website: www.wiley.com/go/provost/science_of_cooking

components and their sources are also causes of controversy. Epidemiological studies show a causative relationship between red meat consumption and significant increase in risk of cardiovascular heart disease and colon cancer. Fat, as well as the products of cooking meat over high heat and smoke, is the likely culprit of health risks. Conversely, several studies suggest that lean red and other meats have a positive impact on human health in providing nutrients, vitamins, and beneficial polyunsaturated fats when included in our diet in moderate amounts. One kind of healthy fat, conjugated linoleic acid, has several positive health benefits including increasing immune functions and providing possible anticancer properties in animal studies. However, for some, the overshadowing philosophical question of the humaneness of raising living beings only to be eaten is a continuing controversy. Despite these issues, over the past decade, yearly meat consumption in the United States has ranged from 110 to 116 lbs per person, slightly up from the 1970s when, on average, a US citizen consumed 104–106 lbs per year. Consumption of meat is not limited to the United States. China and India are also increasing meat consumption in their diets. In 1997, humans were responsible for consuming 235 million tons of meat, and the United Nations predicts that world-wide meat consumption will double by 2050.

The invention of faux meats, products that imitate fish, beef, and chicken, is a new and interesting application of science to meet health and animal use challenges. The energy and environmental resource consumption of raising fish, beef, and birds to meet the predicted 9.6 billion global population by 2050 is a worldwide issue. Three-dimensional printing of meat and extruded chicken with fibers very similar to the authentic meat is within sight. In a blog post, creator of the series *Good Eats*, Alton Brown, enthusiastically wrote about a new mock chicken product that had the consistency and fiber-like quality of authentic chicken [1]. Whether it is creating a new "Earth-friendly" food or sitting down to cook a great steak, understanding the biology, chemistry, and biochemistry of meat and the changes that occur as the food is cooked is an important tool.

Texture, flavor, and the juiciness of meat are the key characteristics used to assess the quality of animal and fish meat. The chemical composition and changes in meat components as the tissue ages, as well as the reactions of these compounds during cooking, are responsible for flavor. A basic understanding of the physiology and anatomy of muscle and knowing the key differences between cuts of meat and the unique characteristics of animal, fish, and shellfish muscle are important for getting the most out of your kitchen.

8.2   MUSCLE MOTORS: HOW MUSCLE WORKS

There are several types of muscle, smooth, cardiac, and skeletal, each of which can be found as a food in some cultures. In most organisms, skeletal muscle performs the same fundamental task: to shorten or contract, creating movement. The flick of a tail, the closing of a shell, and the flexing of a leg are all movements caused by muscle shortening due to the sliding of protein filaments past one another. Muscles are highly organized groups of cells, extracellular protein glue, and connective tissues holding the muscle fibers together. Muscle cells—also called muscle fibers—are the

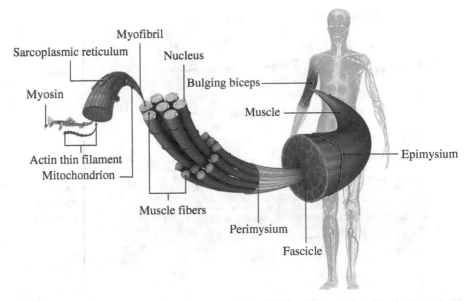

FIGURE 8.1 The structure of muscle and muscle fiber. The organization of tissue from actin to muscle.

basic unit of muscle tissue and collectively make up what we think of as muscles (Fig. 8.1). Muscle fibers are nearly as thin as a thread of hair and often are as long as the muscle itself. Inside the muscle fibers are the proteins: actin, myosin, troponin, and tropomyosin. These proteins are responsible for the contraction and extension of muscle groups and are common among skeletal muscle in fish and mammals. As a muscle develops, the number of these specialized cells stays the same. The increase in size of a developed muscle group is instead due to an increased number of contractile protein filaments in the cells, making a well-exercised muscle grow in size. Unfortunately with more contractile filaments and larger-sized fibers, the meat becomes tougher. Fibers are bundled and held together by connective tissue and filled with long polymerized proteins responsible for shortening the length of a muscle group. In fact, rigor mortis is the stiffening of muscle and loss of extension or relaxation brought about as these key muscle fiber proteins lock in contraction against each other. Let's start our understanding of science and cooking of meat by focusing on the contractile proteins at the center of muscle fibers.

Inside each muscle fiber are long complex polymers of proteins organized into two protein threads called thin and thick filaments. Together these filaments are organized into "sarcomeres." The thin filament is primarily made of actin, a globular protein that polymerizes into long chains forming the core of the thin filament. Wrapped around the actin polymer are other proteins including tropomyosin and troponin (Fig. 8.2). The thick filament is a separate complex of proteins polymerized into a long strand. Myosin is the chief component of the thick filament and acts as a ratcheting motor connecting the thick and thin filaments. Myosin binds to adenosine

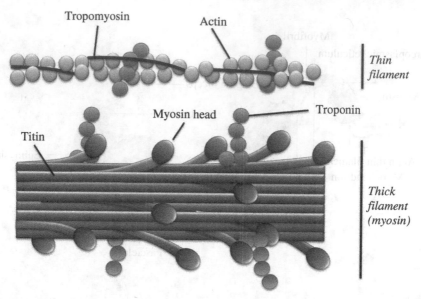

FIGURE 8.2 Muscle complex. Detailed illustration of actin, myosin, troponin, and titin proteins.

triphosphate (ATP), ADP, as well as actin. Titin, the largest known protein with up to 33,000 amino acids and a molecular weight of 3.8 million atomic mass units (many proteins range between 10 and 250 thousand atomic mass units), makes up a third filament binding to actin thin filaments. Titin acts as a spring and anchor for the thin and thick filaments to slide together.

Muscle contraction happens as nerves signaling to muscle fibers release calcium from containment within the muscle fiber (Fig. 8.3). In the beginning of the cycle, myosin is bound to ADP (i.e., ATP) but not actin. Before calcium binding, the troponin and tropomyosin combine to mask the interacting site between actin and myosin. Calcium causes troponin and tropomyosin to bind to actin by exposing the actin–myosin binding site. Once the actin binding site is available, myosin forms a temporary bond, also called a cross-bridge, with actin. The interaction between actin and myosin induces myosin to change its shape, and then myosin will release an ADP molecule. The combination of these two events triggers myosin to tilt much likes a wrist curling toward the forearm. Such ratcheting motion pulls the actin thin filament about 10 nm along the thick myosin filament (don't forget titin is holding actin in place) shortening the distance between the ends of both filaments. In muscle physiology, this is called the power stroke.

Following the ratcheting of myosin, the two proteins remain fixed in place until ATP binds to myosin causing myosin to release from actin. ATP binding to myosin causes the interaction between actin and myosin to break, allowing myosin to reposition and extend itself toward the next actin binding site. After release from actin, myosin-bound ATP is converted to ADP and Pi. Finally calcium is transported back into its space

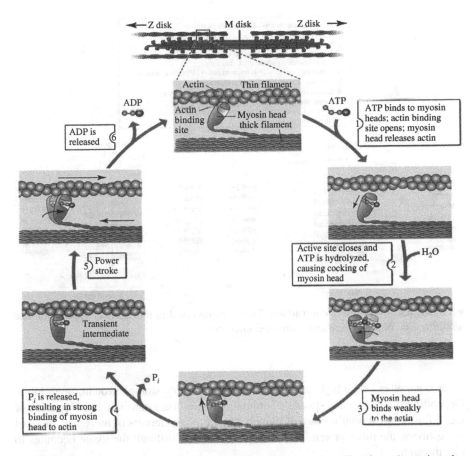

FIGURE 8.3 Mechanism of muscle movement. The myosin walks along the actin using the hydrolysis of ATP. Only one myosin head is shown. Voet Voet and Pratt, figure 7.32. Reproduced with permission from Voet Voet and Pratt 4th Ed.

inside the muscle cell. This happens not at one actin–myosin interaction site but thousands of interactions for each set of filaments, and each muscle fiber is filled with these filaments. Therefore the true motor in our muscle is a molecular motor controlling the sliding of filaments past each other, shortening the fiber and, consequently, the muscle group. True strength comes from this event happening over and over again. Think about hundreds to thousands of these events happening per sarcomere; this provides strength to the muscle (Fig. 8.4).

8.3 MUSCLE ORGANIZATION

Sets of thick and thin filaments called sarcomeres are bundled together to form myofibrils. Together a group of myofibrils make up a muscle cell (muscle fiber). Included in the muscle cell are the organelles typical of a mammalian cell: nucleus, mitochondria,

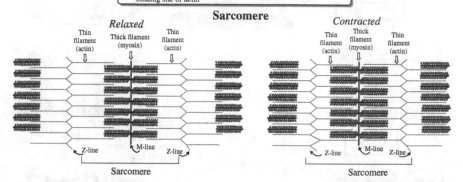

> **Sliding filament theory—how muscle contracts**
> 1. Nerves trigger calcium release exposing sites on actin
> 2. Myosin binds to actin
> 3. The power stroke of the cross bridge induces a sliding of thin filaments past the thick filaments
> 4. During ratcheting, ADP is released and ATP binds to myosin
> 5. ATP-bound myosin disconnects from actin
> 6. Myosin converts ATP to ADP reenergizing myosin to bind to actin
> 7. Calcium ions are transported away, hinging the myosin binding site of actin

FIGURE 8.4   Sarcomere contraction. The steps involved in muscle fiber contraction and an illustration of the relaxed and contracted sarcomere.

other organelles, and a high concentration of the oxygen storage protein, myoglobin. Depending on the age and species of animal, muscle fibers can contain many thick or much fewer and shorter myofibrils, which impact the tenderness of meat. The more of these fibers, the more protein polymers and the more difficult the tissue becomes to bite through.

Inspect a piece of beef or pork meat cut across the grain to see how muscle fibers are organized (or review Fig. 8.1). The elongated arrangement of muscle fibers is the grain of meat. Cut against the grain and you can see into the end of how muscle fibers are grouped together. Long muscle fibers are bundled together into a fascicle. Fascicles are further bundled together to form muscles. Special connective proteins, collectively called the perimysium, glue the fascicles together. The key protein of the perimysium is collagen; this is important for gelatin dishes, emulsions, and other cooking uses. Surrounding the muscle (bundles of fascicle) is a protective fibrous protein sheet called the epimysium. At the end of each muscle group are connective tissues called tendons, also made of a different form of collagen. The amount of connective tissue and collagen and the thickness of the fascicles are factors influencing the tenderness of a cut of meat. Feel the difference by rubbing the uncooked meat in your hand. Tender meat with less connective tissue, collagen, and smaller myofibrils will be noticeably more smooth, where a tough cut of meat will be much more firm and uneven and grainy. We cut against the grain of the meat to create shorter less connected muscle fibers, making it easier to chew.

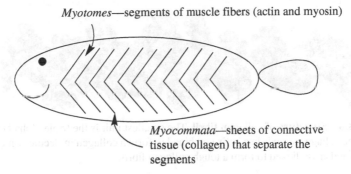

FIGURE 8.5 Muscle structure of fish.

The flaky quality of fish skeletal muscle gives us insight into the different muscle needs between land and water animals. A terrestrial animal must not only propel itself but also support its weight as it walks, climbs, jumps, and generally fights gravity. Fish (teleost), on the other hand, are suspended in a buoyant environment and therefore do not require strong muscles or well-connected muscle tissue. Like land mammals, fish muscle cells are filled with thin and thick actin and myosin fibers contracting and relaxing with the availability of calcium and ATP. However, fish do not possess the elaborate organization of beef and meat of other land animals. Individual muscles of a fish are organized into individual blocks of collagen sheaths in the shape of a "W." Each of these muscles overlaps the other allowing for the undulating nature of a swimming fish. The coordinated contraction of each nestled into the next (via the thin and thick fiber contraction) allows for the swift flick of the tail and contortion of the fish body in water. These individual sheaths, called myotomes, are obvious as individual flakes of a cooked fish (Fig. 8.5).

Shellfish (animals with shells that live in water) are not technically fish, but they also have less defined muscle organization and rely on the contractibility of muscle fibers in a similar way to finned fish and land animals. With the exception of crab, most crustaceans have a head and a body region, and it is the abdomen or body portion that contains most of the muscle fiber used for food and cooking. Expanded to include octopus, cuttlefish, and squid, shellfish muscles are composed of thick groupings of muscle fibers connected to exoskeletons for support and mobility.

8.4 TENDER CONNECTIONS

The thickness of the muscle fiber is due to the number of thick and thin filaments. This is one of the factors that make muscle tissues tough or tender. A tender piece of meat is easily chewed and cut. Meat with a high quantity of long interconnected protein fibers will be harder for teeth to cut than a cut of meat with fewer of these proteins. To test for tenderness of meat an instrument called the Warner-Bratzler tenderometer is used to measure the shear forces of muscle and meat. Essentially this instrument is a

FIGURE 8.6   Structure of collagen fibril. The smallest unit is the triple alpha helix shown above. Three collagen strands in a helix combined to form a collagen molecule, which are then overlapped and cross-linked to form a tough collagen fibril.

guillotine connected to a computer to measure the pressure required to cut through the sample. Fibers and connective tissues combine to make a cut of meat tough or tender.

Connective tissue holds muscle fibers and groups of fibers in place. It covers the muscle, acts as glue within the muscle cell, and fixes muscles to skeletal elements. Most of the connective tissues are made of a few kinds of proteins. Two of the major connective proteins are collagen and elastin. The space within and around muscle cells is filled with proteins that act as a kind of mortar holding bundles of cells in place. The epimysium is a protein membrane also known as a silver skin. This is the thin, tough white and silvery colored connective tissue made of different kinds of collagen. Tendons, informally called sinews, are dense connective tissues attaching muscle to the bone. Similar to another connective tissue and cartilage, tendons are primarily made of collagen (80% of total protein) and elastin (2–4% by weight). While not particularly tasty itself, understanding the composition of connective tissue can help you cook a more tender cut of meat, prepare a tasty soup, or even make a nice dish of pho.

Collagen is a diverse family of proteins, each with a slightly different amino acid composition. There are two main forms of collagen found in muscle, types I and III. Collagen is made of three long, insoluble helical protein chains, each wrapped around three other collagen strands to form a triple helix much like braided hair (Fig. 8.6).

There are a few special characteristics of collagen that impact meat and cooking. First, each collagen protein is shaped into a long narrow helix. The overall shape is reminiscent of a slinky toy pulled nearly straight. The tight collagen helix is a particularly difficult shape for proteins to assume because of the severe turns between atoms in the protein backbone causing atoms to spatially clash. However, collagen has a special repeating amino acid sequence of glycine/proline followed by any amino acid (often written as Gly–Pro–X; Fig. 8.7). This amino acid sequence allows for the special shape and stability of each collagen strand. Because the side chain of glycine is very small (Fig. 8.7), the peptide backbone can easily twist upon itself and permits the protein to form into a long extended coiled chain. The high abundance of proline in collagen is very important as it stabilizes the collagen triple helix. The enzyme prolyl-4-hydroxylase is responsible for adding the stabilizing O—H group to the proline, and it is this O—H group that strengthens collagen triple helix (Fig. 8.8). The reaction catalyzed by oxidase prolyl-4-hydroxylase requires vitamin C. Scurvy is a collagen-related disease that occurs when people are deprived of citrus fruit, a

Glycine
(Gly)
Proline
(Pro)
Hydroxyproline
(HPro)

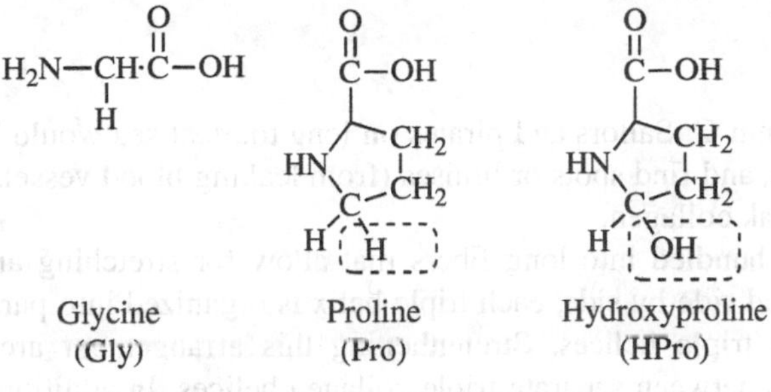

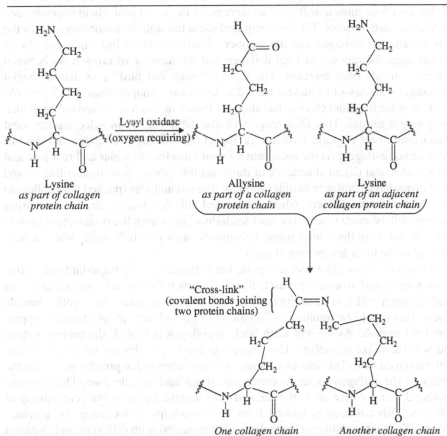

FIGURE 8.7 Key amino acids of collagen.

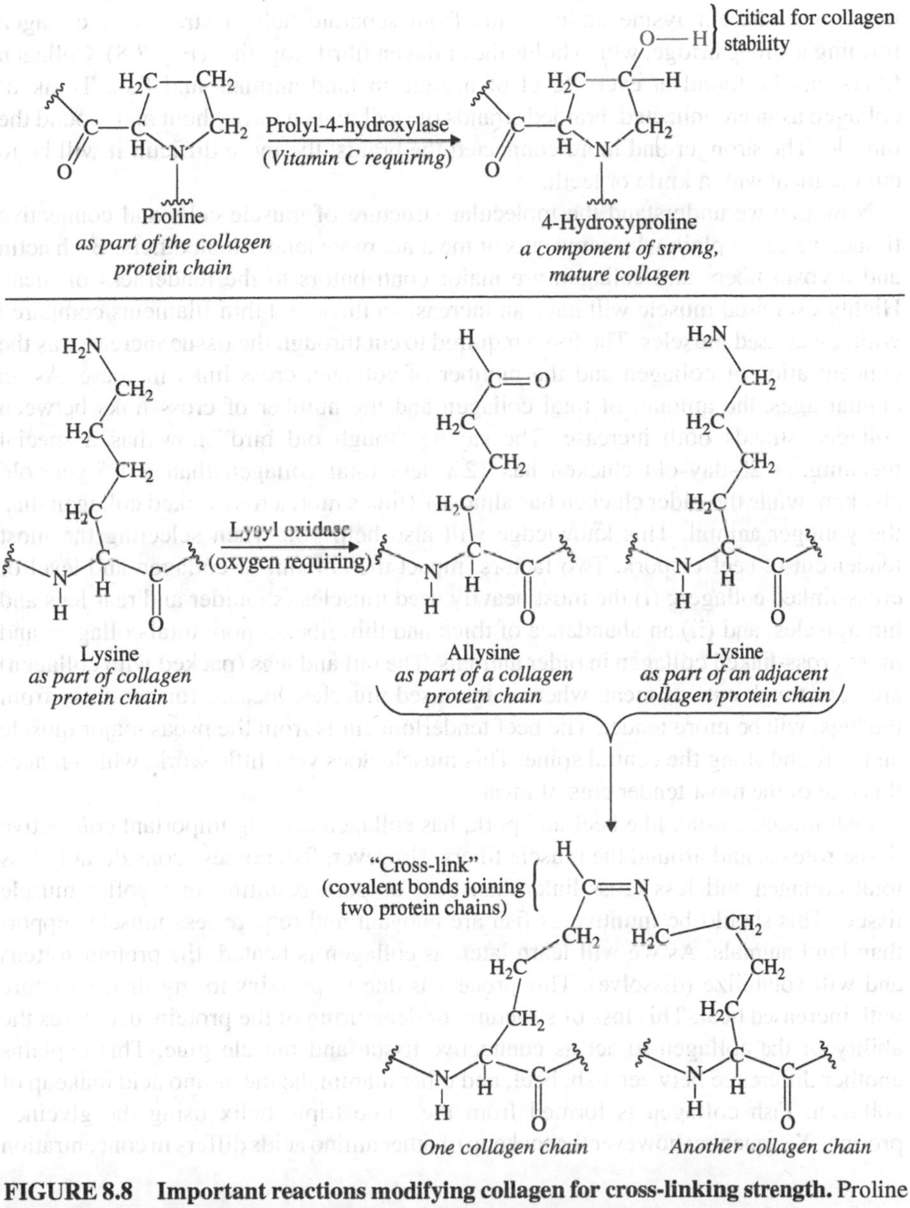

Critical for collagen stability
Prolyl-4-hydroxylase
(Vitamin C requiring)
Proline
as part of the collagen protein chain
4-Hydroxyproline
a component of strong, mature collagen
Lysyl oxidase
(oxygen requiring)
Lysine
as part of collagen protein chain
Allysine
as part of a collagen protein chain
Lysine
as part of an adjacent collagen protein chain
"Cross-link"
(covalent bonds joining two protein chains)
One collagen chain
Another collagen chain

FIGURE 8.8 Important reactions modifying collagen for cross-linking strength. Proline and lysine modification and cross-linking reactions are shown.

source of vitamin C. Sailors and pirates on long tours at sea would lose teeth, have bleeding gums, and find spots or bruises (from leaking blood vessels) on their skin, all due to a weak collagen.

Collagen is bundled into long fibers that allow for stretching and great tensile strength. Packed side by side, each triple helix is organized into parallel bundles of other collagen triple helices. Strengthening this arrangement are cross-links or covalent bonds between separate triple collagen helices. In addition to glycine and proline, collagen of mammal and fish contains the amino acid lysine. An oxygen atom requiring enzyme found in and around connective tissue called lysyl oxidase links two adjacent lysine amino acids from separate helical strands of collagen forming a strong bridge, which holds the collagen fibrils together (Fig. 8.8). Collagen fibers can be found at every level of muscle in land animals and fish. Think of collagen as interconnected, braided strands of steel woven throughout and around the muscle. The stronger and more connected the braids, the more difficult it will be to cut the meat with a knife or teeth.

Now that we understand the molecular structure of muscle cells and connective tissue, we can explain why some cuts of meat are more tender than others. Both actin and myosin fibers and collagen are major contributors to the tenderness of meat. Highly exercised muscle will have an increase in thick and thin filaments compared with lesser used muscles. The force required to cut through the tissue increases as the concentration of collagen and the number of collagen cross-links increase. As an animal ages, the amount of total collagen and the number of cross-links between collagen strands both increase. The saying "tough old bird" now has a special meaning. A 20-day-old chicken has 12% less total collagen than a 1.5-year-old chicken, while the older chicken has almost 12 times more cross-linked collagen than the younger animal. This knowledge will also help you when selecting the most tender cut of beef or pork. Two factors impact the amount of collagen and level of cross-linked collagen: (i) the most heavily used muscles (shoulder and rear legs and hip muscles) and (ii) an abundance of thick and thin fibers, more total collagen, and more cross-linked collagen in older animals. The tail and legs (packed with collagen) are very tough cuts of meat, whereas less used muscles, located further away from the legs, will be more tender. The beef tenderloin cut is from the psoas major muscle and is found along the central spine. This muscle does very little work, which makes this one of the most tender cuts of meat.

Fish muscle tissue, like beef and pork, has collagen serving important connective tissue roles in and around the muscle fibers. However, fish possess considerably less total collagen and less cross-links than beef or pork resulting in a softer muscle tissue. This should be intuitive as fish are buoyant and require less muscle support than land animals. As we will learn later, as collagen is heated, the protein softens and will solubilize (dissolve). This process is due to proteins losing their structure with increased heat. This loss of structure, or denaturing of the protein, decreases the ability of the collagen to act as connective tissue and muscle glue. This explains another difference between fish, beef, and other mammals: the amino acid makeup of collagen. Fish collagen is formed from the same triple helix using the glycine–proline–X sequence; however the makeup of other amino acids differs in concentration between fish and beef, and this impacts the stability of the collagen. The point at which human collagen melts and denatures (with a different total amino acid composition than beef or fish) is 118°F (48°C). Beef and pork collagen denatures at 98.6°F (32°C), and fish collagen will begin to melt and denature at an even lower temperature of 68°F (20°C). Thus it takes less heat and less time to cook and soften the tissue of fish compared with beef or pork.

8.5   RED OR WHITE MEAT

The color of meat is much more complex than picking a chicken leg or a breast. Muscle fibers are organized into different categories depending on the amount of myosin, cellular content, and potential to metabolize food and produce ATP. The myoglobin and mitochondrial content of the cells is responsible for the different colors and taste of each tissue. We will focus on two kinds of muscle fibers: red fibers called type I (i.e., slow oxidative or slow-twitch) muscle fibers and white muscle fibers called type II (i.e., fast glycolytic or fast-twitch) fibers. There is a third intermediate muscle fiber type, a hybrid of both type I and type II fibers. Type I red muscle fibers use a wide variety of food macromolecules to produce energy in the form of ATP, including carbohydrates, amino acids, and fat. Type I cells require oxygen and additional enzymes to meet the metabolic demand and produce a sustained supply of ATP. These components both add a red color to the tissue and, when cooked, provide several interesting, savory flavors to the meat. Type II or fast-twitch muscle fibers primarily produce ATP from aerobic metabolism of carbohydrates. These fibers, also called white or glycolytic muscle fibers, are rich in glycogen (i.e., "animal starch") and the enzymes of the glycolytic pathway (see Chapter 1). As a result of the high concentration of glycogen in type II muscle fibers, these tissues hold more water and are considered juicier. Thus, endurance muscle groups will need a longer sustained ATP production for muscle contraction and will have a predominantly red color. Slower-moving or less used muscles require less ATP except for short bursts of activity and will have more type II fibers and appear white (Table 8.1).

The red/brown color of meat comes from the predominance of type I muscle fiber rich with metal binding proteins called hemoproteins. Lighter-colored muscles are primarily composed of type II fibers that lack appreciable levels of these proteins and appear pale or white. Hemoproteins (a type of metalloprotein) bind an iron cation within a carbon cage known as a porphyrin ring. The iron bound within the porphyrin is called heme, and this reddish-brown molecule is surrounded and held in place by the amino acids of a protein. Together the metal, heme, and protein are known as hemoproteins and have a wide range of interesting chemistry.

Type I muscle fibers have a high mitochondrial content. Mitochondria contain many colorful hemoproteins involved in oxidative metabolism of carbohydrates, amino acids, and fats. A family of hemoproteins in the mitochondria that give some of the dark brown/red tint are the cytochromes. These proteins are responsible for the transfer of electrons ultimately to oxygen in the generation of ATP.

TABLE 8.1   Properties of Muscle Fibers.

	Type I Muscle Fibers	Type II Muscle Fibers
Alternative names	Slow-twitch, slow oxidative, red fibers	Fast-twitch, fast glycolytic, white fibers
Physical role	Endurance and aerobic muscle work; small generated force and smaller in size	Fight or flight and anaerobic muscle work; large generated force and fast movement
Color	Red	White or pale
Myoglobin and mitochondrial content	High	Low
Glycogen content	Low	High
Fat content	High	Low
Flavor/water content	High/low	Moderate/high

Two additional major hemoproteins found in animals are hemoglobin and myo-globin. Hemoglobin is an iron-containing protein located in the red blood cells and is responsible for transporting oxygen throughout the circulatory system. Myoglobin also binds oxygen; however myoglobin is limited to the muscle fiber where it stores O_2 needed for muscle cell metabolism. Myoglobin is a monomeric (single-chain) protein made mostly of alpha helices (Fig. 8.9). As a storage compartment of oxygen in muscle, it is critical for the exercise capacity of red muscle. The more oxygen a muscle fiber can contain, the more metabolism and ATP the cell can produce during long bouts of muscle use. The level of myoglobin can increase with exercise. Deep-diving animals have such a high myoglobin content, which gives their tissue a dark red color. Both hemoglobin and myoglobin are deep red in color when the porphyrin ring iron is bound to oxygen (Fig. 8.9). However, hemoglobin has almost no impact on the color of meat as most of the hemoglobin is bled from the tissue prior to packing. While there is some hemoglobin in fish postharvest, myoglobin is the dominant pigment in muscle enriched with type I fiber.

The *myoglobins* and *cytochromes* are also excellent sources of *bioavailable* iron for humans. Iron is an essential mineral for the human diet—since humans also use the heme- (and therefore iron)-containing proteins hemoglobin and myoglobin to transport and store oxygen. It is not well understood, but *is* well known, that animal sources of iron are more digestible in humans. Plants do contain iron, but we are unable to obtain much of it from eating the plant. The thought is that the *heme* molecule protects the iron and prevents it from being bound up by indigestible plant fiber.

The color of myoglobin and therefore red meat can change depending on what is bound to the myoglobin. Through the heme iron interaction, myoglobin can bind several small molecules including oxygen, carbon monoxide, and water. Each state of myoglobin (oxygen bound, carbon monoxide bound, and whether the iron is +2 or +3) has a different color absorbance that will significantly impact the appearance of the meat. There are three common forms of myoglobin: (i) oxygen-bound myo-globin, which has a distinct cherry red color; (ii) deoxymyoglobin, where iron is in

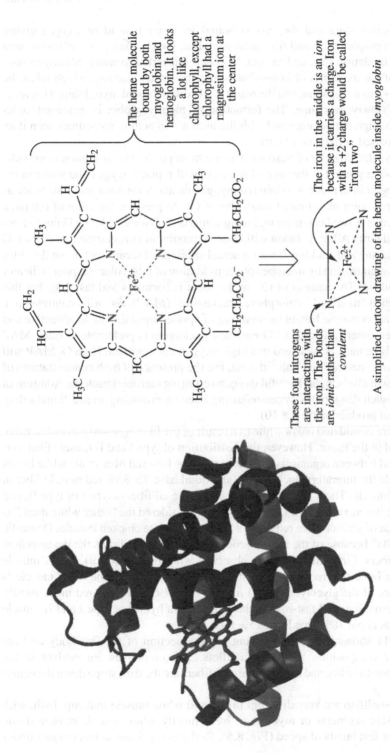

The heme molecule bound by both myoglobin and hemoglobin. It looks a lot like chlorophyll, except chlorophyll had a magnesium ion at the center

The iron in the middle is an *ion* because it carries a charge. Iron with a +2 charge would be called "iron two"

A simplified cartoon drawing of the heme molecule inside *myoglobin*. (Right) The porphyrin ring or hemoprotein bound to an iron (Fe)

These four nitrogens are interacting with the iron. The bonds are *ionic* rather than *covalent*

FIGURE 8.9   Myoglobin. (Left) A ribbon diagram of myoglobin represented with twirly ribbons (i.e., alpha helices) with the heme molecule bound to the protein. There is one molecule of heme for every one molecule of myoglobin. (Right) The porphyrin ring or hemoprotein bound to an iron (Fe) in the center of the heme molecule.

the +2 oxidation state and the iron is bound to water instead of oxygen giving myoglobin a purple color; and (iii) metmyoglobin, which takes place when the iron is in the +3 oxidation state and oxygen has been converted to water. Metmyoglobin is a result of the oxidation of iron—that is, the loss of an electron, which takes the iron from a +2 to a +3 cation, and the result is a brown-colored myoglobin. This reaction occurs slowly over time. The formation of metmyoglobin is increased under conditions of high temperature and with the increase in acidity sometimes seen if an animal is stressed at the time of harvest.

To preserve the red color of fresh meat, some meat producers use carbon monoxide, a gas that, when present at the time of packing, will replace oxygen and water in the +2 iron/myoglobin to produce a deep red myoglobin and meat color that lasts twice as long as the red color of untreated meat. Use of CO to preserve the color of red meat has been used in several countries including Canada and Norway. In the United States the Food and Drug Administration will allow a maximum concentration of 4.5% CO during processing and packing. Because actual spoilage of meat is not about the color but rather contamination by microbes or the oxidation of fats, color is a poor indicator of meat spoilage. The amount of CO in the meat is harmless and tasteless, but this process, called modified atmosphere packaging (MAP), is still controversial. Consumers are concerned about the masking of spoiled meat that looks attractive and red. However, others argue that CO-enriched packaging is preferable to other MAP processes where meat is packaged in a high-oxygen environment. High O_2 MAP will provide a more "natural" red-colored meat, but the presence of high concentrations of oxygen gas will also support harmful oxygen-requiring (aerobic) bacteria. Addition of CO inhibits such dangerous oxygen-requiring bacteria providing an additional safety factor in meat production (Fig. 8.10).

Muscles are considered red or white as a result of the fiber type and amount of myoglobin found in the tissue. However, the distribution of type I and II muscle fiber varies within and between organisms. Chickens possess 10% red fiber in the white breast muscle, while the migratory duck, goose, and quail have 75–85% red muscle fiber in their breast muscle. There is very little muscle where all fiber is type I or type II, and most muscle is a mixture of the two. Pork, while considered the "other white meat," is actually made of 15% or more red muscle fibers than white chicken muscle. Domestic pork is "white" because of the mostly sedentary lifestyle that limits the development of the red fibers. Fine differences are observed with the darker pork leg. A muscle supporting a bone that requires more stamina is darker, where the outer muscle is made of more of the glycolytic type II muscle fiber. Beef, while a red meat, mostly made of intermediate red fast-twitch fiber type IIA (a hybrid of type I and II muscle cells) still has about 10% type I and 15% type II fiber.

Figure 8.11 shows a cartoon showing a cross section of the fish body and the muscle fiber composition from tilapia fillets. The myotomes are evident in the V-pattern along the fillet, and the "red muscle" fibers are the dark stripe down the center of the fillet.

Fish and shellfish are very different in red and white muscle makeup. Fish, with smaller muscle segments or myotomes, have mostly white aerobic muscle tissue arraigned for fast bursts of speed (Fig. 8.5). Red muscle tissue is located just under

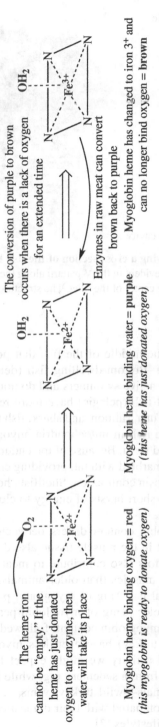

FIGURE 8.10 **The oxidation state of Fe-heme in the muscle.** The different colors of meat are due to the ligand bound to iron and the oxidation state of iron in the heme of myoglobin.

The heme iron cannot be "empty." If the heme has just donated oxygen to an enzyme, then water will take its place

Myoglobin heme binding oxygen = red
(this myoglobin is ready to donate oxygen)

Myoglobin heme binding water = purple
(this heme has just donated oxygen)

The conversion of purple to brown occurs when there is a lack of oxygen for an extended time

Enzymes in raw meat can convert brown back to purple

Myoglobin heme has changed to iron 3^+ and can no longer bind oxygen = brown

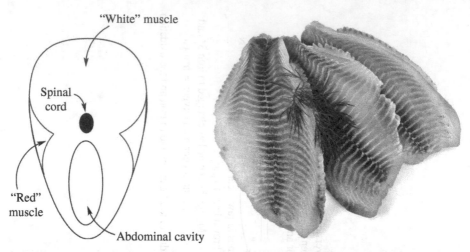

FIGURE 8.11   A cartoon showing a cross section of the fish body and the muscle fiber composition. The myotomes are evident in the V-pattern along the fillet, and the "red muscle" fibers are the dark stripe down the center of the fillet. The structure of the muscles can be seen in the tilapia fillets on the right.

the skin, particularly along the middle of the fish that powers the slower steady swim movements (Fig. 8.11). Bottom-dwelling fish (demersal) drift along with current; most of them are not active swimmers and do not have as much red fiber. Active swimming top current fish (pelagic) have more red muscle fiber and will have a rich taste to the meat. Tuna, salmon, and shark, fish that swim long distances often times at high speed, are rich in mitochondria, myoglobin, and type I fibers and have darker or red-colored flesh. Because of the endurance exercise needed by these fish, the meat is highly marbled with fat, providing energy directly to produce ATP needed for actin and myosin contraction. Shellfish, however, are mostly made of white meat as they require short bursts of energy to close a shell or scuttle to a hiding location.

As we have seen, myoglobin content differs between red and white muscle fibers, but myoglobin content in the muscle tissue also differs by species and the age of the animal. These factors also contribute to meat color. Younger animals have less myoglobin in their muscles than older animals. For example, veal is a very pale brownish pink (veal has 2 mg of myoglobin per gram of meat), while young beef is a cherry red color (8 mg of myoglobin per gram of meat). Across species, beef has the highest myoglobin content, followed by lamb and then pork. Fish and fowl (chicken and turkey) have even less myoglobin. This trend also follows the level of physical activity we might expect from these animals. For example, farm-raised pork is from a sedentary pig, while beef is from the slightly more active cow. Game animals—wild turkey, venison, and so on—have more myoglobin in their muscles compared with their domesticated counterparts due to their more physically active lifestyles [2].

8.6 DEATH AND BECOMING MEAT

There is a point after harvesting the tissue when muscle is considered meat. Fresh muscle is difficult to chew, and while fresh meat possesses some interesting flavors, most of the savory taste associated with meat takes place as the chemical and physical changes of the tissue occur hours and days after the death of the animal. These changes take place in three phases: (i) the continued metabolism by cells in the muscle tissue, (ii) followed next by *rigor mortis*, and finally (iii) the chemical and physical changes in the meat following rigor. When combined with the chemical reactions between the molecules catalyzed by heat in muscle tissue during postmortem and cooking, muscle has changed and is now called meat.

Like cheese and other foods, the enzymes responsible for metabolism and other processes in a cell remain active after the death of the organism. The actions of these enzymes set the stage for important chemical changes in fish and land animal muscle.

One important postmortem change in cells is the carbohydrate metabolism catalyzed by enzymes. Glycolysis (Chapter 3) is a series of chemical metabolic steps converting glucose to pyruvate or lactate and is an immediate source of ATP for muscle cells. In the presence of oxygen, mitochondria containing muscle fibers will continue to oxidize the pyruvate generated by glycolysis to carbon dioxide and water, producing large amounts of ATP. Glycogen, a large, branched polymer of glucose, serves as a reserve of glucose for muscle fibers of animals. Resting glycogen levels will vary depending on the health of the animal and its nutritional state prior to slaughter. Glycogen is quickly broken down to glucose in muscle by a handful of enzymes whose activity is highly accelerated by stress hormones. The glucose resulting from the breakdown of glycogen is further metabolized to create ATP for movement and other cellular needs. Generally fish have much less glycogen than land animals, which means that fish have smaller fuel reserves to continue ATP and pyruvate/lactate production after death (Chapter 3). We will see that the levels of ATP are important for a tissue to proceed to rigor mortis.

The duration of postmortem glycolysis (and thus the production of ATP and other breakdown products) depends on several factors including the level of glycogen prior to harvest. If fish struggle during the catch, they will quickly use their small glycogen stores, leaving only a short time before carbohydrate metabolism is finished and ATP production ceases. Cattle are often rested and fed before slaughter to increase the level of glycogen in the tissue before harvest. Chicken breast glycogen can decrease by nearly half, if the animal struggles during harvest. Stress at the time of harvest clearly influences the level of glucose and ATP in the muscle. Adrenaline, the fight-or-flight hormone, is produced if an animal is stressed and activates the glycogen breakdown enzymes. When combined with increased muscle activity in stressed muscle, the tissue has a very low ATP level and will enter rigor mortis very quickly. This results in cuts of meat known to the beef industry as "dark cutters," because of the dark appearance (Fig. 8.12).

The dark color of the "dark cutters" is due to the change in pH of the glycogen-depleted tissue after harvest. The pH of rested and unstressed muscle at the time of harvest is 7.2. Within 6 h of harvest, glycolysis normally produces enough lactic acid

FIGURE 8.12   Bright red steak.

for the muscle pH level to drop to a pH of 5.6. The lower pH is optimal for myoglobin to bind oxygen and produce the cherry red color upon oxygen exposure, and the low pH also inhibits harmful bacterial growth. But in glycogen-depleted tissue, there is less lactic acid produced in the postmortem muscle, and the pH is 6.0 or higher; this higher pH results in a darker color for oxygen-bound myoglobin. The dark meat is otherwise safe and palatable. Fish will typically have a slightly higher pH of 6.0 several hours after harvest.

There are two additional problems that impact upward of 20% of meat production worldwide and involve metabolism, pH, and quality of land animal meat. Pale, soft, exudative (PSE) and dark, firm, dry (DFD) meat are caused by high stress in the animal prior to harvest and are a major reason for lost meat sales.

PSE is caused by high stress in the animal and leads to increased metabolism by the muscle prior to harvest. The acidic conditions and warm temperatures of the exercised muscle rapidly denature proteins and give rise to mushy, dry meat. Interestingly, certain strains of pigs are more genetically susceptible to PSE. These pigs have a gene encoding for a protein that releases more calcium into the muscle fiber cells during stress. Increased muscle calcium when combined with the lower pH increases the rate of protein degradation and production of mushy tissue.

Stressful conditions also produce DFD conditions in meat. In these muscles the pH doesn't drop far enough. The pH in tissues with DFD is 0.5–1.5 pH unit higher than the expected 5.6. In this case, prolonged stress prior to harvest reduces the glycogen, but the animal recovers from excessive lactic acid production. After slaughter, the tissue does not have enough glycogen to produce more lactic acid and has a lower pH and expected levels of ATP production. This results in meat that enters into rigor quickly, a high pH that produces a darker red myoglobin, and protein that does not denature and tightly holds on to water. DFD meat that is dark in color, tough without protein denaturation, and with less acidic conditions is more likely to spoil with bacterial contamination.

A famous agricultural scientist, Dr. Temple Grandin, whose autism gave her a special insight into how cattle were stressed during handling prior to slaughter, has been key in advancing how cattle are handled. Dr. Grandin has created a system of how to humanly slaughter beef and other animals to reduce stress and create a higher-quality product.

The cellular content and further metabolic breakdown of ATP are critical for the onset of rigor and production of molecules for flavor and taste of meat. ATP is formally called a nucleotide and can be divided into three portions. The adenine portion of ATP is a carbon–nitrogen compound and is considered a base. The adenine is bonded through an oxygen molecule to the simple carbohydrate ribose. Also attached to ribose is a series of phosphate groups (oxygen and phosphorus atoms). Two of these phosphate groups possess energy that is released upon the bond breaking. Many cellular processes including actin and myosin (muscle thick and thin fiber) contraction utilize this energy to produce movement or cellular work.

Without oxygen (called "anaerobic conditions"), the cell produces less ATP from glucose. This metabolic limitation leads to a finite level of ATP for cells after harvest. A "living," functioning cell quickly consumes glycolytically generated ATP. Once the level of ATP is diminished, rigor mortis sets in. As you remember, each contraction of thick and thin filaments utilizes ATP. The relaxation of the contracted muscle filaments requires ATP for myosin to release from actin and for calcium to flow in and out of the cell. Once muscle fibers use the available ATP, the actin and myosin remain locked, and fibers are contracted, unable to relax and extend. Thus rigor mortis, or the stiffness of death, takes place hours to days after death depending on the animal and condition prior to slaughter. More generally rigor is the stiffness when opposing muscle pairs contract and cannot expand, creating a locked, stiff condition. Cold shortening of muscle occurs when fresh meat is chilled prior to rigor. While storing the meat at cold temperatures slows the growth of bacteria, storing the meat at lower temperatures also induces calcium release, shortening (contraction) of the muscle fibers, and shrinking of the meat. Butchers will combat this effect by using electrical impulses and other techniques. The hanging of a carcass stretches the muscle fibers during cooling as a way to combat cold shortening during rigor.

The length of rigor differs with the animal type. Fish with redder muscle fiber will have a longer rigor contraction than fish with more white fiber. Maximum rigor also varies from 6 to 8 h for fish with red muscle and 4.5–6 h for fish with white muscle. Well-fed, farmed salmon will go into rigor at 10 h postharvest and last for nearly 60 h. Rigor will set in beef and pork in 1.0–2.5 h postmortem. The handling of fish, beef, and pork is best left until after rigor has passed. Freezing or cutting the rigor contracted muscle leads to large-scale tears in the muscle fiber leaving space for water to leave and causes the tissue to become soft and mushy (Fig. 8.13).

Eventually rigor is lost and muscle groups extend and loosen. Two factors are responsible for the loss of rigor: the pH and the release of enzymes from the muscle fiber. Over time, the lower pH from glycolysis will create enough lactic acid to denature the contractile proteins, releasing myosin from actin. As muscle cells age, a special set of enzymes called proteases are released from the cells. These proteases bind and cleave the peptide backbone of other proteins. The action of many proteases is enhanced by the acidic conditions of postmortem tissue. These proteases will attack and cleave many proteins including actin and myosin. This enzymatic destruction of protein loosens the muscle fibers and ends rigor (Fig. 8.14).

The changes prior to and during rigor mortis dynamically impact each other and set up transformations after rigor that generate new flavorful molecules. After

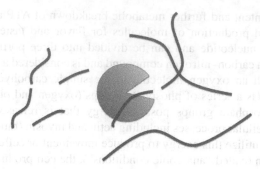

FIGURE 8.13 Cartoon of protease action. A protease, like a pacman, will "chew" proteins into smaller pieces by hydrolysis.

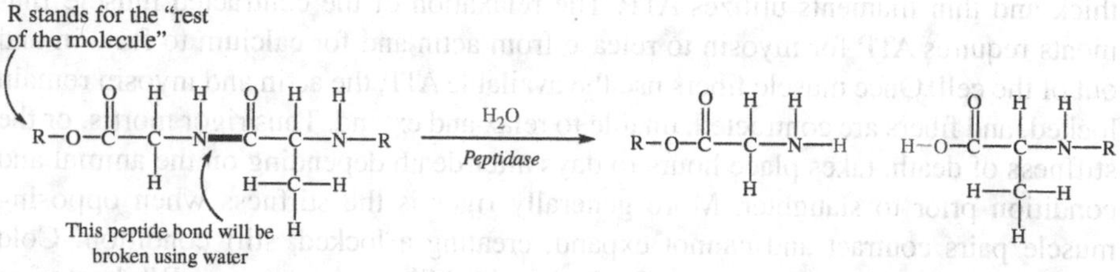

R stands for the "rest
of the molecule"

R—O—C—C—N—C—C—N—R $\xrightarrow[\text{Peptidase}]{H_2O}$ R—O—C—C—N—H   H—O—C—C—N—R

This peptide bond will be
broken using water

FIGURE 8.14 Proteolytic cleavage. Protease hydrolysis of a protein substrate.

reactions with proteins like myosin, ATP is rapidly converted to ADP and AMP. Enzymes in the cell will remove an ammonia group from the adenosine base creating inosine monophosphate (IMP). IMP is further degraded by a series of muscle enzymes to inosine and hypoxanthine. Both inosine and IMP are key components giving meat its "meaty" flavor and smell. IMP, like monosodium glutamate, has an umami flavor and is often added to foods like soup for additional flavoring. Hypoxanthine on the other hand has been reported to have a neutral to bitter taste (Fig. 8.15).

The ribose portion of ATP is also an important flavoring component to meat. Ribose and ribose-5 phosphate (both breakdown components of ATP) increase in aged meat and contribute to the roasted aroma of cooked chicken. Other sugars resulting in glycogen and glucose metabolism are produced during meat aging by a host of different enzymes. These simple sugars including ribose, glucose 6-phosphate, and fructose undergo a diverse set of reactions with proteins during heating meat, creating hundreds of different flavors and aroma molecules.

One last chemical and physical change that occurs with aging muscle takes place with the proteins in and around the muscle fibers. The fall of pH induces proteins to unravel and denature (Fig. 8.16). This results in the loss of interconnected proteins within muscle fibers and begins to soften or tenderize the tissue. Protein denaturation also impacts the ability of muscle to "hold" water. Native folded proteins are organized to keep the water-fearing or hydrophobic amino acids within the globular shape of the protein, away from water. The charged and hydrophilic amino acids are folded to face the exterior of the protein, where these functional groups can form hydrogen

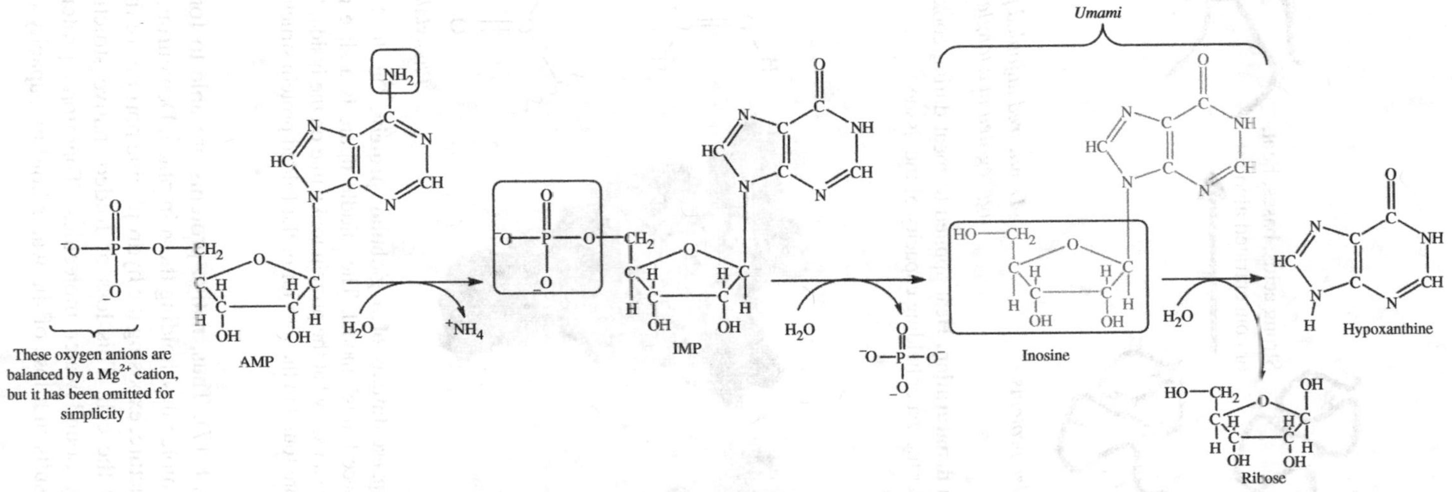

FIGURE 8.15 Production of umami compounds from AMP. The enzymatic degradation of AMP produces inosine, ribose, and other umami-flavored compounds.

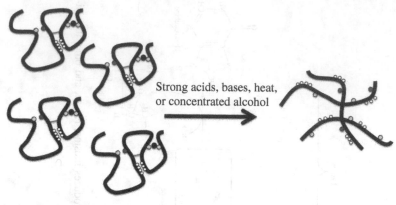

Strong acids, bases, heat, or concentrated alcohol

Native folded proteins

Denatured unfolded protein aggregated in a tangle network

FIGURE 8.16   Protein denaturation. Heat applied to meat during cooking will denature the proteins. This will alter the water holding capacity of the tissue.

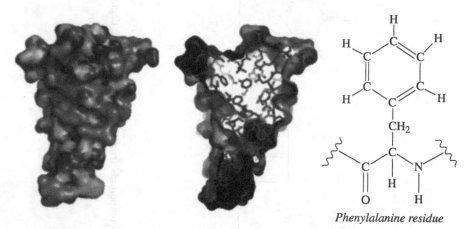

Phenylalanine residue

FIGURE 8.17   Two representations of a globular protein. On the left is the overall globular protein in a "space-filling" model. The middle figure is a slice right through the middle of protein—so we can see what the protein looks like on the inside. Hydrophobic residues like phenylalanine are found in the interior of the folded protein structure.

bonds with water (Fig. 8.17). Thus, native proteins are able to form noncovalent interactions with water molecules, holding them in place. Denaturing proteins by pH opens up the folded structure, exposing the hydrophobic portions of the protein to the watery environment of the cell. This loss of folded, native structure also results in fewer noncovalently bound water molecules. Therefore, protein denaturation by acidic conditions results in a loss of the water holding capacity (WHC) of meat (Fig. 8.18).

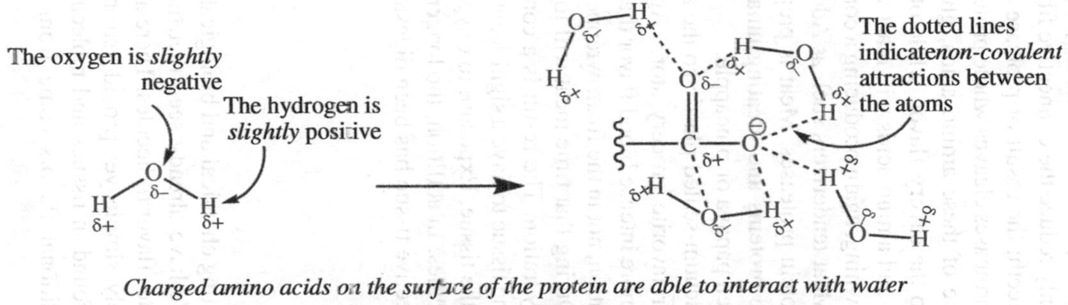

The dotted lines indicate non-covalent attractions between the atoms

The oxygen is slightly negative

The hydrogen is slightly positive

Charged amino acids on the surface of the protein are able to interact with water

FIGURE 8.18 Water holding capacity of proteins.

In addition to acid-induced denaturation of some of the proteins, the proteases released by cells also alter the structure of meat proteins. Proteases play a major role in tenderizing aged meat. Calpains digest the contracting fiber proteins, while cathepsins are proteases that bind and cleave the protein backbone of a number of muscle proteins including collagen and other connective tissues. Together, these proteins degrade many of the connective proteins such as tropomyosin, actin, myosin, and collagen. The proteases not only reduce the connective fibers and myofibrils making the tissue easier to cut with teeth; the result of protease activity are important new flavors. Longer digestion by proteases cleaves whole proteins into small peptides and individual amino acids. Some of these amino acids, glutamate and aspartic acid, produced by proteases add to the savory flavor, while others have a sweet or sour taste. Many of these peptides and amino acids will react with the carbohydrates like ribose and glucose during browning, again producing a complex mixture of the flavor and aroma of cooked meat. Meat tenderizers such as Adolph's Meat Tenderizer are powdered preparations of protein proteases. Meat is prepared by soaking in proteases, cleaving the connective proteins and creating umami flavors. Most of these proteases are from plants, often papaya or pineapple.

Wet aged meat is left in vacuum-sealed plastic so the aging process can occur in the meat department of your favorite grocery store. Aging meat on a foam tray wrapped in plastic leads to a more intense blood flavor than dry aged beef. Dry aging (air exposed in a cool dry location, not in the foam-wrapped tray) results in water loss and can take up to 45 days. During that time meat will lose a significant fraction of its weight in water from dehydration. The result is a concentration of flavor molecules. Aging will also cause the tissue to have a significant protease activity releasing amino acids and tenderizing the tissue. Exposure to oxygen also increases the reaction of oxygen with fat molecules. In addition, the longer process of dry aged meat means that more of the connective tissue has been digested by proteases, yielding a more tender piece of beef.

8.7 FLAVOR

There are a number of interesting chemical and biological reasons for the molecules that create flavor in meat. We have already seen that the metabolism of carbohydrates, ATP, and proteins creates flavor molecules. There are several other interesting molecules that are deceptively simple yet provide strong flavors to meat. These include simple amino acids found in tissues and molecules derived from the food of the animals we eat. Additional flavors come from reactions that take place during cooking.

Shellfish live in a unique environment. The brackish water that many shellfish make as their home requires them to have a special ability to tolerate a wide range of salinities (saltwater concentration). Osmosis is the net movement of solvent molecules, such as water, through a semipermeable membrane from a region of low solute concentration to a region of higher solute concentration. The tissues of most organisms are semipermeable to water (and impermeable to salt ions and other compounds) due to the plasma membrane surrounding each cell. Particles such as salt ions, proteins, and carbohydrates contribute to the solute concentration inside the cell, but the solute concentration inside the cell is still lower than the solute concentration of the saltwater environment, a so-called hypertonic solution. For many organisms, this would mean that the cellular water would, through osmotic pressure, move from the cell into the higher solute containing salt water. This would cause the cells to shrink and die. To survive the saline living conditions, shrimp, prawns, and other shellfish have high concentrations of free amino acids. Specifically glycine and alanine are found in high concentration in the muscle tissue of shrimp. Scallops adjust to the high concentrations of glycine and the complex carbohydrate glycogen. Besides being a way to respond to high salt conditions, meat of these shellfish is particularly flavorful and sweet because of the glycine and some of the simple sugars found in the tissue after glycogen breakdown. Some of the sweetness of shrimp is lost during heating due to chemical changes. Many of the amino acids become entangled by heat-denatured proteins or react with other compounds, changing the structure of glycine.

The flavor and aromas of freshwater and saltwater fish come from naturally occurring compounds, microbial spoilage, reaction with oxygen, and processing reactions. Enzymes called lipases bind and combine unsaturated fats with oxygen and water molecules to create most of the fresh fish smells. Fish fats broken down by lipoxygenases into shorter six, eight, and nine carbon chain compounds now made of aldehydes, ketones, and alcohols each have a unique and short-lived fresh fish aroma and flavor. Freshwater, but not saltwater, fish have enzymes that produce trans-2-hexenal and cis-3-hexenal to give a green, plantlike smell. Other fatty acid metabolites provide plantlike flavor and odor to fish and are found in some vegetables. So the most fresh fish are often compared to the smell of fresh leaves.

Saltwater fish as well as crustaceans and mollusks have a unique flavor and odor associated with them, caused by family of related bromophenol compounds. This odor is missing in ocean farming (Fig. 8.19). Several studies on the organoleptic quality of aquacultural raised fish have reported distinct differences and lower flavor quality in the fish. 2,6-Dibromophenol seems to be responsible for shrimp and crab flavors, while the singly brominated phenols 2-, 3-, and 4-bromophenol enhance the seafood quality of finned seafood (Fig. 8.20). Interestingly, none of these compounds has been detected in freshwater fish, and the bromophenol content in wild seafish is 1000 times higher than that found in farm-raised ocean fish. These differences indicate that fish do not produce bromophenols on their own. Instead, marine worms and algae synthesize the compounds that then are transferred and stored in fatty tissue of predator fish. Including either the algae or low concentrations of bromophenols to the fish food has increased the attractiveness of the fish flavor in farm-raised ocean fish.

Microbial contamination of fish and shellfish contributes to the offensive fishy smell and flavor of older fish. This happens as sulfur-containing compounds are released, phenols are modified by the bacteria, and fatty acids that have putrid aromas and flavors are generated. The strongest flavor and odor caused by microbial growth are due to the conversion of trimethylamine oxide (TMAO) to trimethylamine (TMA) (see Figure 8.21). Saltwater, but not freshwater, fish (except Nile

ing.

FIGURE 8.19 Aquafarming. Farming fish even in the ocean will not produce a "fishy" taste without additional bromophenols in the diet.

OH Br — 2-Bromophenol

OH Br — 4-Bromophenol

OH Br Br — 2,4-Dibromophenol

Br OH Br — 2,6-Dibromophenol

Br OH Br Br — 2,4,6-Tribromophenol

FIGURE 8.20 Simple bromophenols in shellfish. Mono-, di-, and tribromophenol structures involved with ocean fish aroma.

Trimethylamine oxide *an osmolyte used by ocean fish to counteract the effects of salt water* — Bacteria and fish enzymes → Trimethylamine — A source of acid *(i.e., lemon juice, vinegar)* → Trimethylammonium cation Carboxylate anion — *This ionic compound (i.e., a salt) dissolves readily in water and can be washed away*

FIGURE 8.21 Amine compounds and acidic solutions.

perch and tilapia) possess fairly high levels of TMAO, which acts as an additional regulator for osmosis and stabilizes protein structure in finned fish. Deep-sea fish have particularly high levels of TMAO giving strong recognizable smells and tastes. The related sharks and rays use the amino acid breakdown product urea in place of TMAO. Bacteria and some of the fish produce an enzyme that decomposes both urea and TMAO. TMAO decomposes to TMA providing the strong fishy smell of old fish, while urea is converted into ammonia by the bacteria giving older shark flesh the strong odor of cleanser. Cooking fish with an acid such as lemon juice, tomato, or vinegar can reduce the distinctive fish smell. The acid combines with TMA to form an organic salt that is less able to volatilize and be detected. Rubbing your hands with lemon is a good way to remove the fish smell from your hands and utensils in your kitchen!

Sugars like ribose and amino acids produced artificially or naturally occurring in meat are prime targets to react under heat to create an amazing array of new flavor and odor molecules. The special reaction primarily between amino acids, sugars, and lipids under dry, higher temperatures is known as the Maillard or browning reaction. The Maillard reaction was first described over 100 years ago and is responsible for the smells and flavors of cooked meat. The aging of meat and the metabolic changes of proteins, carbohydrates, and ATP all provide the necessary starting materials for the Maillard reaction. Each time you brown a meat like fish, beef, or pork, the reaction between sugars and amino acids creates hundreds of new molecules. For example, ribose from ATP will react with the amino acid cysteine from protein degradation, creating over a dozen different products that can be smelled even when there are only 10 or so molecules per trillion air molecules.

However, meat must be relatively dry to reach the high temperatures necessary for Maillard reactions (around 300°F/149°C), and thus boiled or braised meats will not have some of the nutty and meaty flavor of browned meat. Boiled or poached fish and beef will have a very different flavor than when cooked and browned. The surface of the meat must be heated high enough to evaporate or boil off the water and allow the temperature of the tissue to get above 212°F/100°C.

Maillard reactions are also a challenge for the modern method of *sous vide* cooking. For this modern molecular approach to cooking, food is placed in bags, often with a vacuum to remove air and placed in a controlled circulating water bath. The idea is to cook the entire piece of meat to a uniform temperature. However, the food must be taken out of the bag and quickly browned at the surface to generate the Maillard products. Perhaps the perfect combination of *sous vide* and Maillard browning can be found in this hamburger recipe. First chopped lean meat is formed into a thick burger and cooked in the *sous vide* water bath in a sealed bag to 133°F (56°C). This allows the large beef patty to be evenly cooked from the edge through to the middle without drying or overcooking the edges of the meat. This is followed by a quick dip in liquid nitrogen. A fast exposure of the cooked burger to the low temperature of liquid nitrogen (−321°F/196°C) serves not only as a coolness factor but also freezes only the surface of the meat without cooling the interior. The patty can then be deep-fried or, if you are brave, subjected to a flame from a gas torch for just a minute, creating an evenly cooked and browned modern burger.

8.8 SEARING TO SEAL IN THE FLAVOR—NOT!

Many recipes call for food to be browned on a hot skillet to seal in the juices or lock the flavor into the meat. This old saying could not be further from the truth. The act of subjecting the meat to high enough heat to cause the Maillard reaction will also damage the tissue of the meat. As we will learn in Section 8.9, heat causes the collagen to shrink, disrupting the integrity of the muscle fiber. The inevitable loss of water is detectable by the sounds of water boiling, sizzling during browning, and the leaking of water from the meat after browning. However, browning the meat before roasting does create, not lock in, the flavor molecules.

8.9 STAGES OF COOKING MEAT

The collagen type and content of meat (remember that fish have much less than beef or pork) is reflected in the final temperature for cooked meat. Fish requires a lower temperature as the muscle's connective tissue is more easily degraded than pork or beef collagen. See Figure 8.22 for an interesting comparison between fish and meat cooking. There are predictable phases of cooking meat. At 58–60°C/135–140°F, myosin has already begun denaturing and coagulating. The denaturation of myosin and other muscle fiber proteins leads to a decrease in their ability to hydrogen-bond to water, and the newly exposed hydrophobic portions of myosin will begin to aggregate, firming up the meat. Collagen also begins to denature. However, due to the shape of the long intertwined strands of collagen protein, denaturation causes collagen to shrink.

The combination of myosin denaturation and loss of water binding with the shrinkage of collagen wrapped around the muscle fiber results in a loss of juice and an overall shrinkage of the muscle tissue. Meat heated to this level (medium rare) will be on the cusp of moving from juicy to dry. A slow continual heating of the meat will continue to denature the collagen-containing connective tissue. At temperatures near 140°F/60°C, collagen continues to denature but also slowly decomposes and dissolves into gelatin. Thus, slow cooking at a low temperature can reduce tough, collagen-rich meat into softer, more tender meat as the collagen is dissolved. Continued cooking will drive off the water, raise the internal temperature of the rest of the meat, and dry out the outside portion of the meat. Meat above 165–170°F/74–77°C will have lost most of the water, all of the protein will be denatured, and the myoglobin converted to the brown hemichrome. This is the USDA definition of "well-done" meat.

Collagen is unique among most proteins because in its native state, collagen is an insoluble fibrous protein and when denatured the protein loses some of its three-dimensional structure and becomes water soluble. At about 160°F/71°C, beef collagen will denature and convert into liquid gelatin. Lower temperatures suffice for collagen from pork or fish. If carefully heated, gelatin will form into a gel upon cooling; however if overheated, the long collagen strands will completely denature and aggregate into an insoluble mess. JELL-O™ is the product of collagen conversion to gelatin. Gelatin has a number of culinary uses: it provides the soft, chewable texture of gummy

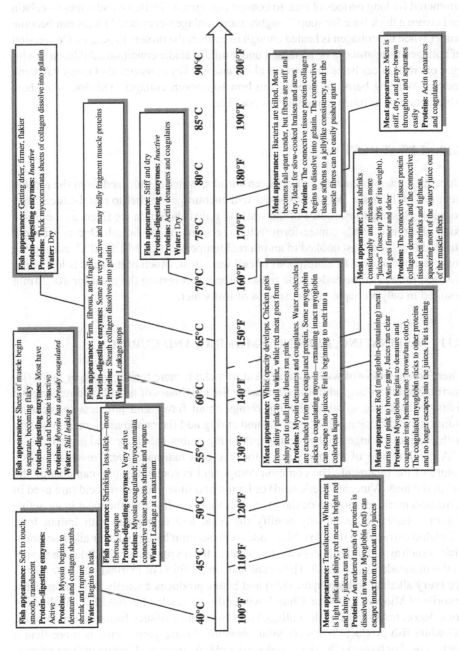

FIGURE 8.22 Stages of fish and meat cooking.

bears, marshmallows, and candy corn and is often used as a stabilizer acting to maintain emulsions much like agar or pectins. Soup bones with connective tissues are often simmered for long periods of time to convert collagen in the tissue and bones to gelatin and create a thick base for soup. Tougher more collagen-connected meats can become tender when the collagen is heated enough to convert the protein to gelatin. Conversion of collagen to gelatin can be increased under mildly acidic conditions. Addition of vinegar or citrus juices helps this process along and is a key component of many barbecue recipes. Winning barbecue cooks learn how to convert collagen with low, slow heat while retaining as much of the juice as possible.

8.10   LET IT REST

Once the proteins denature and collagen shrinks, water is squeezed from the muscle fiber out of the tissue. Most cookbooks will encourage the chef to rest the meat after cooking. Resting allows some of the water to redistribute into the fibers as the tissue cool. Hydrogen bonds can re-form, retaining some of the liquid. One experiment showed that for a roast cooked to an internal temperature of 140°F/60°C, cutting the roast resulted in over 10 tablespoons of cell water lost. Much of the fluid is lost as the tissue is opened at the ends of the muscle fibers. Yet resting the meat for 10–20 min resulted in only a couple of tablespoons of lost water.

8.11   MARINATING, BRINING, SMOKING, AND CURING

There are a number of ways to treat meat to make it more tender, juicy, or more flavorful. Marinades are helpful in decreasing the toughness of meat by immersing the tissue in an acidic solution. Brining brings about flavor and juiciness to meat by osmosis and diffusion, while smoking and curing add flavor and an element of safety with chemical changes to the meat by inhibiting pathogenic microbial growth.

A quick review of the many posts for "the best marinade ever" reveals the true nature of the treatment. The common component is something that can change the pH of the meat. Vinegar (acetic acid) or lemon and lime juice (citric acid) are used in most acid marinades to reduce the pH to less than 4.5. Soaking the meat in an active yogurt culture will eventually acidify the meat and is popular with Indian food (vindaloo curry, tikka masala). Marinades with sodium bicarbonate or alkaline phosphates such as sodium triphosphate (also called tripolyphosphate (TPP)) raise the pH of the marinade from 8 to 10. The Scandinavian tradition of marinating white fish in lye (very alkaline sodium hydroxide) and butter produces a smelly, soft, translucent favorite of Midwest Lutheran Church winter dinners called lutefisk. The alkaline lye soak deconstructs most of the collagen and connective tissues forming a gelatinous, off-white fish product. Whatever your pleasure, ensure your recipe is more than a dash of acid or base to effectively reduce the pH. A squeeze of lemon or lime juice or a pinch of bicarbonate will not significantly bring the pH to the point of action on the meat protein.

The main function of a marinade is to weaken the surface connective tissue by denaturing proteins with acidic or alkaline/basic solutions. The additional ingredients of marinades add to the flavor and salt concentration of the meat. Marinating fish, chicken, beef, or pork has the potential to soften the meat and add flavor, but strangely, marinating can also result in a dry piece of cooked meat. Why would this be? Acid and bases each effectively have the same impact on protein. The addition of moderate amounts of acid or base will change the native structure of the protein and reduce the amount of water held by the charged groups of the protein. At higher concentrations of either acid or base, the protein will begin to degrade and hydrolyze (cleaving the protein backbone into smaller peptides), thus reducing the toughness of the meat.

To explain the action of a marinade, we need to review a little acid–base chemistry. At either end of the pH spectrum, carboxylate groups and the amino side groups of amino acids that make up proteins can change their charge. At lower pH, carboxyl groups will shift from a negative charged ion to a neutral polar compound $(R-COO^- + H^+ \rightarrow R-COOH)$. The amino side charges are relatively unaffected by the lower pH, but at pH above 8.0, they will shift from a positive charged compound to a neutral charged group $(R-NH_3^+ \rightarrow R-NH_2 + H^+)$. The impact of changing the pH below 5.0 or above 8.0 is the denaturing of the protein and loss of water binding capacity of the protein. This happens because the charged group on the protein can bind salt and water through ionic interactions and hydrogen bonds. See Figure 8.23 for a detailed look at these interactions.

One of the problems with marinades is the loss of water or juiciness from the meat. The WHC of meat is a measure of how much water is held in place in meat tissue through noncovalent bonds between water and protein. Most of the water in a muscle fiber cell is tied up with the charged carboxylate and ammonium side chains of amino acids within meat proteins. These two chemical groups, when charged, can interact with other ions and hydrogen-bond with water. This is an effective way to trap water in the meat tissue. However, when the pH is lowered by an acidic marinade, carboxylate ions can become neutral carboxylic acids, while raising the pH with an alkaline marinade turns the ammonium ion into a neutral amine. These neutralized amino acid side chains now have less hydrogen bonding potential. Water, no longer held in place by noncovalent interactions, is free to flow out of a ruptured muscle cell. In addition, as the pH moves to more acidic or alkaline/basic conditions, the protein will begin to denature, exposing the hydrophobic interior of the protein. In this condition, the unraveled proteins also aggregate and hydrophobic interaction potential increases.

Another impact of denaturing protein is the loss of structure within thick and thin myofibril filaments. As the proteins denature and aggregate due to hydrophobic interactions, the myofibrils (thick and thin filaments) tend to shrink, and because these long protein fibers are connected to the cell wall, this shrinking squeezes water out of the cell. A filamental protein called calpain binds the actin and myosin fibers to the cell membrane. When the fibers denature and shrink, they are still connected to the cell wall. Thus, the muscle cell fiber shrinks. An interesting marinade would include a two-step process: first marinade in an enzyme-containing solution (proteases—or tenderizers) to partially denature the connective tissue, followed by an acidic or alkaline/basic marinade to further denature the proteins. The initial protein digestion step would

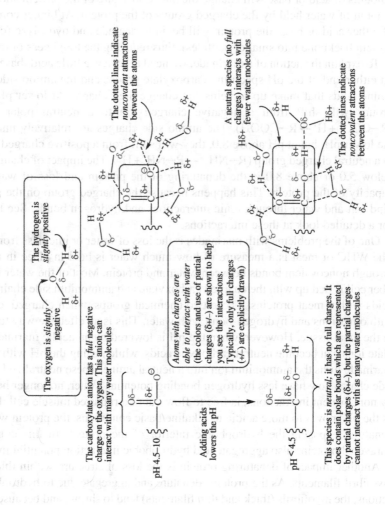

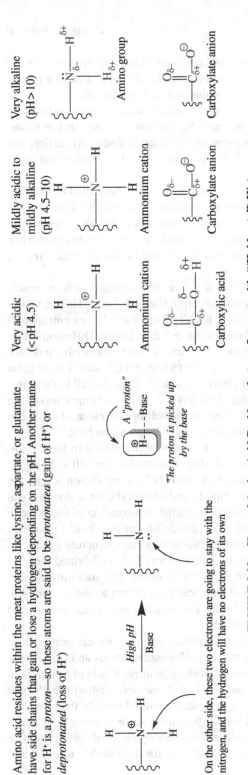

Amino acid residues within the meat proteins like lysine, aspartate, or glutamate have side chains that gain or lose a hydrogen depending on the pH. Another name for H⁺ is a *proton*—so these atoms are said to be *protonated* (gain of H⁺) or *deprotonated* (loss of H⁺)

On the other side, these two electrons are going to stay with the nitrogen, and the hydrogen will have no electrons of its own

FIGURE 8.23 Charged Amino Acid Residues that can Interact with ("Hold onto") Water.

allow you to avoid the squeezing of calpain and collagen. When experimenting in the kitchen, try including an emulsified oil with your marinade. Shirley O. Corriher reports that this leads to a deeper marinade in less time than that without the oil.

Brining is the process of soaking meat in a seasoned saltwater solution and is often used to impart additional flavors and increase the amount of water in the tissue. Brining is a combination of osmosis (movement of water as discussed earlier) and diffusion (of salts, spices, and flavors). There is quite a bit of confusion and some controversy to the science of brining.

As we learned with marinades, some of the water in a cell is associated or "bound" to proteins, while some of the cellular water is considered "free water." Free water is able to move around and be the solvent of the cell. Osmosis is the net movement of water through a semipermeable membrane (e.g., a muscle fiber membrane) from a low to high salt solution, but the complex nature of cellular water makes understanding the brining process complicated.

Most brines will have a 3–10% NaCl solution. The Na^+ concentration in muscle cells is about 12 mM and the intracellular Cl^- concentration is 4.2 mM. This translates to a 0.026% Na^+ and a 0.015% concentration for Cl^-. Thus the salt concentration of typical brine is over 650-fold higher outside of the cell than inside. Diffusion drives ions from a high concentration (the brine) to a low concentration (the muscle). However, muscle membrane is pretty impermeable to Na^+ and Cl^- ions. Under these conditions, the net movement of water by osmosis would be from the cell to the brine. This should lead to a dry piece of meat. Yet, anyone eating a brined turkey on Thanksgiving will tell you how juicy the meat is. A well-brined piece of meat can increase its weight by 10% or more… something has to be going on here.

While the muscle fiber plasma membrane is mostly impermeable to ionic compounds, the concentration of sodium and chloride ions outside the cell (i.e., in the brine) is so great that some of the sodium and chloride ions are driven across the membrane by diffusion. The salt ions then bind to and disturb the three-dimensional structure of muscle fiber proteins. This results in partial denaturation of the protein. Most of this work has been attributed to the chloride anions that bind to positive charges of the myofibril proteins. The altered filament protein structure breaks its contacts with other proteins, allowing the filaments to swell. A well-brined piece of red meat will have a final concentration of 4.5% to nearly 6% sodium chloride. Osmosis (or the movement of water *into* the muscle cell) occurs as the new muscle fiber salt concentration is higher than the brine. The increase in chloride anions also helps the muscle fiber protein bind more water.

Muscle from different organisms or, for that matter, different muscle groups from the same animal will have very different starting salt concentrations and osmolarity. Fish brines are very different from a brine for pork or poultry. A rule of thumb is to start with a lower salt concentration of around 1%. Remember that with a 1% salt solution, it will still be hundreds of times more concentrated outside than inside a typical muscle cell. Protein changes and swelling seem to occur at about 5.5% NaCl in chicken breast and leg muscle. Using different concentrations of salt up to 10% will provide a solid range to test small pieces of meat for their ability to swell and retain water after cooking.

Salting or *curing* is a very old method of meat preservation before the time of refrigeration. The meat was salted (with NaCl) to remove moisture and therefore inhibit the growth of bacteria. With the advent of refrigeration, salt *curing* is unnecessary to preserve meat for months at a time, but salted cured meats are still made for their flavor alone. Often the meat is injected or immersed in brine solutions for a short time—the short curing time means these modern "salt cured" meats must be refrigerated and cooked to ensure meat safety since the curing was not strong enough or long enough to kill bacteria.

Curing evolved as a way to store the harvest of a hunt for longer than could be done with fresh or even cooked meat. There are a variety of different meat cures available, but all are trying to prevent bacterial growth and spoilage of the meat tissue. Many cure recipes use salts or sugar to draw water out of the meat. This leaves the tissue inhospitable for harmful microbes that would thrive on uncooked meat. Some curing recipes allow for beneficial bacteria, such as *Lactobacillus*, also used for cheese production, to grow and produce lactic acid. These cures include sugar to serve as a food for the bacteria to reduce the surface pH to about 4.5. The combination of an acidic pH with low concentration of water and high salt concentration leaves meat dry and able to be stored for long periods of time.

As in brining, salt draws water from the meat. However, unlike wet brining, salt curing requires much higher concentrations of salt. Wet brines range from 15 to 30% salt. Dry brining requires the meat to be packed in a salt mixture. The term corned beef traditionally refers to a brisket (a tough cut of beef from the lower chest of cattle) that has been placed in a barrel of coarse salt granules. Corned comes from the number of kernels or seeds of granular salt used to pack the beef. Over time, more and more water will move from the muscle fiber to the salt cure. The proteins will fully denature and entangle with each other. When combined with the dry texture, this gives dried meat its characteristic firmness and chewiness.

In addition to the use of rock or table salt (NaCl) in curing meat, sodium nitrite ($NaNO_2$), sodium nitrate ($NaNO_3$), and potassium nitrate (commonly called saltpeter or potash) have been used for centuries to inhibit harmful bacterial growth in cured meat. In the sixteenth to seventeenth centuries, cooks realized that using *saltpeter* (which turned out to be potassium nitrate, KNO_3) in addition to salt improved the color, flavor, safety, and storage life of meat. In the 1900s, chemists determined that the active ingredient in saltpeter (KNO_3) was a small amount of nitrite or NO_2. The nitrite reacts with the meat to form NO (nitrous oxide), which binds the iron (Fe) ion in myoglobin—giving it a permanent pink color—and the nitrite also inhibits the growth of bacteria. Meat cured with nitrates and nitrites retards the development of rancid off-flavors due to oxidation of fat during storage and also has the advantage of preventing the growth of *Clostridium botulinum*, the bacteria responsible for botulism.

Early curing used primarily sodium and potassium nitrates at high levels. However, a salt-tolerant bacteria was found to be responsible for converting nitrates to the more reactive nitrites. It is the nitrite ion (NO_2^-) that reacts to form nitric oxide (NO), a short-lived but highly reactive gas. NO slows the oxidation of fat that normally takes place with the iron bound by myoglobin. NO binds very strongly to the heme iron in

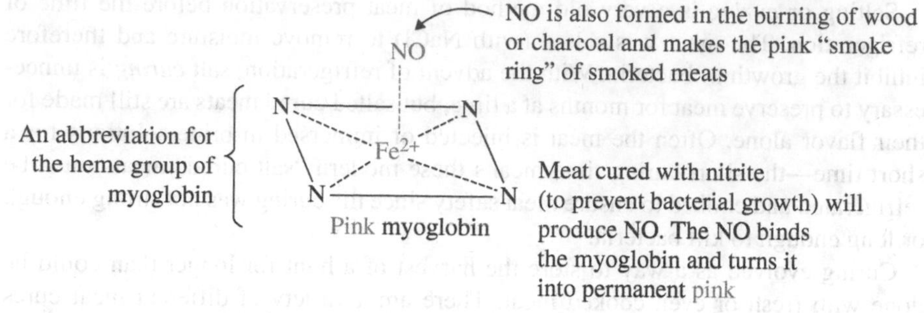

NO is also formed in the burning of wood or charcoal and makes the pink "smoke ring" of smoked meats

An abbreviation for the heme group of myoglobin

Pink myoglobin

Meat cured with nitrite (to prevent bacterial growth) will produce NO. The NO binds the myoglobin and turns it into permanent pink

FIGURE 8.24   Nitrate and meat. The replacement of oxygen by nitrogen compounds and the subsequent binding to iron give smoked meat its pink color.

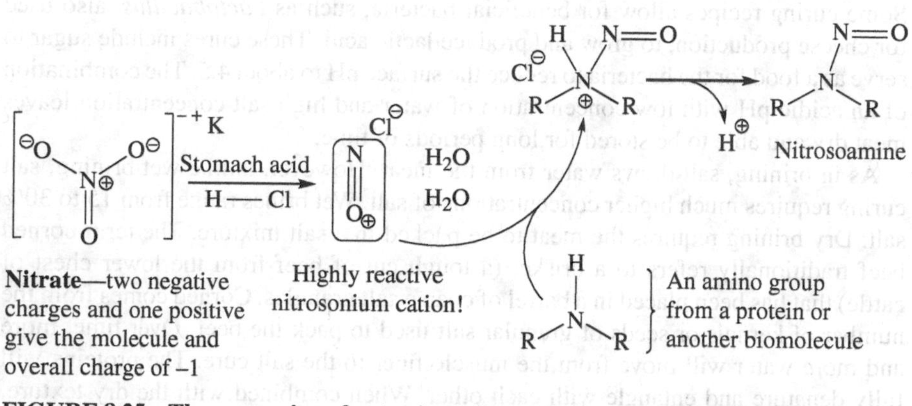

Nitrate—two negative charges and one positive give the molecule and overall charge of −1

Highly reactive nitrosonium cation!

Nitrosoamine

An amino group from a protein or another biomolecule

FIGURE 8.25   The conversion of potassium nitrate to nitrosamine in the acidic stomach.

myoglobin, eventually forming a pink NO–Fe myoglobin complex that remains pink even after cooking (Fig. 8.24). For short-term storage, nitrites are added to meats and sausages for safety concerns. Only for longer-term storage will nitrates be used, and then a culture of the transforming bacteria is included during processing.

There are concerns about including nitrites or nitrates in food. Nitrates and nitrites can react with the nitrogen-containing amino groups of amino acid side chains to form nitrosamines. Nitrosamines are a family of hundreds of closely related compounds, most of which are carcinogenic in animal studies. Nitrosamines bind to DNA, causing mutations, some of which can induce tumor formation. The early studies on nitrosamines and cancer were a little misleading as they used high concentrations of the compound for many days in a row to induce tumor formation. The formation of the nitrosamine primarily happens under high heat or in the highly acidic conditions of the human stomach (pH ~2; Fig. 8.25). Most meats have low concentrations of the nitrates/nitrites, and addition of the antioxidant vitamin C (i.e., ascorbic acid) inhibits nitrosamine formation. Modern regulations limit nitrate concentrations in meat to very low amounts (200 ppm), and the inclusion of ascorbic acid is required for all commercially prepared meats. The National Academy of

Sciences estimates that on average US citizens are exposed to about less than a microgram of nitrosamines per day. Most of the compound included in our diet comes from bacon and beer. The latter is produced during malting. Thus cooking with lower heat and ascorbic acids decreases the formation of nitrosamines, particularly when eating bacon. Limiting your intake of beer is a personal choice.

REFERENCES

[1] Alton Brown on the end of meat as we know it. Alton Brown blog post. Science Wired. Available at http://www.wired.com/2013/09/fakemeat/ accessed on July 5, 2015.

[2] Hedrick, G.B., Aberle, E., Gorrest, M.J. and Merkel, R. ed. (1989) *Principles of Meat Science*, 3rd edn. Kendall/Hunt Pub. Co., Dubuque, IA, pp. 126–131.

9

EGGS, CUSTARDS, AND FOAMS

Guided Inquiry Activities (Web): 5, *Amino Acids and Proteins; 6, Higher Order Protein Structure; 8, pH; 9, Fats Structure and Properties; 12, Emulsions and Emulsifiers; 25, Eggs; 26, Custards and Egg Foams*

9.1 INTRODUCTION

From fluffy omelettes to the French dish of egg white battered Chicken Francaise and delicate meringues, eggs can be found in recipes from breakfast to dinner and dessert. Their perfect combination of water, protein, and fats makes eggs an excellent source of nutrition, a means of thickening, and a source of moisture in baked goods.

The egg houses all the ingredients needed to make a living, breathing creature. It is the essence of life itself. The yolk is a densely packed orb of fuel and raw materials. As Harold McGee unpacks in his tome *On Food and Cooking* [1], the word "yolk" comes from the old English for "yellow," which itself is derived from the Indo-European root meaning "to gleam, to glimmer." This same root gives us our words *glow* and *gold*. Even the ancients recognized the similarity between the egg's yolk and our sun—these yellow orbs are responsible for life.

Historians suppose that eggs have been part of the human diet from the very beginning of our existence. Eggs were cooked into dishes consumed by Romans in the first century, while omelettes can be found in French cookbooks from the fourteenth and fifteenth centuries. In fact, the French monarch Louis XIV was fond of boiled eggs for breakfast, and Parisians were said to marvel as Louis knocked off the small end of a boiled egg with one swift stroke of his fork [2]!

The Science of Cooking: Understanding the Biology and Chemistry Behind Food and Cooking,
First Edition. Joseph J. Provost, Keri L. Colabroy, Brenda S. Kelly, and Mark A. Wallert.
© 2016 John Wiley & Sons, Inc. Published 2016 by John Wiley & Sons, Inc.
Companion website: www.wiley.com/go/provost/science_of_cooking

FIGURE 9.1   A Partridge Cochin cockerel. Sammydavisdog, https://en.wikipedia.org/wiki/Cochin_chicken#/media/File:Partridge_Cochin_cockerel.jpg. Used under CC BY 2.0 https://creativecommons.org/licenses/by/2.0/deed.en

Despite the egg-loving Louis XIV, it was not until the nineteenth century that the chicken and its egg became the fascination of Europeans and Americans alike. In 1834, chickens of the Chinese breed (known then as "cochin") arrived in England as a present for Queen Victoria. These chickens were superior in egg production (and meatier!) than the European and American varieties of the day. The now-familiar bright red comb on the chicken's head was so shocking in 1834 that it inspired chicken mania! When cochins arrived in the United States at the Boston poultry show of 1849, they attracted crowds that numbered in the thousands. This excitement catalyzed what has become known as *The Century of the Chicken* [3], which was also the century of the egg. While farmers were breeding chickens for egg and meat production, eggs were incorporated into cooking like never before. By the 1940s egg production had become industrialized to support the mainstreaming of eggs in the modern diet. Many animals lay eggs, but the chicken egg is by far the most commonly eaten around the world, so this chapter will focus on chicken eggs (Fig. 9.1).

9.2   WHAT IS AN EGG?

Any egg is a type of *cell* that is specialized for *sexual reproduction*—the process by which two parents contribute genetic material (*genes*) to make a new individual (Fig. 9.2). *Germ cells* (e.g., egg and sperm) contain *half* the genetic material necessary to make an individual—the cells are *haploid*. Only eukaryotic organisms undergo *sexual reproduction*, and not even *all* eukaryotic organisms, only the most complex.

Of the two reproductive *germ cells* that combine their genetic material (*genes*) to make a new individual, the *egg* is the larger, less mobile one. The *egg* cell receives the *sperm cell* (the carrier of the other half of the genetic material), accommodates the joining of the two *haploid* sets of *genetic material*, and subsequently divides and differentiates into the new embryonic organism.

Sexual reproduction

Each cell has *half* the genetic material   *Fertilized egg* has a unique
necessary to make a new individual   combination of *all* the genetic material
                                             necessary to make a new individual

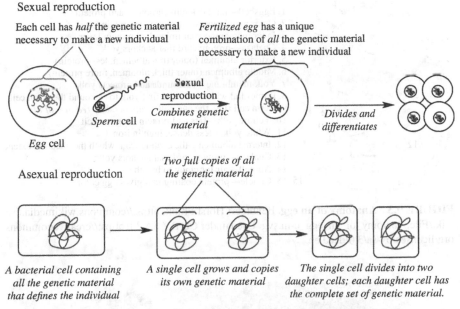

FIGURE 9.2   **A comparison between sexual and asexual reproduction.**

Asexual reproduction is how bacteria, yeast, and other simple organisms multiply.
By definition, the product cells of *asexual reproduction* are exactly genetically identical to the single, original parent—the daughter cells are *clones* of the parent. On the
other hand, sexual reproduction combines two sets of genetic material into a new,
unique individual. Chicken eggs are created by the female hens. The hen will produce the eggs whether they are fertilized or not—mass produced, grocery store eggs
are *not* fertilized. There is no nutritional difference or noticeable difference in
physical appearance between fertilized and unfertilized chicken eggs. Folklore indicates that a "blood spot" in an egg indicates fertilization, but this is incorrect. The
blood comes from the rupture of a blood vessel during formation of the egg and has
nothing to do with fertilization. Blood spots are not harmful nor do they affect taste.
They can simply be removed with a spoon.

A chicken will lay eggs (fertilized or not) until she has accumulated a certain
number of eggs in her nest. If the eggs are removed—perhaps by a predator or a
human—the hen will lay another to replace it and may do so indefinitely.

The familiar chicken egg has a *yolk* surrounded by egg *white*, contained within a
hard shell (Figs. 9.3 and 9.4):

- The yolk: accounts for approximately one-third of the weight of an intact
  chicken egg. Comprised of mostly fats and proteins—it carries 75% of the
  calories and most of the iron, thiamin, and vitamin A. Its purpose is to provide
  food for a developing chick.

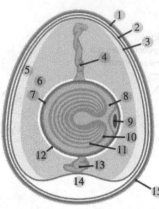

1. Eggshell—hard calcium carbonate and protein
2. Outer membrane—antimicrobial protein layer
3. Inner membrane—antimicrobial protein layer
4. Chalaza—protein cord that anchors yolk
5. Exterior albumen (outer thin albumen, less protein)
6. Middle albumen (inner thick albumen, more protein)
7. Yolk membrane—surrounds and protects yolk
8. Primordial white yolk: the first yolk to surround the germ cell
9. Germ cell (i.e., the egg cell)—*not actually red in real life*
10. Yellow yolk—fats and protein for germ cell
11. White yolk—less dense, high in iron
12. Internal albumen—the coating from which the chalazae extend
13. Chalaza—protein cord that anchors yolk
14. Air cell—air for chick to breathe
15. Cuticle—protein coating that gives egg color

FIGURE 9.3   Anatomy of an egg. Benutzer: Horst Frank, https://commons.wikimedia.org/wiki/File:Anatomy_of_an_egg_c-m.svg. Used under CC BY SA 3.0 https://creativecommons.org/licenses/by-sa/3.0/deed.en

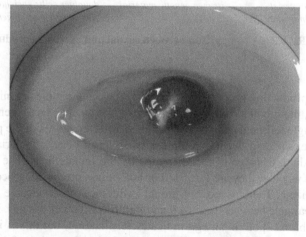

FIGURE 9.4   Raw egg. The chalazae are visible on the yolk. Miya—Miya's own file, https://commons.wikimedia.org/wiki/File:Chicken_egg01_monovular.jpg#/media/File:Chicken_egg01_monovular.jpg. Used under CC BY 3.0 https://creativecommons.org/licenses/by/3.0/deed.en

- The white: accounts for approximately two-thirds of the weight of an intact chicken egg. It is 90% water—the rest being protein. There are only traces of minerals, fatty material, glucose (a sugar), and vitamins. The white provides essential proteins and water and also provides protection to a developing chick.
- The shell: made of calcium carbonate and protein, the shell is riddled with pores (tiny holes) that allow gases to pass in and out of the egg (Table 9.1).

TABLE 9.1 The Composition of a US Large Egg[a].

	Whole Egg	Egg White	Egg Yolk
Weight	55 g	38 g	17 g
Protein	9.9 g	3.9 g	2.7 g
Carbohydrate	0.9 g	0.3 g	0.3 g
Fat	9 g	0	9 g
Monounsaturated	2.5 g	0	2.5 g
Polyunsaturated	0.7 g	0	0.7 g
Saturated	2 g	0	2 g
Cholesterol	213 mg	0	213 mg
Sodium	71 mg	92 mg	9 mg
Calories (cal)	84	20	94

[a] Ref. [1].

9.3 INSIDE AN EGG

An egg contains everything you need to make a chick. Eggs are unmatched as a balanced source of amino acids, and they include a plentiful supply of linolenic acid—an essential polyunsaturated omega-3 fatty acid—as well as several minerals and most vitamins. Eggs also contain *cholesterol*—a *hydrophobic* molecule. Cholesterol is also considered a *lipid* because fatty acids can be converted into cholesterol using many enzymes. In humans, high blood cholesterol does increase the risk of heart disease—a fact that has long made medical professionals recommend limiting egg yolk consumption to 2–3 per week. However, recent studies show egg consumption has little effect on blood cholesterol; rather *saturated fats* have a far more powerful effect on raising blood cholesterol. In addition, the *phospholipids* in egg's yolk interfere with our ability to absorb the cholesterol, so we don't have to count our eggs after all.

9.3.1 The Yolk

The *yolk* (Figs. 9.3, 9.10, and 9.11) surrounds the *germ cell* (Figs. 9.3 and 9.9). The *germ cell* contains half the genetic material needed to make a chick. This *germ cell* is the *haploid egg cell* surrounded by the yolk. The yolk is much larger than the *germ cell*. The yolk membrane (Figs. 9.3 and 9.7) surrounds and protects the yolk and indicates egg freshness.

The yolk is made of about 50% solids (proteins, lipids, and some carbohydrates). There is about twice the amount of lipids as proteins in egg yolk. To the naked eye, an egg yolk looks like a homogeneous yellow solution. However, microscopic analysis shows us something very different. The yolk is actually made of two parts: white (or light) yolk called the plasma (Figs. 9.3 and 9.11) and *yellow (or dark) yolk* called the granule (Figs. 9.3 and 9.10); under a microscope, the yolk appears as islands of semisolid dark yellow granules surrounded by a lighter colored solution. The white yolk is made of mostly lipids like LDL (low-density lipoproteins that we will learn about later) and some proteins and is less dense and especially rich in iron, while the yellow yolk is denser and rich in fats and proteins. The dark yellow granules are made of tightly aggregated proteins enriched in HDL (or high-density lipoprotein), are more protein rich than the plasma, but still contain plenty of lipids including lecithin (also known as phosphatidylcholine). The color of the yellow yolk depends on what the hen eats. Yellow pigments in hen food include xanthophylls, which the hen obtains mostly from corn-based feed.

The high concentration of iron in the white yolk is what prevents it from fully setting when the egg is hard-boiled, while the protein complexes in the granule resist heat treatment and are the reason why it takes higher heat or longer cooking times to turn the yolk solid. But adding salt to eggs before cooking causes yolk proteins to fall out of solution and solidify (denature) much more easily.

Two interesting proteins are found in the yolk. Phosvitin is a protein where almost half of the amino acids are serine. Many of these serines are phosphorylated ($-PO_4^{2-}$). This is highly unusual and the higher content of phosphoserines gives it a high resistance to heat denaturation and proteolytic cleavage. The charges of the phosphorylated serine amino acids give phosvitin a natural ability to bind positive charged metals. The major water-soluble protein is livetin accounting for 30% of the plasma proteins. These proteins are very similar to human antibodies and are a source of allergenic hypersensitivity for some people. Together the three forms of livetin provide passive immunity (the transfer of antibodies from the yolk to the developing embryo).

After water, most of the yolk is made of lipids. Ninety-two percent of the yolk is comprised of different types of lipids. Triglycerides (95%), phospholipids (31%), and cholesterol (4%) make up the main types of lipids in egg yolk with some free fatty acids bound to proteins and other components. Lecithin is the old, traditional name for the phospholipid phosphatidylcholine. This lipid is a minor but important component of egg yolk, making up about 0.9% of the lipids. You will see later with both a water-loving and water-fearing component, it is used in cooking to make hollandaise and mayonnaise among many other foods. Together the proteins and lipids in the yolk are used for their important emulsifying properties.

9.3.2 The White

The egg white is made of mostly water (80%) and a mixture of proteins collectively called *albumen* (Figs. 9.3, 9.5, 9.9, and 9.12). The *albumen* not only nourishes the chick, but it is a biochemical shield against infection and predators. The *chalazae* (Figs. 9.3, 9.4, and 9.13) are dense elastic cords made of *albumen* that anchor the yolk to the ends of the shell and allow it to rotate while suspended in the middle of the egg. *Chalazae* are visible in a raw egg when it is cracked open (Fig. 9.4). We will investigate the egg white proteins more closely in later sections.

9.3.3 Membranes and Shell

The *membranes* (Figs. 9.3, 9.22, and 9.3) line the inside of the shell and are made of antimicrobial proteins. The *shell* (Figs. 9.3 and 9.1) is made of calcium carbonate and protein, and since the developing chick needs to breathe, the shell has

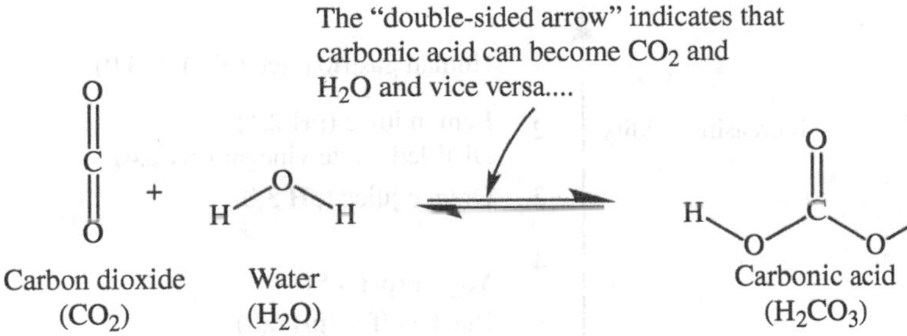

The "double-sided arrow" indicates that
carbonic acid can become CO_2 and
H_2O and vice versa....

Carbon dioxide	Water	Carbonic acid
(CO_2)	(H_2O)	(H_2CO_3)

FIGURE 9.5   Reaction generating carbonic acid.

thousands of tiny *pores* or holes. These *pores* are invisible to the eye. The *cuticle* (Figs. 9.3 and 9.15) is a thin protein coating on the shell. This coating initially blocks the pores to slow the loss of water and prevent the entry of bacteria. When the chicken deposits the proteinaceous *cuticle*, pigment molecules are also deposited into the shell and give the egg its color. The pigments deposited are totally dependent on the type of chicken. White, brown, even blue eggs and yellow spotted eggs—all the colors have to do with the genetic makeup (the breed) of the chicken, for example, Rhode Island Reds lay brown eggs. There is no nutritional difference between white, brown, or even blue eggs. The *air cell* (Figs. 9.3 and 9.14) provides the developing chick with its first breaths of air and is also an indicator of egg freshness.

9.4   EGG FRESHNESS

Although an egg can remain edible for weeks if kept intact and cool, egg quality does deteriorate over time. When an egg is freshly laid, it contains carbon dioxide *dissolved* in the white and yolk. When carbon dioxide is *dissolved* in water (and egg white is 90% water), it is in the form of *carbonic acid* (Fig. 9.5).

When acids are dissolved in water, they make the water *acidic*. This can be measured by the pH scale (Fig. 9.6). On a pH scale, 7 is neutral pH, below 7 is *acidic*, and above 7 is *alkaline*. The more acid in the water, the lower the pH, while adding acid to an *alkaline* solution will lower the pH.

As eggs age, the carbon dioxide dissolved in the white and yolk gradually escapes through the pores in the shell. Since carbonic acid is essentially carbon dioxide and water (Fig. 9.5), as carbon dioxide leaves eggs, the white and yolk lose carbonic acid and become more *alkaline*. This change in pH changes interactions between egg white *albumen* proteins (the proteins interact less) and the *egg white albumen* (Figs. 9.3, 9.5, and 9.9) is consequently runnier.

In addition to carbon dioxide escaping through the pores of the shell, water molecules also escape, making the overall contents of the egg shrink. This shrinking allows air to travel in through the pores in the shell and enlarge the air cell. In a typical refrigerator an egg will lose 4 mg of water a day.

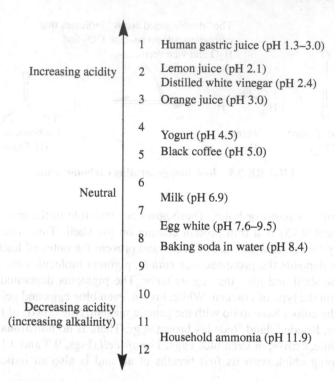

FIGURE 9.6   pH scale, a measure of acidity and alkalinity.

A final indicator of egg freshness is the *yolk membrane* (Figs. 9.3 and 9.7). Because the yolk contains less water than the white, water gradually crosses the membrane from the white into the yolk. In the refrigerator, the yolk gains about 5 mg of water per day. This increase in water makes the yolk swell (enlarge) and the yolk membrane weaken. The grade an egg is given is due to the quality as determined by the way the yolk and white membranes function (Fig. 9.7).

Poaching is the process of cracking a raw egg into a pot of boiling water. The white coagulates or sets in the boiling water before the yolk. Chefs recommend that only the freshest grade AA eggs be used for poaching, otherwise you may find filmy strings of coagulated white floating in the water instead of an intact, perfectly poached egg. Since fresh, grade AA eggs are not too alkaline yet (from the escaping CO_2), the egg white protein—the albumen—is not too runny, and the yolk membrane is tight. These facts along with a gently boiling pot of water help in making the best poached eggs.

9.5   EGG PROTEIN

Egg white albumen is made of mostly water and protein. Ovalbumin is the most abundant protein, and as we will see later, is largely responsible for how egg white cooks. Another interesting protein, lysozyme, is basically the same protein found in

Plate 1 The science behind Cheese.

The Science of Cooking: Understanding the Biology and Chemistry Behind Food and Cooking, First Edition. Joseph J. Provost, Keri L. Colabroy, Brenda S. Kelly, and Mark A. Wallert.
© 2016 John Wiley & Sons, Inc. Published 2016 by John Wiley & Sons, Inc.
Companion website: www.wiley.com/go/provost/science_of_cooking

How does your cookie crumble?

The chocolate chip cookie is a perennial favorite. Chewy, soft, crisp, thick or thin? What is the difference? The science is how the cookie is prepared and the materials used to make the cookie...

Chocolate Chip Cookies

INGREDIENTS

- Fat (shortening, butter)
- Sugar (white, molasses)
- Egg(s)
- Vanilla
- Leavening agent

- Flour
- Water
- Salt
- Semisweet chocolate chips
- Nuts (optional)

Crisp or chewy? Sugar and the protein content in flour will influence if a cookie is crisp or chewy when cooked and cooled.

Water - Just enough from eggs, fats or added liquid can provide the expanding gasses for a puffy cookie or ruin the mix with a runny batter. Egg whites will bind and dry out a cookie and must be balanced by the right amount of water.

Browned, rich flavored cookies require sugars and proteins undergoing the Maillard reaction creating dark brown colors and flavorful molecules.

Leavening agents release carbon dioxide gas to puff the cookie. Thin cookies are made with less leavening agents and a particular mix of flour and fats to spread while the cookie is heated. Baking powder is a typical leavening agent.

Caked cookies need to spread less, do not harden and have a source of gasses to expand during cooking.

Directions

Flour Power
The protein level of flour makes a difference. High protein flour will produce a flat darker cookie while a low protein flour will make for lighter colored and puffier cookies

Fat Spread
Butter is the most popular ingredient for buttery flavor but will make a cookie spread. Reduced fat spreads makes for a more puffy and soft cookie

Preheat oven to 177°C (350°F)

Mix together 2 ¼ cups all-purpose flour and ½ tsp baking powder.

In a separate bowl, combine 1 cup room temperature butter with ½ cup sugar and 1 packed cup brown sugar, whip until light and fluffy.

While mixing, add 1 tsp salt, 2 tsp vanilla, and 2 eggs. Beat until well mixed.

Add the dry flour mixture until just combined, then stir in 2 cups semi-sweet chocolate chips.

Bake 8 to 10 minutes on parchment paper until golden around the edges, but still soft in the

Soda or Powder?
The influence of baking soda (bicarbonate) helps the Maillard reaction and makes the cookie darken with less time and lower heat. Without an acid, baking soda cannot leaven. Baking powder will leaven or raise the cookie by generating carbon dioxide gas.

Sticky Buisness
A greased pan will help spread the dough while cooking making a thin cookie. Denatured protein from eggs will stick to the nooks and cracks of a pan. Either wait for 2-3 min after the cookies are out of the oven or use parchment paper to limit spreading.

Cookie Science

Brown Color and Flavor - Maillard Reaction

The Maillard reaction is the combination of amino acids from proteins and simple sugars. Some sugars like glucose, mannose and ribose react better than others. The more protein and sugar present the more brown and flavorful the cookie. The trick is time and temperature.

- The Maillard reaction requires high heat at the surface of the cookie, and this won't happen until most of the water has evaporated. During early stages of baking much of the energy is used by water to escape from the liquid to gas phase.

- The reaction happens at a greater rate when the pH is more alkali to keep the sugar in the correct open-chain form. This addition of baking soda and reduction of water helps brown your cookie without over heating.

More Protein? It's in the Flour

There are many different types of wheat flour and the main distinguishing characteristic is the amount of protein in the flour. The protein is called gluten.

- All Purpose Flour: Often used for cookies and baking. Protein content ~8-11%

- Bread Flour: also a "hard flour" has a higher protein content needed to capture the gasses of bread. Protein content ~ 12-14%

- Cake Flour: Fine milled flour with less protein used in cakes and cookies. Protein content ~ 7-9%.

High protein flour will absorb more water and help bind the cookie into a tighter form. Low protein flour will not absorb as much water leaving the water to form into steam while cooking giving a raised puffy cookie

Hmmm...

Fats and Sugar

The type of fat the world of difference. A slow melting fat like shortening reduces spreading, while a fat like butter melts quickly...

- Shortening (vegetable) is nearly water-free and has a higher melting point of 117°C than butter 90°C. Cookies made with lard won't spread until the fat melts giving time for the protein to connect and form a solid network during baking.

- Butter his a higher water content ~20% water and low melting point 90°C. The steam made from butter's water will contribute to puff, but the lower melting point of butter will cause spreading well before the cookie sets.

- Margarine has more water than butter. Missing some of the butter flavor it can serve as a nice compromise or part of an experiment!!!

Sugars are essential for the browning Maillard reaction. Sugars contribute to browning and crunchiness.

- Corn syrup is made from corn starch and is broken down to glucose with some maltose. This browns faster and will make the surface of the cookie a little more crunchy.

- Table sugar (sucrose) will not brown as well but makes for hard crystalized crispy cookies

- Brown sugar (light or dark) contains molasses which is mostly a mix of sugars and a few other compounds including fats. These sugars absorb water easily and help soften a cookie.

- Honey is a mixture of fructose and glucose and easily absorbs water keeping the cookie soft when stored.

Did you know?

Plate 2 The science behind Cookies.

Plate 3 The science behind Bread.

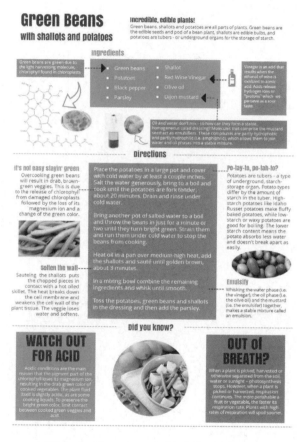

Green Beans
with shallots and potatoes

Incredible, edible plants!
Green beans, shallots and potatoes are all parts of plants. Green beans are the edible seeds and pod of a bean plant, shallots are edible bulbs, and potatoes are tubers - or underground organs for the storage of starch.

Ingredients

Green beans are green due to the light harvesting molecule, chlorophyll found in chloroplasts.

- Green beans
- Potatoes
- Black pepper
- Parsley
- Shallot
- Red Wine Vinegar
- Olive oil
- Dijon mustard

Vinegar is an acid that results when the ethanol of wine is oxidized to acetic acid. Acids release hydrogen ions or "protons" which we perceive as a sour taste.

Oil and water don't mix - so how can they form a stable, homogeneous salad dressing? Molecules that comprise the mustard seed act as emulsifiers. These compounds are partly hydrophobic and partly hydrophilic (i.e. amphiphilic), which allows them to join water and oil phases into a stable mixture.

Directions

It's not easy stayin' green
Overcooking green beans will result in drab, brown-green veggies. This is due to the release of chlorophyll from damaged chloroplasts followed by the loss of its magnesium ion and a change of the green color.

Soften the wall
Sauteing the shallots puts the chopped pieces in contact with a hot oiled skillet. The heat breaks down the cell membrane and weakens the cell wall of the plant tissue. The veggie loses water and softens.

Place the potatoes in a large pot and cover with cold water by at least a couple inches. Salt the water generously, bring to a boil and cook until the potatoes are fork tender, about 20 minutes. Drain and rinse under cold water.

Bring another pot of salted water to a boil and throw the beans in just for a minute or two until they turn bright green. Strain them and run them under cold water to stop the beans from cooking.

Heat oil in a pan over medium-high heat, add the shallots and sauté until golden brown, about 3 minutes.

In a mixing bowl combine the remaining ingredients and whisk until smooth.

Toss the potatoes, green beans and shallots in the dressing and then add the parsley.

Po-lay-to, po-tah-to?
Potatoes are tubers - a type of underground, starch-storage organ. Potato types differ by the amount of starch in the tuber. High-starch potatoes like Idaho Russet potatoes make fluffy baked potatoes, while low-starch or waxy potatoes are good for boiling. The lower starch content means the potato absorbs less water and doesn't break apart as easily.

Emulsify
Whisking the water phase (i.e. the vinegar), the oil phase (i.e. the olive oil) and the mustard (i.e. the emulsifier) together, makes a stable mixture called an emulsion.

Did you know?

WATCH OUT FOR ACID
Acidic conditions are the main reason that the pigment part of the chlorophyll loses its magnesium ion, resulting in the drab green color of cooked vegetables. The plant fluid itself is slightly acidic, as are some cooking liquids. To preserve the bright green color, limit contact between cooked green veggies and acid.

OUT of BREATH?
When a plant is picked, harvested or otherwise separated from the soil, water or sunlight – photosynthesis stops. However, when a plant is picked or harvested, respiration continues. The more perishable a fruit or vegetable, the faster its respiration rate. Plants with high rates of respiration will spoil sooner.

Plate 4   The science behind Green Beans.

Hot Sauce & Guacamole

The secret of hot stuff!!!
It is all about... capsaicin. Capsaicin is a pungent compound that binds to your pain receptors (nociceptors). While similar to compounds found in vanilla, ginger and pepper, capsaicin stimulates a totally different center in the brain and brings the pain.

INGREDIENTS
Mango-Pineapple Hot Sauce

- Mango
- Pineapple
- Ginger
- Habañero
- Sugar & Salt
- Lemon & Lime Juice
- Cumin
- Xanthan Gum

The citric acid in lemon and lime juice adds some sour flavor and inhibits browning of the fruit.

Why Guacamole?
Capsaicin is non-polar/hydrophobic and therefore not water soluble. Washing down hot food with cold is also just spreads capsaicin to more pain receptors. Oily or fatty foods like whole milk will dissolve and help remove the offending compound. Avocados are high in fats and provide a tasty way to pull the capsaicin out of the way.

Due to their high levels of capsaicin, habaneros (or other) will add heat that's matched with sweeter tastes including...

Xanthan gum is a complex polysaccharide and a little of it goes a long ways to make your hot sauce thick and stick to those chips.

Directions

Lose the seeds
Most of the capsaicin is in the placenta not the seeds. The thick coating on the seeds has alkaloid compounds which are bitter and will detract from your hot sauce.

Slice and remove the pepper seeds, but be careful to leave the placenta (the ribs) behind. Blanch the fruit of the pepper in boiling vinegar. Reserve the cooled vinegar.

Add all ingredients except xanthan gum (2 cups mango, 1 cup canned or fresh pineapple with some juice, ¼ tsp cumin, 1 tsp sugar and 1 tbsp molasses, ⅓ cup each lemon and lime juice, 1 tsp ginger and salt to taste). Blend or puree until smooth.

Make it Stick
Most sauces will separate over a short period of time. Xantham gum (a flavorless polysaccharide made by fermentation of three simple sugars), will bind water and "hold" water in place.

Add vinegar, salt, lime or lemon juice to taste. If you desire more heat, add more peppers.

Sprinkle ¼ tsp xanthan gum and pulse mix with immersion or stand blender. Allow the gum to hydrate before adding more. It may take more than an hour to fully thicken...

Hot Sauce Basics
A simple hot sauce is made by blending hot peppers with vinegar. Fermentation helps extract and create new flavors. Tabasco hot sauce is made this way.

Acid Helps
The lowered pH of hot sauce due to lemon and lime juice as well as vinegar acts as a preservative. Few strains of bacteria are able to easily grow.

Guacamole

INGREDIENTS

- Avocados
- Juiced Lime
- Salt
- Cayenne
- Cumin
- Red Onion
- Roma Tomatoes
- Cilantro
- De-seeded Jalapeño or Publano
- Garlic Clove, Minced

More Acid
Once opened, the avocado releases polyphenol oxidase. This enzyme reacts with oxygen and the phenols from the cell wall to create a brown, bitter product. This "little bit of nasty" is no actual natural defense mechanism of the fruit. Besides tasting good, lime juice helps slow down the enzyme activity.

Skin and scoop the avocado flesh and immediately mix with lime juice. Mash the mixture but do not blend to a smooth paste. Allow the "chunks" of avocado to hit your pallet and create a contrast to the blended paste mixed with the spices.

Add salt, cumin and cayenne and fold in with onions, tomatoes, cilantro and garlic.

Serve after one hour at room temperature.

More Pungent
When speaking about food, "pungent" means "heat" or a "strong unpleasant flavor or aroma". The heat from peppers is one kind of pungency, but garlic and onions also contain pungent compounds. Onions and garlic produce sulfur containing compounds that do not bind to the pain receptors.

Did you know?

Salsa microbes
Many of the compounds used to make salsa have antimicrobial properties. Saccharomyces cerevisiae, Staphylococcus aureus, Bacillus cereus, and Escherichia coli are all inhibited by one or more of the components in most salsa.
This helps the popular garnish act as a preservative while tasting... amazing

Salsa means Sauce
There really is no difference between hot sauce and salsa. Salsa translates to sauce in Spanish. Thus all sauces are salsas.

BUT... picante means "hot spicy"

Since the early 1990's salsa is the most sold and used condiment in the United States - even more than ketchup!

Plate 5 The science behind Hot Sauce.

Lemon Souffle
with Creme Anglaise

This recipe uses eggs in two ways
The high protein content of the egg white is used to make a foam in which denatured protein stabilizes trapped air bubbles. The egg yolk is used to flavor the souffle and also provide the rich pourable custard of the creme anglaise

Lemon Souffle ingredients

- Butter
- Sugar
- Flour
- Milk
- Eggs plus egg whites
- Lemon juice
- Lemon zest
- Salt

Only the egg whites are used for creating the egg white foam. Addition of extra egg white adds more protein for a stronger foam.

This recipe uses a ramekin - a small porcelain dish with tall sides. The porcelain transfers and holds heat, while the tall sides provide support to the souffle while it is still rising.

The terpenes that give lemon its characteristic odor are highly concentrated into the peel. The zest and juice will be used to flavor this souffle base.

Directions

Separate
Before beginning, the cook must separate the egg yolks from the egg whites. Egg whites are 90% protein, while egg yolks contains all the egg's fat.

Denature the protein
Agitating the egg white by beating adds energy which unfolds the 3-dimensional structure of the proteins. With their hydrophobic interiors exposed, the proteins coagulate into a stiff solid, trapping air bubbles.

Expand the Air
Beating the egg white trapped many air bubbles between the coagulated protein. Heating those air bubbles causes them to expand. And the souffle rises right out of the dish!

Melt butter in the saucepan and add the flour. Cook for 1 to 2 minutes and whisk in the milk. Add sugar. Cook for 2 to 3 minutes. Remove from the heat and let cool. Whisk in the egg yolks, 1 at a time. Add the lemon juice and zest and reserve.

Put the egg whites and a pinch of salt in a mixing bowl. With an electric mixer, beat the egg whites to stiff peaks. Using a rubber spatula, scoop one-third of the egg whites into the lemon mixture. Gently and quickly fold the whites in. Repeat this process 2 more times with the remaining whites. Be sure that the whites are fully combined and there are no streaks. Carefully spoon the mixture into the chilled ramekins until they are about three-quarters of the way full. Place the filled ramekins on a sheet tray and into the preheated oven. Bake the souffles until the souffles have risen straight and tall over the rims of the ramekins, 13 to 15 minutes.

Make a Roux
Cooking flour in fat coats the wheat proteins (i.e. glutenin and gliadin) with the greasy fat molecules and prevents them from mixing with water. Instead, the water is used to get the starch molecules found in wheat and add extra thickening power to the souffle base

Thicken the Yolks
When heating, the added energy unfolds the 3-dimensional structure of the proteins in the food. But unlike egg white, the protein in egg yolk (40% of the eggs total!) is diluted with fat. The fat molecules interfere with protein coagulation, which yields and very soft solid.

Creme Anglaise

Creme Anglaise ingredients

- heavy cream
- vanilla bean
- sugar
- egg yolks

The loose texture of this pourable custard owes itself to the lower protein and high fat content of egg yolks

Extract for flavor
Heating the vanilla bean with the cream extracts the flavor molecules from the vanilla into the cream

Place the heavy cream in a small saucepan. Split the vanilla bean down one side and scrape the seeds out and put them in the heavy cream. Whisk in half the sugar. Place the pan on a medium heat and bring it to a boil.

While the cream is heating, whisk together the egg yolks and remaining sugar. When the cream comes to a boil, pour half of it into the egg and sugar mixture, whisk it and then pour it IMMEDIATELY back into the saucepan. Whisk to combine. Bring the sauce to a boil and immediately remove it from the heat and chill.

A cream, not a custard
Like the souffle base above, heating unfolds the 3-dimensional structure of the proteins in the egg yolk. The presence of fat molecules in the yolk and the fat and water in the cream interfere with protein coagulation. The network of coagulated protein is so weak and fragile that a solid never forms. Instead, we get the thickened "egg cream" of that is creme anglaise

Did you know?

FAT DESTROYS FOAM
Every spec of egg yolk must be removed for a proper egg white protein foam to form. The exposed hydrophobic regions of the denatured protein must cluster around air bubbles, not hydrophobic fat molecules!

FOAMS ARE FRAGILE
Egg white foams must be handled gently. Folding of the custard mixture into the foam is necessary to prevent the escape of air bubbles.

Plate 6   The science behind Lemon Souffle.

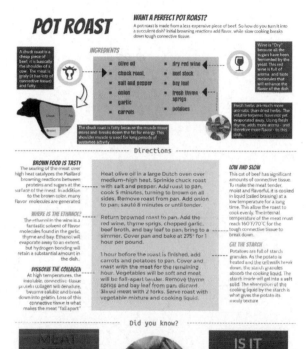

Plate 7   The science behind Pot Roast.

Great Gravy

Sauces and Roux

What is Gravy Anyway?

The term 'gravy' is very culturally dependent. For Italian-influenced food, gravy can mean a tomato based "sauce" while in other cultures, gravy is a thickened juice from some cooked meat or a cream-based thickened sauce.

BASIC INGREDIENTS   White Sauce

- Milk or cream
- Flour
- Butter
- Salt & Pepper
- Herbs & Spices (optional)

A basic white sauce (more formally Béchamel, meaning medium thick sauce in French Cooking) is a classic starch based sauce that becomes a cream sauce when cream is added!

The fat in butter, milk or cream adds to the thickness of the sauce due to milk globules or butter fat. However the fat also serves to limit the interaction between flour particles - coating the carbohydrate polymers as they swell preventing the sauce from being too thick, lumpy or gooey.

Rule of thumb: per cup of starch-based gravy/sauce

Thin: 1 tablespoon
Medium: 2 tablespoons
Thick: 3 tablespoons

Flour serves as a starch source, some gravies, sauces or roux can use grain starches such as cornstarch or rice starch (use 1.5 as much of these for every 2 tablespoons of flour).

Directions

IT IS ALL ABOUT THE STARCHES

Starches are large glucose polymer molecules. Amylopectin is branched and amylose is a long unbranched polymer. Both types of starch interact with water through hydrogen bonds and act as long, loosely connected tangles giving the sauce its thickness. The straight chained amylose forms a thicker sauce than the less efficiently packed amylopectin.

Amylose forms a more firm opaic gel while amylopectin forms a clear glossy sauce.

BRANCHED OR UNBRANCHED

Grain starches like flour, rice or corn are higher in amylose than other starch sources. These starches will firm into a moldable gel when cool. Sauces made with these starches will thicken just below the boiling point of water and are re-heatable without thinning. However, they lose ability to hold water during and after heating.

Root and waxy starches, such as arrowroot, potato and tapioca starches, are mostly amylopectin (up to 99%). These sauces will cool to a clear glossy gel at a lower temp than grain starches. This starch freezes and thaws well.

FLOUR STARCHES ARE THICKER

Adding flour to a hot butter will denature (kill) the enzymes in the flour that degrade the carbohydrate - the longer the carbohydrate the thicker the sauce?

Basic white sauce:

Simmer 2 cups whole milk with 1 bay leaf and a 1/4 sliced onion. Minced garlic clove (1-2 cloves) and crushed pepper corns are optional. Once a simmer is reached, remove from heat, strain the milk, and set aside while making a base.

Melt 4 tablespoons butter and add 4 tablespoons all purpose flour. Stir over low heat for 2-3 minutes but do not brown. Remove from heat.

Vigorously whisk the strained milk into the flour paste. Heat until a simmer is reached and continue to whisk for 3-5 minutes.

To make a Cream Sauce, replace 1 cup milk with 1 cup heavy cream and simmer the final mixture for 10 minutes.

To make a cheese (i.e. alfredo) sauce use the basic recipe and add 1/4 cup of Parmigiano-Reggiano, Gruyere, or other aged hard cheese with 2 tablespoons lemon juice or dry white cooking wine or 1/4 teaspoon tartaric acid.

Season with turmeric, nutmeg, cayenne or dry mustard alone or in combination to taste.

Brown Pan Gravy:

After roasting or frying turkey, chicken or beef, remove the meat from the pan and remove all but 4 tablespoons of fat from the pan. Mix in 24 ounces of a chicken, vegetable, or beef broth along with 8 ounces of red wine. Whisk while simmering and continue for 2 - 5 min to slightly reduce the volume (~20%). Scrape the tasty bits from the pan while mixing.

Add 4 tablespoons flour (for a medium thick gravy) and whisk continuously to combine while simmering. Add salt to taste.

Roux:

For mac and cheese, chowder soups, and of course sauces, you need to make a flour/fat base called a roux. There are three versions of roux, white, blond and brown. Each version depends on the extent of the Maillard reaction you allow to happen. A darker the roux will not thicken as well as a white or blond roux.

Heat oil, butter or clarified butter. Some recipes call for heating the butter until water mostly gone (bubbles decrease).

Add flour or another starch source to the butter while whisking constantly. Add cold liquid to a hot roux or a hot liquid to a cold roux to give enough time to disperse the starch before full swelling of the sugars.

STARCH GRANULES

Structurally, starches are small semicrystalline granules. Starches in the granules are organized in layers like the tight rings of an onion. Swelling of starch takes place in the outermost layer as warm water begins to bond with the carbohydrate.

GO SLOW

Slowly heat a starch while stirring - it takes time for the water to bind to the carbohydrate - too fast and a shell of water binds tightly to the flour mass that is impenetrable. The result, is a solid lump of gel surrounding dry starch!

ACIDS, SAUCES AND CHEESE

Acids reduce the thickness of a sauce by helping to reduce the length and connections of the starches - add any acid components last to keep the starch as thick as possible.

Adding cheese can make a stringy mess instead of a sauce. Adding cheese to the roux prior to adding milk or cream helps as the starch and fat coat the casein proteins before they can aggregate into strings.

Adding acids to a cheese sauce will limit the interactions between casein proteins. The lower pH will neutralize the negative charged proteins. Thus they will not interact as well with each other or calcium, making a smoother sauce.

Lemon juice (citric acid) and wine (tartaric acid) are both used in cheese sauces. Wine just tastes better!

Did you know?

I Broke my Sauce

A broken sauce happens when instead of a thick cream or gravy, you have an unthickened and watery mess. You broke your sauce. What happened and can you fix it?

Wait a bit - sauces can fall apart into fat and water if they are left too long. Adding an emulsifier like a xanthan gum help to hold water in place.

Too HOT! At higher temperatures the starches will hydrolyze (become cut into small pieces) especially if there is an acid around.

Choose well - Pick the right starch for your need - temperature and time of use

But those LUMPS!

We have all done it, added flour to a hot liquid and were too slow to mix. A lump of rock hard jellified, start your gravy over again mess. Starch binds water tightly leaving dried granules on the inside. Of what we all think of as lumpy gravy!

This phenomena happens because the carbohydrates bind so many water molecules (shells and shells of water) before allowing water to penetrate the starch granule.

Whisking is key - but slowing down the process is important. Make a cold slurry of flour while mixing and slowly add to the hot roux or liquid to keep granules apart and the lumps at bay.

Plate 8   The science behind Great Gravy.

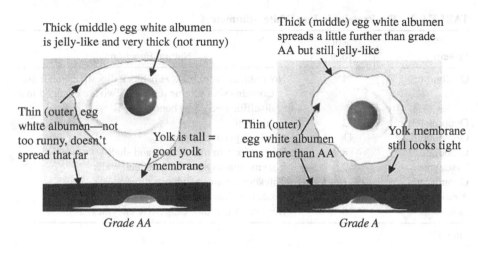

Thick (middle) egg white albumen is jelly-like and very thick (not runny)

Thin (outer) egg white albumen—not too runny, doesn't spread that far

Yolk is tall = good yolk membrane

Grade AA

Thick (middle) egg white albumen spreads a little further than grade AA but still jelly-like

Thin (outer) egg white albumen runs more than AA

Yolk membrane still looks tight

Grade A

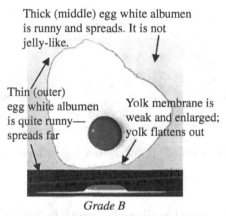

Thick (middle) egg white albumen is runny and spreads. It is not jelly-like.

Thin (outer) egg white albumen is quite runny—spreads far

Yolk membrane is weak and enlarged; yolk flattens out

Grade B

FIGURE 9.7   Egg grade is a function of egg white and yolk membrane.

human tear ducts. In both cases the protein serves to act in an antibacterial function. The protein will cleave protein–carbohydrate bonds on the surface of bacteria and fungi cell walls. This leaves the bacteria damaged and unable to survive. Lysozyme has been used as preservatives stopping microbial growth in meat, fruits, and vegetables. In the egg white, it stops the microorganism contamination that may pierce the egg shell (Table 9.2).

Proteins are *amphiphilic* molecules—they have both *polar* (hydrophilic) and nonpolar (hydrophobic) parts. In nature, we find proteins in water-based environments. Because not all parts of the protein love the water (some parts are *hydrophobic*), the protein *folds* in a three-dimensional way that buries the *hydrophobic* parts on the inside of the structure and exposes the *hydrophilic* parts to the outside, where they can interact with the watery environment. It is easy to obtain egg white protein, and so egg white proteins were among the first proteins studied for their molecular

TABLE 9.2   The Proteins in Egg White Albumen[a].

Protein	% of Total	Natural Function
Ovalbumin	54	Nourishment for chick, *may* block digestive enzymes. Contains six cysteine residues. Two are engaged in a disulfide (—S—S—) bond
Ovotransferrin	12	Binds iron
Ovomucoid	11	Blocks digestive enzymes
Globulins	8	Plug defects in membranes and shell
Lysozyme	3.5	Enzyme that digests bacterial cell walls
Ovomucin	1.5	Thickens albumen, inhibits viruses
Avidin	0.09	Binds the vitamin biotin
Others	10	Bind vitamins; block digestive enzymes…

[a] Ref. [1].

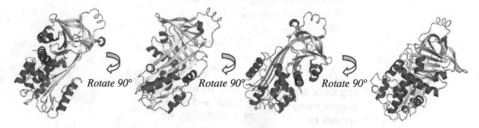

FIGURE 9.8   The protein ovalbumin (from chicken egg white) as viewed rotating around a vertical axis.

structure. Modern biochemistry has allowed scientists to "see" how proteins fold at a molecular level. What they found was quite beautiful (see Fig. 9.8).

But how can Figure 9.8 be depicting a protein? There are no visible amino acids or peptide bonds, just twirly ribbons and curvy arrows. What are these images depicting anyway?

KEY CONCEPT

Most proteins fold into three-dimensional structures made of α-helices, β-sheets, and loops each held together by hydrogen bonds. The protein folds in order to hide hydrophobic parts and expose hydrophilic parts to water.

In the cartoons of Figures 9.8 and 9.9a, a red twirly ribbon is an α-*helix* ("alpha" helix), while a yellow curvy arrow is a β-*strand* ("beta" strand) and many curvy

(a)   (b)   (c)

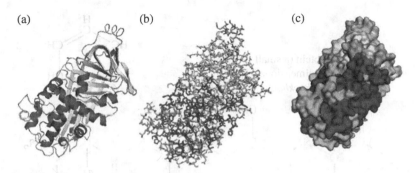

FIGURE 9.9 The same orientation of the ovalbumin protein molecule represented as a cartoon (a) and then as using *amino acid residues* (b)—the amino acids joined by peptide bonds and a (c) *space-filling* representation of ovalbumin.

(a)   (b)   (c)   (d)   (e)

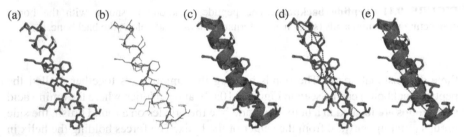

FIGURE 9.10 A single α-*helix* from ovalbumin drawn five ways: first as *amino acid residues*— amino acids joined by peptide bonds (a), and then in (b) a thin line is tracing the three-dimensional pattern of the peptide bonds between each amino acid (called the *peptide backbone*), which makes the helical ribbon shape. In (c), a cartoon ribbon has replaced the thinly traced line. In (d) and (e), we see that hydrogen bonds (dashed lines) hold the amino acid residues in this special helical shape.

arrows lined up lengthwise next to each other make a β-*sheet*. The squiggly green parts are called *loops*. If we represent a structure of the molecule ovalbumin using the same color scheme as you see in Figure 9.8 and 9.9a, but instead we replace the squiggles, twirly ribbons, and curvy arrows with atoms, the structure looks like Figure 9.9b. In Figure 9.9b, those same *helices* (ribbons), *sheets* (arrows), and *loops* are depicted using *amino acid residues* (the amino acids joined by peptide bonds). The last version of ovalbumin (Fig. 9.9c) is called a space-filling model. Here the surface of each atom (the edge of where the electrons are found most of the time) is shown in this presentation to allow researchers to get a feel for what happens at the surface of a protein.

Let's look more closely at the amino acid atoms making a *helix* in Figure 9.10. The twirly ribbons in Figure 9.10c and e are α-helices and the loops that join them are really cartoon representations of how the *peptide backbone* of the protein looks in

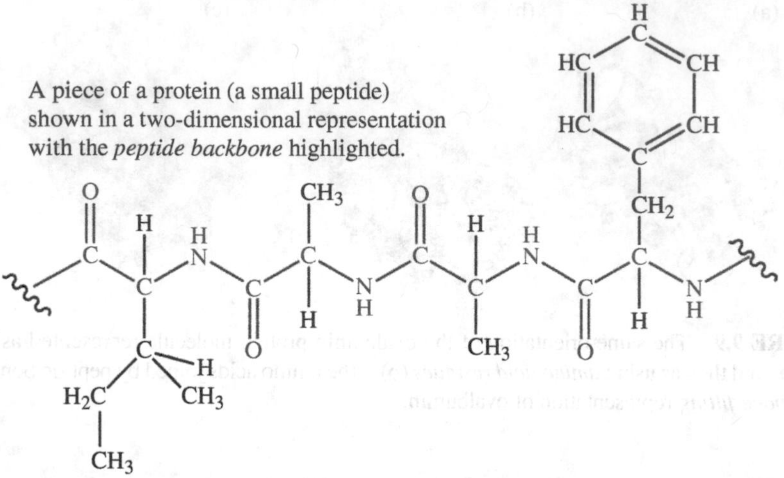

A piece of a protein (a small peptide)
shown in a two-dimensional representation
with the *peptide backbone* highlighted.

FIGURE 9.11 Peptide backbone. The peptide backbone is shown with the bonds connecting the amino acids (side chains alternating above and below) the backbone.

three-dimensional space. The bonds holding the amino acids together (called the peptide backbone) can be seen in Figure 9.10a, b, and d. Notice where the amino acid side groups are located in a helix—outside like the bristles on a round brush, the side groups pointing outward from the center of the helix. The forces holding the helix in place are the hydrogen-bonded atoms within this backbone shown in Figure 9.10d and e. The *peptide backbone* is the $C=O$, alpha carbon, and nitrogen of each amino acid residue along the protein chain (Fig. 9.11).

In ovalbumin (as with pretty much any protein) some groups of amino acids fold into α-*helices*, some into β-*strands*, and some into *loops*. The *primary sequence* of the amino acid residues determines which structure they form. Figure 9.12 depicts how an α-*helix*, for example, is created by an arrangement of *amino acid residues* held together by *hydrogen bonds*—remember *hydrogen bonds* are weak, noncovalent bonds that form between partially positive and partially negative atoms. β-*sheet* and *loop* structures are also held together by *hydrogen bonds*.

From Figures 9.7 and 9.8, we can see that the protein ovalbumin has many α-helices and β-sheets all *folded* on top of and around each other to make a *globular* structure. In fact, when you fill in all the atoms using a type of *space-filling* representation, we can see the lumpy, three-dimensional *globular* looking protein (Fig. 9.8c). The α-helices, β-sheets, and loops also interact *with each other* by hydrogen bonds. β-sheets are stabilized by the same atoms of the peptide backbone; however, the arraignment is very different (Fig. 9.12). Unlike in a helix, the hydrogen bonds are between the same atoms of the peptide backbone, but between two strands rather than within the same strand (Fig. 9.13). Each amino acid will form hydrogen bonds with neighboring chains.

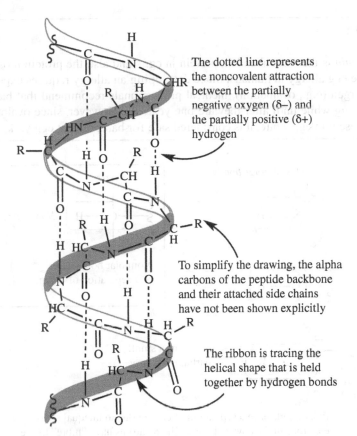

The dotted line represents the noncovalent attraction between the partially negative oxygen (δ−) and the partially positive (δ+) hydrogen

To simplify the drawing, the alpha carbons of the peptide backbone and their attached side chains have not been shown explicitly

The ribbon is tracing the helical shape that is held together by hydrogen bonds

FIGURE 9.12   Backbone of an α-helix. The helix is in a right-handed turn where hydrogen bonds form between every four amino acids in the same chain.

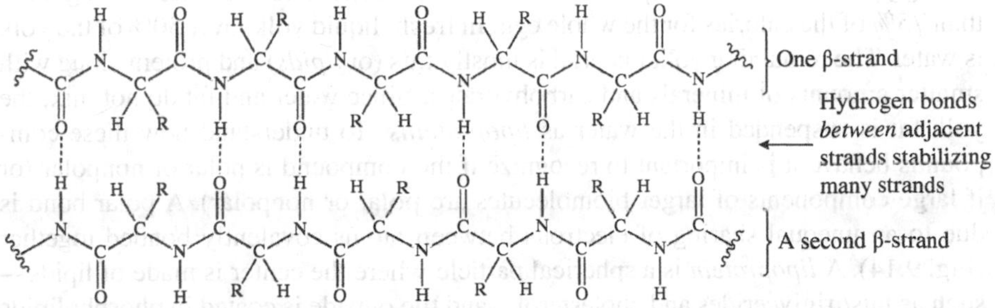

One β-strand

Hydrogen bonds *between* adjacent strands stabilizing many strands

A second β-strand

FIGURE 9.13   β-sheets. In β-sheets, the peptide backbone is extended rather than coiled in a helix. In this form, hydrogen bonds form between the amide and carbonyl groups of neighboring strands.

BOX 9.1

Ovalbumin is the most abundant protein in eggs and also the protein most likely to cause egg allergy in humans. Since developing an allergy requires exposure to the allergen (e.g., ovalbumin), medical professionals recommend that babies do not eat egg white until after they are one year old. However, since ovalbumin is only present in egg white, it is considered safe for babies to eat egg yolk.

Examples of *polar bonds*

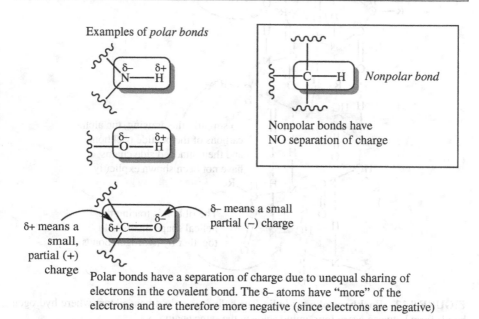

Nonpolar bonds have
NO separation of charge

δ– means a small
partial (–) charge

δ+ means a
small,
partial (+)
charge

Polar bonds have a separation of charge due to unequal sharing of electrons in the covalent bond. The δ– atoms have "more" of the electrons and are therefore more negative (since electrons are negative)

FIGURE 9.14   Polar bonds.

9.6   EGG FATS

The egg yolk contains some protein but also all of the fat and cholesterol and greater than 75% of the calories for the whole egg. In fresh, liquid yolk, over 50% of the yolk is water. The remaining yolk material is mostly fats (or *lipids*) and protein along with smaller amounts of minerals and carbohydrates. Since water and fat do not mix, the yolk fat is suspended in the water as *lipoproteins*. To understand how these compounds behave it is important to recognize if the compound is polar or nonpolar (or if large components of larger biomolecules are polar or nonpolar). A polar bond is due to an unequal sharing of electrons between atoms covalently bonded together (Fig. 9.14). A *lipoprotein* is a spherical particle where the center is made of lipids— such as fats/triglycerides and cholesterol—and the outside is coated in phospholipids and proteins. Phospholipids and proteins are amphiphilic—they have both hydrophobic and hydrophilic parts. *Hydrophobic* groups of atoms are comprised of nonpolar bonds and *do not* interact with (literally, "fear") water. *Hydrophilic* groups of atoms are comprised of polar bonds and/or contain charged atoms: these groups of atoms interact with (literally, "love") the polar bonds of water via their charged or partially charged atoms (Fig. 9.15). See Chapter 1 for a review of *hydrophobic* and *hydrophilic* (Fig. 9.16).

The proteins and phospholipids of lipoproteins give egg yolk *emulsifying* properties. *Emulsifiers* can mix with water and fat and bring the two phases together in a fine, cream-like, stable mixture. When fats/oils and water form a stable mixture (i.e., one that is not separated into two phases), it is called an *emulsion*.

In its simplest form, mayonnaise is an emulsion of oil and vinegar, where eggs are the *emulsifier*. Mayonnaise does not separate into the oil phase (hydrophobic) and the water phase (the vinegar) (hydrophilic) because the amphiphilic molecules in egg stabilize the emulsion. The yellow color of egg yolk is due to small amounts of carotenoid pigments ingested by the hen in her feed. Carotenoids are light capturing pigments made by plants. Learn more about carotenoids in Chapter 7.

9.7   COOKING EGG PROTEIN

When a protein *denatures*, energy (e.g., in the form of heat) added to the mixture of proteins makes the molecules move around. Add enough energy and the proteins move and groove enough to break all the *hydrogen bonds* holding the α-helices, β-sheets, and loops together, and the three-dimensional *globular* protein structure unravels (Fig. 9.17). When the protein unfolds, all the *hydrophobic* parts that were buried inside the protein are exposed to the watery environment, the hydrophobic parts *hate* the water, and instead they'd like to find another hydrophobic place to be. The exposed hydrophobic parts of a protein join together with exposed hydrophobic parts of other proteins, clumping together in a process called *coagulation*. The *unfolded, coagulated* protein is now a large, aggregated complex that can't stay dissolved in the water, so the coagulated protein *solidifies*, trapping water molecules between the unfolded proteins.

The proteins of egg white each have a unique denaturation temperature that is dictated by their specific amino acid sequence and three-dimensional fold (Fig. 9.18). Ovotransferrin is the least stable of the egg white proteins and in a pure form will denature at 57°C/135°F. Globulins, ovalbumin (the most abundant egg white protein), and lysozyme denature in their pure forms at 72°C/192°F, 71.5°C/191°F, and 81.5°C/179°F, respectively. The mixture of proteins in egg white results in a composite denaturation temperature of approximately 93°C/145°F with coagulation above 99°C/151°F. The network of denatured and coagulated egg white protein is stabilized by interaction of exposed hydrophobic areas and disulfide bond formation between protein chains.

The abundance of ovalbumin makes its denaturation very important to cooking egg white. Ovalbumin actually unfolds in two steps. The half-unfolded ovalbumin is called S-ovalbumin; it is surprisingly heat stable and difficult to denature. Over time, most of the ovalbumin in freshly laid eggs gradually converts to the S-ovalbumin form—in fact, by the time an egg reaches the grocery store shelf, approximately 50%

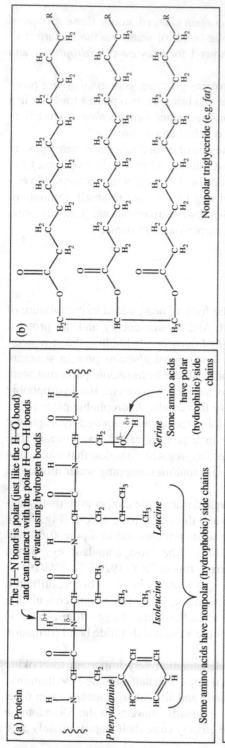

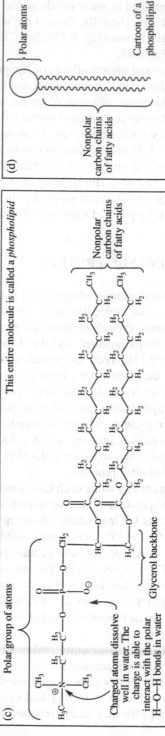

FIGURE 9.15 The major components of lipoproteins. (a) Protein contains hydrophobic (nonpolar) and hydrophilic (polar) groups of atoms. Proteins are *amphiphilic* and can mix with water by folding in a way that exposes the hydrophilic parts to water. **(b)** Triglycerides are nonpolar (hydrophobic) and do not mix with water. **(c)** Phospholipids are a modification of a triglyceride in which one of the fatty acid chains is replaced with a very polar group of atoms; consequently, phospholipids are *amphiphilic*. **(d)** A simple cartoon representation of a lipid.

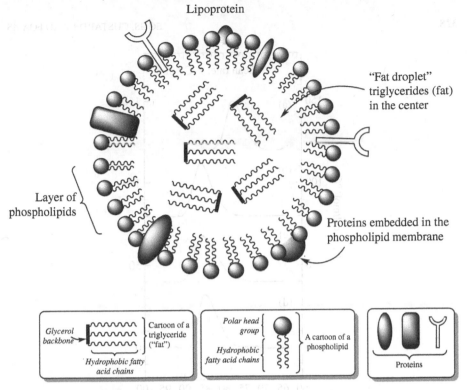

Lipoprotein

"Fat droplet" triglycerides (fat) in the center

Layer of phospholipids

Proteins embedded in the phospholipid membrane

| Glycerol backbone | Cartoon of a triglyceride ("fat") | Polar head group | A cartoon of a phospholipid | Proteins |

Hydrophobic fatty acid chains

Hydrophobic fatty acid chains

FIGURE 9.16 Cartoon of a lipoprotein. A lipoprotein is coated with the *amphiphilic* phospholipids and proteins, while the interior is made of *hydrophobic* fats.

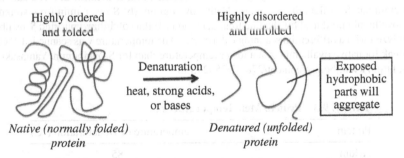

Highly ordered and folded

Highly disordered and unfolded

Denaturation
heat, strong acids, or bases

Exposed hydrophobic parts will aggregate

Native (normally folded) protein

Denatured (unfolded) protein

FIGURE 9.17 Protein denaturation at the molecular level.

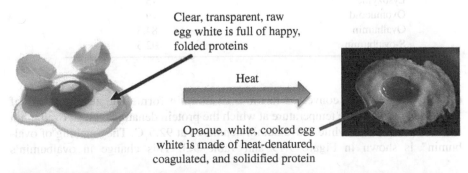

Clear, transparent, raw egg white is full of happy, folded proteins

Heat

Opaque, white, cooked egg white is made of heat-denatured, coagulated, and solidified protein

FIGURE 9.18 Egg protein denaturation. Protein denaturation caused by heat—as seen with the unaided eye.

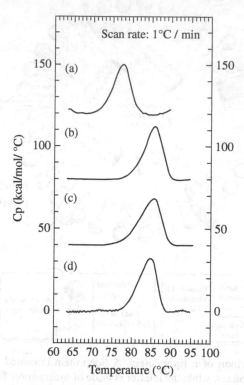

FIGURE 9.19 Thermostability of native ovalbumin and S-ovalbumin. The stability of ovalbumin and S-ovalbumin to heat. (a) Native ovalbumin, (b) S-ovalbumin from stored egg, (c) S-ovalbumin produced by alkaline treatment, and (d) the solubilized purified S-ovalbumin as analyzed with a differential scanning calorimeter. The temperature was scanned at 1°C/min. The peak for native ovalbumin is at a lower temperature than for S-ovalbumin. Yamasaki, JBC 2003, http://www.jbc.org/content/278/37/35524.full

TABLE 9.3 Protein Melt Temperatures[a].

Protein	Temperature of Denaturation (°C)
Avidin	85
Egg globulins	92.5
Lysozyme	75
Ovomucoid	79
Ovalbumin	84.5
S-ovalbumin	92.5

[a] Ref. [4].

of its ovalbumin has converted to the S-ovalbumin form. The accumulation of S-ovalbumin raises the temperature at which the protein denatures. Fresh ovalbumin denatures at 84.5°C, while S-ovalbumin denatures at 92.5°C. The "melting of ovalbumin" is shown in Figure 9.19 and Table 9.3. This change in ovalbumin's three-dimensional structure makes it more difficult to denature and results in "runnier" whites when cooking [5].

All this egg protein coagulation happens well below the boiling point of water (212°F/100°C). Overcooking an egg with too much heat and for too long can evaporate out the water molecules that were trapped among the unfolded protein—the result is a rubbery solid. You may have noticed this effect if you have ever prepared scrambled or fried eggs in a too-hot skillet!

Denatured proteins coagulate into a network stabilized by interactions between hydrophobic regions of the unfolded protein and disulfide bonds that form between the protein changes (Fig. 9.20). In the cooking of a whole egg, the white will always set before the yolk. This interesting fact makes possible the soft yolks of delicate poached eggs, sunny-side up fried eggs, and baked eggs. While egg white protein begins to set at approximately 93°C/145°F with coagulation above 99°C/151°F, egg yolk lipoproteins only begin to denature around 95°C/149°F and finally coagulate by 70°C/158°F. The large amount of fat—relative to protein—in the egg yolk makes it that much harder for the denatured proteins to find one another and stick their exposed hydrophobic parts together; therefore, higher temperatures are required.

9.8 CUSTARDS

Heat *denatures* (or unfolds) proteins, and when proteins *denature*, the exposed hydrophobic parts of the protein join together with exposed hydrophobic parts of other proteins, clumping together in a process called *coagulation*. We observe this process in the cooking of an egg—the clear, runny white becomes a white solid mass of denatured and coagulated protein.

When we add other liquids and/or molecules like starch, sugar, and/or fat (from milk, butter, cheese, etc.) to the eggs, we *dilute* the protein mixture and consequently *raise the temperature* at which the proteins begin to coagulate. *Dilution* of the proteins surrounds the protein molecules with many more water (and/or sugar or fat) molecules, which makes it harder for the proteins to find one another and stick their hydrophobic parts together (*coagulate*), so we have to raise the temperature, which makes the molecules move around that much more rapidly and find each other. When the coagulated protein network does finally form in these diluted mixtures, the solid matrix of coagulated protein is tender and fragile—the large networks of coagulated proteins are filled with water, sugar, and/or fat molecules—for example, as in a custard.

If you heat a custard preparation too high or too fast, the proteins will *curdle* instead of *thicken*. In curdling, proteins denature and form hard, tight lumps of coagulated protein that exclude the water, sugar, and/or fat molecules (in custard making, the effect is called *syneresis*; Fig. 9.21). Because custards are so sensitive to overheating, they are often cooked in an oven while sitting in a *water bath*. The water bath keeps the temperature constant during cooking. The boiling point of water is 212°F/100°C, while a typical oven temperature is 325–350°F. No matter how hot you make the oven, the water will only ever reach 212°F/100°C, upon which it will boil

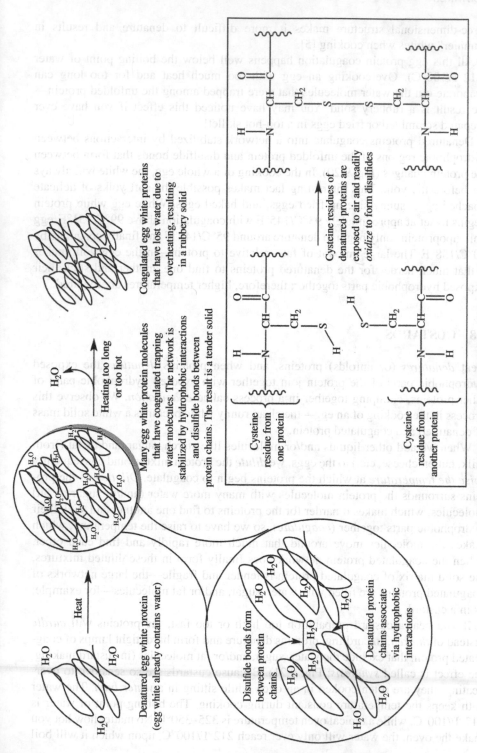

FIGURE 9.20 Coagulation of egg white protein.

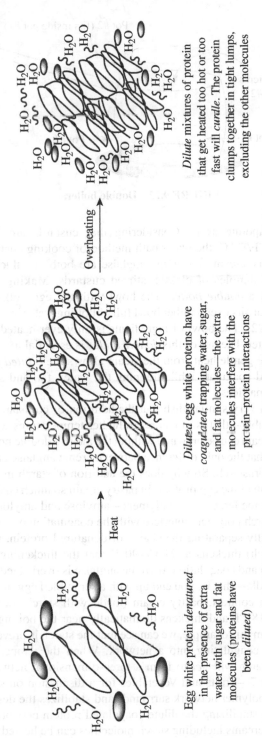

FIGURE 9.21 Egg protein coagulation diluted with water, sugar, and fat molecules.

Egg white protein *denatured* in the presence of extra water with sugar and fat molecules (proteins have been *diluted*)

↑ Heat

Diluted egg white proteins have *coagulated*, trapping water, sugar, and fat molecules—the extra molecules interfere with the protein–protein interactions

→ Overheating

Dilute mixtures of protein that get heated too hot or too fast will *curdle*. The protein clumps together in tight lumps, excluding the other molecules

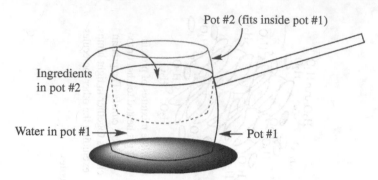

FIGURE 9.22 Double boiler.

and eventually evaporate away. Considering that custards are best cooked at approximately 180°F/82°C, the water bath method of cooking custards works particularly well. Pastry cream and crème anglaise are both egg-thickened pourable creams and good examples of classic, stirred custards. Making crème anglaise requires the use of a *double boiler*. The bowl with the egg yolks, vanilla, milk, cream, salt, and sugar sits over another bowl full of boiling water in a variation on the *water bath* (Fig. 9.22). The bowl of ingredients is only ever heated with the steam from the boiling water. The use of this *double boiler* is essential to successful crème anglaise because the gentle heat from the steam allows the *diluted* mixture of protein to denature and coagulate while trapping the water, fat, and sugar molecules among the protein matrices.

Pastry cream is a very similar dish, but the addition of cornstarch makes pastry cream more heat stable—that is, it can reach higher temperatures without curdling. Therefore, pastry cream can be prepared without the aid of a double boiler. Examining the recipes reveals that the only difference between crème anglaise and pastry cream is the addition of cornstarch. So why does the addition of starch make the denaturation and coagulation of the egg proteins in pastry cream so much easier? The answer lies in the effects of the large starch polymers—amylose and amylopectin. At lower temperatures the starch polymers interfere with the coagulation of the denatured egg proteins by physically separating the chains of denatured protein. Crème anglaise (made without starch) thickens at 71°C/190°F, and the thickening is evidence of protein denaturation and coagulation. If crème anglaise is overheated to 82°C/180°F, the mixture will curdle—that is, you end up with a scrambled egg mixture instead of a smooth cream. In contrast, pastry cream (made with starch) does not begin to thicken until 82°C/180°F and thickens dramatically near the boiling point of water (100°C/212°F). From this example, we can see that the starch is preventing the denatured protein from coagulating into a network. When the temperature is finally high enough, the starch will *gel* to form a three-dimensional network of polymer chains (see Chapter 7—Fruits and Vegetables for a discussion on starch gelation), and this additional polymer network surrounds and stabilizes the denatured protein. Starch is so good at stabilizing the dilute coagulated protein network that stirred or baked custards and creams including starch molecules can be heated to much higher temperatures without the risk of curdling, although the starch does change the final taste and texture.

Baked custards—such as flan or cheesecake—operate on the same principles. Recipes that use no cornstarch must be carefully heated in a water bath to ensure slow formation of the delicate network of denatured and coagulated egg protein diluted with fat and sugar. Recipes with cornstarch can be baked without a water bath.

If you are using cornstarch in your baked or stirred custard, you should be aware that raw egg yolks contain an enzyme called *alpha amylase*, an enzyme that breaks starch molecules into individual sugar molecules. When making a custard or cream with added starch, the primary ingredient is egg yolks, but the *alpha amylase* can totally ruin the product by cleaving the starch molecules. The solution is to heat the raw egg yolks above 180°F/82°C before adding them to the starch; this high temperature denatures and inactivates the alpha amylase enzyme.

9.9   EGG WHITE FOAMS

Denaturing a protein always uses some kind of *energy* to break the *hydrogen bonds* holding the protein together and subsequently unravel its globular structure (Figs. 9.23 and 9.24). The unraveled protein now has exposed *hydrophobic* regions that can stick together (coagulate) with exposed *hydrophobic* regions of other proteins. *Agitation* can also denature proteins—specifically the kind of aggressive agitation that introduces large amounts of air into the protein mixture, creating a *foam*. When cooks use a whisk or wire beater to beat egg white, the protein will (over time) stiffen into a white foamy semisolid. This stiffening and change in color from clear to white mirrors what we see in heat *denaturation* of egg white protein. When egg white protein is *denatured* by whipping or beating air into the mixture, the *denatured* proteins *coagulate* and trap air bubbles in addition to water. The *denatured* proteins

FIGURE 9.23   An egg white foam (creative commons license). https://www.flickr.com/photos/kt_kuksenok/5403419666

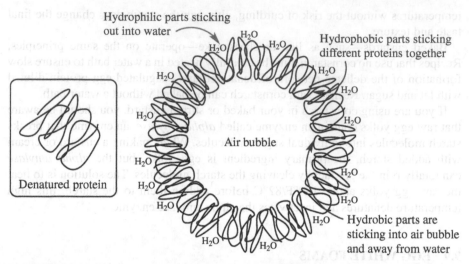

FIGURE 9.24 **Denatured egg white protein coagulating around an air bubble.** This creates a foam.

cluster around and stick their exposed hydrophobic portions into the air bubbles. The large numbers of trapped air bubbles give the *coagulated* protein matrix a soft, semisolid texture. The semisolid foam of *denatured* protein and air bubbles can be mixed with other ingredients and then heated. Heating accomplishes additional protein *denaturation* within the foam—especially of ovalbumin. Ovalbumin is not particularly susceptible to *denaturation* by agitation, and when the foam is heated, ovalbumin finally denatures, hardening the foam even further. Heating expands the air bubbles in the foam, which causes considerable rise, and it also dries out the foam as water molecules evaporate. When the proteins finally set enough from the heat, the gas bubbles are trapped and can't expand anymore. The result is a puffy/tall, hardened yet airy solid. Meringue cookies like the one shown here are classic examples of a cooked egg white foam (Fig. 9.25). The structure of the cookie is created almost entirely by egg white protein.

As we saw with custards, addition of other molecules dilutes the proteins and makes it more difficult to form the coagulated protein network that stiffens, thickens, or solidifies. In particular, when making an egg white foam, *fat molecules* are particularly destructive. Addition of any fat to a foam stabilized by denatured protein is disastrous to the foam. The denatured protein preferentially clusters around the fat molecules (since fats are very hydrophobic) and the air bubbles escape or never form in the first place (Fig. 9.26). In fact, instructions for creating an egg white foam begin with separating the egg yolks from the egg whites. Since the yolk contains all the fat within the egg, removing this fat is essential to creating an egg white foam. This is demonstrated in the preparation of a soufflé. A soufflé is made by first beating egg white into a stiff, glossy foam and then *folding* in a "base" to add flavor to the whites, and the subsequent mixture is then baked. The base typically contains egg yolks, milk, butter, flour, sugar, and flavoring. Because egg white foams are created by the

FIGURE 9.25   A lemon tart topped with meringue (creative commons license). https://flic.kr/p/egrcsh

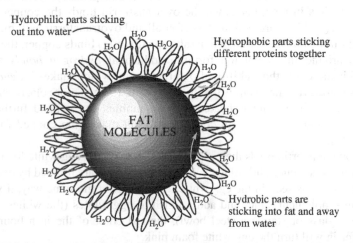

Hydrophilic parts sticking out into water

Hydrophobic parts sticking different proteins together

FAT MOLECULES

Hydrobic parts are sticking into fat and away from water

FIGURE 9.26   Protein denaturation by agitation in the presence of fat molecules.

trapping of air bubbles by denatured protein, the base must be carefully folded in so as not to destroy the foam.

When a soufflé hits the oven, it rises dramatically as the heat expands the air bubbles within the foam. Simultaneously, the heat is further denaturing the protein, which prevents the air bubbles from escaping. The soufflé base typically contains starch and protein to reinforce the walls of the air bubbles that exist in the egg white foam—this gives the soufflé a stronger structure that can stand up straight out of the dish.

Interestingly, it has been known for centuries that egg white proteins form better foams when beaten or whipped in a copper bowl. The secret lies with the protein

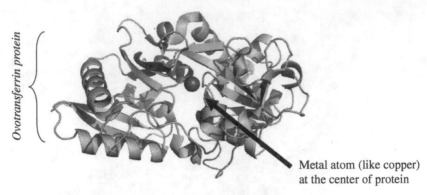

FIGURE 9.27   Ovotransferrin binding a metal at the center.

ovotransferrin (also known as *conalbumin*). Remember that ovotransferrin is present in egg white in order to bind iron—that's its job (Fig. 9.27). Well, when a cook beats egg whites in a copper bowl, the ovotransferrin binds the copper from the bowl! Copper and iron are both metals after all and ovotransferrin will bind copper as easily as iron. When ovotransferrin (aka conalbumin) binds copper, the structure organizes around the metal ion forming additional *hydrogen bonds* and *ionic bonds*. This increases the stability of ovotransferrin, which makes it *denature at a higher temperature*. The result is that the egg white foam is more elastic and stable. When the egg white foam is heated, the air bubbles can expand further before the copper-bound ovotransferrin coagulates—this makes the cooked foam taller or puffier [6].

When ovotransferrin binds the copper from the bowl, the egg white foam takes on a yellowish-golden hue. The golden yellow copper atoms are trapped by ovotransferrin, and the color is seen in the egg white foam itself. In the same way, if you crush up an iron supplement tablet and add a pinch to egg whites (the whites should be beaten in a glass or stainless steel bowl), the red color of the iron bound by the ovotransferrin will turn the egg white foam pink.

If you don't have a copper bowl (they are expensive and hard to clean), then addition of acids like cream of tartar (i.e., tartaric acid) or beating the egg white over a bowl of hot water can improve the formation of an egg white foam by encouraging protein denaturation. In each case, using acid or heat disrupts the *hydrogen bonds* holding the protein structure together causing denaturation. The acid-induced denaturation of egg protein is similar to the formation of curds in the acidification of milk that occurs when making yogurt or cheese.

The addition of acid unravels the protein by adding protons (H^+) that change the charges of some amino acid residues—like glutamate, shown in Fig. 9.28. The change of charge disrupts the noncovalent interactions that hold the protein structure together. The normal, folded, three-dimensional protein structure has water-loving (hydrophilic) parts on the outside and water-hating (hydrophobic) parts on the inside of the folded protein, but addition of acid unfolds the protein by disrupting

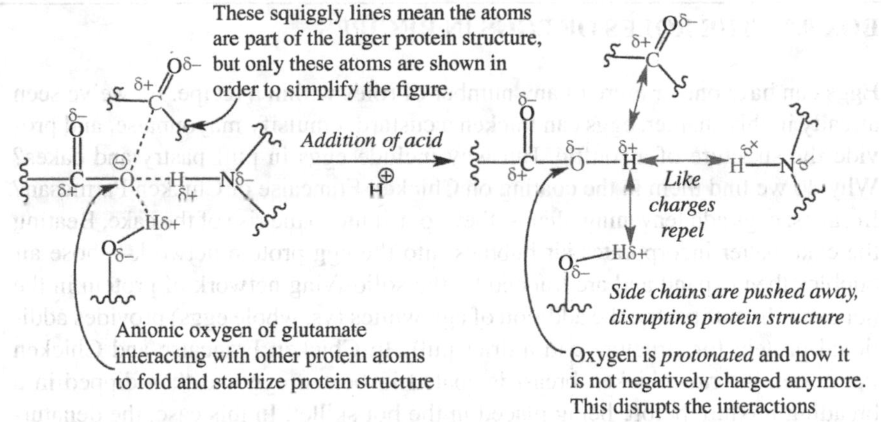

These squiggly lines mean the atoms are part of the larger protein structure, but only these atoms are shown in order to simplify the figure.

Addition of acid

Like charges repel

Side chains are pushed away, disrupting protein structure

Anionic oxygen of glutamate interacting with other protein atoms to fold and stabilize protein structure

Oxygen is *protonated* and now it is not negatively charged anymore. This disrupts the interactions

FIGURE 9.28 Effect of pH on protein coagulation. Several amino acids will ionically interact depending on pH. Tartaric acid is used to reduce the aggregation of egg proteins.

the stabilizing interactions. The acid *denatured* protein *coagulates* with other unraveled protein by sticking the exposed hydrophobic areas together.

9.10 EGG PASTEURIZATION

Heating eggs at 140°F/60°C for 5 min or at 190°F/88°C for 1 min is essential in order to kill *Salmonella*. *Salmonella* is a type of bacteria that can be found inside raw eggs. The bacteria come from the chicken that laid the egg—not necessarily from poor handling of eggs in the grocery store or farm. Since the bacteria are living inside the egg, washing the surface of the egg has no effect on removing them. Since *Salmonella* bacteria are living organisms, they prefer warm temperatures to grow; bacterial growth is slowed at cold (refrigerator) temperatures and the bacteria are killed by hot temperatures. There is no way to tell if an egg is harboring *Salmonella*—you simply have to prepare egg dishes in order to kill *Salmonella* were it present.

Unfortunately, pasteurization does affect the quality of the foam prepared from the pasteurized egg whites such that whipping a foam takes longer for pasteurized egg whites than for raw egg whites, and pasteurized foam is less stable. The brief increase in temperature during pasteurization denatures ovotransferrin, which has the lowest denaturation temperature (57°C/135°F) of all the egg white proteins, and also causes partial denaturation of ovomucin and lysozyme [7]. Lysozyme is a very alkaline protein, and so it binds tightly to acidic proteins in egg white, including ovomucin—it is the ovomucin–lysozyme network that denatures during pasteurization. To improve foaming after pasteurization, metal ions can be added to stabilize the ovotransferrin, and/or the denatured ovomucin–lysozyme network can be removed.

BOX 9.2   THE ROLES OF EGGS IN RECIPES

Eggs can have one or more of any number of roles within a recipe. As we've seen already in this chapter, eggs can thicken a custard, emulsify mayonnaise, and provide the structure of a soufflé. But why include eggs in puff pastry and cakes? Why do we find them in the coating on Chicken Francaise or Chicken Parmesan? In cakes, eggs add leavening, that is, they contribute to the rise of the cake. Beating the cake batter incorporates air bubbles into the egg protein network—those air bubbles then expand and are trapped by the solidifying network of protein in the hot oven. In puff pastry, the addition of egg whites (vs. whole eggs) provides additional protein for structure and a drier puff. In Chicken Francaise and Chicken Parmesan, the raw chicken breast is coated in beaten egg and then dipped in a breading mixture before being placed in the hot skillet. In this case, the denaturation and coagulation of the egg protein help glue the breading to the meat.

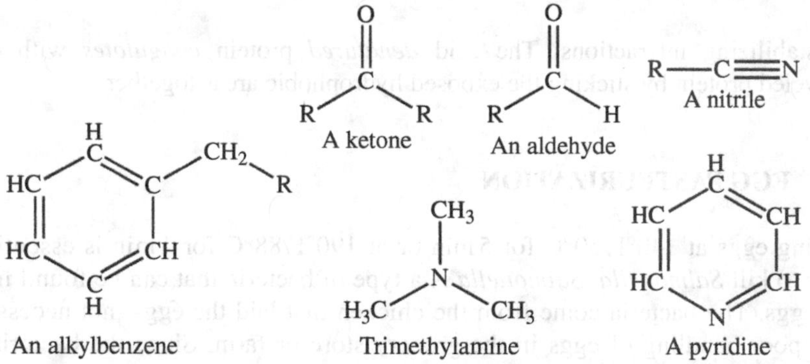

FIGURE 9.29   Classes of odor molecules produced by cooking eggs.

9.11   HEATING EGG PROTEIN CAUSES CHEMICAL REACTIONS

What comes to mind when you think of hard-boiled eggs? Perhaps, the unpleasant and complex sulfurous odor of cooked eggs? The strangely greenish-yellow yolks that come from overcooking? What about the complex odor of well-cooked eggs? As we have seen in Chapter 2, odor is due to volatile molecules that are released into the air, and cooking eggs causes the release of a variety of molecules that we see and smell.

Much of the flavor and sensory perception of eggs comes from chemical reactions that produce gaseous or volatile molecules. Cooked whole eggs produce 141 unique, identifiable odor compounds. Most are nitriles (an organic compound with carbon–nitrogen triple bonds) and alkylbenzenes (benzene rings with one or more carbon atoms), and another 33% were either ketones or aldehydes from fatty acids. The Maillard reaction between sugars and amino acids produces pyrazine and thiazole compounds giving another strong egg-like flavor and odor (Fig. 9.29).

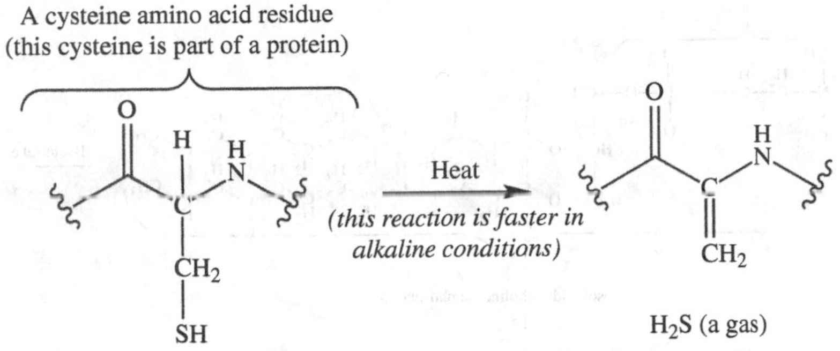

A cysteine amino acid residue
(this cysteine is part of a protein)

FIGURE 9.30   The reaction of cysteine and heat to produce H_2S.

Heating proteins causes them to *denature* and *coagulate* into a solid, but heating also causes chemical reactions within the proteins. In particular, the characteristic "eggy" odor of hard-boiled eggs can be attributed to the degradation of cysteine amino acid residues within the albumen protein to produce H_2S (*hydrogen sulfide*; Fig. 9.30). The sulfurous odor of H_2S is what we associate with cooked eggs. In high concentrations, H_2S can be objectionable. The cysteines within the egg white albumen protein react more readily to produce H_2S when the protein has unfolded (*denatured*). This exposes cysteines that were previously buried inside the protein and allows them to react.

The reaction to produce H_2S also happens more readily in *alkaline* conditions (pH > 7), and older eggs inevitably turn alkaline due to loss of carbon dioxide. Some cookbooks recommend that you add lemon juice or vinegar to the water when hard-boiling eggs in order to reduce the "eggy" smell. The acid of the vinegar or lemon juice is supposed both to lower the pH and reduce the formation of H_2S (Fig. 9.31).

The breakdown of egg white protein upon heating also produces nitrogen-containing compounds including ammonia, methylamine, and dimethylamine. The latter can be further altered to form the fishy smelling trimethylamine (see Chapter 8 for TMA). When the egg white is more alkaline (pH around 9–10), heating the egg produces a number of pyridine types of molecules (Fig. 9.29). These strong-smelling, nitrogen-containing, cyclic molecules are only partially soluble in water making the escape of their distinctive fishy odor slow and continual. The yolk contains more fats than proteins and upon cooking produces odor molecules from the breakdown of fatty acids including ketones, aldehydes, and short-chain fatty acids. Bacterial breakdown of phospholipids such as phosphatidylcholine inside a contaminated egg or inside the human gut can produce choline and eventually trimethylamine (the fishy odor) in aged eggs (Fig. 9.32).

The reaction to produce FeS happens when the eggs are heated for extended periods of time. Boiling eggs in water for 15 min is a sure way to make iron sulfide. Sometimes when hard-boiling eggs, a green gray discoloration can occur at the interface of the yolk and white and can even (in some cases) discolor the entire yolk.

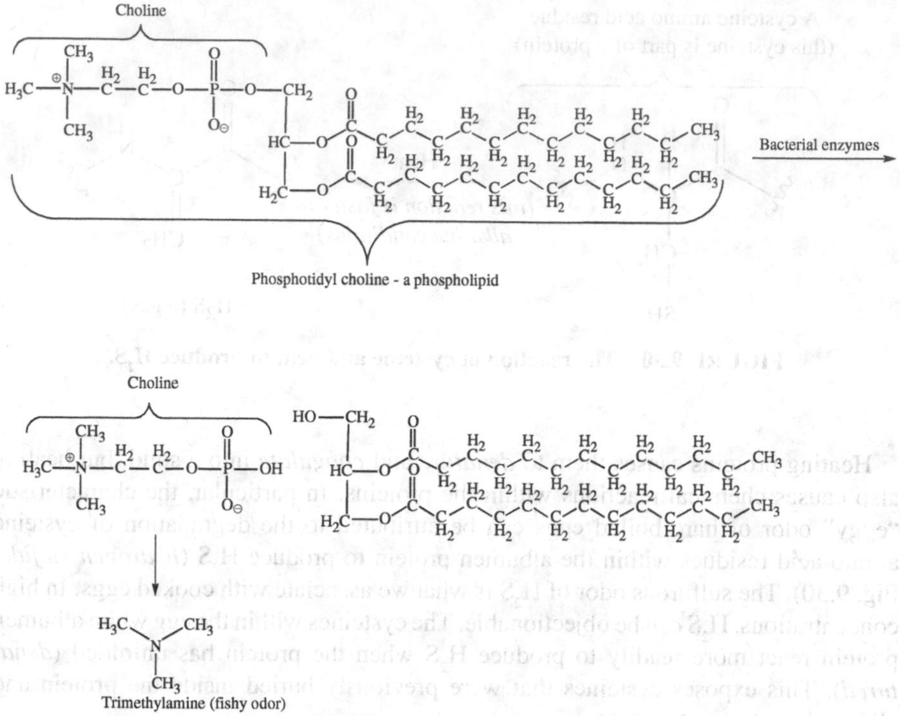

Choline

$H_3C - \overset{\oplus}{N} - \overset{H_2}{C} - \overset{H_2}{C} - O - P - O - CH_2$ ... Phosphotidyl choline - a phospholipid → Bacterial enzymes →

Choline

Trimethylamine (fishy odor)

FIGURE 9.31   Formation of fishy trimethylamine in aged eggs.

Green-gray line
at interface

Bright
yellow
yolks

Green-gray
discoloration
throughout

FIGURE 9.32   Green eggs and iron sulfide.

This greenish color is caused by the compound *iron sulfide* (FeS). The source of the
sulfur (S) in FeS comes from the H_2S released by albumen cysteines, while egg yolk
is particularly iron (Fe) rich. The H_2S gas migrates from the white into the yolk—
forming the green iron sulfide (FeS; Fig. 9.32).

REFERENCES

[1] McGee, H., ed. (2004) *On Food and Cooking*. Simon and Schuster, Inc., New York.

[2] Kenneth, F. K. and Kriemhild Coneè, O., ed. (2000) *The Cambridge World History of Food*; Volume I. Cambridge University Press, New York.

[3] Smith, P. and Daniel, C. (1975) *The Chicken Book*; Little, Brown and Company, Boston.

[4] Donovan, J.W., Mapes, C.J., Davis, J.G. and Garibaldi, J.A. (1975) A differential scanning calorimetric study of the stability of egg white to heat denaturation. *J. Sci. Food Agric.* 26: 73–83.

[5] Huopalahti, R., López-Fandiño, R., Anton, M. and Schade, R. (2007) *Bioactive Egg Compounds*. Springer Science & Business Media, New York.

[6] Harold, J.M., Sharon R.L. and Winslow, R.B. (1984) Why whip egg whites in copper bowls? *Nature* 308: 997–998.

[7] Lomakina, K. and Míková, K. (2009) A study of the factors affecting the foaming properties of egg white—a review. *Czech J. Food Sci.* 24: 110–118.

10

BREAD, CAKES, AND PASTRY

Guided Inquiry Activities (Web): 6, Higher Order Protein Structure; 7, Carbohydrates; 9, Fats Structure and Properties; 18, Starch; 27, Bread

10.1 INTRODUCTION

Few of us can walk by a bakery without being drawn to the aromas and visuals of freshly baked breads, pies, or cakes. At times, you might wish to bite into a chewy, dense wheat bread. On other occasions, you might be drawn to a flaky, buttery croissant. Perhaps on a third occasion, you might choose a light and fluffy piece of cake. Flour and water provide the basis for all of the foods, yet they have significantly different textures, aromas, and mouthfeel due to additional ingredients, treatment of, and structure of the dough or batter. In this chapter, you will learn about flour and its molecular components, the process of making a yeast-based bread, quick bread, sponge cake and pastry, and variations in other ingredients that help to yield the final baked product with the desired texture, aroma, and flavor. The flour that is used in most baked goods comes from wheat. Not only is wheat readily available, but it produces a flour that has the components and properties that are ideal for different types of baked goods, ranging from leavened breads to flaky pie crusts. Thus, most of this chapter will center upon wheat-based flours and baked goods. However, due to the fact that there are many individuals that are afflicted with celiac disease, we will also discuss some other flour options and characteristics of the disease itself.

The Science of Cooking: Understanding the Biology and Chemistry Behind Food and Cooking,
First Edition. Joseph J. Provost, Keri L. Colabroy, Brenda S. Kelly, and Mark A.Wallert.
© 2016 John Wiley & Sons, Inc. Published 2016 by John Wiley & Sons, Inc.
Companion website: www.wiley.com/go/provost/science_of_cooking

10.2 WHEAT-BASED FLOUR, WHERE IT COMES FROM AND ITS COMPONENTS

All flours come from a grain that is milled or ground into a fine powder; however, different types of flours are produced when each component of the grain or various types of wheat are milled. First, let's talk about the structural components of a grain.

A grain (or seed) consists of three major structural components: the germ (or embryo), the endosperm, and the bran (Fig. 10.1). The endosperm makes up most of the bulk of the grain and contains cells that are filled with carbohydrates and proteins. Carbohydrates, in the form of starch, make up about 70–80% of the endosperm. The protein component (about 7–15%) consists of both soluble and insoluble proteins. The sprout of the plant emerges from the part of the grain called the germ or embryo. The germ is nutrient rich and contains fats, vitamins, and mostly soluble proteins. The bran makes up the outer shell of the grain; it contains some minerals (~7%), soluble proteins (~16%), and carbohydrates but mostly consists of the structural, insoluble carbohydrate called fiber. In a seed, fiber has a primary role of protecting the endosperm and germ components.

You might be asking, what do the structural components of a seed have to do with flour? Different types of flours are distinguished by the component of the grain that is milled during the production process. For example, white flour comes almost solely from the endosperm portion of a grain of wheat, while whole wheat flour

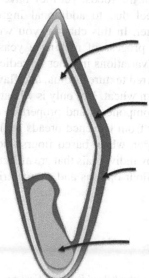

Endosperm
• *Starch (complex carbohydrates—flour)*
• *Limited proteins*
• *B vitamins*

Aleurone
• *Outermost layer of endosperm*
• *Lining made of specialized cells*
• *Contains most of nutritional whole grain components*

Bran—Just under the hull/husk
• *Insoluble carbohydrates (fiber)*
• *Trace minerals, phenolic compounds*
• *B vitamins*

Germ
• *"New plant" embryo*
• *Nutty sweet flavor*
• *Vitamin E, folic acid, thiamine, and metals*

FIGURE 10.1 Anatomy of a grain. Flour is primarily composed from starch and proteins found in the endosperm.

TABLE 10.1   Protein Content and Use of White, Wheat-Based Flours.

Flour Type	Protein (% of Total)	Primary Use in Cooking
Cake	7.5–8.5	Cakes, quick breads, muffins, and pancakes. Produces a tender crumb
Pastry	8–10	Pie crusts and pastries
Instant	9.5–11	Sauces and gravies
Bleached, all purpose	9.5–12	General use flour for all cooking and baking purposes but contains a little too much protein for the best pie crusts, muffins, and pancakes and too little flour for the best yeast breads
Bread flour	11.5–12.5	Yeast breads, pasta, pizza
Durum wheat (semolina)	13–13.5	Pasta

contains the ground bran, germ, and the endosperm. You have probably seen or used other types of flours as well, including all-purpose, bread, cake, semolina/durum, pastry, and instant. These white, wheat-based flours are distinguished by their protein content (Table 10.1).

Wait a minute, protein content! How do you get flours of different protein content when these flours all come from wheat? Each component of a wheat grain contains a different amount and type of protein (soluble and insoluble). Moreover, differences in wheat type, growing climate, and soil conditions will all impact the amount of protein present within a grain of wheat. The endosperm of spring-sown wheat is more brittle, disintegrates readily on milling, and tends to have a higher protein content; thus spring-sown wheat produces "hard" flours. "Soft" flour is usually obtained from winter wheat, as these flours have a protein content of less than 10%.

Try this at home. Place a few tablespoons of bread flour, all-purpose flour, and cake flour in different unlabeled dishes. Feel the texture of each. Can you tell the difference between the softer flours (low-protein content) and the hard flours (high-protein content)?

The "all-purpose" white flour that you purchase in the grocery store is usually a blended flour to give a more consistent product in terms of protein content. Most flours typically do not contain any other additives, aside from drying agents to prevent the flour from forming lumps when exposed to moisture in the air, unless the flour is designated as "self-rising." Self-rising (self raising in some countries) flour contains a raising agent like baking powder; these flours are sometimes called for in quick bread, biscuit, or cake recipes.

There is clearly a lot to learn about the protein in flour. However, you can't consider protein in the absence of carbohydrates. Let's learn more about the carbohydrate and protein components of a flour and how each impacts the characteristics of a loaf of bread, pie crust, or cookie.

10.3 CARBOHYDRATES IN FLOUR

10.3.1 Starch

You already know a lot about carbohydrates and their importance in cooking, food, and nutrition. To refresh your memory, glucose is a carbohydrate (it contains carbon, hydrogen, and oxygen atoms) in which five of the carbons and an oxygen are most commonly arranged in a "ring" or cyclic structure (Fig. 10.2). When glucose is not bonded to any other carbohydrate molecules, it is a monosaccharide.

However, most of the glucose molecules in flour don't exist in this monosaccharide molecular form; they are present in the form of a glucose polysaccharide that is called starch. You have likely heard of the term "starch," used the term "starch," and maybe even try reducing "starch" from your diet, but have you ever thought about what starch looks like at the molecular level?

Starch is composed of two different types of glucose polysaccharides, amylose and amylopectin, both of which consist of glucose molecules that are covalently bonded to one another via glycosidic bonds to make long molecular chains of glucose (Fig. 10.3). A single molecule of amylose may contain up to several thousand glucose molecules that are bonded together via $1 \rightarrow 4$ glycosidic links. An amylopectin polysaccharide can contain many more glucose molecules (on the order of a million) that are bonded via $1 \rightarrow 4$ glycosidic links and $1 \rightarrow 6$ glycosidic bonds.

A starch granule in flour, which contains many individual molecules of amylose and amylopectin that make up 98–99% of the total molecular components, has a size of 2–40 µm and consists of 10–20% amylose and 80–90% amylopectin. Amylose has a more linear three-dimensional structure, while the $1 \rightarrow 6$ glycosidic links in an amylopectin molecule result in a branched structure that kind of looks like a bristled brush. The different polysaccharide molecules interact loosely via intermolecular interactions with themselves, with one another, and with any small amount of protein that is present within the flour. However, because of the differences in three-dimensional structure, amylose molecules pack more tightly against one another and any interacting protein molecules than amylopectin. Thus, amylose molecules and any interacting proteins that are present in the starch granule are more resistant to degradation by enzymes, since the enzymes have a difficult time accessing the molecules to break them down (Box 10.1).

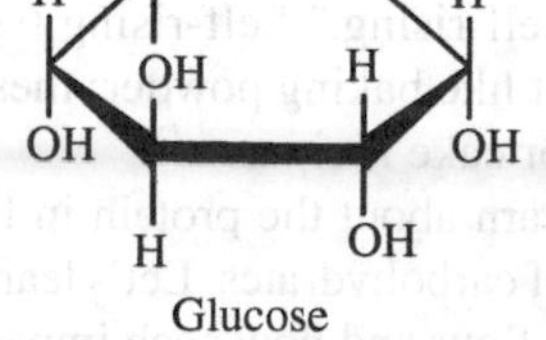

Glucose

FIGURE 10.2 The structure of glucose.

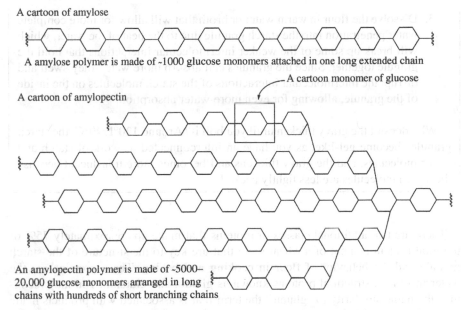

A cartoon of amylose

A amylose polymer is made of ~1000 glucose monomers attached in one long extended chain

A cartoon monomer of glucose

A cartoon of amylopectin

An amylopectin polymer is made of ~5000–20,000 glucose monomers arranged in long chains with hundreds of short branching chains

FIGURE 10.3 Starches of flour. Amylose and amylopectin are both polymers of glucose monomers. The branched amylopectin has two types of glycosidic bonds and provides many "ends" for the storage and quick release of glucose by digestive enzymes.

BOX 10.1 STARCH, FLOUR, AND GRAVY

The amount and type of protein in a starch granule along with its location are critical to cooking. You see this when you make a flour-based gravy or sauce. When cold water is added to flour, it is absorbed by a small amount of water soluble protein that is present, but it can't penetrate the amylose and amylopectin carbohydrates due to the tightly packed structure of the carbohydrates. Once the proteins absorb water, they become sticky and cause the starch granules to clump together. Can you say "lumpy gravy"? Once a large lump has formed, the molecules near the center of the lump become protected from the outside, so the lump will not break down, even when you add more water or try to whisk it away. What are some possible ways to prevent the problem?

1. Use a low-protein flour (e.g., Wondra™). Less protein decreases the likelihood that the proteins will absorb all of the water, allowing some water to penetrate and interact with the carbohydrates.
2. Initially dissolve a little flour in a large amount of water with rapid mixing. If there is enough water and stirring for water to penetrate the starch granule, they begin to swell and absorb enough water that they will not clump (but they will separate from the water if you leave the mixture to sit in a bowl for a few minutes).

3. Dissolve the flour in warm water or broth that will allow for more complete water penetration into the starch granule due to the heat of the water, which will break up some of the weaker intermolecular interactions that hold the granule together. Since the granules can absorb more water, they swell and disrupt the intermolecular interactions of the starch molecules on the inside of the granule, allowing for even more water absorption.

Why doesn't the gravy thicken until you boil it? Around 120°F/49°C, the starch granules become gel-like, as you have an interconnected network of starch and water molecules, and the gravy thickens and becomes more translucent because the starch molecules are less tightly packed.

There are two additional classes of proteins that make up approximately 75% of the total protein content of a wheat grain that are key to the structure of the starch granule and the behavior of flour in cooking and baking. These proteins are the water-insoluble, structural proteins known as gliadin and glutenin. You may recognize the name similarity to "gluten," the term that is associated with products made from wheat-based flours. In the following section, you will learn that gluten is not a protein but is formed when the two classes of proteins, the gliadins and glutenins, interact with each other (Fig. 10.4).

10.4   WHEAT PROTEINS AND GLUTEN FORMATION

How significant is the impact of gliadin and glutenin proteins in baked goods? Huge! The behavior of these proteins can lead to a flaky, tender, or dense, chewy pie crust. Interactions of the proteins can result in a loaf of bread with a tender and light crumb or a dense, heavy, chewy texture. If you are a baker, it is very useful to have some knowledge of the structure and behavior of these proteins so that you can treat them appropriately for the baked result that you desire.

The gliadins are proposed to be small, tightly coiled proteins that fold in a compact, spherical three-dimensional structure. Each gliadin protein molecule can fold and is stable in a monomeric form due to the presence of internal (i.e., within the same molecule) intermolecular interactions such as hydrogen bonds and electrostatic interactions (aka ionic interactions). A single variety of wheat may have over 40 different types of proteins that are classified as "gliadins" that have small differences in sequence and structure.

Although glutenin proteins also have coiled, helical structural components, the structures are much less tightly packed and individual glutenin molecules interact with other glutenin protein molecules to form much larger protein aggregate complexes. These complexes have some very distinctive structural and protein sequence features that contribute to their physical properties and behavior. Like what? The complexes have a large center section (called a domain) containing 440–680

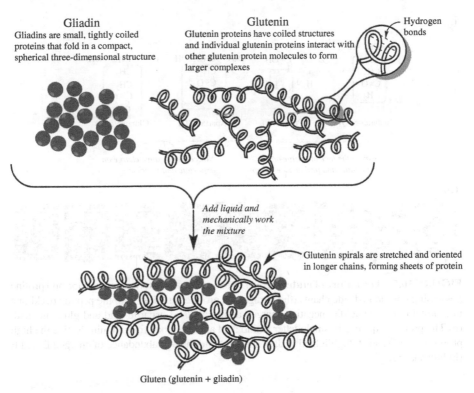

Gliadin
Gliadins are small, tightly coiled proteins that fold in a compact, spherical three-dimensional structure

Glutenin
Glutenin proteins have coiled structures and individual glutenin proteins interact with other glutenin protein molecules to form larger complexes

Hydrogen bonds

Add liquid and mechanically work the mixture

Glutenin spirals are stretched and oriented in longer chains, forming sheets of protein

Gluten (glutenin + gliadin)

FIGURE 10.4 The muscle of flour. Two of the key proteins (not the only proteins) in flour, glutenin and gliadin, combine to form gluten, called the muscle of flour.

amino acids that is formed from short, repeating amino acid sequences with a high relative proportion of the uncharged side chains of amino acids, glutamine and pro-line (abbreviated Q and P), and a low relative proportion of negative charged amino acids (abbreviated D and E). See Figure 10.5 to analyze the high quantity of Q and P versus D and E amino acids in the amino acid sequence for glutenin [1].

The glutamine residues have two important roles. They act as an important nitrogen source for a germinating grain of wheat, and, structurally, the amide side chain of glutamine (R—NH—R') allows for formation of hydrogen bonds between different glutenin molecules and between glutenin and water (Fig. 10.6). However, even though glutenin can hydrogen-bond with water, the glutenins remain insoluble due to the low relative proportion of charged amino acids within this center domain of the protein complex.

Are there any other structural domains within the glutenin complex that impact protein behavior? Yes, the glutenin central domain is surrounded by N- and C-terminal domains that contain a high proportion of the amino acid, cysteine (Fig. 10.7). Remember cysteine? It has a sulfur-containing side chain and readily forms strong, covalent, disulfide bonds with other cysteine residues in oxidizing environments (Fig. 10.8).

(a)

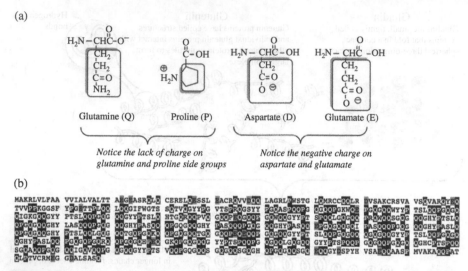

(b)

```
MAKRLVLFAA VVIALVALTT AEGEASRQLQ CERELQSSSL EACRQVVDQQ LAGRLPWSTG LQMRCCQQLR DVSAKCRSVA VSQVARQYEQ
TVVPPKGGSF YPGETTPLQQ LQQGIFWGTS SQTVQGYYPG VTSPRQGSYY PGQASPOOPG QGQQRGKWQE PGQQQQWYYP TSLQQPGQGQ
QIGKGQQGYY PTSLQOPGQG QQGYYPTSLQ HTGQRQQPVQ GQQPEQGQQE GQWQQGYYPT SPQQLGQCQQ PROWQQSGQG QQGHYPTSLQ
QPGQGQQGHY LASQQOPGQG QQGHYPASQQ QPGQGQQGHY PASQOOPGQG QGQHYPASQQ PEGQGQOQI PASQQOPGQG QQGHYPASLQ
QPGQGQQGHY PTSLQQLGQG QQTGQPGQGQ QPGQGQQTGQ GQQPEQGQQE GQGQQGYYPT SLQQPGQGQQ QGGQGQQGYYP TSLQQPGQGQ
QGHYPASLQQ PGQGQQGGQQ PGQGQGHPEQ GKQPGQGQQQ YYPTSPQQPG QGQQLGQGQQ GYYPTSPQQP GQQQPGQGQQ QGHCPTSPQQ
SGQAQQPGQG QQIGQVQQFG QGQQGYYPTS VQQPGQGQQS GQGQQSGQGH QEGQGQQSGQ EQQGYDSPYH VSAEQQAASP MVAKAQQPAT
QLETVCRMEG GDALSASQ
```

FIGURE 10.5   The nature of glutenin. (a) The amino acid structures of glutamine and proline possessing noncharged side chains (the nitrogen on proline is tied up with the peptide backbone in a protein) and the acidic, negative charged amino acids of aspartic acid and glutamic acid. (b) The protein sequence in single letter amino acid abbreviations for glutenin. Notice the high proportion of Q and P (highlighted in dark shading) and the low abundance of charged D and E (lightly shaded).

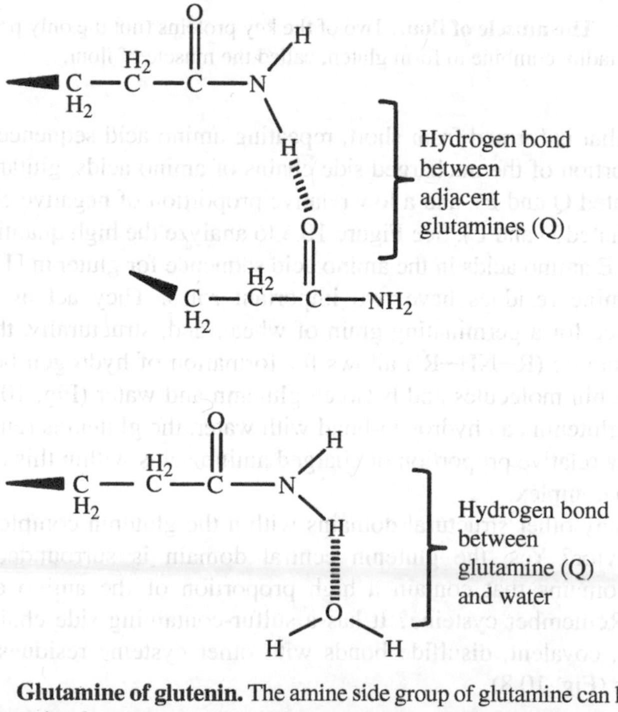

FIGURE 10.6   Glutamine of glutenin. The amine side group of glutamine can hydrogen-bond with another glutamine forming cross-links between strands, within strands and with water.

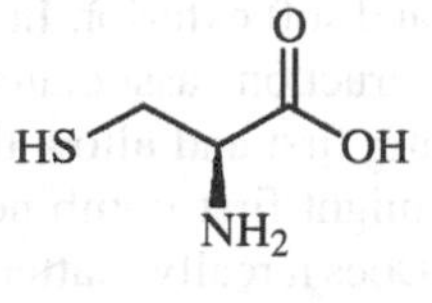

FIGURE 10.7   Cysteine.

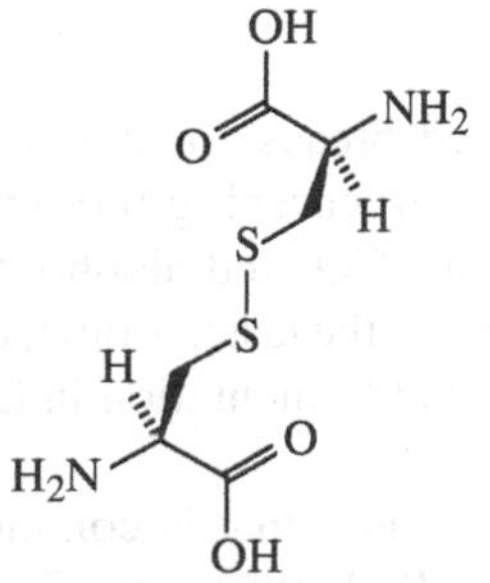

FIGURE 10.8   Cysteine disulfide. Two cysteine amino acids can bond through the S atoms. In bread this cross-links proteins making them more stable and mechanically rigid.

When you have dry flour or a "just-combined" flour/water mixture, the gliadin molecules and glutenin protein complexes do not significantly interact with one another, starch molecules or water molecules; they have a random and disorganized arrangement. However, once a flour/water mixture begins to be worked, through stirring, mixing or kneading, the gliadins and glutenin complexes begin to interact with one another via intermolecular interactions. How and why does this happen? Well, as you work with the dough, the gliadin and glutenin proteins are stretched out, literally. They no longer are balled up in a spherical structure because the mechanical action of stirring or mixing extends and stretches the proteins out, kind of like a slinky. As they are stretched, the proteins have more opportunity and space to interact with other protein molecules; correspondingly they align against one another in orderly, stretched, interacting protein sheets that have a new name: gluten. Thus, gluten is the complex network of interacting gliadin and glutenin proteins that has properties distinct from either of the individual proteins including strength, elasticity, and stretchiness. Gluten is of the utmost importance in baked goods. To make a good bread, you desire gluten formation. In a pie crust, you want to avoid the formation of gluten. Let's apply and expand our basic understanding of gluten to something that you can really chew on: bread.

10.5   YEAST-RAISED BREAD

The ingredients and steps of baking a fresh loaf of bread are pretty simple: mix flour, water, yeast, and salt; knead the dough; let the dough rise; and bake. Yet, anyone who has made bread can attest that a perfect loaf is not easily attained. The bread may be dense, flavorless, or have a pale and soft exterior. In addition, there is some variation in ingredients and the specific instructions associated with a basic white bread. Your mother might mix yeast with water first and allow the mixture to sit (or proof) for a few minutes, while your uncle might first combine yeast with the flour, then add water to the yeast/flour mixture. Does it really matter? Yes! Let's begin our discussion about bread by learning more about the organism that is key to it all: yeast.

10.5.1 Yeast

Humans have been eating raised breads for about 6000 years; however, we only began to understand the process of leavening through the work of Louis Pasteur in the 1850s. Now, we know that the CO_2 and alcohol that are generated during yeast fermentation are key components to the taste, aroma, and texture of a bread. You have heard about yeast, metabolism, and fermentation in Chapter 4, but let's review some of the basics again.

Yeast is a living organism, a fungus that is commonly used in baking and in the brewing of alcoholic beverages. Its Latin name, *Saccharomyces cerevisiae*, means brewer's sugar. In bread dough, yeast consume starch and sugar that is present in flour and convert it to CO_2 and alcohol through the process of fermentation. These yeast products give a baked loaf of bread its texture, flavor, and aroma. How do the yeast accomplish this? Can yeast just begin "chewing" up the starches and producing CO_2 immediately when added to flour? Not exactly. Yeast actually can't break down starch in its gigantic, polysaccharide form; it is too compact and tightly packed for the yeast enzymes to access it for metabolism. Thus, although there is a lot of potential, because the glucose molecules are stored in the form of the starch polysaccharide, the yeast can't initially use it to support its own growth and reproduction. Fortunately, you might recall that flour contains soluble enzymes called amylases. When activated in the presence of water, the amylases carry out hydrolysis reactions, and use water to break the starch down into smaller molecular compounds: maltose, glucose, and other dextrins (smaller glucose polysaccharides) (Fig. 10.9). These smaller sugars can serve as a food source for the fermenting yeast. Isn't it great that flour contains precisely the enzymes that yeast need to thrive?

Once yeast have a useable food source, they begin to carry out glycolysis, an anaerobic (i.e., no oxygen is required) metabolic process that is essential for the yeast organism's growth and reproduction. In Chapter 4, you learned about the importance of glycolysis to all organisms, as it is the primary metabolic pathway whereby organisms can generate the energy source molecule, ATP. After an organism carries out glycolysis, however, it needs to do something with the products of the pathway, by breaking them down further or by repurposing them into molecules that are essential for the function of the organism. Yeasts have the enzymes necessary to accomplish both of these tasks in the process called fermentation (Fig. 10.10). In fermentation, yeast break down the pyruvate generated by glycolysis via two reduction reactions, converting it first to acetaldehyde and carbon dioxide and then ethanol. NADH is used as the reducing agent and is converted to NAD^+. Thus through fermentation, the yeast generates ethanol and carbon dioxide gas and regenerates the NAD^+ that is

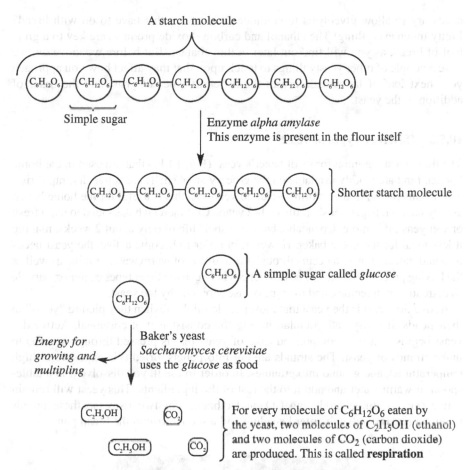

FIGURE 10.9   Yeast metabolism. Yeast utilize its enzyme, amylase, to break down complex starches to the simple sugar, glucose. Glucose and not starch is then converted to carbon dioxide and other compounds via respiration.

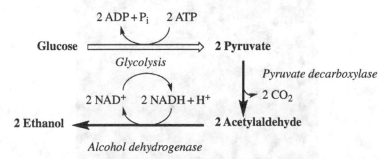

FIGURE 10.10   Ethanol production in yeast. Continued respiration (glucose metabolism to pyruvate and carbon dioxide) requires a regeneration of NAD^+ from NADH. Yeast use alcohol dehydrogenase for this step. The by-product (from the yeast's perspective) is ethanol.

necessary to allow glycolysis to continue. What does this have to do with bread? Pretty much everything! The ethanol and carbon dioxide products are key to a great loaf of bread, as you will find out later in this chapter. But before we move on, we have a couple of other yeasty things to touch upon that may affect how you approach your next loaf of homemade bread: the different types of yeast and the order of addition of the yeast.

10.5.2   Different Types of Yeast

The three most common forms of baker's yeast (Fig. 10.11) that are used in the home kitchen (and are readily available on the grocery store shelf) are active dry, rapid-rise, and instant yeast. All these forms are dried versions; this is useful for the home baker, as they have a refrigerator shelf life of 1–2 years. Commercial bakers tend to use a fresh or wet yeast; it is more dependable but has a shelf life of only about 2 weeks, making it less ideal for the home baker. However, in order to become active, the yeast needs heat and moisture, which occurs through the addition of warm water or milk, as well as the baking process (for heat). The differences between the three types center on particle size, treatment in recipes, and the type of rise provided by the yeast.

Active dry yeast is the yeast that most people think of when they picture "yeast" in their heads. It is dry and granular, having the consistency of cornmeal. Active dry yeast begins as a wet, compressed cake of yeast that is pressed through screens to make strands of yeast. The strands are cut into pellets, dried (via exposure to high temperatures), and ground into granules. In order to use it, you dissolve it in a tablespoon of warm water and add it to the rest of the ingredients. This yeast will remain active (i.e., continue producing CO_2 and ethanol) for two rises, so these breads require a rise immediately after kneading and a second rise in the bread pan.

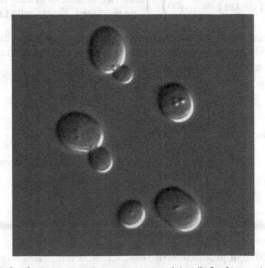

FIGURE 10.11   Baker's yeast. *Saccharomyces cerevisiae* (baker's yeast) in DIC microscopy. "*S. cerevisiae* under DIC microscopy" by Masur.

FIGURE 10.12   Proofing yeast. Baker's yeast incubating in sugar and water.

Rapid rise yeast is dried at lower temperatures and is milled into smaller particles than the active dry version. Because of this, it doesn't need to be dissolved in water and can be added directly to the dry ingredients. Typically, rapid-rise yeast has other additives (like enzymes) that also help to make the dough rise faster. With this yeast, you only need to do one rise. In other words, right after kneading, you can shape the dough into the bread pan, let it rise, and toss it into the oven for baking. However, what you save in time, you lose in flavor and structure. Your final loaf will be fairly bland and commercial tasting. If substituted for either the active dry or the instant yeasts, you will see a difference in the final bread product in the rise (or fall) of the bread or in its taste/texture.

Instant yeast is also known as bread machine yeast. It is milled into finer particles than active dry yeast, so it doesn't need to be dissolved in water. However, unlike rapid rise, it will give you two separate rises and can be used interchangeably with active dry yeast.

10.5.3   First Steps: Yeast Proofing

Most home bakers use the active dry version of yeast, thus "proofing" the yeast by combining with a warm liquid, along with a little sugar or honey is a necessary step for a good rise (Fig. 10.12). Following a resting period of 5–10 min, you should see a foam or bubbly liquid, as the yeast begin to ferment the added sugar if the yeast are active. The added sugar, although not absolutely necessary, jump-starts the activation of the yeast. If there are no bubbles or foam, something has gone wrong—the yeast may have expired, the liquid may be too cool to awaken the yeast, or too hot and it may have killed them. The optimal temperature for the liquid (which is typically water, milk or some combination of the two) during proofing is 105–110°F. For rapid-rise yeast that is added to the flour first, you will want your liquid to be at a slightly higher temperature (120–130°F/49–54°C). Once the yeast are activated, you are ready to add the liquid to the other ingredients (typically flour and salt), which is where the yeast and the molecular components of the flour come together to yield a great loaf of bread.

10.5.4   Next Steps: Mixing, Kneading, and Gluten Formation

Mixing may seem like an unimportant afterthought relative to the yeast and kneading processes. However, there is more to mixing than simply combining the physical ingredients. When water is added to flour, the water is absorbed by the starch granule.

What does this mean? The glutenin and gliadin proteins on the exterior of the starch granules generate intermolecular interactions with the water molecules (and vice versa), thus the proteins become hydrated. Remember that the proteins were already involved in intermolecular interactions with themselves to create their three-dimensional, compact, spherical structures. By involving water in the intermolecular interactions, the overall three-dimensional structure of the proteins is loosened and less compact. This loosening allows for new intermolecular interactions to form with neighboring protein molecules. As these new interactions are created, so are the first strands of gluten. However, that is not all! As the proteins become hydrated, so do the starch granules that the proteins are a part of. Thus, the individual starch granules that were separate entities in dry flour become sticky and bind together; you have experienced this with your eyes and hands if you have ever mixed a bread dough. And that is still not all! As water gets absorbed by each starch granule, the amylase enzymes now have the water needed to break down starch into smaller molecules, specifically maltose and glucose. The yeasts that were activated during proofing can use these sugars as a food source and begin to carry out the processes of glycolysis and fermentation. Wow! Who would have thought that all of this is happening during the mixing of a bread dough: hydration, enzyme activation, yeast fermentation, and the beginning stages of gluten formation? However, this is only the beginning of the process. Let's bring together some concepts as we learn more about kneading.

Have you ever tasted bread that has been kneaded for one minute and one that has been kneaded for 10 min? Not only is there a difference in the final product but also in how the dough feels to the touch. What happens at a molecular level during kneading that promotes the formation of gluten, which impacts dough texture, rising, baking, and taste? As mentioned previously, during mixing, the starch granules bind together and the proteins change structure as they begin to have intermolecular interactions with water and each other.

During kneading, the "bound" starch granules are moved apart, which forces the proteins to unwind and become even more linear (while still kinked and coiled), allowing for interactions with other protein molecules in different and more ways via both intermolecular interactions (like hydrogen bonds) and covalent bonds (like disulfide bonds; Fig. 10.13). As the dough continues to be kneaded, these interactions are broken and reform in new and different ways, which eventually results in glutenin

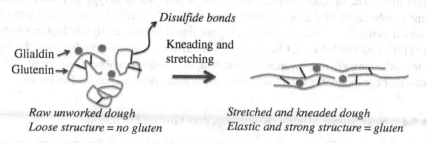

FIGURE 10.13 Gluten forms during kneading dough. Stretching and kneading flour dough lengthens the proteins where they can then cross-link with each other forming long, stronger gluten complexes.

protein molecules that are interacting with one another end to end, forming long, coiled protein chains that might be a few hundred glutenin molecules in length, kind of like a slinky (Fig. 10.14). This arrangement leads to a nice elastic dough, so it stretches but it doesn't break unless you stretch it apart really far. As the dough is stretched, the proteins uncoil and lengthen; when you release the stretch, the dough bounces back to its original shape as the proteins recoil and reform the disulfide and hydrogen bonding interactions with the neighboring glutenin proteins. You can recognize when this happens during kneading; the dough becomes less sticky, smoother, and more elastic, yet it remains firm.

You might be thinking, what are the gliadin proteins doing as the glutenin proteins are stretching out into these long coils? The gliadins contribute to a dough property called plasticity. Remember that the gliadins are typically compact in their native structural state, so they actually interfere with or limit the intermolecular interactions between glutenins. You may be thinking, but doesn't this reduce the strength of the gluten network? Yes, it does. If the network is too tight (i.e., the proteins are held too strongly or tightly together via intermolecular interactions), then the dough wouldn't stretch, it would not deform under the pressure of your kneading hand, and it wouldn't mold into a ball for rising. Thus, the gliadins help to prevent a gluten network that is too strong.

We have a couple of questions left to address about gluten. What additional factors (besides kneading time) affect the quality of the gluten network? Can you break a gluten network, once it has been created? What happens when the gluten network becomes too strong?

10.6   CONTROL OF GLUTEN FORMATION

There are multitudes of ways to control gluten formation that do not require an excessive amount of time or money. A relatively easy way to control gluten formation is by adjusting the type of flour that you use for a particular type of baked good. Since bread flour has a higher protein (i.e., more gliadin and glutenin) content than all-purpose flour, you will create a stronger gluten network with the same amount of kneading and ingredients in a dough made from bread flour versus all-purpose flour. Secondly, the presence of oxidizing agents or air enhances gluten formation. This may seem a bit unusual, but there is a scientific basis for this statement. Because disulfide bonds are more likely to form in an oxidizing environment and the formation of gluten involves the creation of disulfide bonds, you will have a stronger gluten network if more air (i.e., oxygen gas) is incorporated into the dough. While on the subject of disulfide bonds, aged flour also makes a better yeast-raised bread than fresh flour, as the thiol groups are more oxidized in aged flour (again, aged flour has been exposed to oxygen for a longer period of time). The disulfide bridges are more likely to form, leading to a stronger gluten network. The fourth factor has already been addressed: the kneading time. The more kneading, the better the gluten network to a certain point. If you are hand kneading, your hands and arms will tire long before you have reached the point of kneading too long. However, overkneading can happen, particularly if you are

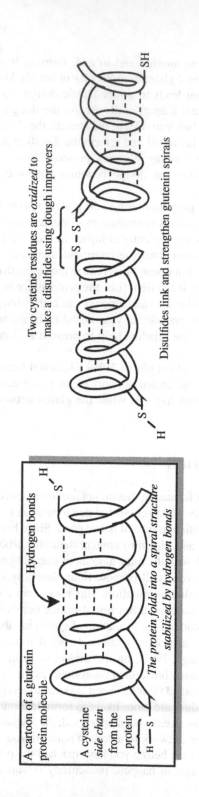

A cartoon of a glutenin protein molecule

Hydrogen bonds

A cysteine side chain from the protein

The protein folds into a spiral structure stabilized by hydrogen bonds

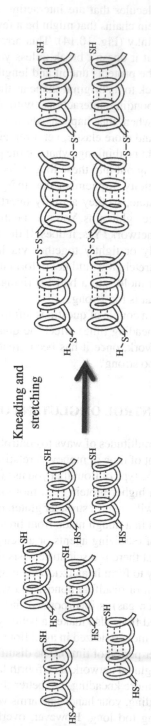

Two cysteine residues are *oxidized* to make a disulfide using dough improvers

Disulfides link and strengthen glutenin spirals

Kneading and stretching

FIGURE 10.14 Molecular details of gluten formation. The cross-linking of gluten comes from the disulfide bond between strands of glutenin. The high glutamine (Q) content supports the helix by forming many hydrogen bonds.

letting an electric mixer do the work for you. When dough is overkneaded, the gluten network is so tight, it feels very dense and tough, it tears apart when it is stretched, and is even difficult to knead because it doesn't flatten and fold easily.

If you are using unbleached flour, the dough color might even change when you reach the point of overkneading, as the excessive oxidation causes the flour to bleach. One thing that you can do to prevent overkneading and reduce your kneading time is a technique that was developed by a French breadmaking authority named Raymond Calvel in the 1970s. In this technique, called autolysis, the flour and water are first mixed and allowed to rest before kneading. As the mixture rests, the protease enzymes in the flour begin to break down/cleave the glutenin and gliadin proteins into smaller pieces. These smaller proteins are more easily straightened and aligned during the kneading process, so an organized gluten network is created more quickly. The autolyzed bread dough requires a much shorter kneading time.

Clearly, gluten formation is important. The question is, why is it important? The formation of a strong, elastic gluten network is key to the leavening or rising step of bread baking; this is where the fermenting efforts of the yeast and your kneading efforts come together.

10.7   THE RISING BREAD

Any bread baker visually knows what happens when bread rises; it gets bigger, often doubling in size after sitting in a warm location for about 60–90 minutes. You already know that it has something to do with the yeast, but what exactly is happening at the molecular level? During the rise, the yeast really go to work, reacting with the sugars produced from the breakdown of starch (carried out by the amylase enzymes in the flour), releasing carbon dioxide (CO_2) and ethyl alcohol. Both of these molecules (and alcohol byproducts) are really important to bread; the CO_2 is particularly key to the rising phase. Why? Remember the CO_2 is a gas. Gases can bubble out of a solution, similar to what you see in a carbonated beverage. However, a gas can also be trapped inside a mixture. If your bread dough has a strong gluten network, the sheets of gluten become blown up with the CO_2 gas like a balloon. As the gas molecules are captured, the dough rises. However, in order for the CO_2 to be captured, the gluten sheets need to be robust enough to not break as they expand and plentiful enough to capture the CO_2 gas molecules as very small bubbles. If the gluten sheets are weak, the CO_2 will burst out of the gluten network, rather than being captured, and the dough will not rise properly. If the CO_2 is captured in large, rather than small bubbles, you have gaping, large holes in the final loaf. Thus, formation of a strong, elastic, and plastic gluten network actually allows a bread to rise. With production of a greater amount of and smaller gas bubbles, you will have a tenderer product because of the interruption in the gluten network with air pockets.

Bread rising typically happens in a warm space. Why is there a temperature dependence? If the yeast are not happily duplicating, growing, and fermenting, you will have less CO_2 production over a period of time. This is why refrigerated bread dough requires a rising time of 8–12 h; it just takes longer for the yeast to produce

---

BOX 10.2   RAISING YOUR DOUGH

It might be convenient to allow your dough to rise overnight. However, the slower rise time does lead to production of different flavor molecules than a more typical rise. These stronger, often more sour flavors are preferred for some breads or in some individuals. A slow rise also encourages the development of a more open, irregular crumb structure to the bread due to the redistribution of the dough gases that occurs when the cooled dough warms.

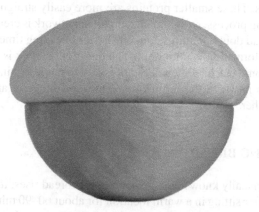

FIGURE 10.15   Rising bread dough. Dough rising with the production of CO_2 gas.

the same amount of CO_2 at 39°F/4°C relative to 73°F/23°C. The optimal temperature for bread rising is 27°C (81°F). Yeast actually reproduces more rapidly at 95°F/35°C, but the yeast, which may give your bread a less desirable bitter flavor, releases bitter metabolites (Box 10.2).

Some bakers do not disturb a rising bread dough, while others incorporate a turning step into the dough rise. Turning involves delicately folding the dough over several times at one or multiple points during the rise. Turning the dough stretches it and builds strength to the gluten network, as it helps to align any misaligned gluten sheets. In addition, turning removes some CO_2 from the dough, an excessive amount of which inhibits yeast activity, to ensure the maximum rise and production of flavor molecules. What are these flavor molecules? You know that one product of fermentation is ethanol, but as the yeast proliferate, they also synthesize proteins, aldehydes, ketones, diacetyl, and other aromatics that contribute to that scent and flavors of a raised bread dough and final product. We will discuss the specifics of the flavor molecules later in the chapter.

When is a bread raised enough? Most cookbooks tell the baker bread has raised enough "when it is doubled" but it is sometimes hard to know when enough is enough. Rising is complete when you can poke your finger into the dough and the hole doesn't immediately close (Fig. 10.15). At this point, the gluten has been stretched to the limit of its elasticity. If you try to raise the bread any more, the gluten network will break and the bread will deflate, releasing all of the captured CO_2. Rising time for a typical dough that sits in front of a sunny, draft-free window is 60–90 min.

10.8 THE PUNCH AND SECOND RISE

In many bread recipes, after the first rise, you knock down or punch the dough and let it rise again. Is this a step needed to relieve aggression or does it have some purpose? To start, the punch is not actually a punch but more of a push down, also known as degassing. As the degassing term implies, when pressure is put onto a raised dough, some of the CO_2 bubbles are expelled from the dough and those that remain break into smaller gas bubbles. These smaller bubbles can be blown up again during a second rise, giving a final bread product a fine, light texture. The yeast cells, the sugars, and the products of fermentation are also redistributed throughout the dough during the punch, which leads to better growth during the second rise. Interestingly, the temperature of the dough is also equalized during the punch. The inside of a raised dough is warmer than the outside because of the activity of the yeast; in equalizing the temperature the second rise is more evenly distributed in a punched dough relative to one that is not punched. Finally, by punching down the dough, the gluten network has a chance to relax before it is stretched again, so the dough will be easier to shape. Thus, there are scientific and culinary reasons why you need to punch down the dough; although it seems like anyone could do it, there is some technique associated with the process (Fig. 10.16).

For a final bread product with a fine crumb (like a sandwich bread), first, push your fist quickly but gently into the center of the dough. Then, pull the edges of the dough to the center. Take the dough out of the bowl, pat it, turn it over, and do two or three kneadings to release any additional air bubbles. For a coarser crumb with better chew, turn the dough (gently folding it over) to reactivate and redistribute the yeast and sugars without pressing out as much air.

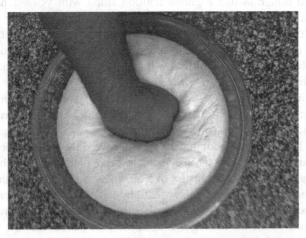

FIGURE 10.16 Punching down dough.

FIGURE 10.17 Appret. Slightly slicing the surface of a growing dough to limit tearing by allowing a "pleat" to stretch.

Once a dough has been punched accordingly, it is shaped into the form of the final loaves and then left to sit and rise in a warm place. This second rise is sometimes called "finishing" or "appret" as it is the final opportunity for the yeast to do their work before they succumb to the heat of the oven during baking. Some bakers slit the tops of the shaped dough before or after the second rise, which ensures that the gluten network is not stretched to its limits during baking, reducing the presence of unattractive tears in the crust (Fig. 10.17).

10.9 BAKING

You may view the baking process as simply "cooking" the dough, but there is so much more to it!

First, let's talk about why the bread expands (even more) during baking. Remember that the raised dough is like a gluten balloon filled with yeast-generated carbon dioxide and air that was introduced into the dough during mixing and kneading (Fig. 10.18). Your experiences (in the kitchen and in life) tell you that gases expand when heated: a balloon left in a warm car explodes, the air pressure in your car tires decreases in the winter. The scientific law that tells us that gases expand when heated is known as the ideal gas law: $PV = nRT$.

The letters in the ideal gas law stand for: pressure (P), volume (V), amount in moles (n), a gas constant (R, which doesn't change in value), and temperature (T). The behavior of most gases is consistent with this equation, including the gases present in rising dough. Thus, if the amount of gas that you have is the same in the prebaked and the baking dough (i.e., n is constant), then the increase in temperature that occurs during baking causes an increase in the overall value on the right-hand side of the equation. If the right-hand side of the equation increases in value, then the terms on the left-hand side the equation must, together, account for this temperature increase. The gas pressure inside the baking loaf does increase but so does the volume of the bread. How does this happen? As the gas bubbles expand, the dough, which is still soft during the initial stages of baking, is pushed aside and visibly expands. If the

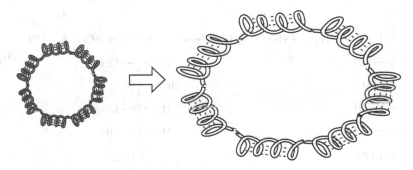

FIGURE 10.18 Expanding gases are held in place by stretched gluten protein cages.

starch and gluten networks are too rigid, the bread doesn't expand because the gluten network cannot move.

When does the bread become more rigid? In other words, what happens to make the bread fully "cooked"? During baking, the starch granules in the bread's interior absorb water, swell, and create a permanent solid structure that is a spongy network of starch and protein filled with tiny air pockets called alveoli. Bakers use the term "crumb" to describe the characteristics of this network. In contrast, the outer surface (also known as the crust) develops a dry, dense texture. Clearly, we have different things going on in the crumb versus the crust; let's look at what is happening in each component.

As the dough heats, it becomes more fluid due to the increased movement of the molecules, the gas bubbles expand, the gluten stretches, and the dough rises. In addition to the expansion of gas bubbles, at about 60°C/140°F, more gas is generated in the expanding bread due to vaporization of the yeast-produced ethanol and water; this expands the dough by as much as half of the initial dough volume. This expansion stops when the crust becomes too firm and stiff to be pushed around anymore, at about 90°C/194°F. At 100°C/212°F, the water in the bread's interior and on the exterior fully evaporates. In the interior, the generated steam is redistributed into the bread because it cannot escape. However, on the bread's surface, the water fully evaporates, resulting in a drying and hardening of the bread's crust. As the temperature continues to rise in the bread's interior to about 150–180°C/300–350°F, the starch begins to "gel" into a more solid state due to the absorption of water vapor and resulting swelling of the starch granule. Moreover, the gluten proteins, which are now denatured due to high temperatures, form even stronger intermolecular and disulfide bond interactions with one another. In essence then, the starch and protein network becomes much tighter and stronger and the gas bubbles have no more space for expansion. Pressure builds as the gas bubbles try to expand until they rupture, turning the structure of the loaf from a network of separate gas bubbles into an open, porous network. During the end phases of baking, the starch continues to form an even more solid, gel-like structure. However, even after baking and during cooling, the starch granules continue to become more firm, making the bread easier to slice when cool than when piping hot.

What about the crust? The crust's color and aromas are a result of Maillard reactions. You have heard about Maillard reactions before. These reactions, discovered by Louis Camille Maillard in 1911, occur between sugars and amino acids (Fig. 10.19). Because

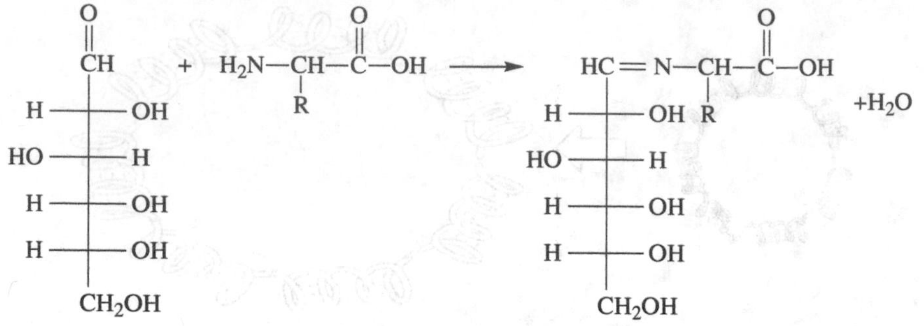

FIGURE 10.19 Maillard reaction browning bread crusts. Sugar, shown on the right in the straight chain form, will condense with the nitrogen of an amino acid releasing water starting the first step of the browning reaction.

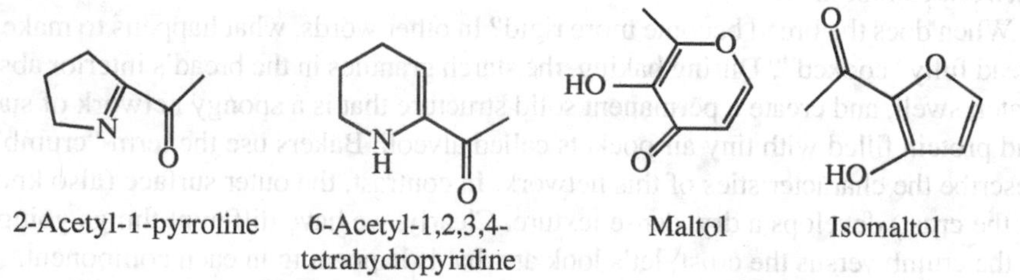

| 2-Acetyl-1-pyrroline | 6-Acetyl-1,2,3,4-tetrahydropyridine | Maltol | Isomaltol |

FIGURE 10.20 Aroma compounds from baked bread.

there are 20 different amino acids and hundreds of different sugars, the Maillard reactions are best described as a group of reactions rather than a single one. The first step of any Maillard reaction occurs between the carbonyl group of a sugar molecule and an amino group, resulting in the formation of a new covalent double bond between the original molecules and a molecule of water. The reactive amino group may be part of a free amino acid or an amino acid that is still part of a protein chain.

What types of Maillard reactions occur in bread and what do they do? Well, they certainly contribute to the brown color of a bread crust, but they also contribute substantially to the flavor and aroma of the bread. You already know some of this chemistry, but let's look at a few of the hundreds of Maillard reaction products that are generated in the dough, crust, and crumb of a great loaf of bread.

In an early scientific study on the aroma compounds of wheat bread, 372 compounds were identified in the crust and 86 in the crumb of a wheat-based white bread. In the crumb, non-Maillard molecules including aldehydes, alcohols, and ketones predominate, while in the crust, the Maillard products predominate, including pyrazines, pyrroles, and furans. Wow! all of these aromatics (Fig. 10.20), it is no wonder that bread smells so good!

Two Maillard products associated with bread flavors and aromas are 2-acetylpyrroline (roasty, bread crust-like) and 6-acetyl-1,2,3,4-tetrahydropyridine (roasty, cracker-like flavor), while maltol and isomaltol contribute to the freshness of baked bread.

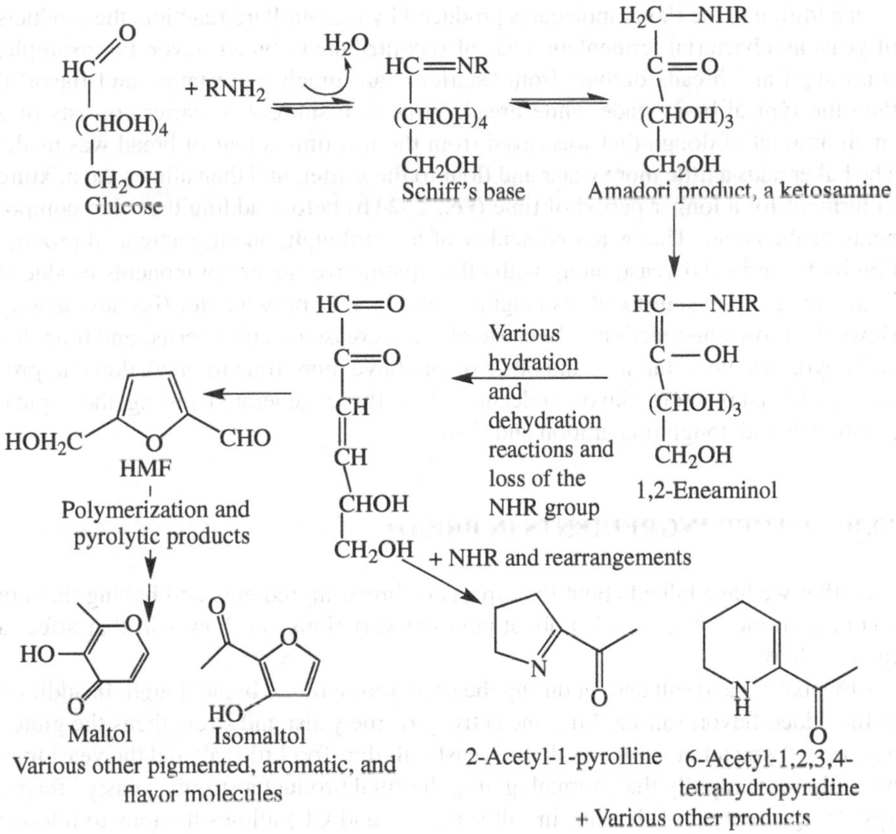

FIGURE 10.21 The browning reactions forming brown crusts of bread.

In short, the aldehyde group (CHO) of the sugar and the amine group of a free amino acid or amino acid that is present within a protein react to form a compound called a Schiff's base (it has the C=N) group. Schiff's bases are not stable compounds, so this compound rearranges itself to a more stable compound called an Amadori compound. All of the chemistry up to this point is fully reversible and is the same chemistry for all Maillard reactions. What happens next, however, varies widely depending upon the exact conditions of the reaction. The Amadori compound can rearrange, condense with another compound, isomerize, polymerize with another like compound, cyclize, or undergo degradation reactions. It is the products of the Amadori intermediates that give us the extensive array of color, flavor, and aromatic compounds associated with baking or freshly baked bread. However, one of the major pathways produces a compound called a 1,2-eneaminol, followed by 5-hydroxymethyl-2-furaldehyde. From 5-hydroxymethyl-2-furaldehyde, the brown, melanoidin pigments associated with a brown crust form. A more minor pathway produces a 1,3-enediol to eventually produce various alpha-dicarbonyls. A third pathway is called the Strecker degradation, which produces aldehydes (Fig. 10.21).

In addition to the flavor molecules produced by the Maillard reaction, the products of yeast and bacterial fermentation can also contribute to bread flavor. For example, sourdough and breads derived from "starters" are much more tangy and flavorful than the typical homemade white bread. What is a starter? A starter consists of a small amount of dough that was saved from the last time a loaf of bread was made. The baker adds a little more water and flour to the starter, and then allows the mixture to ferment for a longer period of time (i.e., 2–24 h) before adding the other components of the bread. The increased acidity of a sourdough, due to lactic acid production by bacteria and yeast, along with other distinctive flavor components produced in the dough, give sourdough its tanginess and lack of brown color (because acidity slows the browning reactions). Because of the increased resting period and time, the carbohydrates, fats, starches, and other sugars have more time to break down to produce/yield many more flavor molecules than those generated during the typical (shorter) bread dough preparation and rising.

10.10   OTHER INGREDIENTS IN BREAD

Now that we have talked about the aspects of bread ingredients and baking that are common to most bakers, what about possible variations and how will that affect a finished loaf?

Most bakers add salt and fat during the preparation of the bread dough. In addition to the added flavor, salt modifies the activity of the yeast and strengthens the gluten network. If you add too much salt, the yeast will die. Too little salt and the yeast may multiply more rapidly than normal, giving the final product a strong "yeasty" flavor. The charged nature of the ions in salt (i.e., Na^+ and Cl^-) allows the ions to interact with and stabilize any charged amino acids that are present within the gliadin and glutenin proteins. This will decrease any possible charge repulsion between individual protein molecules (e.g., two positively charged groups repelling one another), allowing the two salt-bound protein molecules to bond (covalently and via intermolecular interactions) more extensively.

Fats inhibit formation of the gluten network, increase the volume and lightness of a dough, and slow the staling process. Fats like to bind to other fats and other hydrophobic species. The glutenin and gliadin proteins do have hydrophobic components, thus the fat molecules bind to the proteins, preventing them from interacting with one another. However, fat also helps to stabilize the walls surrounding the gas bubbles, preventing them from rupturing, which lead to a lighter, more voluminous bread. If you have ever tried to crunch your way through a 2-day old baguette loaf, you likely know that fat prevents bread staling. What happens when bread stales anyway? Although staling is an issue of water loss, the loss of water is due to a change in the starch. With time, the amylose and amylopectin molecules in a starch begin to associate with one another and thus stop interacting with water. The term used to describe this phenomenon is retrogradation and it occurs in most any starch-based food including potatoes and pastas. During the first phase of staling, which begins to take place within a few hours of baking, amylose molecules begin to associate and crystallize with one another, while in the second stage of staling, amylopectin molecules retrograde. These phenomena result in a more crystalline and solid structure of the starch-based food, and loss of the water molecules that were involved in intermolecular interactions with the starch granule.

Staling can be reversed (in part anyway) by "melting" the crystals through heat and moisture. Fat protects against retrogradation, as it coats the starch granules, preventing the amylose and amylopectin from interacting with one another. By placing your bread in the freezer, you also prevent the crystallization of the starch molecular components and water migration.

Although a small amount of sugar is important in "jump-starting" the yeast, in a dough containing a substantial amount of sugar, the sugar is really problematic for gluten development. Because sugars bind to water, water is unable to form intermolecular interactions within the gluten network, yielding a weakened gluten structure. Thus, a rich dough (containing a high sugar and fat content) are quite fragile and relatively soft. Some bakers overcome this challenge by kneading in the fat and sugar after the gluten network has been established within the dough.

10.11 GLUTEN AND CELIAC DISEASE

Most grocery stores and college cafeterias now have a large number of products and areas that are designated as gluten free. You (yourself) or your friends may have adopted a gluten-free diet for medical or nonmedical reasons. What is the medical basis behind celiac disease? How does a baker prepare a high-quality, tasty baked dough without gluten?

Celiac disease is an autoimmune disease. In any autoimmune disease, the body recognizes a molecule as foreign and utilizes its immunological system to attack the foreign molecule. In most cases, the immune system can recognize harmful foreign species (e.g., bacteria) from nonharmful species (e.g., foods, human biological molecules), thus helping to protect us from harmful foreign species. However, in the case of celiac and other autoimmune diseases, the immune system triggers an abnormal response against a molecule or organ. In celiac disease, gluten is the recognized foreign molecule. Thus, when a person with celiac disease eats gluten, the lining of his or her intestine becomes inflamed and damaged, leaving the intestine unable to properly digest and absorb nutrients from other foods.

An estimated 1% of people worldwide are thought to have celiac disease; a gluten-free diet is the only treatment for those afflicted, which is problematic given that gluten is present in so many common and nutritious foods. Gluten is found in wheat (of course), barley, rye, bulgur, farro, and some suggest oats, all of which are a good source of vitamins, minerals, and fiber to one's diet. Thus, in order to eat a healthy, nutritious diet, a celiac-afflicted individual must avoid most breads, crackers, cereals, pastas, pastries, and a number of processed foods but can indulge in eggs, meat, fish, fruits, and vegetables.

In addition to those individuals diagnosed with celiac disease, you may have friends who are gluten sensitive or have made a dietary decision not to eat gluten.

What should you do if you would like to serve bread at your next dinner party? The easy response is to go to the gluten-free aisle in the grocery store. But maybe you would like to experiment with baking a gluten-free bread. How can you achieve a chewy, light texture without gluten?

Often gluten-free flours are supplemented with (or you can add) xanthan or guar gum or other emulsifiers. These products can contribute elasticity (the gums), stabilization of, and reduction in the diffusion of the gas bubbles (the emulsifiers). It is tricky, but with some experimentation with different flours, gums, and emulsifiers, you can still bake a great loaf of bread.

10.12   MUFFINS AND BATTER BREADS

Perhaps on a given morning, you desire the tender structure and sweet taste of a blueberry muffin or a lemon poppy seed loaf rather than a piece of chewy, toasted sourdough. Although still "bread," these breads are a lot more like cake, they are quickly prepared (hence the name quick bread), use chemical leavening agents, and have a completely different texture from yeast-based breads. In muffins and batter-based breads, you do not want to generate the chew of a good gluten network but a nice, tender, porous crumb; the ingredients and the techniques help to accomplish this.

The ingredient list for a muffin is expanded relative to a yeast-based bread and includes more sugar and fat, eggs, baking soda or baking powder or a combination of the two, and in some recipes buttermilk, sour cream, or milk. In addition to providing richness, sweetness, and body, the "extra" eggs, fats, and sugars all interfere with gluten formation by interacting with the gliadin and glutenin proteins themselves, thereby preventing the molecules from interacting with one another to form the gluten network. However, even in the presence of these interfering molecules, simply stirring the mixture can generate a gluten network. Therefore, any muffin man knows that you should combine the dry ingredients, add all of the wet ingredients in one swoop (often making a "well"), and fold the wet into the dry with a rubber spatula only until the dry ingredients are barely wet and the batter is still lumpy (Fig. 10.22).

Using this method, your muffins will be tender and have a nice porous interior but will stale quickly due to the lack of an even distribution of the fat to coat the starch particles. However, if you make a good muffin, the entire batch will be eaten within a day so staling won't be a problem.

10.13   CHEMICAL LEAVENING AGENTS

If you have ever forgotten to add the baking powder to a batch of muffins, you know the importance of a leavening agent to the final product. Your unraised muffin probably felt as heavy as a pool ball and certainly did not have a porous, honeycomb-like structure. Under the right conditions, chemical leavening agents produce carbon

FIGURE 10.22 The wet and the dry. Keeping the two sets of components of a cake or muffin mix separate until needed is a secret to avoiding excess gluten formation.

FIGURE 10.23 Baking soda versus baking powder. Baking soda is a pure sodium bicarbonate, while baking powder has other components added to the sodium bicarbonate.

dioxide gas, the same gas bubbles that are released by yeast during fermentation. The two most commonly used chemical leavening agents are baking soda and baking powder. Although they both produce carbon dioxide, they do so under different conditions and over a different time frame, thus if you try to interchange them in a recipe, you will probably be underwhelmed with the resulting product (Fig. 10.23).

10.14   BAKING SODA

Baking soda causes the reddening of cocoa powder, hence the name devil's food cake. So what is baking soda all about? Baking soda consists of a single, basic chemical called sodium bicarbonate ($NaHCO_3$), which produces carbon dioxide, water, and a salt in the presence of an acid (H^+) (Fig. 10.24)

Wait a minute! An acid? What type of acid is available and safe to use in the kitchen? Buttermilk or sour cream (lactic acid), molasses (various acids including acetic, propionic, aconitic), lemon juice (citric acid), or cream of tartar are the most common additives to a baking soda leavened quick bread or muffin. The great thing about this reaction is that it will cause the production of millions of tiny carbon dioxide gas bubbles in the thick muffin batter. If these expanding gas bubbles can be captured in the structure of the batter as it becomes firm in the hot oven due to its transition from a foam into a sponge, you will have a tender, porous baked good. The problem with this reaction is that it is initiated immediately when the bicarbonate and acid are exposed to liquid water. Thus, as soon as you add the wet to the dry, carbon dioxide begins to be produced. Thus, you need to get that batter into the oven as quickly as possible so that it reaches a firm state before all of the carbon dioxide gas is lost to the air (Box 10.3).

Interestingly, baking soda is a base, thus the acid not only assists in production of CO_2 but also helps to neutralize the mixture so that your final product doesn't have the bitter taste associated with alkalinity (basicity). Thus, make sure that you are adding the exact amount of baking soda that is called for in a recipe. Baking soda is probably

$$NaHCO_{3(s)} + H_{(aq)}^+ \longrightarrow H_2O_{(1)} + CO_{2(g)} + Na^+$$

FIGURE 10.24   Sodium bicarbonate reaction in acid.

BOX 10.3   TARTAR AS AN ACID

Cream of tartar is a white powdery substance, whose most common use is in the stabilization of beaten egg whites in a meringue. Given its slightly acidic nature (a pK_a of 4.3) and solid structure, it reacts relatively slowly when compared with a stronger aqueous acid like lactic acid (pK_a 3.9). In a reaction with baking soda (sodium bicarbonate) carbon dioxide is readily liberated for a quick production of expanding gas (Fig. 10.25).

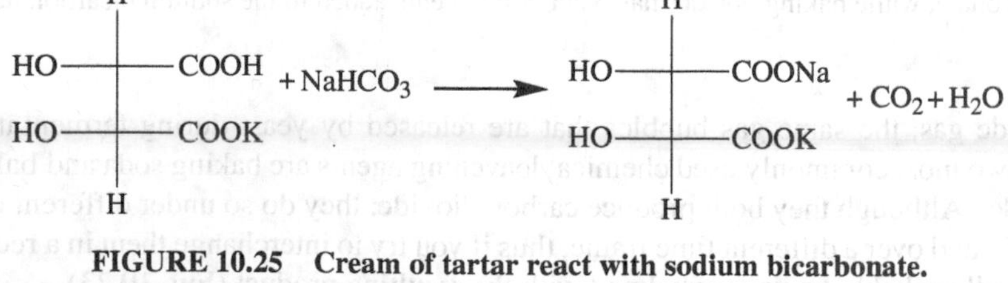

FIGURE 10.25   Cream of tartar react with sodium bicarbonate.

---

the only leavening agent used in your favorite chocolate chip cookie recipe (it causes the cookies to brown due to Maillard reactions and spread out), while in other recipes (e.g., buttermilk pancakes) baking soda is used in combination with baking powder.

10.15 BAKING POWDERS

Baking powder contains a mixture of three components: baking soda (sodium bicarbonate), one or more acids (e.g., cream of tartar, monocalcium hydrogen phosphate, dicalcium hydrogen phosphate, sodium aluminum sulfate, or sodium aluminum phosphate), and a filler/diluent. What is a filler? Given the baking powder is primed to produce carbon dioxide as soon as it gets wet, it is prone to losing its effectiveness particularly under warm, humid conditions. Cornstarch is added as a filler to keep the molecules of acid and bicarbonate physically separate from one another. However, even with the presence of cornstarch, baking powders will degrade over time and have an expiry date (that should be paid attention to). You have probably figured out several perks associated with the use of baking powder in a recipe: (i) you don't need an acid within the ingredients list of your recipe; (ii) baking powders are preneutralized, thus they do not have an "off" taste that might sabotage your recipe. However, there are even more advantages associated with a baking powder due to its double acting nature.

What is a double acting baking powder? Double acting baking powders actually have two acid additives; acid 1 (e.g., monocalcium dihydrogen phosphate) is soluble in the mixture and immediately reacts with the available sodium bicarbonate to produce carbon dioxide, much like the reaction and behavior that you see with baking soda. However, as we talked about earlier, in order for these gas bubbles to be retained within the structure of the baking batter, the batter needs to set relatively quickly in the oven. In reality then, only a portion of the gas bubbles will actually be retained to leaven the final product, and the work of acid 1 really just gives the batter some body before it cooks. The second acid (e.g., sodium aluminum sulfate) in a baking powder only becomes reactive in the presence of heat because it is insoluble (or slightly soluble) in the batter mixture at room temperature. Thus, when the muffins are in the oven and the batter is not yet set enough to hold onto the expanding carbon dioxide bubbles being produced by acid 1, acid 2 dissolves and begins to react with the remaining sodium bicarbonate, producing more carbon dioxide that will be able to be retained within the quick bread structure as it sets. Thus, a recipe that utilizes baking powder will typically yield a larger rise than a recipe that uses baking soda, even if the baking is delayed by 15–20 min. Since much of the reactivity of acid 1 is complete within a few minutes (~2–5 min), most of the rise is due to the work of acid 2.

10.16 BAKING SODA VERSUS BAKING POWDER

Some recipes call for one, some recipes call for the other, some recipes call for both. How do you decide? It depends upon the desired characteristics of the final product. Baking powder will give you a better rise, but if you want a flatter or browner structure, as seen in a chocolate chip cookie, baking soda will be used as the leavening agent. The baking soda will cause the cookies to brown and spread out on the pan and not up into the air. Because of the rise, baking powder cookies will taste more cake-like than a baking soda cookie. If you are a fan of the cake-like cookie and would like to carry out an experiment in the kitchen, remember that you will need to add more baking powder to a baking soda-based cookie recipe, given that baking soda comprises only about a third of the same amount of baking powder.

If a recipe calls for both baking powder and soda, the baked good is actually being leavened by the baking powder, which contains the correct proportions for all of the acids to react with all of the basic bicarbonate, while baking soda is added to neutralize any acid (like buttermilk) that is present within the ingredients list to prevent a sour overtone. What is the point of buttermilk in the first place if you don't need an acid? Why not use milk? In addition to providing the milk proteins that interfere with gluten formation, which are common to both milk and buttermilk, buttermilk is thick. When you add a thick liquid to your batter, then you don't need to add as much flour, also reducing the development of gluten.

10.17 CAKES

Cakes are similar to quick breads and muffins in many ways. Most cakes have a tender, porous, and moist texture; they have a similar basic ingredients list including flour, eggs, sugar, fat, and a chemical leavening agent; however, a cake is much lighter, finer, and spongy than a muffin (some cakes are formally called sponge cakes). You already know that "lightness" is generated by the retention of gas bubbles within the molecular structure of a food, but what does this structure look like in a cake and does the structure come from the ingredients or their treatment? In short, a cake is an egg foam. However, you need a lot more than eggs to make a great cake; let's take a closer look at the ingredients list and techniques with an eye toward the science.

10.17.1 Generation of the Foam

You already know about foams and egg foams from Chapter 9, where you learned what happens when you whip egg whites to make a meringue. The egg foam for a sponge cake needs to be more robust than that of a meringue; it has to withstand the weight of the cake and any fillings after all. Thus, there is more to a cake than just the egg foam. In fact, the generation of the foam begins right at the start through the incorporation of air bubbles into the batter. You probably have the first step on a cake recipe memorized, "cream butter and sugar together until light and fluffy." Although you might want to skip this step, it is actually one of the most important steps in cake baking as it strongly influences the final texture. By beating the butter and sugar, you incorporate air bubbles into the mixture that will eventually expand in the heat of the oven (along with the steam and baking powder-generated carbon dioxide bubbles) to make a properly risen cake that is light and fluffy, along with evenly distributing the sugar. How exactly does

> **BOX 10.4 A LIGHTER SHADE OF PALE**
>
> What does it mean when a material becomes lighter in color or pales? A white material is white because it reflects all light. Light is able to be reflected if the particles present within the material are approximately the same size as a light wave (400–700 nm). With more beating, the air bubbles that are incorporated into the dough get increasingly smaller. As they decrease in size, the dough become whiter because more of the light is reflected off the dough.

this incorporation happen? The solid sugar crystals, along with your mixing beaters, carry air molecules into the semisolid (crystalline/liquid) room temperature structure of the fat. The air bubbles are retained in the fat, thus, as it is creamed and the color of the mixture becomes lighter and the volume increases (i.e., it becomes fluffier) (Box 10.4).

The next step typically involves addition of the eggs and other liquid-based ingredients (e.g., dairy products, flavorings). Obviously, the eggs are important to development of the foam, but they also provide moisture, richness, and flavor to the cake. If you don't recall what an egg foam looks like, remember that eggs are made of protein molecules. As you begin to beat eggs, they undergo mechanical stress that causes the proteins to denature. The hydrophilic parts of the proteins organize themselves to interact with other water-based molecules, while the hydrophobic parts of the proteins stick to the fats so that they can avoid the water. With continuous beating, the incorporated air becomes enclosed in a protein/fat bubble with the hydrophobic parts directly toward the inside where it can interact with the nonpolar air molecules (like oxygen), while the hydrophilic parts are directed toward the outside where they can interact with the hydrophilic sugars. With more beating, air continues to be incorporated and the air bubbles within the foam get smaller, which will make the cake more tender (Fig. 10.26).

10.17.2 Flour

Although there are recipes for flourless cakes, almost all cakes contain flour. A flour is important to a cake, as it helps to stabilize the egg foam and prevent it from collapsing in the oven. However, you don't want to generate gluten nor do you want to pop all of your air bubbles and break down the egg foam. Therefore, it is best to gently fold the flour (and other dry ingredients) into the eggy, voluminous batter with a rubber spatula, particularly with an unleavened cake (see in the following). As soon as all of the dry ingredients are incorporated, you stop folding. In a cake that utilizes a leavening agent, you may fold in the flour (as aforementioned) or mix your batter slightly to generate a limited amount of gluten. The gluten development will make the batter a little more elastic for the capture of some of the leavening agent-generated carbon bubbles and you will have a well-raised cake. The problem with gluten generation is that your cake will not have as tender a texture.

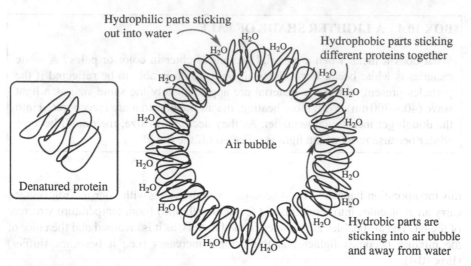

FIGURE 10.26 Proteins and foam. The denatured proteins will form a cage around air mixed with protein and fat (not shown here).

10.17.3 To Use a Leavening Agent or Not?

Many, but not all cakes use leavening agents. A Genoese sponge cake provides an excellent example of a cake that does not use a chemical leavener. As was discussed previously, if you cream sugar and eggs properly, you will generate a stiff egg foam that can retain the incorporated air bubbles. Sugar is necessary at this point (as opposed to just starting with the eggs) because it increases the viscosity of the eggs. Without the sugar, the speed of beating that would be necessary to create a stable egg foam would not be achieved without using heat to assist with the denaturation of the egg proteins. The careful folding of the flour and melted butter into the mixture is key to ensure that you do not destroy your foam and to give it additional strength. Even in the absence of a leavening agent, the cake rises a little during baking due to the expansion of the air bubbles and steam generation that is captured in the denatured egg and flour protein structure. The key aspect of a nonleavened cake are the air bubbles; they must be small and you must generate a lot of them, as measured by the stiffness of the egg foam prior to addition of the flour/butter. The fat also begins to "melt" the egg foam relatively quickly, thus this is a cake that you want to bake as quickly as possible.

Baking powder (as opposed to baking soda) is the most common leavening agent for a cake because you desire a good rise and do not desire browning reactions.

The key thing about the use of baking powder as the leavening agent is that you need to keep the generated carbon dioxide bubbles small. This can be promoted in a number of ways: (i) preparing a "thicker" batter (relative to a Genoese sponge cake) and (ii) controlling the baking conditions. If you are using a double acting baking powder, you will get some generation of some small carbon dioxide bubbles as soon as the "wet" ingredients are added to the "dry" ingredients. A substantial increase in carbon dioxide production (due to the "slower acting" acid) will occur when the temperature of the batter reaches approximately 120°F/50°C; at this temperature the previously made CO_2 bubbles will grow much larger. Large gas bubbles, particularly when generated in a runny batter, may come to the surface of the cake and be lost like a popped balloon. If a lot of gas is being lost in this way, there will be less carbon dioxide in the cake and your cake may not rise adequately. At about 140°F/60°C, protein denaturation occurs, which stabilizes the batter and retains the air bubbles in the batter. These bubbles can expand to a certain point, but if there is still a substantial amount of new carbon dioxide being produced from the baking powder, eventually the bubbles will burst. With all of these competing factors, many of which are out of your control, some bakers think that making a sponge cake without a leavening agent is actually less complicated and more likely to be successful than sponge cakes that contain leaveners. You will have to judge for yourself which type of cake is preferred to bake or eat.

10.17.4 The Woes of a Collapsed Cake

If a cake is undercooked, it will collapse. As you know from the discussion previously, baking stabilizes the structure of the cake by increasing protein denaturation and the formation of more protein–protein intermolecular interactions (in the egg and flour proteins). How do these interactions help with stabilization? As the proteins that surround the air bubbles denature, they unfold and stretch out. When a protein stretches or linearizes, it has more of an opportunity to interact with other protein molecules that are also denatured. When the denatured protein molecules form intermolecular interactions and covalent cross-links (e.g., disulfide bonds) with one another, they become much more rigid and less fluid in their behavior. Thus, a baking cake makes a transition from a viscous liquid batter into a more solid, sponge-like structure and the foam structure that had been holding the air bubbles is strengthened enough to withstand the increasing pressure associated with the expansion of the gas and vaporization of water. If baked properly, the structure of the cake should hold even as the cake cools, the air bubbles contract, and the vapor condenses. Why does a cake collapse in the middle? Cakes are more cooked along the edges because that is the section that interacts with the metal pan; the pan conducts heat and structurally supports the edge of the cake. This is why a cake is gooier in the middle and more crusty/crisp along the edges (Fig. 10.27).

10.18 PASTRIES: FLAKY PIE CRUSTS AND PUFF PASTRIES

What did your grandmother always tell you when you were making pie crusts? Handle the dough as little as possible, and now you know the reason why. Mixing, working with, and kneading dough generates gluten, and you definitely do not want gluten if you desire a flaky, tender pie crust. However, there are a few other aspects of preparing a pie crust or pastry that make these baked goods quite distinct from a yeast-based bread, a muffin, a cookie, or a cake. The ingredients for a pie dough are simple: flour,

FIGURE 10.27   Pie crust.

water, and butter. You don't need an electric mixer or even a whisk, just two knives or a pastry cutter. Simple? Hmmm... most bakers would tell you otherwise.

In a typical bread dough, flour and water are the primary ingredients used to make the thick, hopefully nonsticky, mass that we call a dough. The fat or any other ingredients are not present at a very high concentration. In pastry dough (by contrast), the high-fat content is necessary for the flaky texture that we associate with a pie crust. Why? The fat separates the starch particles present within the flour from one another and prevents water from accessing the starch, thereby preventing gluten formation. This separation begins right at the beginning of the preparation of a pie crust. In most recipes, you first "cut" the cold fat into the flour, preferably one with a low-protein content, using two knives, a pastry cutter, or your food processor until the pieces of fat are pea sized. In order for the fat to be distributed evenly among all of the starch granules, it needs to be in small pieces. The smaller the pieces, the more separated the granules, the less likely water will access the granules (remember that water and fat don't mix) and the flakier the crust. You might ask, why can't you just melt the butter and mix it into the flour such that it is evenly distributed? Remember that butter is made of fat, milk proteins, and water. When melted butter is added to flour, the water from the butter penetrates the starch granules and the granules stick together before the molecules of fat can surround them. Therefore, you end up with water-filled starch aggregates (i.e., gluten alert!) rather than individual starch particles. Moreover, when cold butter goes into a hot oven, the water evaporates immediately. When warm, soft butter goes into a hot oven, the butter water seeps into the structure of the crust, resulting in a soggy mess. It is the same rationale for using cold butter to make a crust, rather than one at room temperature.

Once you have your fat-separated starch granules, you add the moisture to make a cohesive (but not a pretty) mass. Adding a cold liquid is important (remember the aforementioned point about warm butter), but the exact liquid will vary depending upon the recipe. It may be icy water, cold buttermilk, or even vodka (see Box 10.5); the

BOX 10.5   PIE CRUST LIQUIDS

When making a pie crust, the goal is to limit gluten formation. You can limit its formation by using a low-protein flour (e.g., a pastry flour), you can limit formation by not kneading the dough, and you can also limit formation by the type of liquid that you use to get the dough to coalesce into a mass. Recall that water promotes gluten formation by entering the starch granules and forming intermolecular interactions with the gliadin and glutenin proteins, causing them to unfold and to begin to interact with one another. Thus, the more water, the better the chance of gluten formation. Eighty proof vodka consists of 60% ethyl alcohol (CH_3CH_2OH) and 40% water. Interestingly, the ethanol does not readily interact with the gliadin and glutenin proteins, thus it doesn't help them to unfold and promote gluten formation. Thus, by using vodka or some other alcohol, you can add a little more moisture to make the dough easier to work with without worrying about the gluten that is forming because of the extra water. For those vodka fans out there, a vodka-prepared crust will actually not have any alcohol taste at all. The alcohol evaporates out of the crust in the heat of the oven.

key part about the cohesion mass is that you should not try to make it homogeneous and pretty. The more moisture, the more chance for gluten development, thus typically a recipe will suggest that you sprinkle in the liquid in tablespoon amounts until you have just enough for the dough to coalesce. You basically want to get just enough moisture into all of the dry ingredients so that you can make a disk-shaped mass that can be wrapped and placed into the refrigerator or freezer. Because you have already coated the starch granules with fat, the starch granules will be unable to swell, and the starch proteins will be unable to interact with one another and with water. At colder temperatures, proteins also absorb less water (because solubility is temperature dependent). Keeping everything as cold as possible will reduce the hydration of the granules and further prevent gluten development. However, if you start to knead or work with the dough extensively or the dough begins to warm, the gluten that will make the crust chewy and tough will begin to develop.

In the next step, the pie crust dough (and you) has a chance to relax. The dough goes into the refrigerator; you might go to the couch. In addition to making the dough cold, the rest period gives the protein molecules a chance to relax, such that if any were stretched out during mixing, they have the chance to shrink back without forming the intermolecular interactions that stabilize gluten formation.

Once the dough is in the fridge, you might be thinking that you are home free. You have followed all of the directions, so now you just need to roll it out, put it into the pan, and bake it. Not so fast! During rolling, you stretch the dough and flattens the fat chunks into sheets. With any stretching, some gluten develops, thus you need to work to keep the gluten at a minimum. How? The pie dough also needs to be kept cool during rolling so that the fat doesn't start melting on you. Some professional baking kitchens have a marble or metal countertop for rolling pastry doughs. Metal is the best surface for conducting heat, followed by marble, and then very distantly come wood or plastic countertops. Why do the properties of the rolling surface matter? Metal and marble will stay as cold as the pie dough because both will remove the rolling pin-generated heat energy readily.

In addition, during rolling, you quickly learn if you added the correct amount of water or generated gluten in your pie crust preparation by the elasticity of the dough. Elastic, sticky dough that readily bounces back to its original shape and size when rolled out has a high gluten content and excess water. Add a little extra flour to the work surface during rolling to try to get the dough to the right consistency. Drier pastry dough will be very stiff and won't hold its shape upon rolling; you may even have chunks breaking off during rolling. You will need to add a little more water to these doughs in order to get a dough that can readily be rolled to approximately 3–4 mm thick.

Is it ready for the oven? Not yet. Once made and rolled, it is important to let the dough stand for another 10–20 min. The dough has just undergone a lot of mechanical stress, which will (again) promote gluten formation due to the stretching of the proteins. As the dough relaxes, the rolled sheet will contract and any stretched proteins that had not yet formed fully into gluten will relax. The amount of contraction will depend upon the elasticity of the dough; the more elastic the dough, the more contraction you will see. If the dough is not allowed to contract on the counter, the dough will contract during the baking, potentially spilling your pie filling contents over the edge of the crust. You want the dough to keep its shape when cooked and you actually want to know what that shape will be!

10.18.1   Baking the Crust

You have your pie crust rolled out on the counter and you have a choice of an aluminum foil, glass, or metal pie plate. Which one should be chosen? Any shiny metal pan will reflect heat away from the pie; your pie will take longer to cook. Heat will readily pass through a glass dish and your pie crust will cook and brown quite evenly. Of particular note, avoid the flimsy, aluminum foil pan at all costs. Your crust will not cook evenly and the pans are so thin that they will not retain any heat, so the crust will cook slowly. You know this if you have ever picked up aluminum foil wrapped bread directly out of the oven with your bare hands. If you hold onto the edges of the foil, you will not get burned.

You know during baking that your crust gets "cooked," but by now, you also know that there must be some science behind the baking process. During baking, you want the crust to become firm, crisp, and brown. This state is achieved by baking at a high enough temperature that the dough heats up quickly, which causes the internal moisture (from added water or butter) to be converted to steam and evaporate out of the crust. If this works well, as the steam evaporates, it will make layers in the crust, giving it its characteristic flakiness. But, as you learned during the dough preparation, if the butter melts slowly, that water will seep into the starch granules, leaving you with a soggy crust. You have the same "soggy crust" problem when your crust is cooked with a filling; you know the characteristic soft, soggy bottom crust of Aunt

Ruth's pumpkin pie. How do you avoid this when you want to actually eat a filled pie and not just a pie crust? Prebake the crust. You can throw the pie crust in the oven, filled with dried beans or pie weights to ensure that the crust doesn't develop air pockets, and bake it until the edges are barely golden (~15–20 min), then brush an egg wash onto it. This technique is known as blind baking a crust. Then, you add the filling and finish baking. The protein layer of the egg will prevent the filling from seeping into the crust. You can also make that bottom crust a little more firm and browned by baking your pie on the lowest oven rack.

There you have it. All that you needed to know to make a great pie crust. The trouble with it is that you might make a great pie crust on one occasion with the perfect balance of structure and flakiness, then on another occasion, it doesn't turn out quite right. If you have already figured this out, pie crusts are finicky things; one extra teaspoon of water, a humid day, or variation in the protein content in your flour can all be the culprits of a crust disaster. However, with knowledge and practice, you will be able to make the necessary adjustments to make a great crust every time.

10.18.2 Puff Pastry

A puff pastry is a dough that has flaky layers. Baklava provides the perfect example of the flakiness of a puff pastry; it is so flaky, flakes of it drop onto your plate as you take a bite (Fig. 10.28). Interestingly though, although puff pastry is flaky, each layer of the pastry has a high gluten content. Hmm, how can this happen when you just learned that gluten makes pie crusts chewy, not flaky? A layer of fat separates all of the gluten-rich pastry layers. Thus, when it is baked, the moisture in each fat layer is converted to steam, which causes separation of each layer, each of which has its own gluten network, none of which is connected.

In order to achieve this type of structure, you prepare a dough of flour, water, and sometimes a little bit of fat, and mix it minimally. You roll the dough into a square or

FIGURE 10.28   Saint Honore cake with chocolate and raspberry. Notice the flaky layers of cooked dough. The lack of interactions is due to "shortening" or fat that breaks up the cross-linking of proteins in flour.

FIGURE 10.29   Folding puff pastry.

rectangle and let it rest to allow the gluten to relax. You pound out a large piece of butter (by hitting it with your rolling pin) and place it on the dough (Fig. 10.29).

The dough is turned onto itself (over the top of the butter), is rolled out again, and the steps are repeated five more times. Throughout the process, you develop layers of gluten that are separated by layers of fat, 729 and 728 layers to be exact, respectively.

As you can imagine, if the butter-generated steam is critical to the preparation of a puff pastry, you need to keep everything cold. Don't try a recipe for puff pastry on a hot day; you will have better results making it in your friend's kitchen who doesn't turn the heat on in the winter. You could also store the rolled pastry in the refrigerator during the resting periods. After the final turn, the dough rests for about an hour, then is rolled out to about 1/4 inch/6 mm and baked in a very hot oven. The escaping steam causes the layers to puff to a height that is at least four times the starting height of the unbaked pastry.

REFERENCE

[1]   Anderson O.D., Greene F.C., Yip R.E., Halford N.G., Shewry P.R., Malpica-Romero J.M. (1989) Nucleotide sequences of the two high-molecular-weight glutenin genes RT from the D-genome of a hexaploid bread wheat, *Triticum aestivum* L. cv RT Cheyenne. *Nucleic Acids Res.* 17: 461–462.

11

SEASONINGS: SALT, SPICES, HERBS, AND HOT PEPPERS

Guided Inquiry Activities (Web): 13, Flavor; 14, Cells and Metabolism; 19, Plants; 20, Plants and Color

11.1 INTRODUCTION

We will learn about the molecular structure and behavior of simple inorganic salts, organic compounds that make up spices and herbs, and the more complex biomolecules that are found in hot peppers. We will discuss and investigate how these compounds impact food and how they are best prepared and used and look closely at the history and use of hot peppers.

As you sprinkle salt on your popcorn, have you ever thought about the historical impact of those tiny grains on humans and our livelihood? Humankind's historic search for salt, spices, and herbs has caused wars, sent peoples to discover new lands, and encouraged the development of new methods to access and acquire particular spices. Seasonings not only enhance the flavor of food, but they have helped to define some cultures and significantly changed the ways in which food can be prepared and stored safely. We all know that the addition of a small amount of a compound or mixture can enhance the flavor of food.

Humankind's historic search for salt, spices, and herbs has caused wars, sent peoples to discover new routes to access the seasonings, and changed health and food. There are many historical tales of seasonings being used to mask the flavor and odor of tainted foods. In 1939, biochemist J.C. Drummond in *The Englishman's*

The Science of Cooking: Understanding the Biology and Chemistry Behind Food and Cooking,
First Edition. Joseph J. Provost, Keri L. Colabroy, Brenda S. Kelly, and Mark A.Wallert.
© 2016 John Wiley & Sons, Inc. Published 2016 by John Wiley & Sons, Inc.
Companion website: www.wiley.com/go/provost/science_of_cooking

Food: Five Centuries of English Diet tells of the medieval recipes used to mask the flavor and odor of rotten meat [1]. In *Food and Cooking in Roman Britain*, Marian Woodman describes the difficulty in storing food and use of seasoning to mask the lack of freshness of food [2]. However, food historians will argue these tales as myth and point out that those who could afford spices and herbs were those who were least likely to eat spoiled meat and food. In fact, some seasonings are now known to reduce the incidence of spoilage and contamination, rather being used to "cover" it up. Salts are dry and inhibit bacterial growth in cured meat and some spices and herbs even behave as antibiotics. Thus, it is entirely plausible that people who found certain flavors attractive were more likely to use them in cooking and were less likely to become sick from microbial pathogens. These people would teach the use of spice and pass on the genes for the seasoning taste receptors along with a heightened desire for spiced food. Hot peppers have grown in popularity to rival most herbs and spices, have a very interesting science, and have earned their own place in books on cooking. This may have been particularly important in hot and humid climates where refrigeration was at a premium. Salt, spices, and herbs all play an important role in health and the taste of our food. Let's start our discussion with the most simple, most utilized, and perhaps most important seasoning of all, salt.

11.2 SALT: FLAVOR ENHANCER AND A DRIVING FORCE OF HISTORY

Salt, whether mined from the ground or dried from the sea, is a critical component of human health and has been integral in shaping much of the world's history. Many modern roads were initially paths created by animals to salt licks. The discovery and harvest of salt (and spices) created trade routes, resulting in global power shifts and colonization around the world. Romans used salt as part of a soldier's pay; salt is the root of the term "salary" [3]. In order to understand how salt is indispensable in cooking, baking, and human health, you have to understand its chemistry and molecular structure.

11.2.1 Chemistry of Water and Salt

The chemical definition of a salt is an ionic compound formed by a reaction of an acid and a base. However, commonly, salts are compounds that are composed of cations (positively charged ions) and anions (negatively charged ions) whose charges balance one another out. Common examples of monovalent (singly charged) salts include table salt, sodium chloride (consisting of an equal number of Na^+ and Cl^- ions), and potassium chloride (K^+ and Cl^- ions). An example of a divalent salt is magnesium chloride, which consists of one Mg^{2+} and two Cl^- ions. In Chapter 1, we discussed polyatomic ions that can make up a salt, such as that found in sodium sulfate, which consists of two Na^+ and one SO_4^{2-} ion). In the solid state, a salt forms into a well-organized, three-dimensional network called an ionic or crystal lattice, where each ion is surrounded by ions of opposite charge (Fig. 11.1). This ionic attraction

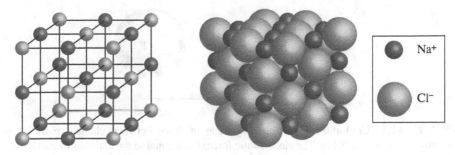

FIGURE 11.1   Ionic lattice of sodium chloride.

Sodium chloride (i.e., table salt) is an ionic compound. It is made of two different types of atoms that are held together by a positive-to-negative attraction called an *ionic bond*

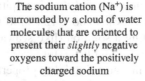

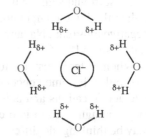

The sodium cation (Na⁺) is surrounded by a cloud of water molecules that are oriented to present their *slightly* negative oxygens toward the positively charged sodium

The chloride anion (Cl⁻) is surrounded by a cloud of water molecules that are oriented to present their *slightly* positive hydrogens toward the negatively charged chloride

FIGURE 11.2   Salt dissolves in water. The polar nature of water helps to disrupt the attractive force between ions in a salt crystal.

holds the cations and anions in place, allowing for the formation of large crystals due to the presence of a repeating geometric pattern.

When you dissolve table salt or some other salts in water, the ions dissociate from one another (Fig. 11.2). This means that the cationic and anionic components separate from one another. Why? Remember that water molecules are polar, where the oxygen atom has a partial negative charge due to its attraction or affinity for electrons, while the hydrogen atoms have a partial positive charge. In solution, the water molecule surrounds the salt ions due to its large dielectric constant.

What in the world is a dielectric constant? A dielectric constant tells you how likely two ions will come apart in a particular solvent. Water has a large dielectric constant because water counters the attraction between the cation and anion of the salt, since water is polar. Basically, water reduces the force (defined by Coulomb's law; Fig. 11.3) that holds the ions together because it is attracted to and surrounds

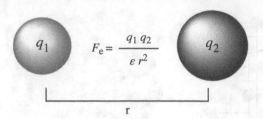

FIGURE 11.3   Coulomb's law. The magnitude of force between charged particles is described by Coulomb's law. The electrostatic force, F_e, is equal to the constant times the two charges all divided by the square of the distance times ε. ε is the dielectric constant and for water it is 80.4 and for benzene is 2.2.

the cationic and anionic components of the salt. In a salt solution, shells or cages of water surround salts, with the negative or positive pole of water combining to neutralize the ionic compound. This arrangement shields the attraction of positive and negative ions (e.g., Na^+ and Cl^-) from each other, keeping the ions in solution and the salt dissolved.

When water is lost due to evaporation, the ions lose some interactions with water molecules, become concentrated, and begin to form into ion clusters (consisting of interacting cations and anions). With increased loss of water, these ion clusters get large enough to form into crystals and precipitate out of the aqueous solution. You may be thinking, do different salts have different precipitation properties? Yes, every salt has a different solubility in water and thus precipitates more or less readily at different water concentrations. In water above 104°F/40°C, KCl is more soluble than NaCl. Thus, by carefully heating a solution (like seawater) that contains both KCl and NaCl, you can remove NaCl from the mixture by precipitation while leaving the bulk of the KCl still dissolved in water. Solubility properties of salts are quite important in the kitchen, so we will come back to this later in the chapter.

11.2.2   Sodium and Health

You have probably seen a "reduced sodium" can of soup in the grocery store or been told to watch your salt intake by your doctor. What do sodium and salt have to do with human health? Everything! Salt is essential for human life. Sodium is found in the blood, in the lymphatic system, and in and around cells throughout the body. Nerve cells use the concentration/charge gradient of sodium and potassium ions to fire signals throughout the nervous system. Sodium and potassium ions are essential in the formation of the high⊠energy molecule ATP from ADP. A 180 lb adult has nearly 0.2 pounds of sodium; on average, we have 0.1–0.2% sodium in our body. The recommended daily allowance of sodium for adults is 2000–2300 mg, about one half teaspoon of table salt each day, most of which comes from our food and drink. Although this seems like a lot of salt, most of the salt that we ingest does not come from the salt shaker, but from highly processed, canned, and frozen foods or salty snack foods. Diets high in sodium significantly contribute to high blood pressure, heart disease, and stroke. Combined, these diseases kill more people in the United States than all cancers combined. A recent study of over 3000 participants [4] found that a moderate four weeklong decrease in sodium intake from 9–12 to 5–6 g/day decreased blood pressure in those with and without high blood pressure problems. The decrease in sodium intake and blood pressure was accompanied by an increase in the kidney enzyme, renin. While long⊠term increased levels of renin have negative potential impact on several health issues including diabetes and vascular and renal disease, the benefit of sodium reduction is thought to outweigh such negative impact. A recent study of over 3000 people found that moderate levels of sodium intake did not translate into a greater risk of hypertension or coronary disease, while in this same 2011 study, lower sodium intake was associated with higher heart disease mortality [5].

11.2.3 Use of Salt in Cooking

The culinary use of salt is vital. Salt improves the way we see, taste, and smell food. Salt in food helps to keep proteins from aggregating into a clotted mess. Sodium chloride can stabilize oil⊠in⊠water emulsions, reducing the separation of oil and water. We have already discussed the preservative role that salt plays in food. Although excessive salt intake is detrimental to human health, cooks cannot ignore the benefit and role of salt in cooking.

Over 5000 years ago, humans began to use salt to enhance the flavor of and preserve food. How do we taste salt, and how, biologically, does it enhance flavor? Proteins on the surface of taste bud cells transport Na^+ into the taste bud, which initiates a signal to the brain that you taste a salty food. However, if you have ever added salt (or forgotten to add the salt) to a cookie recipe, you know that salt does more than just cause a food to taste salty. Low concentrations of salt suppress bitter flavors, thereby allowing other flavors to come through the palate. At high concentrations, salt can increase umami or savory flavors by decreasing sour and sweet flavors. For example, a mixture of table sugar (sweet) and urea (bitter) was found to be equally bitter and sweet by most tasters. However, addition of sodium to the mixture made it seem overwhelmingly sweet. A salad made with a bitter green like spinach or arugula can be made sweeter through the addition of salt. This property of salt may help you to recover and fix a seemingly bitter and ruined meal!

In addition to its contributions to taste, salt impacts the behavior of proteins during cooking and baking. As you know, during cooking, proteins often denature in the presence of acid or heat. The denatured, unraveled protein molecules get tangled up with other protein molecules, forming large protein aggregates that then precipitate. You can recognize this as clots or curds of protein (like you see in curdled or sour milk). How does salt impact protein denaturation? In low salt conditions, salt ions will interact with the protein molecules, preventing their aggregation and forming an insoluble (precipitated) complex. This property comes in handy when you are making a meatloaf or meatballs. Have you ever wondered why you add an egg? The egg works as a binding agent, by helping the meatloaf to hold its shape. This binding property happens to be due to the egg white proteins. What does this have to do with salt? Well, you have to heat up a meatloaf when it cooks to temperatures that are high enough to denature the egg white proteins. However, a meatloaf doesn't have an eggy⊠looking exterior or interior like a fried egg. The addition of salt to the egg white proteins helps the proteins to remain in the solution (even in acidic or heated conditions) and not aggregate, thus enhancing their ability to bind the food together.

What about baking? Why does homemade bread that doesn't contain salt have a horrible taste? In bread, salt plays a different role with proteins. If you recall from Chapter 10, gluten is a complex network of proteins that is responsible for creating a stretchy bread dough due to the formation of cross⊠links with other wheat proteins. These cross⊠links govern the texture of the final bread product and cannot form until water hydrates wheat flour. The proteins that make up gluten have many positively charged amino acids; in solution these charges cause the protein molecules to repel each other. The negatively charged chloride ion, provided by table salt, binds to the positive charges in the proteins, allowing them to come close together and to form the cross⊠links and connections that strengthen the dough. However, the same interaction (between protein molecules and sodium chloride ions) slows down the hydration of gluten proteins because the interaction of proteins with ions reduces interactions between protein molecules and water (since the protein charges are now neutralized by the salt). Bakers will sometimes begin to mix their dough without salt to reduce this effect and then add the salt following hydration of the flour.

Emulsifiers help two solutions that would normally remain separated (like oil and water) and will help maintain two immiscible solutions as one homogeneous mixture. Table salt can be used as an emulsifier; however, other more complex salt compounds such as sodium citrate, phosphates, and tartrate are more commonly used to emulsify foods such as cheese and dairy products. In some foods, table salt will enhance the water binding to proteins, forming the gel and keeping phases from separating.

Do you add salt to your pasta water? The old tale of adding salt to increase the temperature of or make the water boil faster actually has a bit of scientific truth to it; however, in reality, it is not very accurate. It is true that adding salt or other compounds to water will elevate the boiling point of the solution. The boiling or melting point of a solution is called a colligative property. Any substance, whether salt, sugar, or cinnamon, dissolved in water will alter both the boiling and melting point of the solution. The more substance dissolved, the more the melting or boiling point will change. How much is the change and does it matter?

The equation for boiling point elevation is shown below, where ΔT_b is the change in boiling point, i is the van't Hoff Factor, K_b is a constant for the solvent, and m is the molarity or concentration of the solution

$$DT_b = iK_b \times m$$

The van't Hoff Factor, i, is equivalent to "2" for table salt since it dissociates into two ions in water (Na^+ and Cl^-). Thus, for a water and salt solution, the only factor that will impact the change in boiling point is the mass of table salt dissolved in the water (m). The more the salt, the greater the change in boiling point. Because m is a positive value, ΔT_b will be positive and the boiling point will increase. What does this mean for the cook? When you add salt, more heat is required to bring the mixture to a boil than pure water. It will also take longer to boil the same volume of pure water as salt water. However, the boiling saltwater solution will have a higher temperature; therefore, food cooked in salted, boiling water could cook faster than food cooked in unsalted water.

Unfortunately, validation of the old tale ends here. Let's imagine that you add 533 g (about 2 1/4 cup) of salt to one gallon of water. The boiling point will increase by a whopping 2.5°C. In other words, the teaspoon of salt that you may add to your water does not change the boiling point by a significant amount. At this point the salt will have a greater impact on the taste of the food than the few degrees of water temperature.

However, the addition of a pinch or two of salt to water is important in cooking! Salt increases the volatility of some compounds, making it easier for these compounds to escape the boiling water and enhance the flavor and aroma of the cooked food. Salt water saturates the starches in boiling pasta, thus enhancing the flavor of the pasta. So, keep adding salt to your boiling water, and it does make a difference in the flavor of the final product.

11.2.4 Too Many Kinds of Salts

A trip to the gourmet grocery store or a close inspection of various recipes may lead you to think there are different kinds of salt: smoked salt, sea salt, table salt, flake salt, kosher salt, etc.—the list is long and confusing. However, in chemistry, salt is any salt, consisting of anions and cations. KCl, NaCl, and magnesium sulfate are all salts. However, in cooking, salt takes on a slightly more focused definition. The culinary definition of salt is sodium chloride, table salt. The various salts that are described in recipes or are present on the grocery store shelf are all sodium chloride; the difference is in the preparation and presence of additional components (contaminants).

How is salt made? Around the world, salt is primarily produced by two techniques: mining or evaporation of seawater. Most table salt comes from rock salt that is mined from the earth. Large, ancient, underground salt formations can be mechanically removed with mining equipment and explosives, ground into small pieces, and dissolved in water. Alternatively, water can be pumped through mines, which dissolves the salt, making a saturated brine solution that is pumped from the caverns. The salt solution is evaporated to concentrate the salt, and the salt crystallizes and precipitates out of the solution. In windy and warm conditions, the brine is left to concentrate in open pans. Modern evaporation techniques use vacuum chambers to reduce the water content. The brines contain a number of contaminating minerals and other compounds; these contaminants are removed by selective precipitation of the contaminants using a variety of reagents.

Seawater, which contains approximately 3.5% NaCl, can provide a nearly inexhaustible source of salt. The simplest method to harvest salt from seawater is to pump the seawater into smaller drying ponds where the water evaporates in the wind and sun. The evaporated seawater is rinsed with a pure, saturated saltwater solution to rinse off impurities, without dissolving the salt crystals. This is a simple process that can easily be re☒created if you live near a saltwater body. In short, filter a clean, nonpolluted gallon of seawater (to remove rocks and debris), place it in a shallow pan, and evaporate the water using the heat of an oven or the sun. A gallon of sea-water will produce a little less than a cup of salt. If you heat the final product in an oven at a high temperature, you can evaporate the last drops of water and kill any remaining microorganisms. Is there any difference in the sun evaporation versus the oven evaporation methods? Maybe. Rapid evaporation and the addition of small salt crystals (to initiate the crystallization process) will yield the small granular crystals that we associate with table salt. Slow evaporation allows the salt crystals to grow larger and become irregular in size. These salt flakes are sought after due to their sharp edges, which allows the salt to stick to cooking surfaces, producing a crunchy, salty finish to foods. Crystal size and shape are two ways of distinguishing different types of salts. Let's look at some other differences that lead to the different types of salts that you may see on the grocery store shelf.

11.2.4.1 Table Salt Table salt, consisting of small, regularly shaped grains, is slow to dissolve because of the shape and purity of each grain or crystal. Iodine is often added to table salt (you may have seen "iodized" and "uniodized" versions) to reduce the incidence of goiter, a problematic thyroid disease associated with prob-lems in mental development, which, prior to iodized salt, plagued the United States and other countries. Glucose is used in very small amounts to stabilize the iodine for long☒term storage. To keep the salt free flowing, additional additives are used at low concentrations, such as other sodium, calcium, aluminum, phosphate, and silicon salts. These salts help to absorb any water that may be present in the salt or in the air, thus preventing clumping and caking.

11.2.4.2 Kosher Salt According to Jewish dietary law, meat must be treated with salt to draw out the blood. Given that salt already adheres to the Jewish dietary law, there are no additional dietary or production requirements for Kosher salt. However, most pro-ducers either do not include any additives (iodine or other) or just add anticaking agents. Kosher salt is typically produced by slow evaporation techniques, which produce large, thin sharp☒edged salt flakes. The thinner and jagged☒edged flakes quickly dissolve and more readily stick to foods than the table salt granules; thus cooks sometimes prefer this form of salt.

11.2.4.3 Canning or Pickling Salt Canning or pickling salt is produced in the same manner as table salt; however it doesn't contain any iodine or anticaking or anti-clumping agents; these additives are often insoluble at the high salt concentrations required for the canning processes. Both canning and pickling salts are produced and milled to form a small fine☒grain crystal that more quickly dissolves than the larger cubed table or rock salt.

11.2.4.4 Rock Salt Rock salt is the product of raw, undissolved, crushed, or large crystal sodium chloride formation. Because it doesn't go through a purification process, it contains contaminants of minerals and other compounds. Consequently

FIGURE 11.4   Gourmet salt. Three types of salt with sea salt on the right.

(as you might predict), rock salt is a less expensive salt that is usually not used for food and cooking, except when making homemade ice cream using a maker that requires use of a salt/ice mixture to freeze the delightfully creamy and rich dessert. Rock salt is not used in the salt grinder that you might see on your dinner table; salt grinders contain large crystals of purified salt.

11.2.4.5   Gourmet Salt   The diversity and custom flavors found in gourmet salts have caught on with chefs and at home cooks alike (Fig. 11.4). Gourmet salts are flake salts formed from local water sources, which often contain regional or added contaminants (organisms, other salts and minerals) that give the salt a unique color or taste (organisms, other salts and minerals trapped in the salt flakes). Some gourmet salts are made with added flavorings. The most famous and perhaps interesting artisan salt is made by an ancient French technique. Traditional fleur de sel (flower of salt) is made from salt beds in west central France. Minimally disturbed, the salt flakes grow by dehydration on the surface of the pond and are collected by workers who scrape this layer of salt before the crystals sink to the bottom of the container or ponds. Unrefined and often not washed, these salt flakes contain a range of minerals and even small amounts of algae, all of which contribute to the flavor. Is fleur de sel safe to eat? Yes, this salt is safe for consumption, but given the high cost (approximately $30/lb), it is most frequently used as a garnish or is sprinkled to finish a dish.

Himalayan pink salt is crystallized rock salt that comes from the Pakistan Himalayan mountains, where a high mineral content leads to its characteristic color. The lava or coral particles added to make Alaea Hawaiian sea salt gives the salt its red or pink brown color. You can buy or make numerous other flavored salts, which provide unique and interesting flavors. Common flavored salts include garlic, celery seed, and lemon, but if you check out the spice section at your local grocery store, you will find numerous others that will add character to a dish.

11.2.4.6 Sea Salt Like gourmet salt, sea salt is produced from solar heating or thermal evaporation of seawater. Since sea salt is less purified and refined than table salt, it is off⊠colored (sometimes gray) with large pyramid⊠shaped crystals with sharp edges. Because of the size and shape of the flakes and its expense, sea salt is best used to stick to the surface of a prepared food; its delicate structure will dissolve quickly in the mouth, providing a crunchy, salty sensation. However, there is little evidence to support that sea salt tastes differently or is healthier than other salts. Why might there be a health benefit? The idea is that there is less sodium per tablespoon in sea salt relative to table salt, due to the presence of potassium and calcium salts. However, both table and sea salts contain about 40% sodium by weight. Moreover, the additional minerals found in sea salt are often included in our diet from other sources. The true benefit of sea salt is the quick dissolving, crunchy mouthfeel that is present when the salt is used as a finishing salt for a dish, added immediately prior to serving.

11.3 HERBS AND SPICES

While some argue that salt is a spice, it is not. Salt, herbs, and spices are all seasonings. However, herbs and spices are the products of plants (and you know from our discussion about salt that salt definitely does not come from plants). Table 11.1 shows some common herb–food pairings. Simply defined, herbs are the leaves of a plant, while spices are harvested from the rest of the plant (i.e., the root, stem, bark, seed, or plant fruits). Some plants, like cilantro and dill, produce both spices and herbs, while others like basil produce herbs or spices, respectively. In general, herbs are grown in more temperate climates, while spices grow in warmer or tropical zones. Therefore, NaCl is an inorganic mineral and does not fit into the description of herbs and spices.

Like salt, herbs and spices have influenced mankind in many ways. Historians find links to trade routes, changes in political power, and geopolitical conflict based on

TABLE 11.1 Herb–Food Suggested Parings.

Food	Suggested Herb or Spice Combination
Beef	Bay leaf, marjoram, nutmeg, onion, pepper, sage, thyme
Lamb	Curry powder, garlic, rosemary, mint
Pork	Garlic, onion, sage, pepper, oregano
Chicken	Ginger, marjoram, oregano, paprika, poultry seasoning, rosemary, sage, tarragon, thyme
Fish	Curry powder, dill, dry mustard, marjoram, paprika, pepper
Carrots	Cinnamon, cloves, dill, ginger, marjoram, nutmeg, rosemary, sage
Corn	Cumin, curry powder, onion, paprika, parsley
Green beans	Dill, curry powder, marjoram, oregano, tarragon, thyme
Potatoes	Dill, garlic, onion, paprika, parsley, sage
Squash	Cloves, curry powder, marjoram, nutmeg, rosemary, sage, cinnamon, ginger
Rice	Chives, green pepper, onion, paprika, parsley

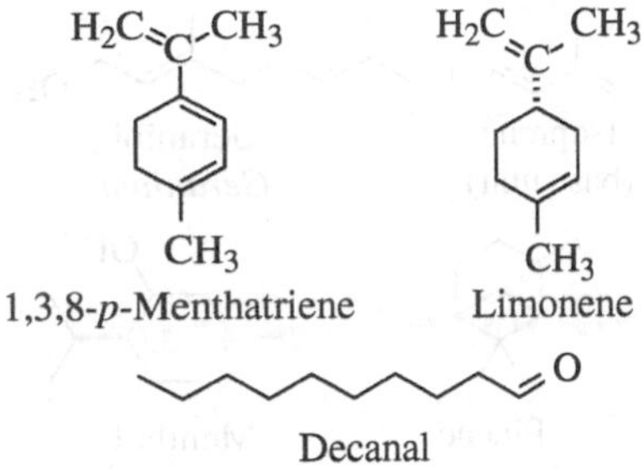

FIGURE 11.5 **Aroma compounds of cilantro and parsley.** The two ring compounds 1,3,8‑*p*‑menthatriene and limonene are responsible for the smell of parsley, while the long carbon chain decanal gives cilantro its odor.

access to herbs and spices. The earliest evidence for the use of herbs or spices comes from ancient humans who wrapped their food in leaves; presumably, they found the food to be more flavorful. However, there is also historical evidence for the use of spice and herbs to preserve food, as perfumes, in religious ceremonies, and for medicinal purpose. Because of the myriad of uses, human desire for access and control of herb and spices drove colonization and expanded exploration routes.

The aroma and flavor associated with most herbs and spices are due to what chemists call volatile organic compounds. Volatile organic compounds are primarily made of carbon atoms (organic); they are not very polar or ionic, and they have a low vapor pressure and low water solubility. Simply put, this means that such molecules have few interactive forces with water and have enough energy at relatively low temperatures to escape from a liquid into a gas phase, where the aroma can reach our nose.

Let's look at two herbs as an example: cilantro (coriander) and parsley. If you look at the two leafy herbs sitting on the refrigerated shelf in the grocery store, they can be easily confused (although a mnemonic might help you remember that the "c"ilantro has curved leaves, while "p"arsley has pointy leaves). However, crushing a leaf of either plant, which allows the volatile chemicals to escape into the air, will immediately tell you which herb you have. Cilantro's aroma and flavor mostly come from a family of carbon compounds called decanals (Fig. 11.5). Decanals are 10 carbon‑containing molecules with an aldehyde functional group on the first carbon. Some of the decanals also have a double bond within the carbon chain. The compounds associated with the aroma of parsley also lead to a complex scent, but the principal compounds found are 1,3,8‑*p*‑menthatriene and limonene (Fig. 11.5). 1,3,8‑*p*‑Menthatreine provides parsley with its floral scent, while limonene is the same compound found in oranges and lemon adding to the complex aroma bouquet of parsley.

If you compare all three compounds, you will notice that none of them are charged and all of them are hydrophobic and nonpolar. The structures of the two parsley compounds are very similar, the main difference being the placement of the double bonds in the ring structure, while the shape of decanal is quite distinct. Nevertheless,

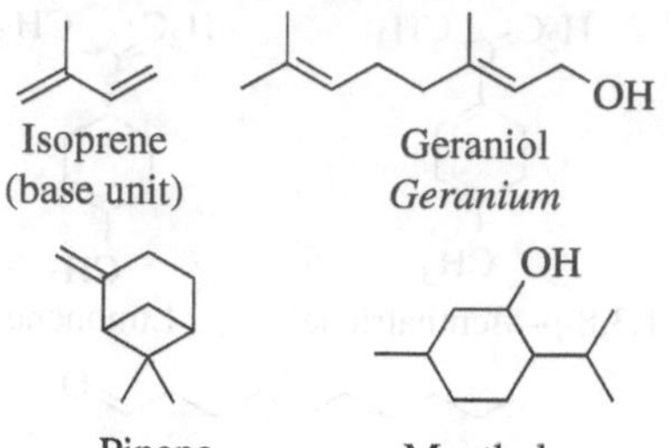

FIGURE 11.6 Terpenes of spices and herbs. The terpene is a base unit used by the enzymes in plants to produce an amazingly diverse set of compounds including those shown here.

as you might predict, all of these compounds are poorly soluble in water and have the high vapor pressure that is characteristic of volatile aromatic compounds. The compounds smell differently because you have receptors for taste and smell in your nose that detect the subtle differences with great discrimination and at very low concentration.

The flavorful and aromatic compounds found in herbs and spices are often called essential oils. This is an appropriate term because these numerous organic compounds that are found in the herb or spice more easily dissolve in oil than water. Some of these organic compounds are less volatile than others, providing a longer‑lasting flavor and aroma due to the fact that they will remain mostly in the liquid form. What makes a compound more or less volatile? The chemical shape and functional groups contained within the molecule (see Chapter 1). Let's talk more about chemical shapes and functional groups that are important in herbs and spices here: aldehydes, ketones, alcohols, amines, esters, ethers, terpenes, and thiols.

11.3.1 Terpenes

The most volatile and aromatic molecules found in herbs and spices fall within a family of compounds called the terpenes. Terpenes, found both in plant and animal cells, are comprised of a diverse set of carbon structures that are built from smaller five carbon units called isoprenes or isoprene units (Fig. 11.6).

Terpenes are organized by the number of isoprene units combined to make the compound (Table 11.2). What is the function of terpenes in animal and plant cells? In animal cells, an isoprene unit is the chemical building block for important steroid molecules, including cholesterol, testosterone, estrogen, and steroid hormones (e.g., corticosteroids). In plants, terpenes play a more secondary role but are very common. The blue smoky haze of the Appalachian Mountains forms due to terpene secretion by the pine trees.

Terpenes are common and found in an interesting number of examples beyond cooking. In animal cells, isoprenes are the building blocks for cholesterol, testosterone, estrogen, and sterol hormones including corticosteroids. In plants, terpenes

TABLE 11.2 Isoprenes in Herbs and Spices.

Isoprene Unit	Name	Formula	Use/Example
2	Monoterpene	$C_{10}H_{16}$	Citral, thymol (mandarin orange), menthol, pine, geraniol
3	Sesquiterpene	$C_{15}H_{24}$	Chamomile, cinnamon, clove, ginger
4	Diterpene	$C_{20}H_{32}$	Vitamin A, rosemary
6	Triterpene	$C_{30}H_{48}$	Lanosterol, cholesterol

play a secondary role for the plant but are very common. The blue smoky haze of the Appalachian Mountains are formed by terpenes secreted by the pine trees. Monoterpenes also serve as seeds of cloud formation. Terpenes found in herbs and spices are fairly volatile and are the primary component of essential oils. The volatile nature of these compounds is why you immediately smell a strong odor quickly associated with the herb or spice upon heating. Terpenes are also fairly reactive, especially with oxygen; when these reactions occur, the new compound is generated. The combination of terpenes with oxygen is a chemical change that creates a new compound that may not be detected by the receptors in your nose. Thus, the aroma may and is the reason why such scents seem to "disappear" after a while in the air or after aging in oxygen-rich environments. Generally speaking, the larger the terpene compounds, the less volatile the compound is. This is beneficial in cooking because larger, less volatile terpenes will remain in the food during cooking, providing a more enduring taste to your food. Rosemary and ginger both contain less volatile, larger terpene molecules, allowing for their lingering taste and smell in a roasted turkey or batch of gingerbread cookies. Examples of these less volatile, larger compounds include the taste and smell of cooked rosemary and ginger.

11.3.2 Phenols

Thousands of compounds in herbs and spices contain or are derived from a group of molecules called phenols. Phenols are compounds that have a benzene ring (a six-carbon ring system with alternating double bonds) attached to a hydroxyl group (–OH; Fig. 11.7).

When several phenol groups are bonded together, a polyphenol is created. Polyphenols have diverse biological and chemical uses; they are used as dyes and in the generation of plastics. In herbs and spices, some polyphenols come from a component of the plant cell wall called lignin, while others are used in defense against herbivores or disease.

One type of polyphenol that is particularly important in food and drink are the tannins (Fig. 11.7). Another complex family of flavor compounds, tannins are polyphenols derived from bark, stems, and woody plant material, which provide a pucker-like feeling called astringency. Several spices contain tannins including tarragon, cumin, vanilla, cinnamon, and cloves.

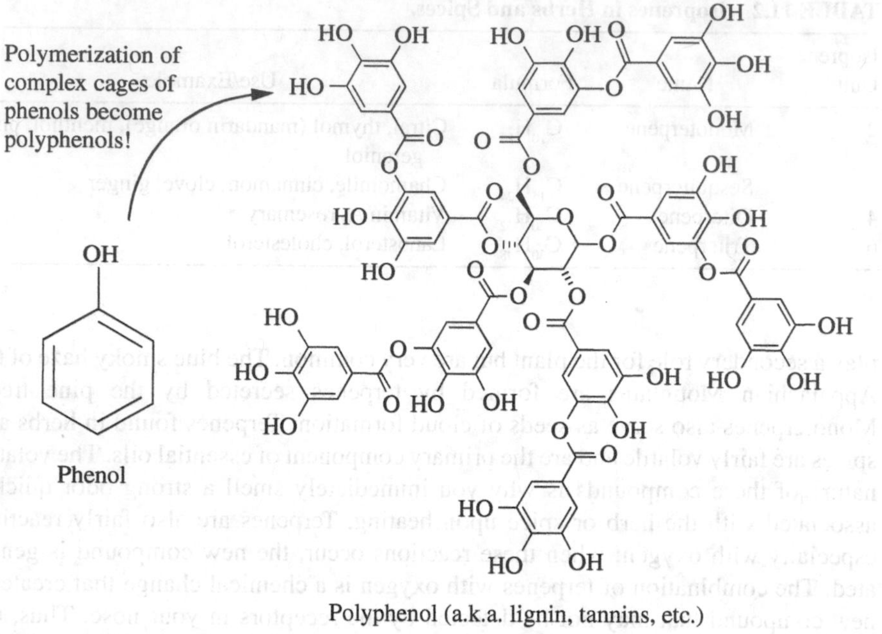

FIGURE 11.7   **The creation of polyphenols.** Lignin is shown on the right as an example of a polyphenol.

Oregano, cumin, thyme, bay, and cinnamon are all examples of spices or herbs that contain phenol groups. An extraction and analysis of these five herbs and spices using a sensitive mass spectroscopy analysis found 52 different phenolic compounds [6]. Rosmarinic acid, first found in rosemary plants, is common to all five of these herbs or spices and is found at very high levels in oregano, rosemary, and thyme (Fig. 11.8). Caffeic acid, found in coffee, is another polyphenol compound identified in all five herbs or spices, although it is found in lesser amounts in cinnamon, cumin, and bay (Fig. 11.8). Caffeic acid is a key intermediate in the production of lignin and is found in nearly all plants. A third phenolic compound, chlorogenic acid, is found at relatively similar levels in each of the five herbs or spices (Fig. 11.8). Chlorogenic acid is produced by the modification of caffeic acid and is important in lignin biosynthesis. In addition to its presence as a flavorant and odorant in herbs or spices, it is also found in coffee beans and some fruit.

11.3.3   Esters

Remember that you have already learned about several chemical functional groups in Chapter 1, including the esters. The presence of functional groups in a compound leads to different chemical characteristics and unique biological activities. Esters are commonly produced from the reaction of carboxylic acids and alcohols (Fig. 11.9).

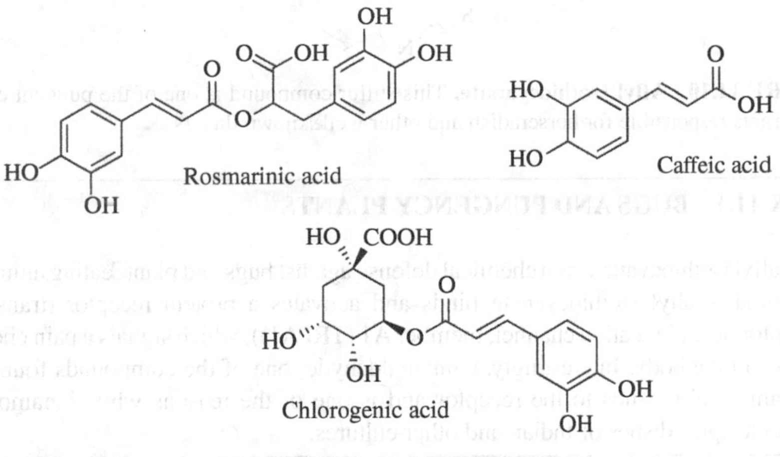

FIGURE 11.8   Spices as phenols.

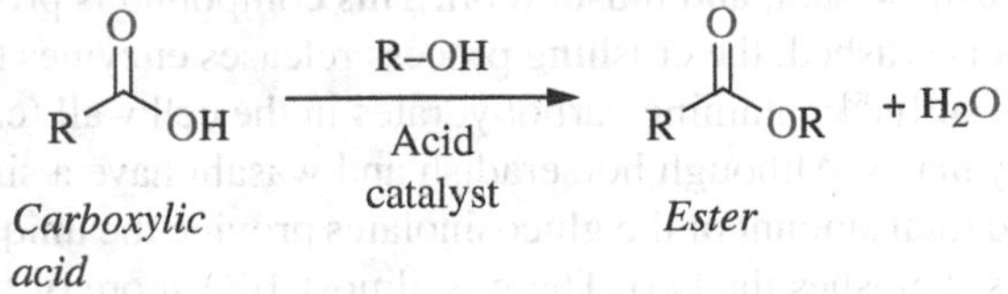

FIGURE 11.9   Formation of esters from organic acids. Many odorants are generated as esters by the loss (dehydration) of water from a carboxylic acid and combined with another carbon‑containing compound.

Esters tend to offer a fruity taste and aroma to our food and drink. Pine, cinnamon, and jasmine are a few spices that contain high concentrations of esters.

11.3.4   Pungent

A flavor family that is not defined by the chemical structure of its flavorants and odorants, but by the sensation of heat, "hotness," or unpleasantness that they bring upon us, is appropriately called pungency. In general, pungent compounds do not bind and activate food and odor receptors, but compounds in this family interact with receptors that signal pain or thermal events. The more formal definition for this type of perception is chemesthesis—the activation of senses in the mouth, nose, or throat for pain, touch, heat, or cold. The cooling sensation of menthol is a chemesthesis event, as well as the heat sensation of wasabi. Horseradish, mustard, wasabi, ginger, pepper, and chilies make up the herbs and spices of this flavor family. Later in this chapter, we will spend some time focusing on chilies and capsaicin. In the meantime, the classification and chemical structure of these compounds can be placed into four categories: thiocyanates, alkylamines or alkaloids, and everything else.

Horseradish, cabbage, wasabi, and mustard all contain a chemical functional group called a thiocyanate (Fig. 11.10). Allyl isothiocyanate is the pungent compound

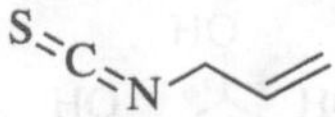

FIGURE 11.10   Allyl Isothiocyanate. This sulfur compound is one of the pungent classes of odorants responsible for horseradish and other well‑known flavors.

BOX 11.1   BUGS AND PUNGENCY PLANTS

Use allyl isothiocyanate as a chemical defense against bugs and plant‑eating animals. In humans, allyl isothiocyanate binds and activates a protein receptor (transient receptor potential cation channel, member A1 (TRPA1)), which signals a pain chemo sensor in the body. Interestingly, cinnamaldehyde, one of the compounds found in cinnamon, also binds to the receptor and is one of the reasons why cinnamon is added to spicy dishes of Indian and other cultures.

---

found in horseradish, wasabi, and mustard oil. This compound is produced when the root of each plant is crushed; the crushing process releases enzymes that catalyze the breakdown of the sulfur‑containing carbohydrates in the cell wall (called glucosino‑lates) to isothiocyanates. Although horseradish and wasabi have a similar flavor, the various forms and total amount of the glucosinolates provide the unique flavor profile of wasabi that distinguishes the two. There is almost 10% more of the allyl isothio‑cyanate in wasabi than in horseradish! However, because of the similarity, the less expensive and easier to obtain horseradish is often tinted green and used as "wasabi." Don't let the color fool you though: 10% more allyl isothocyanate makes a world of difference to the receptors in your nose, throat, and mouth!

Dried mustard seeds or powders are not very pungent because the drying process halts enzyme activity. However, once hydrated with water, the enzymes are able to produce allyl isothiocyanate, leading to the pungency that you associate with mus‑tard. It can, however, take several minutes to hours for the enzymes to make enough of the isothiocyanate for detection. For example, if mustard is mixed with an acidic solution such as a citric acid or vinegar, the enzymes will function, but at a much slower rate, leading to a less pungent dish. In addition, isothiocyanates are fairly unstable and break down quickly. Although the addition of acid reduces the rate at which the enzymes produce isothiocyanate, the lower pH substantially prevents the allyl isothiocyanate from breaking down in your recipe. Extended exposure of the isothiocyanates to heat also increases breakdown and formation of a nonpungent product. What is to be learned from this discussion? The cook who thrives on the preparation and consumption of pungent dishes should wait until the end of the cooking period to add the mustard or horseradish (Box 11.1).

The second group of chemical structures that define the pungent family of flavors is the alkaloids. Alkaloids are a large diverse family of carbon‑based compounds that contain a nitrogen base. In plants, alkaloids are important in the development of the plant, fruit, and seed. However, while they have interesting chemistry and biology, most alkaloids are not flavorants. If, however, you enjoy a blackened grilled salmon

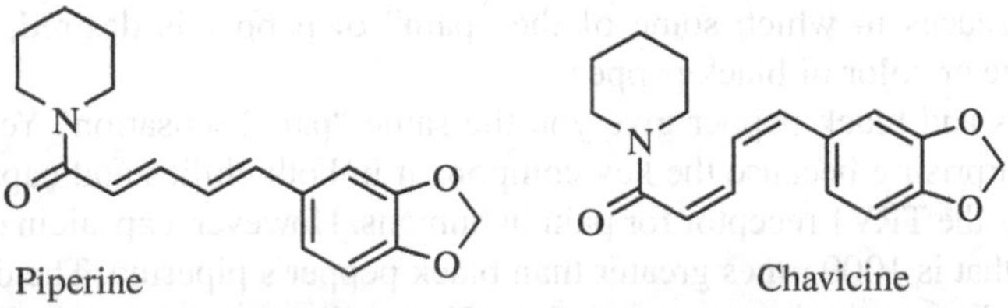

Piperine   Chavicine

FIGURE 11.11   Pungent alkaloids. Shown here are isomers of two alkaloids. There are several isomers, which each bind and activate the receptors responsible for sending pungent pepper odor to the brain.

or spicy salsa, then two alkaloids are critical to enlivening your taste buds. The alkaloids piperine and capsaicin are responsible for the pungency of black pepper and chili pepper, respectively.

Piperine, produced by black peppers of the fruit of *Piper nigrum*, acts by binding and exciting the receptors (TRV1) for pain nerve cells (Fig. 11.11). If pepper has ever caused you to feel pain, now you know why! Piperine acts similar to, but has a greater efficacy than allyl isothiocyanate. Three different piperine isomers are found in the pepper fruit: chavicine, isochavicine, and isopiperine. If you remember, isomers are compounds with the same atomic makeup or molecular formula, but they possess a different organization or structure. Chavicine also has a strong bite and aroma, but it, as well as the other isomers, slowly degrades, while piperine remains stable and pungent. The pepper berries include other aromatic volatile compounds, adding to the aroma and pungency of pepper. Terpene, limonene, and linalool all combine with piperine to give fresh black pepper its woody floral taste with a bite of pain.

Did you ever wonder why fresh cracked black pepper tastes and smells so different from ground pepper, especially ground pepper that has been left in a shaker for a long time? The difference in taste and aroma is all due to the chemistry and biology of the pepper plant and its compounds. The mature fruit or berries from the pepper vine are dark red and contain a single seed. The blanched and dried berries are left to age in the sun, which ruptures the cell wall of the berries, allowing for enzymatic and Maillard browning reactions to take place. These reactions produce dark-colored polyphenols and other volatile compounds in the intact berry, now called a peppercorn. When you grind peppercorns in your mill, many of the volatile compounds are released into the air, resulting in that "peppery" aroma. Although aged ground pepper still contains many piperine and terpene compounds, a significant portion of the more volatile compounds will have evaporated over time. This is the reason why whole peppercorns are often used for longer forms of cooking or preserving instead of ground pepper.

Have you ever seen white pepper? White pepper is derived from the same *P. nigrum* vine berries as black pepper; however the outer fruit layer is removed by bacterial decomposition in water. This process results in the loss of most of the terpene aromatic compounds but allows for the retention of much of the pungent piperine molecules. This milder preparation of pepper used for its less aromatic flavor and aroma while still adding pungency to a food. White pepper is often used in foods, salads, and cream sauces in which some of the "pain" of pepper is desired, but not the stronger taste or color of black pepper.

Do chilies and black pepper give you the same "pain" sensation? Yes and no.... This isn't surprising because the key component in both chilies and capsaicin binds and activates the TRV1 receptor for pain in humans. However, capsaicin does so with an efficacy that is 1000 times greater than black pepper's piperine. The difference in pungency or hotness in the many varieties of hot chili peppers is primarily due to the level of capsaicin in each pepper. What makes capsaicin so powerful? Notice the nitrogen atom in the middle of the molecular structure of capsaicin. Capsaicin, like piperine, is a relatively large molecule that has some polar functional groups; it is even able to hydrogen bond with water! This characteristic makes capsaicin much less volatile, which you are likely grateful for if you've ever touched a sliced jalapeno and then your eyes! We will talk more about hot peppers later in this chapter.

There are lots of other compounds in the "pungent" family, in which a few compounds are worth noting. Gingerol (found in ginger) closely resembles capsaicin but does not have the nitrogen base of an alkaloid; rather, it is a modified phenol. While pungent, gingerol is rated less pungent than pepper. However, age and heat cause the degradation of gingerol to another compound called shogaol. Shogaol happens to be twice as pungent as its parent, gingerol. Thus, dried ginger has a more pungent flavor than fresh ginger. A lesser known spice compound called paradol is found in the seeds of Guinea pepper. The compound is a phenol, is similar in structure to gingerol, and is rated with the same pungency as piperine. Paradol has an interesting property in that it activates a process called thermogenesis, a biochemical metabolism that burns fat to produce energy. The length of the carbon chain of paradol seems to be critical for its fat burning ability. In a study in which mice were fed a high fat diet, the shorter the chain on the compound, the lower the weight gain in the mice that were fed the compound (Box 11.2).

BOX 11.2   COOKING WITH HERBS AND SPICES—NOW IT SHOULD MAKE SENSE!

Cookbooks and advice articles on herbs and spices recommend a few approaches to using herbs and spices:

- Crush leaves and grind spices immediately before adding to your dish.
- Use whole spices or herbs while cooking when you desire slow release of flavor (e.g., adding whole cloves or bay leaves early in a recipe).
- Add ground spices no more than 15 min prior to the end of cooking time.
- Add more ground, dried, or preprepared herbs and spices than fresh.
- Dried herbs and spices are best used when cooking with oil.
- When using fresh herbs and spices, chop and grind the leaves, seed, or root to small pieces to release the flavor.

Now that you understand the basics of the components of herbs and spices, you can recognize the reasons behind each of these hints. More volatile compounds evaporate over time or react with oxygen; thus fresh herbs and spices have more potency and should be used more sparingly than their dried counterparts. In whole herbs and spices, the cells need to be broken before the enzymes can be released that make the flavorant molecules, so cooking for longer periods of time is effective. Because many of the flavorful compounds in herbs and spices are poorly soluble in water, a better liquid for mixing is oil. Application of these hints and your newly gained knowledge of the science behind the hints will certainly help you better understand what and how to best approach the kitchen pantry as an experimental cook and baker.

11.4 A CLOSER LOOK AT A FEW HERBS AND SPICES

If you haven't already gathered this, the biology and chemistry of herbs and spices are pretty interesting. The impact of flavor, the evolution of the originating plants, and the biological impact of herbs and spices are the subjects of many fascinating books. Here, let's focus on a handful of herbs and spices that have an interesting scientific story and play a significant role in the kitchen.

11.4.1 Vanilla

Vanilla is one of the world's most popular flavorings, finding its way into food, beverages, perfumes, and even pharmaceuticals! Vanilla, which originated in Mexico and Central America, comes from the vanilla orchid, a vine that produces vanilla beans as a dried seed pod of its fruit. The value of the bean pods was recognized by the Aztecs who used them to flavor their drinks made with powdered cocoa beans, ground corn, and honey. However, due to the hermaphroditic character of the plant (this means that the plant has both male and female reproductive organs), the plant flower requires pollination to set the fruit. This characteristic was problematic during first attempts to cultivate vanilla outside of Mexico and Central America given that the natural pollinator of the flower was not native to other tropical areas where vanilla was first transplanted (Fig. 11.12).

Due to the great culinary value of the vanilla, cultivars of the plant were brought around the globe in the early to mid⊠1800s, including the West Indian island of Réunion, where the vine would grow but the pod would not develop. Although this gave the Central American growers a lock on the much⊠desired vanilla flavor, scientists continued to strive to find a way to cultivate the plant outside of Central America. In 1836 a Belgian botanist discovered the importance of the *Melipona* bee for pollination; however it wasn't until 5 years later, in the West Indies, that a 12⊠ year⊠old slave discovered and developed a hand pollination method for the vanilla flower. This method is still used today. Thanks to 12⊠year⊠old Edmond Albius, who won his freedom for the development of this process, vanilla vines can now grow and fruit in many tropical areas around the world.

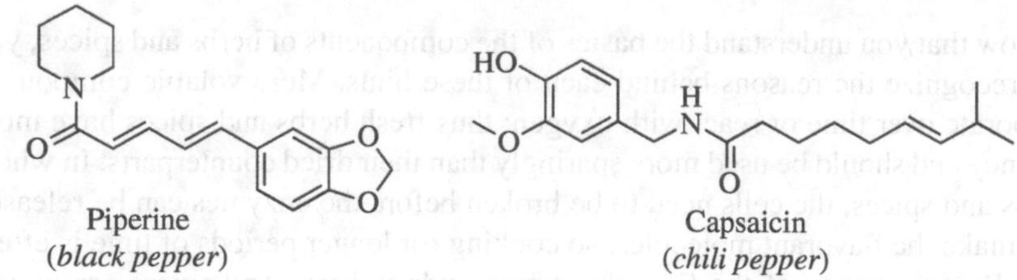

Piperine
(*black pepper*)

Capsaicin
(*chili pepper*)

FIGURE 11.12   Hot or not? Two pungent alkaloids give food a hot flavor but act by very different receptors. Capsaicin but not piperine stimulates our pain receptors, giving a hot feeling.

FIGURE 11.13   Vanilla bean pod. An open, close up image of a vanilla pod. Notice the small black seeds held within the pod. These seeds are used to extract vanillin compound for cooking and baking.

Madagascar and Indonesia are the world's largest producers of vanilla. Madagascar and the West Indian island of Réunion (previously called Bourbon) produce Bourbon or Madagascar vanilla. These pods produce the rich flavor that you most often think of as "vanilla." Tahitian vanilla is derived from a plant hybrid that is grown in the Philippines. While Tahitian vanilla has a desirable flowery and fruity flavor, it is susceptible to breakdown by heat. The sensitivity of Tahitian vanilla to heat is particularly crucial, as part of the curing process of the vanilla bean is to heat the pod, which promotes the browning reactions necessary to form mature vanilla flavors. Mexican and Indonesian beans also have a more subdued vanilla flavor and smoky or wine like aroma than do the Madagascar/Bourbon pods.

When you cook with a vanilla pod, rather than vanilla extract or flavoring, the food has a much more interesting and complex flavor. Why? Most of the vanilla flavor resides in the sticky material inside the pod, as well as in the small black bean seeds. How do you work with a vanilla bean pod in the kitchen? Slice down the length of a bean pod, scrape out the sticky black material and seeds, and include the combination of scraped seeds and the bean in the recipe (Fig. 11.13). This is particularly delicious when you are making a dish comprised of milk or cream. Because the compounds that

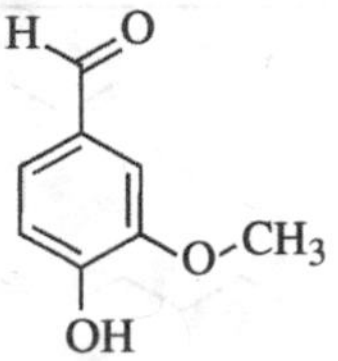

Vanillin

FIGURE 11.14 Vanilla flavorant. Vanillin is one of the key compounds responsible for the flavor and aroma of vanilla.

provide the flavor and scent of vanilla are more soluble in fat and oil than water, the fats in milk solubilize the vanilla flavor molecules, leading to wonderful concoctions like vanilla milk or vanilla bean ice cream. An interesting additional use of unused sliced pods is to submerge the opened, uncooked pods into a closed container of table sugar; this creates a rich, vanilla scented sugar that is worthy of baking and candied treats.

Which molecules give vanilla its characteristic vanilla flavor, and how are these flavoring molecules produced? During the aging and browning process of the bean pods, some of the glycoside components of the seed and plant cell walls are converted to vanillin, the molecule most responsible for vanilla flavor and aroma. Remember that glycosides are sugars that are covalently bonded to other sugars or functional groups via a glycosidic bond. There are many other compounds detected and responsible for part of the flavor of vanilla, but vanillin is responsible for most of the flavor and aroma. The worldwide demand and expense for vanillin far exceed (by about 10 fold) the capacity of the plant to produce the flavor. Therefore, synthetic vanillin accounts for most of the vanilla flavoring market and is produced at one hundredth the cost of the natural product (Fig. 11.14).

What is the difference between a vanilla that is made naturally and synthetically? Natural vanilla extract is a complex mixture that includes vanillin extracted from alcohol soaked vanilla beans or processed beans that are repeatedly washed over with alcohol. Pure natural vanilla extract is best characterized by its sweet fruity, spicy flavor, and aroma. Most vanilla is a synthetic production of vanillin, which contains added sugar and other compounds. This is still pure vanilla; it is just not naturally produced by the plant. However, regardless of whether vanilla is artificial or a pure vanilla extract, the compound is very volatile, so you should add it later in the cooking process to avoid evaporation and loss.

11.4.2 Coriander and Cilantro

Part of the carrot family and a native Middle Eastern plant, coriander and cilantro are two widely used herbs grown and utilized broadly around the world. Cilantro and coriander come from the same plant; the leaf is used as a herb, while the spice comes from the fruits or seeds of the plant and is typically ground. Coriander and cilantro provide a great example of how herbs and spices are distinguished. Coriander is a spice, as it comes from the fruit of the plant. Cilantro is a herb, since it is the leaf of

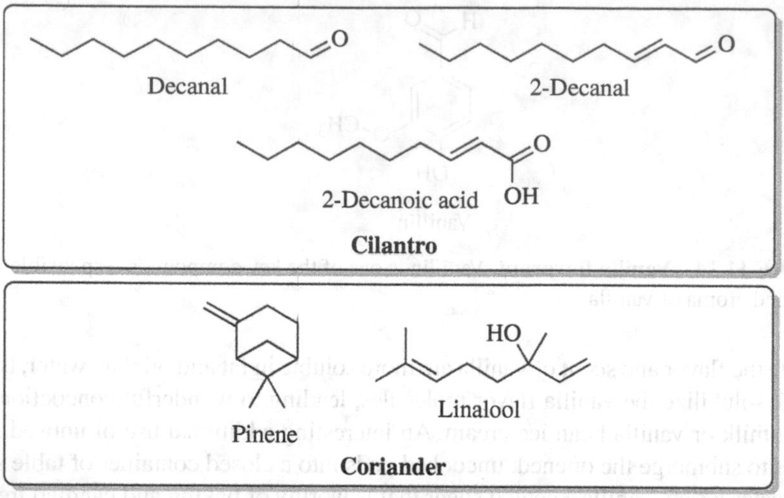

Decanal

2-Decanal

2-Decanoic acid OH

Cilantro

HO

Linalool

Pinene

Coriander

FIGURE 11.15 Compounds of the coriander plant. Cilantro and coriander, while from the same plant, are a herb and a spice, respectively. The compounds responsible for their unique characteristics are shown here.

the plant. If you have ever tasted or smelled cilantro and coriander side by side, you know that there is a distinct difference in flavor and aroma. These differences are due to the distinct molecules that are present in the fruit relative to the leaf of the plant. Cilantro leaves contain 41 different volatile compounds including decanal and similar isomer compounds [7]. Coriander flavor and aroma come from the terpene flavor molecules, linalool and pinene, which give the spice a fruity, pine⊠ or sage⊠like flavor and odor. Linalool is a branched carbon chain with an alcohol (OH) functional group. Pinene contains a complex carbon ring system and is the molecule also found in pine resin, pine oil, and lemon oil. Mixed with cumin, ground coriander seeds provide the base for many Indian culinary dishes (Fig. 11.15).

Cilantro deserves a bit more discussion because the flavor of this herb is polarizing. Some people love cilantro and mix it with their homemade salsas and Mexican food. Other people hate cilantro, claiming it tastes like soap. Why is there such a love/hate relationship? Let's take a closer look at the compounds found in the herb and inspect our DNA.

Why the soapy taste? At the molecular level, some of the compounds in soaps are structurally similar to the decanal⊠based flavor molecules of cilantro. The aldehyde component of the decanals is also structurally similar to the odor molecules that some bugs produce as a defensive weapon. Harold McGee, in a post for *The New York Times* [8], posits that these similar compounds can remind people of experiences with soap, earth, or even bugs. It is no wonder some people have strong negative feelings about the herb!

What is the connection between cilantro and our DNA? Fourteen to twenty⊠one percent of people with Asian, European, and African ancestry report detection of a soap⊠like flavor and a corresponding dislike of cilantro. In contrast, only about 3–7%

of people from South Asia, Central America, and the Middle East (where cilantro is heavily used) did not like the taste of cilantro [9]. In a recent study, scientists found a genetic change (a change in a single DNA base) in the chromosomes of about 10% of the population that is linked to a dislike for cilantro. This change in one nucleotide (the chemical building block of DNA) is called a single nucleotide polymorphism (SNP, pronounced "snips").

SNPs are not uncommon; they are found about every 300 nucleotides on a chromosome. SNPs are part of what brings about genetic diversity in humans, plants, and most organisms. Since the human genome has about three billion nucleotides, there are about 10 million SNPs in our genes. Although 10 million seems like a lot, most of these single mutations do not affect us, as the bulk of the nucleotides that make up our chromosomes do not code for proteins. However, when an SNP happens in a part of the chromosome that does code for a protein, the proteins that these genes code for may have some very unique characteristics.

There are two SNP variants linked to perception of cilantro. How were these variants detected? In one study containing over 14,000 people, individuals were asked whether they detected a soapy cilantro taste and had their genes sequenced. In this study, a connection was made between two SNPs and a group of genes on chromosome 11 that had a single mutation that codes for olfactory receptors [10]. The OR6A2 gene was altered in nearly half of the participating European descendants and codes for a receptor that is highly sensitive to aldehyde⬚containing compounds. Another study found a link between a dislike for the herb and three different genes. What is the take⬚home message? Small genetic differences in chromosomes can alter the structure and function of proteins. These genetic differences may be the reason why there is a difference in the perception of flavor and odor between two individuals.

11.4.3 Cinnamon

One of the most popular spices or herbs, cinnamon, is broadly used in the cooking of sweet and savory foods, beverages, and candies or is sprinkled on a piece of toast. Cinnamon is a spice that comes from the inner layer of the bark of tropical evergreen trees and shrubs from the *Cinnamomum* genus, which consists of 250 trees and shrubs. Given the different species used to make the spice, the term "cinnamon" doesn't fully capture the different characteristics of molecules present in each preparation. Consistent among all cinnamons is the main flavor ingredient, cinnamaldehyde (Fig. 11.16), while the minor components of the spice will vary from source to source.

What are the different types of cinnamon? "True cinnamon" is derived from the *Cinnamomum verum* tree that is native to South India and Sri Lanka and has also been transplanted to grow in Madagascar. Because of the historic importance of the spice, true cinnamon is called Sri Lankan (or Ceylon—Sri Lankan's former name) cinnamon. A second species of tree, *Cinnamomum cassia*, used to make the spice called "cassia cinnamon" (also called Chinese, Padang, Saigon, or Batavia cinnamon), grows more widely in Vietnam, India, and Indonesia. The *Cinnamomum burmannii* tree is the source of burmannii cinnamon, which is also called Indonesian

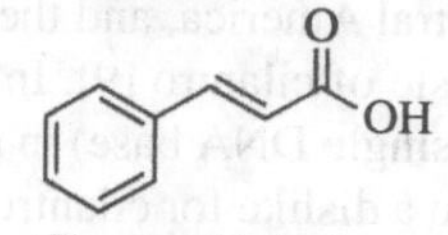

Cinnamaldehyde

FIGURE 11.16   Cinnamon flavorant. Cinnamaldehyde is the main compound responsible for the flavor of cinnamon.

or korintje cinnamon. In order to understand what makes these three types of cinnamons different from one another in taste and use, we need to talk about how cinnamon produced.

Cinnamon is obtained by stripping the inner bark from the shoots of 2–3 year old stems of the tree. The inner bark is dried and curls in a characteristic way that we associate with cinnamon sticks. The cinnamon trees are cut or pruned to allow for the growth of new shoots for the next crop of bark. The work is difficult and requires skilled peelers. Sri Lankan or true cinnamon is paper thin and forms into a single curl (or quill), while cassia and burmannii cinnamons are thick and curl into a double curl/quill. Moreover, only true cinnamon will have many thin layers rolled into its single quill. Once ground into powder (in the absence of chemical analysis), it is nearly impossible to distinguish between true and cassia cinnamon. Cinnamon powders typically come from low grade and chipped bark; the leaves and low grade bark can also be distilled or solvent extracted to harvest cinnamon oil.

Is there a difference in taste between the different types of cinnamon? Sri Lankan cinnamon, which was used in many early European desserts and Mexican recipes, is slightly sweeter and has a mild flavor. Cassia and burmannii cinnamons have a stronger, almost peppery flavor, are the "cinnamon" that you buy in the grocery store, and are the cinnamon spice aromas and flavors that you associate with gum and apple pie. In other countries, cassia cinnamon is distinguished from others by labeling the spice as cassia, not cinnamon.

Since there is a difference in flavor in the different types of cinnamon, you can likely surmise that the different cinnamons must have a different chemical composition. The dried bark of any form of cinnamon or cassia contains 0.5–3% volatile oils that provide most of the flavor and aroma of the spice. The key molecular component, which makes up between 75 and 90% of all the compounds in cinnamon oil, is cinnamaldehyde (Fig. 11.16). Cinnamaldehyde is a modified phenol compound (Fig. 11.17), where a short carbon chain containing an aldehyde has replaced the phenol OH. The compound can be detected at very low concentrations (0.1–0.5 of the total percent of food); upon binding to its receptors, it provides a pungent sensation in addition to the sweet taste.

All cinnamons contain cinnamaldehyde. True cinnamon also contains volatile terpene compounds, including the pine scented pinene and the sweet, floral compounds linalool and eugenol, which are found in many plants that smell of cloves and honey. In contrast, cassia cinnamon also includes a small amount of vanillin, higher concentrations of tannins, and only trace amounts of eugenol. These molecular differences

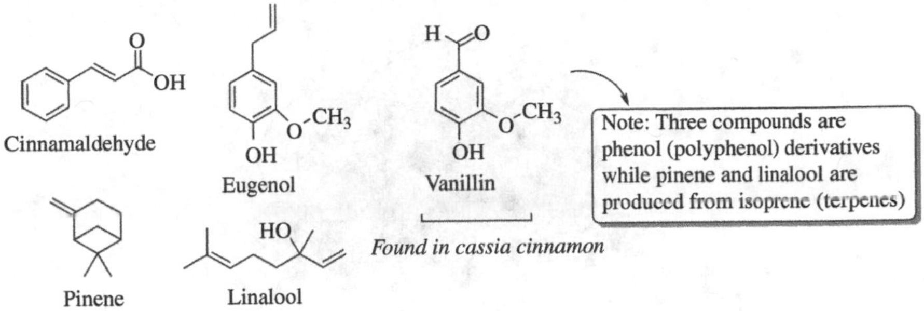

FIGURE 11.17 Compounds of cinnamon. While both true and cassia cinnamons have cinnamaldehyde, true cinnamon has more eugenol than cassia, while cassia cinnamon has small amounts of vanillin.

BOX 11.3   CINNAMON CHALLENGE

Coumarin is found in all forms of cinnamon, although in cassia cinnamon the concentration is 60 or more times higher than in true cinnamon. Coumarin, a naturally occurring fragrant compound found in many plants, is used in perfumes and as a precursor compound to the anticoagulant warfarin (coumadin). Coumarin has a modest and incompletely tested antidiabetic property where it may help to lower blood sugar and lipid levels. However, at higher levels, coumarin is toxic; it causes liver failure in a genetic subset of the population and, upon exposure to animals at high concentrations for long periods of time, may have carcinogenic activity. An adult weighing 130 lbs would have to consume 2 g of cassia cinnamon a day to approach the toxic level of coumarin that has been set by the German government. While other countries have not set a minimum or maximum intake standard, the European Food Safety Authority recommends a coumarin daily intake limit of 0–0.1 mg/kg of body weight per day or up to 50 mg/kg of food. As of 2014, the US Food and Drug Administration (FDA) does not limit the amount of coumarin used in cinnamon-flavored products. However, the USDA does prohibit coumarin as a food additive. Current recommendations are that those who take in high doses of the spice in food or as a health supplement should take caution, as several of the animal studies are difficult to interpret and extrapolate to humans.

---

partially explain the difference in flavor (with true being sweet and mild and cassia being more potent and peppery) between the two (Box 11.3).

11.4.4 Saffron

Saffron is a very interesting spice that, perhaps, you have never used (or even heard of). Why is it interesting? For starters, it is the most expensive seasoning, selling for $1500 and $2500 per pound. Why is it so expensive? The spice comes from a flower, the *Crocus sativus*, which is grown in limited regions of the world; only three threads of the spice

FIGURE 11.18   The delicate spice saffron.

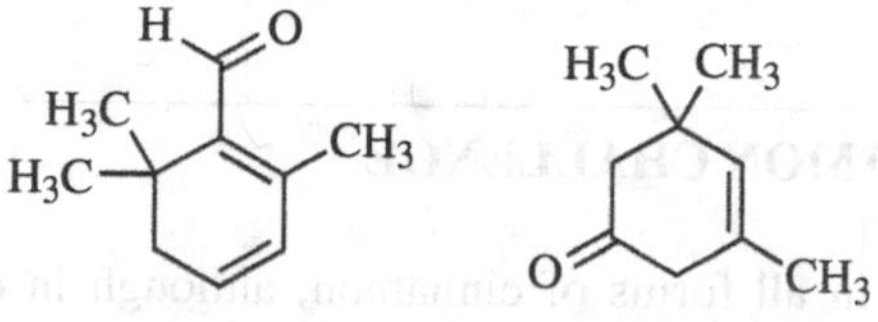

Safranal   β-Isophorone

FIGURE 11.19   Volatile compounds of saffron. The two key components of authentic saffron.

are produced by each flower, and harvesting of the spice is performed by hand! Are you intrigued? Let's learn more about this interesting and expensive spice (Fig. 11.18).

Saffron is sold as a thin red thread; the threads are the stigmas of the crocus flower, which are picked by hand. Given the small number of saffron threads per flower, it takes about 75,000 plants to produce a pound of the spice; one acre of saffron plants yields only about 10 pounds of the spice annually. Iran, Spain, and Portugal are the main producers of the spice; however the flower is also grown in India. The high cost and low availability of saffron have led to a significant counterfeit market for the spice. True saffron has a unique smell, the threads will turn a cup of water yellow, the resulting water will have a bitter taste, and upon addition of baking soda, the water will remain yellow. Many counterfeit saffron will turn water red or brown with an increase in pH. American or Mexican saffron, which comes from a daisy flower, does not impart the flavor of true saffron. However, turmeric is used in many Indian dishes as a saffron substitute, providing a similar color and flavor to the food.

Like the other spices and herbs we've studied, saffron is a complex mixture of volatile substances; over 150 unique compounds have been identified in the spice. Some of the volatile compounds that contribute to the aroma and flavor are produced while drying the stigmas. Analysis of the volatile compounds produced just during the aging process identified 23 different compounds. Although some of these may have been breakdown products produced during the testing process, this is another example of the complex nature of herbs and spices. The two key volatile components are safranal (which makes up ~70% of the volatiles) and beta‑isophorone (Fig. 11.19). Both of these small organic compounds have very little water solubility, which you know makes a good volatile fragrance.

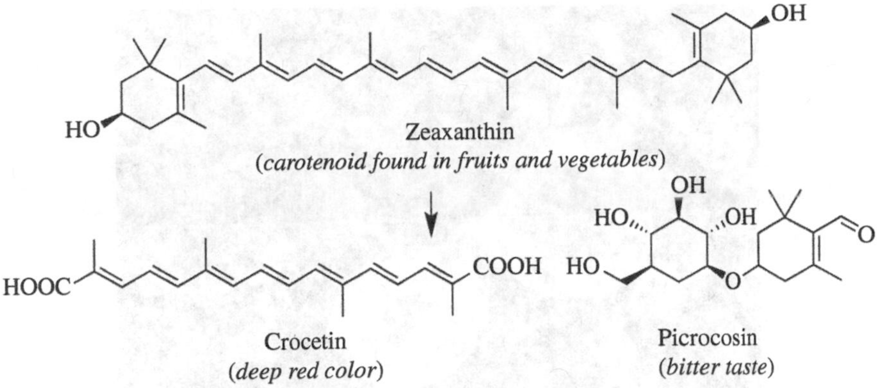

FIGURE 11.20 Saffron color agents.

The deep red color of saffron is due to the presence of the fat⊠soluble carotenoids crocin and crocetin. These strongly colored pigments are formed by the breakdown of a compound called zeaxanthin. A by⊠product of zeaxanthin degradation is a bitter compound, picrocrocin. Picrocrocin, a glucose derivative and water⊠soluble flavor molecule, is the telltale compound found in true saffron. Interestingly, the other by⊠product of this reaction is the aromatic safranal. Why do the stigmas have to be dried to yield these flavor molecules? As we have seen before, the drying process breaks open the cells, releasing the enzymes responsible for these reactions (Fig. 11.20).

Saffron is common in Middle Eastern and Spanish **dishes** such as rice, risotto, and paella. Fortunately for those cooks who are on a tight budget, a pinch of thread is all that is needed for sufficient flavor and development of the characteristic yellow color. Often, the spice is steeped in warm water or milk for 30 min prior to use; this process draws out the color and assists in generating a homogeneous mixture of flavor in the finished dish.

11.4.5 Nutmeg and Mace

Originally grown in the Indonesian islands, the tropical evergreen *Myristica fragrans*, commonly known as the nutmeg tree, produces both mace and nutmeg. Nutmeg is made from the seed or pit, while mace comes from the webbing that covers the shell of the pit. Growing nutmeg trees requires great patience. The nutmeg tree is dioecious; this means that there are separate male and female trees. Only the female trees can produce fruit; to complicate matters further, it takes nearly 8 years to identify a tree as male or female. To overcome this complication and reduce the risk of an all⊠male orchard, cuttings are used to clone the female plants, and 10 female trees are transplanted for each male tree (Fig. 11.21).

In the United States, nutmeg is used in potato⊠based dishes, cookies, pastries, sausages, and, during the holidays, eggnog. The savory flavor of nutmeg is used in many Indian, Middle Eastern, and Indonesian dishes. If you have never cooked with mace, you would find it to have a nutmeg⊠like aroma, but it is more pungent and savory.

FIGURE 11.21   The nutmeg fruit.

Although it was once a highly sought‐after spice, it is now used as a dominant flavor in spice doughnuts and spice cakes. Both spices have some of the same terpenes as cloves having woody and floral notes. Nutmeg and mace are often included with a mix of other spices including cinnamon, cumin, and vanilla. The compounds contained in both spices are also found in cloves and are terpene based with woody and floral tones.

Off and on in recent years, the potential for nutmeg to produce a drug‐like, hallu-cinogenic high moves through communities. However, consumption of the amount of nutmeg needed to produce the high will result in severe side effects including vomiting, diarrhea, nausea, kidney and central nervous system issues, and irregular cardiac rhythms. The nutmeg molecule that is believed to contribute to the halluci-nogenic effect is myristicin. Chemists and biologists interested in this question believe that the drug is either acting directly in the body or must undergo a chemical change to exhibit its activity. Understanding which processes of these is occurring is called the mechanism. There is scientific evidence that myristicin is chemically converted to a psychedelic amphetamine called MMDA in the liver. However, other studies show that myristicin is chemically converted to other nonhallucinogenic compounds in the liver. Other studies show that myristicin itself binds to and stimulates the receptors that are activated by other psychedelic drugs such as amphet-amines, serotonin, and dopamine. It is not clear which of these possible mechanisms might be at work in humans, given that these studies were conducted in rodents or petri dishes. However, either way, the dangers and severe unpleasant reaction to myristicin are nasty and can lead to an emergency room visit.

11.4.6 Curry

The term curry is used for both the dish and the powder that is used as a base seasoning for the dish. Distinct from the seasonings that we have discussed thus far, curry powder is actually a mixture of herbs and spices, most of which do not come from either of the two plants with the name "curry." Curry leaf comes from the *Murraya koenigii* plant, has a lime–lemony taste with woody overtones. Although most curry powders do not contain curry leaf, Indian curry powder mixtures often include curry leaf, or the leaf is added directly to Southern Indian cuisine on its own. The unrelated curry plant comes from *Helichrysum italicum*, a plant that is similar to flowering sage bush and smells of curry but is not included in curry dishes or powders. Interestingly (and perhaps confusing), neither is a traditional component of most curry spice mixes. These plants have a large concentration of alkaloids. However, Indian curry powder mixtures often include the leaf, or the leaf is part of Southern Indian cuisine on its own. The unrelated curry plant appears similar in some ways to flowering sage bush and smells of curry but is not involved in curry dishes.

The term curry is more appropriately used to describe a variety of dishes from around the world, including foods from India, Pakistan, Sri Lanka, Singapore, Thailand, and Japan. Like dialect and cultural variation, there are as many kinds of curry spice mixtures as there are villages. In most cases, the recipes for curry powder start by toasting or browning the spice. The traditional spices in a curry powder include turmeric, coriander, cinnamon, allspice, and cumin, to name a few. Northern curry powder, like the Punjabi style of curry known as garam masala, tends to be sweet and contains black pepper, cardamom, and coriander as its main components.

Thai curries, unlike the Southern Indian dishes, do not contain curry leaves, are sold as a powder or paste, and are typically identified by the color of the powder/paste: red, green, and yellow. Thai curry paste tends to be particularly spicy, where the level of spice depends upon on the chili being used. Red Thai curry is made with red chilies, garlic lemongrass, ginger, and shallots. Green Thai curry paste is similar to red curry but has a green chili pepper and includes coriander and cumin. Yellow Thai curry gets its distinctive color from the addition of turmeric and cumin and is a little sweeter and creamier than the green and red versions due to the addition of coconut milk.

A northern Africa/Arabic curry called Ras el hanout is a blend of black pepper, cardamom, sea salt, ginger, cinnamon, mace, turmeric, allspice, nutmeg, and saffron; this blend is common to Moroccan and Arabic cuisines.

Madras curry sauce is a British version of a hot Indian⊠inspired curry paste, containing chili powder, turmeric, cumin, and cinnamon, which is added for its pungent and savory flavor that contributes to the "heat" of a Madras curry. Given that we are talking about the "heat" associated with a curry, let's talk more about the chilies and capsaicin that give many other dishes "heat."

11.4.7 Chilies, Capsaicin, and Heat

The fruits of the flowering plant *Capsicum* include the mild green pepper and the hottest peppers ghost pepper or bhut jolokia, Carolina Reaper, and the Trinidad moruga scorpion. The primary compound contained within capsicum plants is capsaicin; however the name of the fruit that produces the compound is not universally accepted. At times "chili" is used to describe the pepper, believed to be derived from chili con carne, a tomato⊠based dish that is made with the pepper. Eventually,

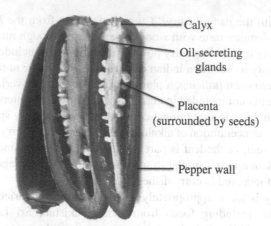

Calyx

Oil-secreting glands

Placenta
(surrounded by seeds)

Pepper wall

FIGURE 11.22 Anatomy of a pepper. Most of the "hot" compound, capsaicin, is found in the placenta. The seeds are filled with bitter☒tasting cell wall material and are coated by the oil glands with capsaicin.

the name for the food and the pepper was shortened to chili. Chile, the name of the South American country, and consistent with the Spanish "e" ending, has also been used as a name for hot peppers. However, perhaps the most historically correct way to refer to a pepper is chili. A Spanish physician and botanist, Francisco Hernández de Toledo, in *Four Books on the Nature and Virtues of Plants and Animals for Medicinal Purposes in New Spain*, used the Aztec native language to describe white habanero peppers as "arbol chili" in 1615. We will use this historical common name, chili or pepper here.

Chilies belong to a larger family of flowering plants including tomato, potato, and petunia plants and over 2700 other species called Solanaceae (nightshade). Plants are organized from this larger family into smaller subsets (or genus). Chili is in the *Capsicum* genus, which includes 22 wild and more than three domestic species. Most hot peppers lie in the *Capsicum annuum*, including bell, anaheim, banana, jalapenos, cayenne, and some of other commonly used peppers. The breadth of taste and heat that lie within the *C. annuum* species is somewhat surprising. A few familiar peppers do belong to a unique species: Tabasco and Thai (*Capsicum frutescens*), habanero, and Scotch bonnet (*Capsicum chinense*).

11.4.7.1 Anatomy of a Chili

Have you ever thought about a chili as a hollow container for seeds (Fig. 11.22)? That is essentially what it is! Inside of the hollow pod, the thin☒shelled seeds are attached to the glands and placenta of the fruit. Chilies are mostly water (70% or more); the dry mass consists of fibrous, soluble, and insoluble complex carbohydrates with a significant concentration of glucose and free amino acids that provide flavor that is hidden behind the heat of a chili. Although there are volatile oils and other fats that also contribute to the flavor and aroma of the fruit, most of the characteristic colors, flavors, and aromas come from molecules called the carotenoids and capsaicinoids.

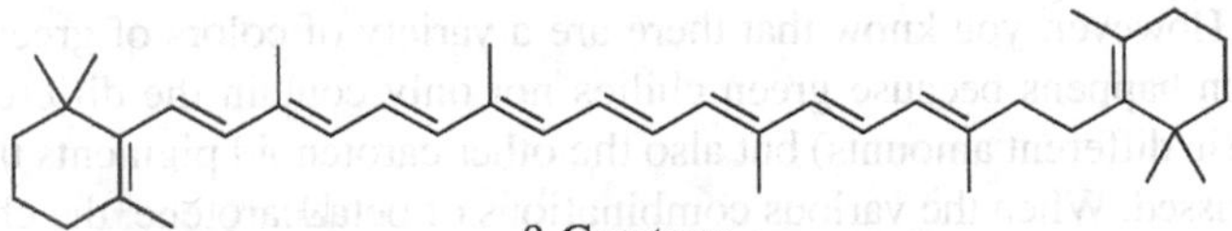

β-Carotene

FIGURE 11.23 Beta☒carotene. The compound responsible for many of the colors in plants and vegetables. Metabolism of this compound generates many different colors in vegetables and fruits.

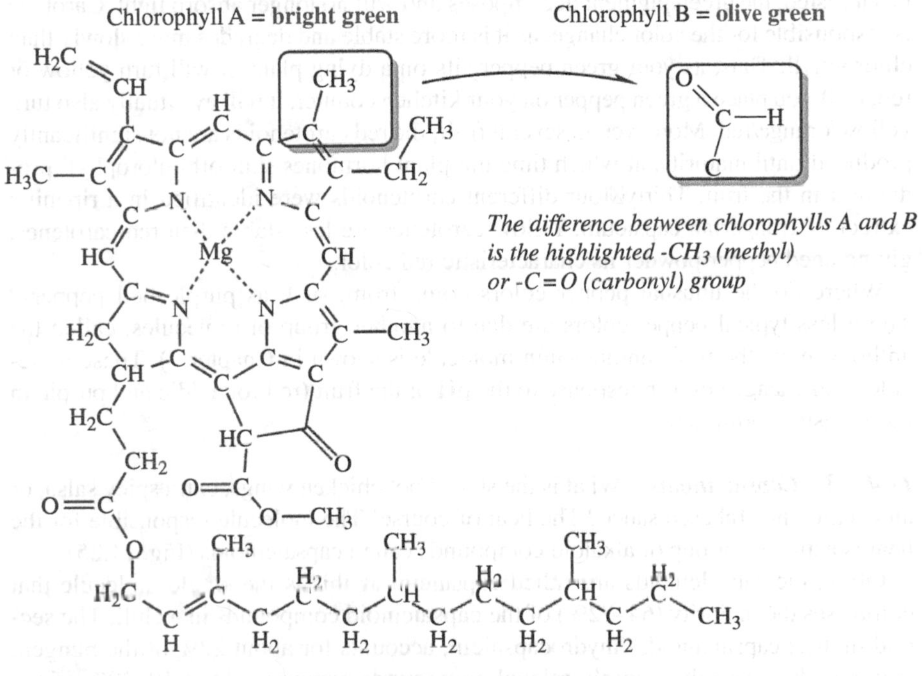

Chlorophyll A = **bright green**   Chlorophyll B = **olive green**

The difference between chlorophylls A and B is the highlighted -CH₃ (methyl) or -C=O (carbonyl) group

FIGURE 11.24 Chlorophyll.

11.4.7.2 Color Have you ever seen or tasted an orange bell pepper? The varied colors of raw and powdered peppers are highly valued for the aesthetic component that these vivid colors bring to a food. The pigments that give plants and select microorganisms yellow and red colors are the isoprene and phenol☒based compounds called the carotenoids. There are over 20 different carotenoids in the fruit of chilies.

Beta☒carotene is responsible for much of the yellow☒orange color peppers, while the red color of cayenne, red bells, and even some red spices like paprika comes from the less common carotenoids, capsanthin and capsorubin (Fig. 11.23).

What about green peppers and plants? As you may recall from high school biology, chlorophyll gives plants their green color. While there are several different types of chlorophyll (a, b, c, d) with slight structural differences, all of the forms have the four☒ ringed structure shown in Figure 11.24. This structure causes chlorophyll to absorb blue and red light. Thus, chlorophyll reflects green light, which is the light and color that we see. However, you know that there are a variety of colors of green in chilies. This variation happens because green chilies not only contain the different types of chlorophyll (in different amounts) but also the other carotenoid pigments that we have already discussed. When the various combinations of beta⊠carotene, the chlorophylls, and other carotenoids come together, red, blue⊠green, and blue lights are all absorbed, leaving a spectrum of green light to be reflected from different types of chili fruits.

Have you ever seen a pepper fruit change color while on the plant in your garden? Chlorophyll is an unstable compound, but when the plant stops producing it or if fruit is harvested, the green pigment decomposes and will no longer absorb light. Carotene is responsible for the color change, as it is more stable and degrades more slowly than chlorophyll. Thus, as your green pepper sits on a dying plant, it will turn yellow or red, or if you place a green pepper on your kitchen counter, it will eventually also turn yellow/orange/red. Moreover, in several fruit, the red carotenoids are not significantly produced until maturity, at which time the plant hormones shut off chlorophyll production in the fruit. Thirty⊠four different carotenoids were identified in a ripening extract of Hungarian capsicum. Yellow carotenes are less stable than red carotenes, giving aged pepper powder its characteristic red color.

Where do the unusual pepper colors come from, such as purple bell peppers? These less typical pepper colors are due to another group of molecules, called the anthocyanins (the basic anthocyanin molecule is shown in Chapter 7). These molecules can change color in response to the pH of the fruit (red for acidic and purple in more basic conditions).

11.4.7.3 Capsaicinoids

What is the star of hot chicken wings, extra spicy salsa, or an eye⊠tearing Tabasco sauce? The heat of course! The molecule responsible for the heat is actually a group of alkaloid compounds called capsaicinoids (Fig. 11.25).

Often, the capsaicinoids are called capsaicin, as this is the single molecule that comprises the majority (64–72%) of the capsaicinoid compounds in a chili. The second highest capsaicinoid, dihydrocapsaicin, accounts for about 22% of the pungent compounds. Five other closely related compounds comprise about 10–20% of the capsaicinoids. A close inspection of the compounds shows that they are similar in molecular structure to other flavorants, such as piperine, gingerol, and vanillin, particularly in the ring component of the molecules. In fact, capsaicinoids are part of a group of molecules called vanilloids (e.g., gingerol, vanillin, and capsaicin) due to the presence of the phenol group. However, the carbon "tails" provide great diversity in the signaling of each compound to our brain; the longer, more hydrophobic tail of capsaicin is chemically very different from the OH and straight chain found in gingerol. Because of these structural differences in particular areas of the molecules, each compound binds to entirely different receptors and sends very different signals to our sensory system.

Perhaps you have noticed a difference in taste when you include (or don't include) jalapeno seeds in your homemade salsa. Many people associate the heat of a chili with the seeds. However, capsaicin is produced in the placenta of the fruit. Thus, while some of the compound finds its way to the thin delicate seeds, the seeds are not the source of capsaicin, nor do they make positive contributions to taste, as they have

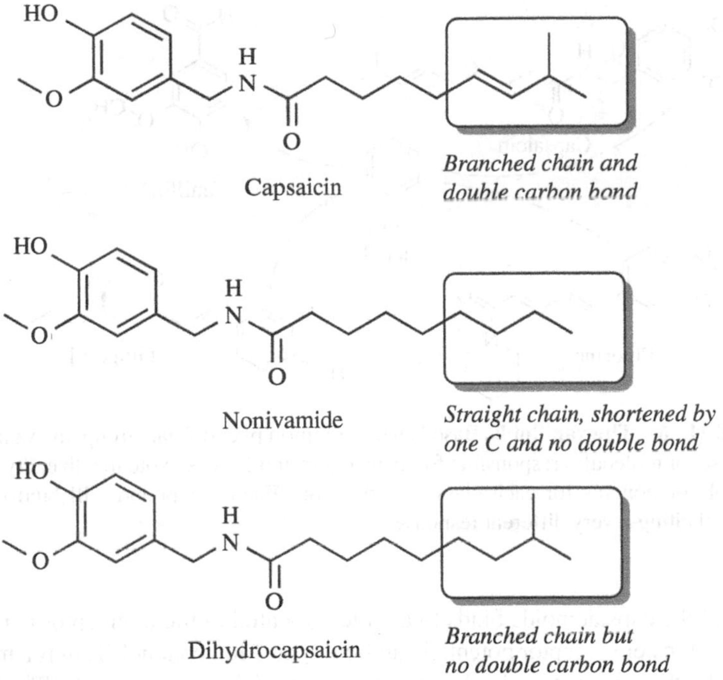

FIGURE 11.25 Capsaicinoids. Capsaicin is the parent compound of the capsaicinoids. Carefully inspect the differences in the tails; the remaining structures are all identical.

membranes with various tannins and polyphenols and are often bitter tasting. Thus, the best way to appreciate the heat, retain the flavor, and reduce the less appealing, bitter overtones of a chili is to remove the seeds and retain the rest of the chili in your dish. Perhaps you have also noticed that the same type of chili (like a jalapeno) may have different levels of heat. The more stress from heat or dry conditions during the growing process increases the capsaicinoid level in the fruit. The fruit also increases the production and secretion of capsaicinoids as the color begins to change and the fruit begins to wither (Fig. 11.26).

When you bite into a chili, the sensory impact is obvious: pain. However, there is little to no real biological damage that occurs when you eat chilies. Capsaicin (and related compounds) interacts and binds to a pain‑sensitive nerve, not a taste receptor! More precisely, capsaicin compounds bind to a class of protein receptors located on the surface of specialized sensory nerves called nociceptors. Nociceptors, found on the skin and mucous membranes, are typically activated by a noxious stimulus that provides a damage signal to the body. Damaged tissue associated with trauma, heat, or chemicals leads to the activation of nociceptor nerve fibers, which provides a signal to the person experiencing the event to withdraw, stop, or get away from the event. Thus, capsaicin does not cause damage; it just sends a signal that pain or heat is being experienced. Let's take a closer look at the biology of this process.

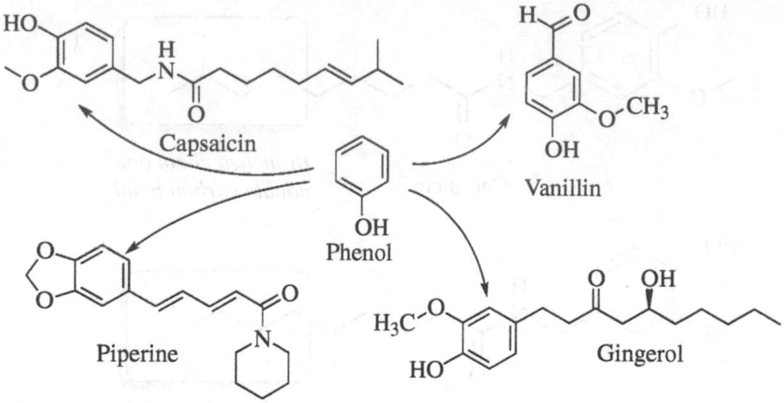

FIGURE 11.26 The vanilloids. Based on the common phenol head group, the vanilloids are a diverse set of molecules responsible for many odors and flavors. Note the diversity in tails or nonphenol components for each class of compound. Each compound will bind to distinct receptors eliciting a very different response.

Each of the capsaicinoids binds to a protein located in the nociceptor nerve fibers called the transient receptor potential vanilloid type 1 ion channel. That is a mouthful; fortunately, the receptor is also known as the capsaicin receptor or TRPV1. TRPV1 responds to a range of different signals including capsaicin, the pungent compound allyl isothiocyanate, heat above 109°F/43°C, acid conditions below pH 5.3, and even electrical current. At higher concentrations gingerol binds to the TRPV1 receptor, providing the pungency for aged ginger powder. Once stimulated, TRPV1 activates dozens of proteins within the cells and opens the membrane for calcium and other ions to enter into the nerve cell. The result is a slew of biological signals that propagate the activation of pain nerve fibers. As long as capsaicin binds its receptor, the pain signal is sent, and capsaicin binds the TPRV1 receptor very well, with an affinity of less than 700 nM. What does this mean? Humans have the capacity to detect the powerful stimulant in solutions of 10 parts per million, that is, about 3.5 teaspoons of the drug dissolved in an Olympic⊠sized swimming pool. There is plenty of capsaicin in most peppers to be noticed (Fig. 11.27)!

It is perplexing then why, despite the pain and discomfort, so many people find pleasure in eating chilies and, often, the hotter the chili, the better! As you can surmise, the more capsaicin present in the fruit, the more likely the TPRV1 receptor will bind and send its signal of burning pain to the brain. However, even after eating a hot chili, most individuals will eventually want to stop the pain and limit spreading the pain! The key solution to this problem lies in the structure of capsaicin. The compound is fat soluble and not water soluble, so drinking water or soda will only make things worse. In fact, water will spread the hydrophobic oil⊠like capsaicin throughout the mouth to mix and coat other membranes, which will set off more nerve fibers! Soda may make things even worse as the acidic drink seems to synergize the pain signal. The solution is to remove the unbound molecule from the system and to tease the capsaicin from its

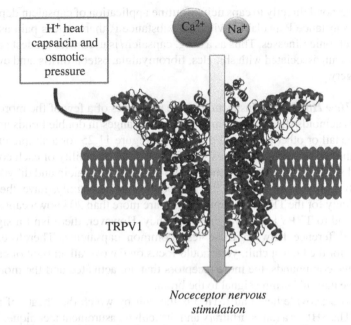

Noceceptor nervous stimulation

FIGURE 11.27 Capsaicin activation of pain nerves. Several signaling cues can activate the membrane receptor TRPV1 including capsaicin. This results in an influx of calcium and sodium ions and sets off pain signals to the CNS.

receptor with a good glass of fat⬚containing milk or oily food that you spit out. Swallowing the now⬚capsaicin⬚soaked milk would only help deliver the compound to the rest of your digestive track, providing you with fun for a long time.

Another piece of advice is don't touch your eyes or any other mucous membrane tissue if you've been cutting chili with your bare hands! Because the capsaicinoids are hydrophobic and primarily nonpolar, the molecules will dissolve in fat. On the skin, a rub in butter will help solubilize the compound into the lipid where it can be washed away.

Do you know anyone who doesn't seem to be bothered by spicy, hot foods? This tolerance for hot food is easily explained at the TRPV1 receptor level. If the receptor is activated for long periods of time or repeatedly stimulated due to the presence of capsaicinoids, enzymes modify TPRV1, shutting down the nerve's ability to send signals to the brain. This property, called desensitization, has been used therapeutically for a long time. Ancient Mayans used capsaicin (chilies) to treat sore throats, while the Aztecs used them to treat toothache. The desensitization caused by capsaicin continues to be used by the pharmaceutical industry. Capsaicin is topically used to alleviate pain and is a key ingredient in the creams that are used as treatment for shingles, muscle aches, and pain patches; it is even sold to reduce the pain of arthritis. A second mode of action utilized by capsaicin to relieve pain is through a peptide called substance P. Substance P is associated with several chronic pains including arthritis; it is responsible for some of the pain signals of different nociceptor nerve fibers that respond directly to capsaicin. Routine application of capsaicin depletes the amount of substance P available; with less substance P, there is less pain associated with some chronic illnesses. Thus as a drug, capsaicin is used with modest success in treating the pain associated with shingles, fibromyalgia, osteoarthritis, and even post-cancer surgery.

11.4.7.4 How Hot Is It? Inspection of the structures of a few of the more than 20 known capsaicinoids shows the seemingly minor changes in double bonds and orientation of the tail or other functional groups (see Figure 11.25 for a simple example). However, these structural differences significantly alter the ability of each compound to bind and send its signal through the TRPV1 receptor. Capsaicin and dihydrocapsaicin, which represent 80–90% of the compounds in the family, have the highest binding affinity for the TRPV1 receptor. There are more than 20 known capsaicinoids that each bind to TRPV1 with varying efficiency. However, there isn't a significant detectable difference between these less common capsaicins. Therefore, if you are looking for the hottest chili, you should focus on the overall amount of capsaicin. The more the compounds, the more receptors that are activated and the more nerves that send the painful burning signal to the brain.

The famous Scoville heat unit (SHU) is the unit by which the "heat" of a chili is measured. The SHU is a rather arbitrary and difficult measurement technique. How did a semiquantifiable scale come about? Parke￼Davis pharmaceutical chemist Dr. Wilbur Scoville was following up on some initial studies performed by a Hungarian physician, Hőgyes. In 1878, Hőgyes had noted that ingestion of capsaicin created a sharp sensation on his tongue and warmth in his gut, followed by belching and flatulence. While Hőgyes' work eventually led to the pharmacological use for the drug, Dr. Scoville was searching for a way to measure the "hot" compound in ground peppers. Why was it necessary? Well, chilies vary significantly in heat from plant to plant or season to season. Thus, in 1912, Dr. Scoville published his one￼and￼a￼half￼page paper "Note on Capsicums," in which he described the process to measure "heat" to further the drug discovery potential for capsaicin [11].

In this historical method, Scoville ground an exact amount of dried chili (a grain or about 64 mg of fruit) in 100 ml of ethanol to dissolve the capsaicin. He then diluted the extract with sugar water. The dilution in which the pungency was no longer detected (by the human tongue) was the measurement of the fruit's capsaicin strength. The heat scale was based on this dilution and later came to be known as the Scoville heat index. However, this taste￼based (organoleptic) test method is problematic, as tasters will wear out, desensitizing their receptors. Thus another more modern method, which utilizes high￼performance liquid chromatography (HPLC), has been devised. Similar to the procedure described by Scoville, a mass of chili is ground in 95% ethanol with sodium acetate. The fluid is then run through fine, silt￼like beads in a steel column that binds and releases the compounds present in the extract differently. Some compounds do not bind with the beads and flow through the column quickly, while other compounds bind tightly and are only released after a long time, or anything in between. This process, called chromatography, separates the compounds present in a complex mixture since each compound flows through the column beads at a different rate. How are these compounds measured or detected? Each compound can be measured by its absorbance of light as the solution flows through a light detector: the larger the absorbance peak, the greater the amount of compound present. Based upon the structure of the compound, the properties of the column beads, and the properties of the fluid (or solvent) that is running through the system, a scientist can predict which peak corresponds to which molecule. Now, scientists can carefully determine the heat of a chili using this standardized technique. SHU (from Scoville's historical method) are still used to measure the heat of chilies. Using the HPLC method, the parts of capsaicinoid present per million of solvent (ppm) is multiplied by 15 to obtain SHU.

For years, the hottest pepper crown was given to the bhut jolokia from India. This pepper was commonly called the "ghost bhut" because of how the pain sneaks up on the consumer or it makes you feel as you have given up the ghost. However, the pepper lost its standing in the Guinness Book of World Records in 2012 to the Trinidad moruga scorpion pepper. Dr. Bosland from the New Mexico State University Chile Pepper Institute using a very standardized approach to testing a batch of plants found the bhut jolokia to have an SHU rating of 1.02 million, with the hottest plant possessing a rating of 1.58 million SHU. The C. chinense variety Trinidad averaged 1.2 million SHU; the hottest individual pepper was found to have just over 2.0 million SHU. To ensure the uniqueness of this pepper, the scientists measured several genes with unique SNPs for each. This showed the C. frutescens bhut was not a variety of the Trinidad pepper. The top spot in Guinness was taken over by the Carolina Reaper 1 year later. Measured in an undergraduate analytical chemistry laboratory at Winthrop University, the average pepper was 1.57 million SHU, while the top hot pepper produced a mouth screaming 2.2 million SHU. To give you some perspective on this heat, pure capsaicin has a rating of 16 million SHU. For better perspective the Carolina Reaper is 440 times hotter than an average jalapeno pepper. Those plugged into the pepper world have heard rumors of new variants that are even hotter being developed and tested. It will be interesting to see how much capsaicin a single pepper fruit can produce!

11.4.7.5 Why? Chemical Warfare!

The idea of eating a chili with two or more million SHU is beyond most people; however the curious (but not crazy) might ask why a plant would produce such a noxious compound? Every molecule that is produced by a cell requires energy. Thus, if, over time, an organism retains the gene that allows it to produce a molecule, there is usually a biological rationale behind it. There are two biological reasons why chili plants produce capsaicin: chemical defense and seed dispersal.

Plants do not have an immune system, nor can they move if attacked or threatened. To defend themselves against being eaten or killed by pathogens, plants have developed an amazing assortment of defense mechanisms to detect and stop invading organisms and animals. The bitter taste of some plants acts as a deterrent to animals biting and eating the plant. Damage caused by insects opens a wound that allows for microbes to enter, damage, and kill the plant. Thus in response to insect or animal damage, plants release volatile repellant compounds such as terpenes and polyphenols to deter further attacks. In other words, the same volatile oils that provide the flavor and aroma of spices act as insect repellants. In addition to repelling further damage, breaking the cell wall in some plants releases enzymes, such as the polyphenol peroxidases (discussed in Chapter 6) that chemically transform the lignin carbohydrates in the cell wall. The modified lignins act as a scar and patch the wounded area, slowing the growth of microbial pathogens already infecting the plant.

Capsaicins also play a very different role in chili plant survival. Most plants need to disperse seeds away from the parent plant to avoid the new plants competing with the parent for sunlight and nutrients. One method that plants use to achieve this is called zoochory. Zoochory occurs when seeds are dispersed through the gut of animals. In fact, zoochory is why the flesh (fruit) surrounding seeds is attractive in color, flavor, and aroma. The fruit acts as an attractant for an animal to eat and spread the seed through its feces. However, this method of seed dispersal is problematic for chilies because the seeds of most chilies are more fragile than seeds found in other fruits. Yet, the chili fruit has considerable sugar content and an attractive color. Let's consider this problem with an example. Pear, orange, and apple seeds have a tough coating. A herbivore with flat grinding teeth would not destroy an apple seed, but a chili seed would be seriously damaged. The conundrum is to attract the right animal to eat and disperse the seed and deter those that would damage the seed and block propagation. Fortunately, there was an evolutionary change that generated a molecule (capsaicin) that would specifically bind to a receptor in animals that warns for burning pain, deterring the animals from eating the fruit and destroying the seeds. Birds, however, that do not have the grinding molars of a mammal have TRPV1 receptors, but capsaicin does not bind to these genetically distant receptors. Thus birds, with a more delicate way to pick and swallow the seed without injury, can spread and disperse the seeds after traveling though the gut.

An interesting study on this problem tested the consumption of fruit and chilies by mice, rats, and birds. All three animals ate the hackberry fruit, while neither the cactus mouse nor the packrat consumed the capsicum fruit. Yet, thrashers (bird) readily ate the hot pepper fruit. Moreover, the seeds eaten by the birds were passed and germinated the same as nonconsumed seeds. However, the rodents did not pass intact seeds, and no plants grew from the nonintact seeds that passed through either the rat or the mouse [12]. It was also found that the birds spread the seeds under its tree where shade helped the plant to grow. This study illustrates that capsaicin is an effective deterrent of ineffective seed dispersers, which increases the evolutionary fitness of the plant.

Another set of studies by the same scientist (Dr. Joshua Tewksbury, University of Washington) highlights a second chemical defensive and evolutionary reason for the capsaicinoids [13]. Recall that the placenta produces capsaicin and some of the compound coats seeds attached to the placenta. Several pathogenic microbes, including fungus, are able to grow inside the pods of punctured or damaged chilies. These fungal plant pathogens produce several compounds that are toxic to the plant host. Since the bite of a foraging insect will leave a wound within which a fungus can grow and bring about disease to the plant, if capsaicin could either detract insects from biting or fungus from growing, the compound would then serve a second chemical plant warfare purpose.

Looking at plants in Bolivia, Argentina, and Paraguay, Dr. Tewksbury and his colleagues found an interesting directional gradient of capsaicinoid⊠producing plants. There were more plants in the southwest than northwest by more than two thirds. Interestingly, the number of insect bites in both groups of plants was the same, indicating that capsaicin was not a deterrent to the insects. However, plants that contained capsaicinoids were half as likely to be infected and rotting due to fungus when compared with those plants with less capsaicin. This study suggests that capsaicin protects seeds from the pathogenic fungus.

Thus, if you want to know "why heat?," the answer is clearly chemical defense. Capsaicin stops the molared mammals from eating and grinding the seeds into oblivion, and the "heat" slows fungal infection, thus sparing seeds from rot. With some humans, however, this evolutionary protection mechanism seems to be lost. We can postulate that we have adapted to the spice as it inhibits fungus on our food, but the organoleptic pleasure of spices is lost when eating a habanero. In fact, one⊠third of humans seek out the pain of eating chili peppers each day. This is nothing new as ancient civilizations in the Old and New World included peppers in food and drink. Clearly our innate aversion to a key biology response to pain has somehow been short⊠circuited. Psychologists have attempted to understand the reason behind the phenomenon of chili consumption. Two hypotheses have been proffered. The first is the social and learned behavior of cultures that consume peppers. Simply put, Mom, Dad, and the relatives eat it and the kids learn to like it too. The second is a "benign masochism." Like the thrill of danger and our fondness for roller coasters, chili heat presents a unique human activity, similar to a trip through a haunted house or a scary amusement park ride. Whatever hypothesis is true for the chili lovers of the world, these pepper heads will continue to fight to claim to have eaten the hottest chili, despite the obvious reasons for avoiding it.

11.5 MEDICAL USES OF HERBS AND SPICES

There are plenty of nonculinary uses of herbs and spices listed in social media outlets, on the internet, or in your great, great grandmother's book of wisdom. Historically, herbs and spices had a distinct antimicrobial activity that kept food from spoiling, which was critical when refrigeration space was nonexistent or at a premium. In addition to fighting microbial growth, several herbs and spices have antioxidant activity, which preserves the food (and the person ingesting the food). There are plenty of medical uses of some compounds to decrease blood pressure, fight cancer, and reduce depression. Let's learn more examples of these medical uses in the following; you may have some medicine in your spice rack without having known it!

11.5.1 Antimicrobial Activity

Prior to refrigeration, people who lived in warm, tropical climates struggled to maintain safe stores of meat and vegetables, much more than those in the cooler climates. Taking advantage of the natural resources growing in these tropical areas, the herbs and spices added to the foods during curing and cooking provided more than an organoleptic purpose. They were added to ensure that the food was safe to eat! In fact, nearly half of the recipes for meat in tropical climates included spices, while less than 5% of those dishes from cooler climates called for the spice. Does this explain the flavoring of Midwestern foods versus those in the southwest corners of the United States? Maybe! Very low amounts (0.05–0.1%) of some herbs and spices inhibit the microbes that cause salmonella, cholera, and food poisoning. Specifically, linalool, eugenol, myristicin, cinnamic aldehyde, allyl isothiocyanate, and vanillin inhibit 50–75% of bacterial growth as compared with nontreated control samples! While the exact mechanism by which the compounds in herbs and spices inhibit bacterial growth is unknown, some evidence points to interference with the bacterial membrane continuity or inactivation of DNA and RNA needed for cell growth and division.

In addition to the bioactive compounds in herbs and spices already discussed, there is a wealth of scientific literature that focuses on the role of herbs and spices in cancer. Some compounds that cause or promote the formation of cancer tumors must first be chemically altered by enzymes in the liver, such as the P450 enzyme system. Compounds in pepper, rosemary, turmeric, and cinnamon stop this process by inhibiting several of the P450 enzymes, thus potentially preventing an accumulation of cancer causing material. Once formed, tumors begin with uncontrolled growth and spread throughout the body in a process called metastasis. Turmeric decreases the signal from the tumor cell that induces its growth. Curcumin and capsaicin inhibit several activation signals to the tumor cell and inhibit the proteins that support cancer cell metastasis. Many herbs and spices also have components that may compete with estrogen for its receptor in human breast cancer cells. Thus, some herbs and spices may reduce the incidence of or progression of estrogen dependent breast cancers.

Although all of these important effects that are imparted by herbs and spices on human health are interesting, they must be treated with a dose of cautious skepticism. Herbs and spices are familiar and considered "natural"; thus many are drawn to use a herbal treatment to achieve a positive health outcome to their ailment. Unfortunately to date, few clinical applications for herbs and spices have been found. Many studies have been performed in culture dishes of bacterial or animal cells, and some work in animal models such as mice or rats show promise; however the transfer of the cell or animal model effect into humans has largely failed. One example is a study that found that cinnamon extracts suppressed the growth of a bacterium (*Helicobacter pylori*) that is the major cause of stomach cancer and some esophageal cancers. Unfortunately, patients with the bacteria who were given high doses of cinnamon extract did not experience an altered bacterial growth or cancer outcome. However, not all of the studies have yielded disappointing results. The curcumin compound found in turmeric blocks inflammation and other enzymes involved in cancer. Early clinical (phases I and II) trials suggest there is an anticancer activity to the compound that has successfully translated to the patient.

Another danger is the thinking that "natural" herbs and spices automatically make them safe. We have already seen examples of the strong side effects of nutmeg and the potential health hazard of high doses of cassia cinnamon. There are many other stories that could fill a book about both the positive and the negative role of herbs and spices. Care and caution with healthy skepticism are always a good spice to include in your decision☒making process.

REFERENCES

[1] Drummond, J.C. and Wilbraham, A. (1939) *The Englishman's Food: A History of Five Centuries of English Diet.* J. Cape, London.

[2] Woodman, M. (1978) *Food and Cooking in Roman Britain.* Corinium Museum, Cirencester.

[3] McGee, H. ed. (2004) *On Food and Cooking.* Simon and Schuster, Inc., New York.

[4] He, F.J., Li, J. and Macgregor, G.A. (2013) Effect of longer term modest salt reduction on blood pressure: Cochrane systematic review and meta☒analysis of randomised trials. *BMJ*, 346: f1325.

[5] Stolarz☒Skrzpek, K., et al. (2011) Fatal and nonfatal outcomes, incidence of hypertension, and blood pressure changes in relation to urinary sodium excretion. *JAMA*, 305: 1777–1785.

[6] Vallverdú☒Queralt, A., Regueiro, J., Martínez☒Huélamo, M., Rinaldi Alvarenga, J.F., Leal, L.N. and Lamuela☒Raventos, R.M. (2014) A comprehensive study on the phenolic profile of widely used culinary herbs and spices: rosemary, thyme, oregano, cinnamon, cumin and bay. *Food Chem.* 154: 299–307.

[7] Potter, T.L. Fagerson, I.S. (1990) Composition of coriander leaf volatiles. *J. Agric. Food Chem.* 38(11): 2054–2056.

[8] McGee, H. (2010) Cilantro Haters, It's Not Your Fault. *The New York Times.* Available at http://www.nytimes.com/2010/04/14/dining/14curious.html (accessed on November 16, 2015).

[9] Mauer, L. and El☒Sohemy, A. (2012) Prevalence of cilantro (*Coriandrum sativum*) disliking among different ethnocultural groups. *Flavour* 1: 8.

[10] Nicholas, E., Shirley, W., Chuong, B.D., Amy, K.K., Joyce, Y.T., Joanna, L.M., David, A.H. and Uta, F. (2012) A genetic variant near olfactory receptor genes influences cilantro preference. *Flavour* 1: 22.

[11] Scoville, W.L. (1912) Note on capsicums *J. Am. Pharm. Assoc.*, 1(5): 453–454.

[12] Tewksbury, J. and Nabhan, G. (2001) Seed dispersal: directed deterrence by capsaicin in chillies. *Nature* 412: 403–402.

[13] Tewksbury, J., Reagan, K., Machnicki, N., Carlo, R.T., Haak, D. and Penanaloza, A. (2008) Evolutionary ecology of pungency in wild chilies. *PNAS* 105: 11808–11811.

12

BEER AND WINE

Guided Inquiry Activities (Web): 3, Mixtures and States of Matter; 4, Water; 7, Carbohydrates; 14, Cells and Metabolism; 15, Metabolism, Enzyme, and Cofactors; 29, Alcohol and Beer Brewing; 30, Beer and Wine

12.1   INTRODUCTION

Beer, wine, whiskey, gin, sake, and tequila: at the heart of each of these and many more liquors is a basic process that shares its science and technology with baking bread, making cheese, and other microbiology sciences. The process has been refined for over 7000 years. Sumerians wrote poems of the effects of wine on cranky teenage princesses, chemists have found trace molecules from beer in ancient Chinese containers, and the Code of Hammurabi included punishments for overcharging tavern customers for drink [1]! Someone, somewhere first found some wild yeast growing in liquid with a sugar or starch source and convinced someone to try it, and an alcoholic drink was born (Fig. 12.1).

For any alcoholic beverage, the basics are simple: water, yeast, a source of sugar, and time. This is alcoholic fermentation. Depending on the conditions (anaerobic and the type of yeast), carbohydrates are primarily metabolized to ethanol and carbon dioxide. While bakers use the carbon dioxide to give rise to their dough, the ethanol is the prized final compound produced by the microbiological factories for alcoholic drinks. The starting source of sugar demands a method to harvest the simple sugars (fruit and grapes) or to convert the complex carbohydrates (starches) from seeds and cereals into simple sugars using a method called malting and mashing. Fermented beverages including wine, beer, sake, cider, and mead involve minimal postfermentation processing

The Science of Cooking: Understanding the Biology and Chemistry Behind Food and Cooking, First Edition. Joseph J. Provost, Keri L. Colabroy, Brenda S. Kelly, and Mark A. Wallert.
© 2016 John Wiley & Sons, Inc. Published 2016 by John Wiley & Sons, Inc.
Companion website: www.wiley.com/go/provost/science_of_cooking

FIGURE 12.1 Sugar and fermentation are common for wine, beer, and distilled spirits.

TABLE 12.1 Alcoholic Beverages Sugar Sources and Processing.

Alcohol	Raw Sugar Starting Material	Additional Processing
Wine	Grapes or fruit	Aging for oxygen and tannin reaction
Beer	Barley, wheat, rice, corn	Added hops and adjuncts for flavor and minimal aging
Mead	Honey, some add fruit or spices	Solids are settled by gravity
Cider	Primarily apples, some other fruits	Pectin is removed by precipitation and solids settled
Sake	Polished white rice	Molds digest starches for yeast, additional alcohol added, solids filtered
Vodka	Potatoes, grains (wheat, rye), fruit	Distillation and rectification (repeated distillation for high alcohol content)
Tequila	Agave cactus, sugars, pineapple	Distilled. Silver—bottled after distillation, Anejo/Reposado—aged in barrels
Rum	Sugarcane products, juice, and molasses	Distilled and aged in oak casks for color and flavor
Whisky	Barley, corn, rye, wheat	Distilled using copper to remove sulfur and aged in oak barrels or casks

and are not enriched in their alcoholic content. Liquor, hard liquor, spirits, or more formally distilled spirits begin with the same basic principle of fermented beverages, that is, a sugar source and yeast. As per the name "distilled spirits," the fermented liquid is enriched in its ethanol content by distillation. Alcoholic beverages owe their flavor and color to the starting compounds, the strain of yeast, and how the fermented mother liquor is processed. Some of the beverages are aged for more complex flavors and others are bottled for immediate consumption (Table 12.1).

12.2   YEAST: METABOLIC ETHANOL-PRODUCING FACTORY

Historic fermentation required some luck to produce alcohol from fruits and grains. Early winemakers unknowingly added yeast from the environment already growing on the grapes and stems or from the feet when grapes were stomped to soften the fruit. Ancient beer production relied on wild yeast blown in from the air or from the barley

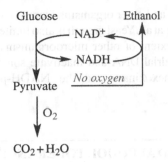

FIGURE 12.2   The presence of oxygen determines the product of yeast metabolism.

(and in some cases bread) to start the process. Yeast is a large family of single-celled eukaryotic fungus. Yeasts exist as individual single cells or large strands of individual cells, a distinguishing characteristic used to make lager or ale beer. There are thousands of species of yeast, some of which produce toxic compounds for humans. The species of yeast in the *Saccharomyces* genus (above species but below family in the biological classification system) have been grown (cultured) and domesticated for its ability to metabolize glucose. Yeasts are facultative anaerobic organisms, which means they can convert food to energy compounds in the presence of oxygen (respiration) or in the absence of oxygen (fermentation).

Oxygen availability is the factor that determines if yeast will convert sugars like glucose to produce CO_2 and H_2O or ethanol and NAD^+ (Fig. 12.2). Fermentation is the anaerobic metabolism of glucose to ethanol, while the simple traditional definition of cellular respiration (or just respiration) is aerobic metabolism of glucose to its most chemically reduced components, CO_2 and H_2O. Fermentation produces fewer high-energy ATP molecules needed for cell division and growth when compared with respiration. Biochemically, the metabolism of glucose is a series of individual metabolic pathways. The conversion of glucose to pyruvate, lactate, or ethanol is considered glycolysis. Further metabolic oxidation of pyruvate to CO_2 and H_2O in the presence of oxygen involves the Krebs cycle, also called the tricarboxylic acid cycle. In the beginning stages of both beer and wine production, oxygen is a critical component to allow respiration so yeast can grow to a high density. The addition of oxygen allows for more ATP to be produced, resulting in a high rate of growth but little ethanol produced. To produce enough ethanol to satisfy consumption of fermented beverages, there must be enough yeast organisms to do the job. The density of yeast cells in the start of fermentation is much too low to make more than a percent or so of alcohol; many more yeast cells are necessary. Thus, the beginning of fermentation is giving yeast food and oxygen to divide. See Chapter 3 for details and review.

Once there are sufficient yeast cells to produce alcohol, sealing the fermentation from further oxygen will shift the cells to produce ethanol. Very little cell division takes place at this stage, as the total ATP production is decreased. Ethanol production offers the cell a chance to convert NADH back to NAD^+ and serves a second purpose. Ethanol serves as a poison to other microorganisms. Three to five percent of ethanol inhibits a range of bacteria and other organisms from growing; higher concentrations will stop most from growing at all. Yeast growth is also limited by ethanol concentration, although not to near the extent of other microorganisms. Higher levels of ethanol will damage the mitochondrial DNA and inactivate some of the enzymes involved in glycolysis including hexokinase and the NADH-producing dehydrogenases (Box 12.1).

BOX 12.1   YEAST AND ALCOHOL TOLERANCE

Most yeast in the *Saccharomyces* genus can tolerate up to 10–20% ethanol by volume before growth is inhibited depending on the specific strain used. That is why most wines or even strong beers are limited in alcohol content. However, scientists have been working to generate a new strain of yeast or ways to help yeast tolerate higher alcohol concentrations and temperature for higher biofuel (ethanol) yield/production [2]. At levels above approximately 20% ethanol, the membranes of yeast cells become porous and the cells die. The new strain of yeast produces a steroid that strengthens the membrane so it can withstand the effects of ethanol production. Another set of scientists found that adding potassium salts helps membrane pumps compensate for the influence of alcohol [3]. While this was done using yeast strains specific for biofuel, it is interesting that changing the culture conditions of the yeast improved the ethanol yield by almost twice compared with/without. What we don't know is how these additions will alter the flavor of fermented beverages or if yeasts growing in such conditions will also generate an additional set of compounds that may impact the final flavor of beer, wine, or distilled spirits.

For those who want more than the ethanol concentration found in beer or wine, further concentration of the alcohol is required. This is most commonly achieved by removing water through distillation. Thus, the total starting amount of simple sugar and the mass of yeast when switching to anaerobic fermentation as well as strain selection each significantly contribute to the final concentration of ethanol in beer or wine. Of course, stopping a fermentation early will result in residual sugars giving a sweet taste, while allowing the fermentation to run to completion (sugar depletion) provides for a dry (less sweet) but higher alcohol content beverage.

In addition to ethanol, additional higher molecular weight alcohols called fusel alcohols are produced, although at low levels. Examples include butyl alcohol, isoamyl alcohol (isopentanol), and isopropyl alcohol (Fig. 12.3). Several of these fusel alcohols provide off-flavors and are also poisons. The danger in amateur distillation is that these dangerous alcohols can be concentrated into the distillate (moonshine can hurt you!). These are produced if the fermentation temperature is too high, the pH is too low, and when there are not enough nutrients available for the yeast (principally nitrogen). Other compounds adding the complex flavor and aroma of fermented beverages depend on the chemical nature of the starting

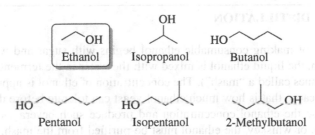

FIGURE 12.3   Fusel alcohols. A number of longer-chain alcohols are produced during fermentation. Ethanol (2 carbon) is the substance commonly called alcohol. All of these compounds are chemically defined as alcohols due to the functional —OH group.

material and the strain of yeast used. Other commonly produced molecules include acetaldehyde (green apple aroma), diacetyl (buttery butterscotch flavor), dimethyl sulfide (sweet corn aroma), sulfur (rotten eggs or burnt matches), and many of the phenolic compounds discussed in the spice chapter.

12.3   ETHANOL

The formal or chemical definition of alcohol is any carbon bound with a hydroxyl group (—OH). For food and cooking, the term alcohol takes on a more generic term for ethanol, also called ethyl alcohol and grain alcohol. Proof is a historic measure of the alcoholic content in a beverage. In the United States the proof of alcohol content is twice that of the percent (weight by volume) of ethanol in solution. A 50 proof alcoholic drink will by these standards be 25% (w/v). Historically, proofing alcohol was a measure used by sailors in the eighteenth century to test their rum rations to see if it was watered down or not. Daily rations of rum were given to sailors who would mix the rum with gunpowder. If the mixture of rum and gunpowder would ignite, the rum was "proofed." However, diluted rum would contain too much water for the gunpowder to catch fire and was "under proof." The relationship between proof and percent is because gunpowder will not burn unless the rum contained at least 57.15% ethanol and was considered to have "100 degrees of proof." Thus, the historic 7/4 times the alcohol by volume as proof is why we do not use exactly one-half percent. Eventually, proof was simplified to be twice that of the percent alcohol content.

Ethanol is a simple organic carbon compound with the hydroxyl group bound to a carbon. Other alcohols include methanol and propanol (1-propanol or 2-propanol; Fig. 12.6). Each of these alcohols have a hydroxyl group which provides a polar nature to the molecule making it a good solvent and why it will mix well with water. The carbon chain provides some hydrophilic nonpolar capacity to dissolve fat-soluble compounds when ethanol is at a higher concentration. Longer-chain alcohols like isopropanol have a higher solvent capacity for such molecules. This is why liquor is used to extract spices and flavor compounds for cooking (Box 12.2).

BOX 12.2 DISTILLATION

The process of making consumable ethanol begins with sugar and water. After fermentation, the liquid ethanol is mixed with the water in the fermentation mixture (sometimes called a "mash"). The concentration of ethanol is approximately 10–12%, because that is how much ethanol yeast can tolerate before they die. In order to raise the ethanol concentration and produce such beverages as vodka, tequila, rum, or whiskey, the ethanol must be purified from the mash. Heating a liquid gives some of the molecules enough energy to leave the liquid and *boil* into a gas or *vapor* (Fig. 12.4). The temperature needed to convert a pure sample of a liquid into a gas/vapor is called its *boiling point*. Water boils at 100°C/212°F and ethanol boils at 78.5°C/173°F, but if you heat a mixture of ethanol and water, it will boil somewhere *between* 78.5 and 100°C. The vapor of the ethanol–water mixture will contain *both* water molecules and ethanol molecules, *but* the vapor of the ethanol–water mixture is more concentrated in the *lower boiling component*— in this case, the ethanol. If we were to capture this ethanol–water vapor and

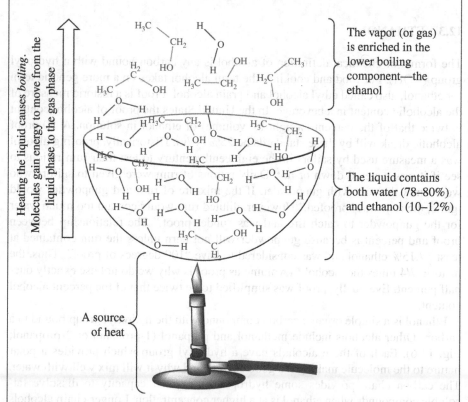

FIGURE 12.4 Ethanol evaporation. The physical change as an ethanol–water mixture is evaporated. This is the first part of distillation.

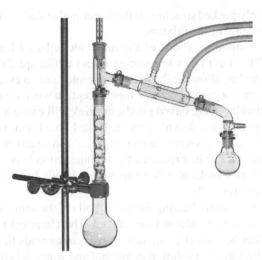

FIGURE 12.5 Fractional distillation apparatus. The mixture is heated where the vapor is cooled in a condensing tube, cooled with water.

condense it back into a liquid, the concentration of ethanol would be higher (Fig. 12.5 shows a common fractional distillation apparatus to distill liquid mixtures). This is the process of *distillation*—vaporization (boiling) and condensation (cooling the vapor back to a liquid). If we continue the process of vaporizing and condensing the ethanol–water mixture over many rounds of *distillation*, we will eventually get a liquid that is highly enriched in ethanol.

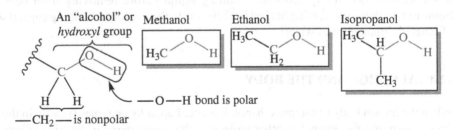

FIGURE 12.6 Alcohol is a polar molecule.

Many of these compounds consist of large nonpolar, hydrophobic regions and are poorly soluble in water. But the carbon chain of ethanol can dissolve many of these compounds for cooking and baking. The hydrogen bond between water and ethanol allows for a loss of volume when the two solutions are mixed together. In liquid water, individual molecules are forming and breaking hydrogen bonds nearly every $1 \times 10^{-5}\,\mu s$. This arrangement means liquid water is very inefficiently packed with an average of 2.3 hydrogen bonds per water molecule. However, ethanol creates a more closely packed structure of the water molecules, essentially shrinking the packing and volume of the mixture.

Because of the nonpolar carbons of the molecule, ethanol has a lower boiling point than water. That is, it takes less energy (heat) to disrupt the interactions between ethanol molecules, allowing the individual molecules to escape their intermolecular interactions into a gas. A common misconception when cooking with alcohol is that heating or flaming food, baked goods, or drink will cause all of the ethanol to "burn off" or evaporate. This doesn't happen. Baked bread retains some of the ethanol for hours even after cooking, something you can smell with freshly baked bread. What is going on can be explained by the interaction between water and ethanol. Water and ethanol molecules form hydrogen bonds between each other and alter the boiling point for both.

There also is a point where heating food will boil off the same amount of ethanol as water, leaving the percent alcohol unchanged. The chemical term for this is an azeotropic point, that is, a specific mixture of two compounds that can't be altered by heating or distillation. As a solution of alcohol and water is heated, ethanol, with its lower boiling point, will evaporate first at its lower boiling point. Thus, the ethanol content does reduce in contact during heating but will not completely "boil off." Part of the reason not all of the ethanol is removed during heating is explained by its azeotropic point. The azeotropic point is the mixture or ratio of ethanol and water where both molecules evaporate at the same rate during a simple distillation. Thus, at the azeotropic point, both some ethanol and water will evaporate at the same rate. The azeotropic point for ethanol and water is about 95/5%. Thus, simply boiling will result in a minimum of 5% of the water. The US Department of Agriculture reports that even after an hour of simmering, 25% of the initial ethanol remained in solution [4]. Eighty-five percent of the original ethanol was found when alcohol was simply added to a boiling liquid before removing from heat. Some of the ethanol remaining behind is likely due to the hydrogen bonding capacity of ethanol. Ethanol loves water!

12.4 ALCOHOL AND THE BODY

When the human body consumes ethanol, it is acted upon by enzymes (mostly in the liver) converting the ethanol to other molecules that cause many of the side effects we associate with alcohol consumption.

Acetaldehyde is the primary metabolite of ethanol. Figure 12.7 shows how both acetaldehyde and acetic acid are produced from ethanol. It is 30 times more toxic than ethanol primarily due to its ability to cross-link proteins. While ethanol is responsible for the feelings of drunkenness (see following text), it is the acetaldehyde that causes the "hangover" and the eventual liver damage. An individual's susceptibility to acetaldehyde toxicity is dependent upon the efficiency of the downstream enzyme, acetaldehyde dehydrogenase, and the availability of NAD^+. A less efficient acetaldehyde dehydrogenase means a greater accumulation of acetaldehyde in the body and greater toxicity.

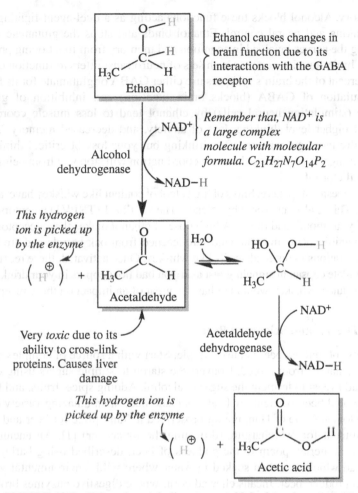

FIGURE 12.7 Ethanol metabolism to acetaldehyde and acetic acid.

Ethanol serves as a drug depressing brain function. γ-Aminobutyric acid (GABA) is a small organic molecule made in the nervous system. GABA is released by one type of nerve cell to inhibit other nerves from signaling. Drugs that mimic or increase GABA signaling are antianxiety relaxing compounds. This happens due to the inhibitory effect GABA has on excitatory-stimulating neurotransmitters. Ethanol increases GABA function by binding to the target for GABA and its receptor and enhances the receptor's ability to take up more GABA into the receiving nerve cell. Ethanol therefore increases the impact of GABA and causes an enhanced inhibition of the target brain function. Alcohol brings about less anxiety, lack of control (need stimulation for muscle control), and other cognitive abilities due to its role on GABA receptors.

The amino acid glutamate is also made and released by neural cells as a neurotransmitter. Unlike GABA, glutamate is an excitatory neurotransmitter. That means glutamate is involved in stimulating neural communication, higher-level thinking, and memory. Alcohol blocks these functions acting as a duel-agent fighting higher-order thinking and neural control. Ethanol binds and stops the glutamate receptor, preventing the nerves involved in cognition and memory from functioning properly.

Proper brain function requires both sets of neurotransmitters to function correctly. Ninety percent of the brain's synapses use either GABA or glutamate for its function. Overstimulation of GABA (blocks brain function) and inhibition of glutamate signaling (stimulating neural cells) by ethanol lead to less muscle coordination, decreased higher-level thinking, loss of anxiety, and decreased memory. You may think you're more courageous after drinking but your loss of critical thinking and memory along with diminished muscle coordination will stop you from being 10 foot tall and bulletproof.

The fun doesn't stop there. Shots of high alcohol content like whiskey have, a burning sensation. This is also answered by science. The vanillin 1 (TRPV1) receptors coat the linings of your mouth and throat. A high dose of ethanol makes these receptors hyperactive to vanillin. Vanillin compounds are leached from oak barrels and are found in low concentrations in distilled spirits like whiskey. Once activated, these receptors can act to stimulate capsaicin signaling just as if you had hot peppers in your drink. Diluted in water or juice, whiskey would not have as great of an impact on these receptors.

12.4.1 The Art and Science of Beer

The process of making beer is fairly simple. Start with a source of complex carbohydrate (e.g., starch) from a seed. Convert the starch to simple sugars using enzymes and then add yeast to ferment the sugar to alcohol. Adding spice, fruits, and hops can create flavored beer. Ancient civilizations made beer in an amazing variety of ways. The Babylonian Code of Hammurabi proscribed a "fair" price for beer and outlined the punishment for overcharging (drowning the innkeeper) [1]. An ancient recipe found in a Sumerian poem to the goddess of beer, described using barley seed to make bread which was then soaked in water where wild, environmental yeast fermented to produce beer. Incans chewed corn, where digestive enzymes broke down the starches for yeast, producing a beer still available today in Peru called chicha beer. Egyptians and Mesopotamians began a technique still used today to use grains as the source of sugars. The process stimulates the seeds to germinate (grow) to produce its own starch-digesting enzymes. Modern crafters and commercial brewers use a mix of the art and science to create an array of types and flavors of beer.

The entire process used by modern brewers is shown in Figures 12.8 and 12.9 and Table 12.2. The starting ingredient for beer is the complex starches found in grain seeds (Fig. 12.10). The grains of rice, corn, or other crops have been used, but barley seed corn is the chief source of sugar for beer production. Barley is a seed grain from the grass family whose seed is arranged in two or six rows at the end of each stem. Like wheat, each seed is covered in a woody husk, lined with living cells (aleurone layer cells), filled with starch-containing endosperm and the plant embryo.

The first challenge is the conversion of the complex starches located in barley endosperm into smaller mono- and disaccharides. Amylose is an unbranched glucose polymer made of thousands of glucose monomers and comprises 20–40% of total

The steps of beer brewing

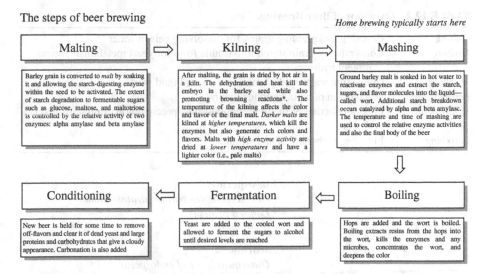

FIGURE 12.8   The process of brewing.

FIGURE 12.9   A simple view of the brewery process.

TABLE 12.2   Overview of Beer Brewing.

Milling	Cracking and grinding grains for dissolving gels in water
Malting	Conversion of grain starch into usable form for yeast metabolism, some protein hydrolysis
Mashing	Converting water-soaked malt into sweet liquid wort
Fermentation	Metabolism of carbohydrates to ethanol
Maturation	Final production of flavor compounds and precipitation of proteins and cell debris
Finishing	Filtering, carbonation, and storage

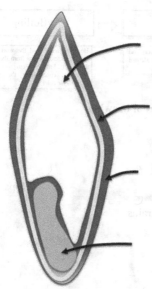

Endosperm
• *Starch (complex carbohydrates—flour)*
• *Limited proteins*
• *B vitamins*

Aleurone
• *Outermost layer of endosperm*
• *Lining made of specialized cells*
• *Contains most of nutritional whole grain components*

Bran—Just under the hull/husk
• *Insoluble carbohydrates (fiber)*
• *Trace minerals, phenolic compounds*
• *B vitamins*

Germ
• *"New plant" embryo*
• *Nutty sweet flavor*
• *Vitamin E, folic acid, thiamin, and metals*

FIGURE 12.10   Anatomy of a wheat kernel. The sugar for beer comes from the endosperm of a wheat kernel. Other cells including the aleurone produced enzymes and hormones to break down the complex carbohydrate found in the endosperm.

barley starch. Amylopectin is a branched polymer of up to 250,000 glucose units and make up 60–80% of total barley starch. Yeast enzymes digest neither glucose polymer. Instead, these complex starches must be separated from the grain, hydrated, and digested into the smaller mono- and disaccharides, glucose and maltose. The entire process is completed as part of the malting step.

12.5   MALTING

The goal of the malting step is to produce enzymes from the grain to digest both starch and proteins to smaller components. Malting consists of three steps: steeping, germination, and kilning. Cracked and milled barley seed is soaked (steeped) in enough water to wake up the cells in the aleurone and embryo. The process takes up to 40 hours when the embryo cells begin to produce plant hormones and will start to sprout (germinate). At this point, water is removed and the seeds are thinly spread to absorb moist air and incubated at 15.5°C/60°F for several days to allow the embryo to begin to develop. During this germinating phase of malting, the embryo will produce hormones such as gibberellin to stimulate aleurone layer cells to produce many different enzymes used to metabolize and digest starch and proteins in the endosperm and the rest of the barley grain. Timing is critical for this stage. Some of the enzymes are lost late in germination and the balance of total protein production versus the types and amounts of each enzyme can change from batch to batch of malt.

Converting the milled grain to a "malt" results in the production of enzymes that break down starch and protein along with a host of other compounds. During this process some of the starch and other nutrients of the grain are used by the sprouting cells. This process typically takes 4–7 days. Shoots and the beginning of roots are produced while the enzymes begin to digest components of the endosperm and the grain itself begins to swell with water content and soften. The process must be stopped at a point where enough enzymes are made to digest the remaining starches and proteins but not left too long where the growing embryo digests too much of the starch.

Once the signs of a growing shoot and root tip from the embryo are observed (commercial malting professionals will also measure the enzyme production in the grain), the process is stopped by heating and air-drying in a kiln. At this point, the grains will be dried to allow the barley malt to be stored for the next step. Drying lasts for 30 or more hours at 80–100°C. Darker beers are heated for longer times and at higher temperatures to allow Maillard and caramelization reactions to occur. Heating and drying is a delicate step that requires careful testing by the malter. The darker the malt, the more the digestive enzymes are destroyed by the heat. If too much moisture remains in the malt, germination will continue and mold may contaminate the malt during storage. The last step is to remove the shoot tips and roots. At this point, malt will appear like a swollen barley seed with a lighter color.

12.6 MASHING

Mixing ground malt with hot water will begin to reactivate the enzymes produced during malting and dissolve most of the starches into solution. Starch processing during mashing can be divided into three phases: gelatinization, liquefaction, and saccharification.

The process of dissolving the long tangles of starches into water is complicated. Gelatinization is the process where by water begins to dissolve starch by interacting via hydrogen bonds with the —OH groups of the starch molecules. Water forms organized shells around the starch causing the starch to swell in size. At some point amylose leaks out of the granules of dried starch and bursts the granules. Barley is a particularly good choice for beer as its starches absorb water at a lower temperature than corn or rice starches; these grains need to be boiled for gelatinization. Thus, it takes less heat to dissolve the barley starch, avoiding further denaturation of the digestive enzymes produced during malting. The result is a gel-like solution thick with the swelled starches. This step is critical for starch digestion to smaller, simpler sugars. Enzymes need the starch to be soluble for proper digestion.

Once the large starch chains are suspended in the water (they have *gelled*), lique-faction can take place. During liquefaction, enzymes break the large chains of starch into much smaller, simpler sugars. Aleurone cells produce two major classes of digestive proteins: starch-digesting and protein-digesting enzymes. The enzymes that degrade starch are similar to those found in saliva and your digestive system. There are three types of starch-degrading enzymes: alpha amylase, beta amylase, and limit dextrinase. Each enzyme uses water to cleave (hydrolyze) starch into smaller components (Fig. 12.11). Alpha amylase will cut the 1,4 glycosidic bond between glucose units on both amylose and amylopectin. Alpha amylase randomly binds and cleaves starches producing much shorter chunks of starch. However, these shortened starches are still too large to be used as food for yeast. Beta amylase also cleaves 1,4 glycosidic bonds, but unlike alpha amylase, beta amylase binds at the nonreducing end of the polymer cutting off two glucose units at a time. Beta amylase thus produces the disaccharide maltose. Partially digested starches include short runs of glucose collectively called dextrin and the trisaccharide maltotriose. Amylopectin contains 1-6 glycosidic branch points that are not digested by either amylase. Another enzyme called limit dextrinase (sometimes called "debranching enzyme") is needed to cleave the 1,6 branches allowing the amylases to continue to digest the starch into maltose and glucose. The end result of this complex digestion is a mixed population of glucose, maltose, maltotriose, and shorter chains of starch called oligosaccharide dextrins.

Protein-digesting enzymes called proteases are also present in malt and activated during the mashing process. These enzymes use water to break the peptide bond of proteins. These proteases, also called peptidases, provide smaller protein fragments called peptides and amino acids from protein degradation. These amino acids pro-vide nitrogen needed for strong healthy yeast growth during fermentation. Dextrins, amino acids, and peptides together contribute to the mouthfeel and taste of beer as well as support a stable foam during drinking. Large, undigested proteins will aggregate and cause a cloudy look or haze to the finished product.

When the large chains of starch have been broken down into very small fragments (i.e., glucose, maltose, maltotriose), saccharification has taken place (Fig. 12.12). The level of starch digestion into the smaller carbohydrates is the fermentability of the mash. The more glucose and maltose produced, the more food there will be for yeast to produce ethanol. A more robust beer will have more body provided by fewer of the shorter and more of the longer starch intermediate compounds. Light beer will have a high-fermentability mash where nearly all of the starch has been digested to glucose and maltose.

During the mashing process, a mixture of malt and water is placed in a metal con-tainer called a mash tun. The tun is heated and held at specific temperatures to allow various classes of enzymes to function. Each temperature stop is called a rest. The mash rest is first held at 40–45°C/104–113°F to allow cell walls to be digested by beta glucanase enzymes. These enzymes break down the cell wall components

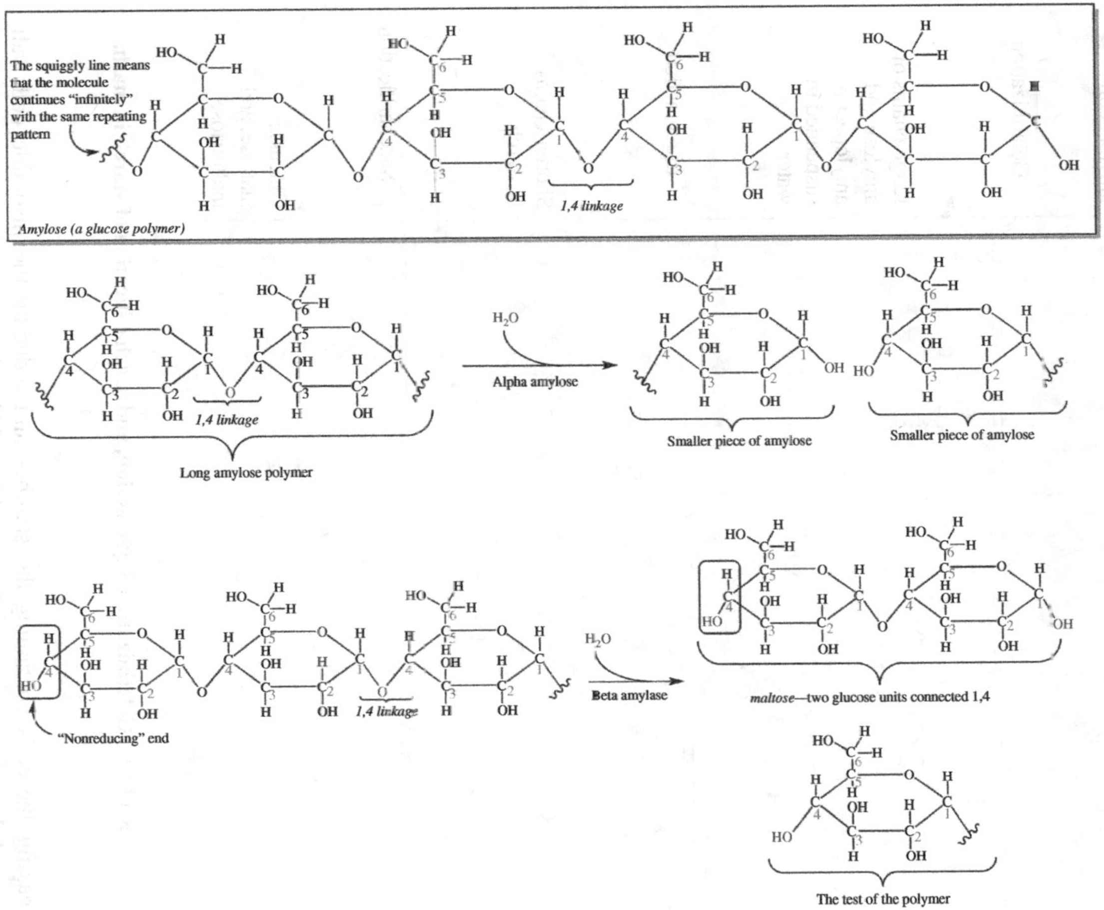

FIGURE 12.11 Alpha and beta amylase reactions.

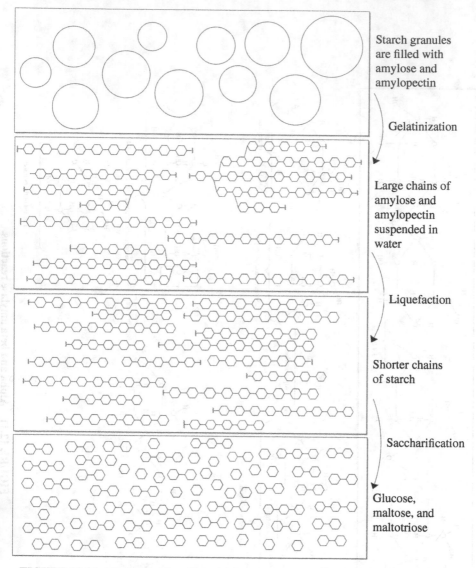

Starch granules are filled with amylose and amylopectin

Gelatinization

Large chains of amylose and amylopectin suspended in water

Liquefaction

Shorter chains of starch

Saccharification

Glucose, maltose, and maltotriose

FIGURE 12.12   Gelatinization, liquefaction, and saccharification of starch in mash.

breaking the cells containing the starches and reducing the insoluble cell wall material that can impart poor flavor and haze if left intact. The protease rest allows for digestion of proteins into peptides and amino acids. Increasing the heat to a final amylase rest can be done at two different temperatures or one for both alpha and beta amylases. A final mashout step will raise the temperature of the mash to slightly above 172°F/78°C. This will denature and inactivate most of the enzymes from the malt as well as most microbial contaminants remaining in the mash.

Darker beers may include removing a fraction of the mash and boiling it to caramelize and add flavor to the mash. Adding a fraction of boiled mash back to the tun is called decoction. Another form of decoction is to boil unmalted adjunct cereals (rice, oats, wheat, sorghum, or corn) to gelatinize these starches. The boiled adjunct starch is then added to the barley mash to allow the enzymes to saccharify the adjunct starch. The result is a low-protein, highly fermentable mash that yields a different taste and higher alcohol content in the finished product. Other adjuncts can also be added at this time to produce different flavors (spice, chocolate, etc.) and increase sugar content (kettle sugars including sucrose or brown sugar).

12.6.1 Wort Processing

Separation of mash solids (grist) from the liquid (wort) happens by filtering through a lauter tun creating a brown sweet liquid called the wort. The wort is then mixed with hops and boiled often in a copper kettle for an hour. Boiling will kill off any residual yeast, bacteria, and enzyme not already lost in the mashing process. Boiling the wort in this fashion will also help remove volatile compounds such as dimethyl sulfide (made from amino acid breakdown during mash heating) that may impart bad flavor to the final product. Wort boiling will also help to convert some of the compounds from hops making them more soluble. Denatured protein will aggregate and mix with some of the cell wall compounds forming a scum called a hot break. The wort is quickly cooled to limit further sulfide products and helps to further denature and aggregate proteins preventing haze.

The act of heating wort also concentrates the sugars and increases the overall density. Remaining compounds from the hops add a required bitterness after which the wort is called a "hopped" wort. A typical mixture of sugars in wort is as follows: maltose (~50%) > dextrins (25–30%) > maltotriose and glucose (10–20%) > sucrose or fructose (1–2%). Maltose and maltotriose are taken into yeast cells using ATP to transport the disaccharides across the membrane, whereas glucose and fructose are passively transported into the yeast cells without expending energy, which impacts the potential ethanol output of the yeast during fermentation. Yeast will grow more slowly due to the increased transport needs of maltose. Once inside the cell, the oligosaccharides will be converted to glucose and used to produce ethanol. Overboiling wort has an additional impact on the final quality of beer; concentrated, highly boiled wort has higher quantities of ethyl acetate and diacetyl. Both compounds produce off-flavors in beer and are considered highly undesirable.

12.6.2 Hops

Resembling a pine cone, hops are the female flowers of the hop vine (*Humulus lupulus*). Hops are relative newcomers to beer brewing. For the first few thousand years, beer makers used herbs or spices to provide flavor and what turned out to be a preservative effect for the fermented beverage. Around 820 AD monks in Eastern Europe began to add the oil-containing cone into their beer. Unspiced or unhopped beer quickly spoils with bacterial or wild yeast growth. Some of the oils from hops inhibit microbial growth and

FIGURE 12.13   Cones of hops on a vine.

provide a unique flavor to the drink. Original ale was unhopped while beer was the term for hopped brews. By the 1600s, nearly all beer was "hopped." Modern commercial production of hops takes place in Germany and in Washington State in the United States (Fig. 12.13).

Like wine grapes, there are a dizzying number of strains of hops whose content is influenced by the environment and earth in which the vine is grown. Hops provide much of the aroma and most of the desired bitter flavor of beer. Located in the lupulin glands of the hop is a yellow powder containing resins. Hops are a complex mixture of compounds including acidic resin oils, tannins, pectins, proteins, waxes, and carbohydrates. The sticky resin of the hop includes α-acids and β-acids. Unaltered versions of both acids dissolve poorly in water. In fact, usually less than one part per million (PPM) of either acid is detected in most beers. Depending on the hop variety, there are low levels of either hop (Noble hops such as the German Hallertau Hersbrucker variety ~3–4% for each acid) or a 3.5 ratio of α-acids to β-acids (14–16 to 4% α-acids for the North American Nugget and Zeus hop varieties). The α-acids are slightly less soluble in water than the β-acids leaving the β-acids to have a larger impact on flavor and taste of beer. Boiling hop acids catalyzes the isomerization of α-acids to iso-acids (Fig. 12.14). Iso-α-acids are much more soluble in beer than their nonconverted α-acid forms. Slightly less than half of the α-acids are converted to the iso-α-acid form and remain in beer after cooling.

There are three main α-acid forms: humulone, cohumulone, and adhumulone. Humulone is the predominant type (48–75%) of α-acid and its isomerized form, isohumulone, contributes to the desirable soft bitterness of beer. Light in the green wavelength will be absorbed by the vitamin riboflavin, which then transfers the captured energy via a reactive oxygen species to isohumulone. The reaction continues and further alters the iso-α-acid. The reaction of light with isohumulone creates a "skunky" or "light-struck" foul odor to the beer. Breaking of the bond by light creates an intermediate compound with a single electron. Such single, unshared electron-containing compounds

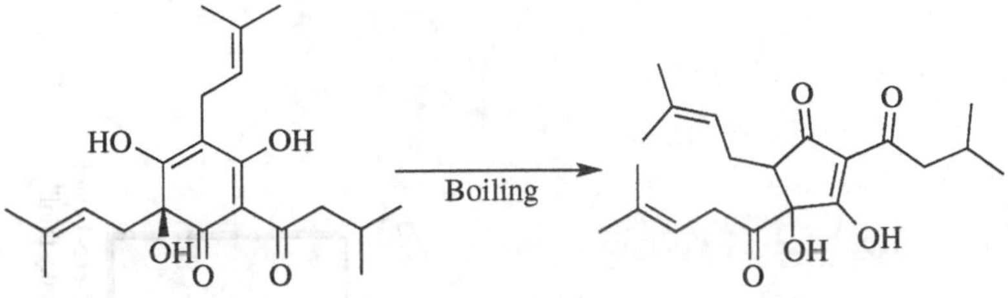

Humalone   **Isohumalone**

(*Alpha acid*)   (*Iso-alpha acid*)

FIGURE 12.14   The conversion of alpha acids by heat. Boiling hops converts the alpha acid humulone to the less bitter isohumulone.

are called free radicals and are very reactive. The iso-α-acid free radical will react with the amino acid cysteine found in the beer to create 3-methylbut-2-ene-1-thiol (MBT), which very much smells like (but is not) skunk spray. Brown or green beer bottles absorb green light inhibiting the reaction while on the shelf. Removing one of the carbon–oxygen double bonds of the iso-acid by adding hydrogen across the bond is called skunk-proofing (Fig. 12.15).

While boiling converts some of the hop acids to a bitter, soluble iso-acid, much of the aroma of the volatile hop compounds are lost during boiling. Brewers will often place hops, pelleted hops, or extract of hop cone back to the wort for the last few minutes of boiling or to the finished, cooled wort. This second hopped or "dry hopped" wort will have a much stronger aromatic component from the hop resins and oils that don't evaporate. Oils include some of the same aroma and flavor compounds found in spices including the terpenes linalool, geraniol, pinene, limonene, and citral providing fruit, citrus, and pine aromas to beer.

12.7   FERMENTATION

The finished wort is a rich, nutrient-dense solution ready to turn into beer. Two different types of yeast strains perform fermentation of wort into beer. There are top fermenters and bottom fermenters. This explains how the two basic categories of beers are organized. Top-fermenting yeasts are used to make ale and bottom-fermenting yeasts produce a lager beer style. Top-fermenting yeasts prefer warmer temperatures (12°C/55°F) and culture faster, often finishing the job in under 8 days. Bottom-fermenting yeasts are much slower fermenters and are best cultured at 4°C/40°F. Of course, the type of starting materials and how the grains are processed, how the wort is produced, and the infusion of adjuncts and hops, together with various strains of each type of yeast, create a rich and diverse beer spectrum for both ales and lagers.

There are hundreds of ale-producing yeasts (top fermenting), most of which originate from the parent yeast strain *Saccharomyces cerevisiae* but may also share genetic identity with other yeasts including strains used for winemaking. Most of

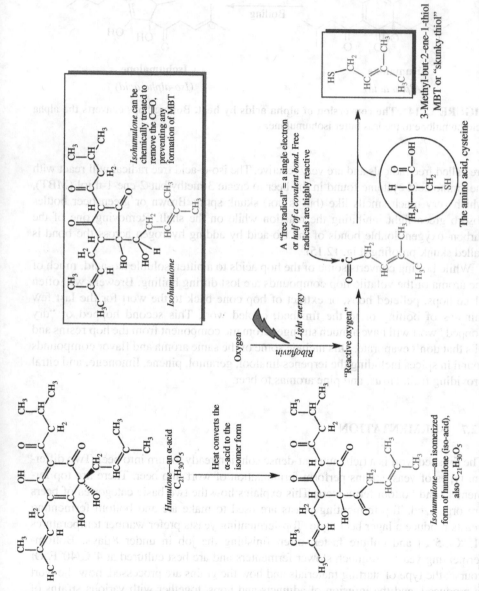

FIGURE 12.15 Skunk-proofing beer. Conversion of isohumulone followed by light-induced oxidation (free radicals) results in foul smelling sulfur compounds.

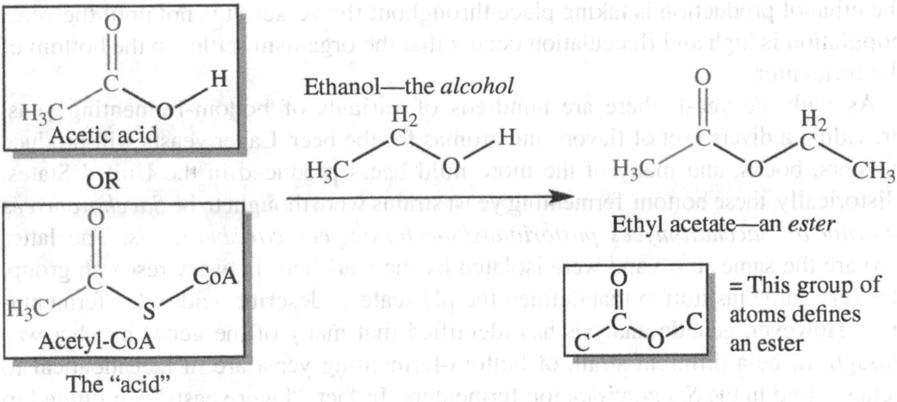

FIGURE 12.16 Formation of esters from acids and alcohol.

top-fermenting strains will clump (flocculate) and rise to the surface as they trap CO_2 gas within the flocculated cells at the end of the culture/fermentation. These strains will also produce a stronger set of flavor molecules including the esters (Fig. 12.16): isoamyl acetate (banana aroma and flavors), ethyl hexanoate (red apple), ethyl acetate (a flowery aroma), and ethyl caproate (a fruity, wine-like aroma). The level of these and other flavor-producing molecules depends on the strain and finely tuned conditions of the fermentation. Ester production takes place inside the yeast. The basic reaction is a condensation reaction between a carboxylic acid and an alcohol. Acetic acid and acetyl-CoA both provide the carboxylic acid for the reaction and ethanol or other longer carbon chain alcohols (butanol, propanol, etc.) react to form a number of final floral and fruity smelling ester products.

Careful brewers will select strains that produce enzymes involved in making acetyl-CoA. To get a larger bouquet of esters, higher yeast growth rates will increase the longer (often called fusel) chain alcohols. Increasing the temperature to encourage faster yeast growth can help this along. After yeast begins to grow at a sufficient rate, lower oxygen levels induce ethanol and fusel alcohol production. This shifts the metabolism from fatty acid production to alcohol leaving some of the acetyl-CoA for the ester reactions. Commercial brewers do this in a number of ways including changing the pressure and temperature of the fermentation, reducing the aeration of wort or extended boiling of wort for a higher-density starting solution. Home brewers can achieve similar effects by ensuring the starter yeast culture will produce the necessary components.

Lagers are often described as clear and crisp, while ales are fruity and complex. The key difference, after the production of wort, is in the species of yeast used for fermentation. Lagers are produced by bottom-fermenting yeast that grow slower and at much lower temperatures. Like the top fermenters, bottom fermenters aggregate or flocculate but do not trap CO_2 and instead sink to the bottom of the fermenter near the end of culturing. The big difference is the rate and metabolism of the two types of yeasts. Bottom-fermenting yeast is more of a misnomer as much of the ethanol production is taking place throughout the vessel. It is not until the yeast population is high and flocculation occurs that the organism settles to the bottom of the fermenter.

As with ale yeast, there are hundreds of variants of bottom-fermenting yeast providing a diverse set of flavors and aromas for the beer. Lager yeasts will produce pilsners, bocks, and many of the more mild beers produced in the United States. Historically, these bottom-fermenting yeast strains were thought to be *Saccharomyces uvarum or Saccharomyces pastorianus/Saccharomyces carlsbergensis*. The latter two are the same strain and were isolated by the Carlsberg brewery research group, the very same institution that defined the pH scale to describe acidity for fermentation. However, genetic analysis has identified that many of the genes in what was thought to be a different strain of bottom-fermenting yeast are in fact identical to genes found in the *S. cerevisiae* top fermenters. In fact, "lager yeasts" are limited to breweries; they are not found in the environment as are other fermenting yeasts. What likely happened is *S. cerevisiae* was crossed and genetically fused with a cold-tolerant yeast sometime before lager beers were brewed in the fifteenth century. A close investigation into the genetic sequence of three unique bacterial strains, *S. uvarum, S. eubayanus,* and *S. cerevisiae*, found that the genes of all three were fused together (most of the genes coming from the parent strain, *S. cerevisiae*) to form what was thought to be a novel species *S. carlsbergensis*. One of the interesting changes was that specific genes involved in sugar metabolism and how the yeast cell used sulfite compounds were transferred from *Saccharomyces eubayanus* to *S. cerevisiae*. These and other traits from *S. eubayanus* and *S. uvarum* allow the once top-fermenting *S. cerevisiae* to grow at colder temperatures and change the flocculation rate and abilities of the cells [5].

Most likely the strains were cultured together allowing fermentation to take place in the colder dark caves of Bavaria, the home of lager brewing. In fact, the German use of "lager" means "to store" and reflects the longer fermentation time of these beers compared with the short time needed for ale fermentation. The end result of "bottom fermenting" is a mild beer without the esters of the ale beers. The mild sulfur odor is due to the sulfite metabolism produced by lager yeast and is appropriate for lager beers.

12.8 CONDITIONING

After the fermentation is over, the beer is considered "green," not carbonated, and unfinished. Conditioning beer requires transferring the young beer from the dead yeast, wort, and hop debris. Yeast is cleared from top-fermenting beer and transferred to a new container. Lagers are traditionally kept at very cool temperatures to stop further yeast fermentation and encourage flocculation and precipitation of fermenting debris. Some add stabilizers or powdered crustacean exoskeleton (chitin) or polyvinylpolypyrrolidone (PVPP) to keep proteins and polyphenols in solution during chilling. Carbonation can be done by directly adding liquefied carbon dioxide, pressurizing a keg with carbon dioxide gas, where the gas will eventually dissolve into the liquid or by inducing a second fermentation. Here, a small amount of yeast is inoculated to clarify green beer with sugar or reserved wort. The mixture is promptly bottled and the continued metabolism of the remaining sugar will create the dissolved carbon dioxide gas.

12.9 OENOLOGY: THE SCIENCE OF WINE AND WINEMAKING

Like beer, the process of making wine requires a source of sugar (grapes or other fruit), water, yeast, time for fermentation to take place, and some postfermentation handling. Unlike beer, the carbohydrate needed for fermentation does not need enzymes or other biological processes to convert complex sugar polymers into glucose and other mono- and disaccharides. Well-grown, mature wine grapes will have plenty of glucose and fructose. However, the handling of wine before, during, and after fermentation is quite a bit different and can be more complex than maturation of most beers (Fig. 12.17).

A simple description of making wine starts with grapes harvested when the sugar content is highest. The grapes are then crushed with the stems and seeds and mixed into a must. Depending on the type of wine, the skins will remain in contact with the juice and yeast is added to begin fermentation. After most of the sugar has been metabolized to ethanol and other products including CO_2, solids are precipitated and the clarified wine is aged until consumption.

FIGURE 12.17   Oenology, the science of wine.

12.9.1   From Grapes to Must, Preparation of Sugars for Metabolism

Grapes are the most common but not only source of sugar for winemaking. Many fruits including apples, pears, and even non-fruits plants such as dandelion can be used to make wine. Using fruit requires additional resources including enzymes to break down the pectin from cell walls. For this chapter, we will focus on grapes.

Like beer, winemaking is an ancient discovery. There is abundant evidence for wine production throughout the ancient world, BCE 3000–4100 in China and the Middle East areas. The simplest and most likely origins of grape wine were from environmental yeast landing and metabolizing the sugars from the fruit. Wine as a preserved source of social lubrication has been in favor since.

The formal term for crushing the grapes, stems, and seeds is *maceration*. At their peak, grapes will consist of 20–30% sugar. Most of the sugar comprises the monosaccharide glucose with a significant portion of the sugar being fructose. Glucose is transported into the yeast and metabolized at a higher rate than fructose, but both provide the metabolic carbon needed for fermentation. Grapes also provide micro- and macronutrients essential for yeast growth. Macronutrients such as lipids, proteins, and complex carbohydrates are needed to build new yeast cell material and micronutrients such as vitamins and minerals, which are crucial for enzyme activity inside the yeast.

Wines are often termed "varietal," which means that a particular wine is made from one type of grape variety rather than a mixture of grapes. The genetic makeup of the grape heavily influences the nature of the finished wine. The grape can be simply divided into the exocarp (skin) and mesocarp (flesh) (Fig. 12.18). The berry, its skin, seeds, and stems are all crushed together. The initial crushing of the grapes produces water, sugar, and organic acids in a clear juice called the *free run*. The pulp is what remains after the *free run* is removed by crushing. The juice remaining in the pulp is found in the skin, seed, and stems but lacks many of the flavor compounds. Further crushing or "pressing" of the pulp results in greater extraction of terpenes, tannins, thiols, and other more bitter tasting components. The combination of the free run and any other liquid pressings is

Exocarp (sking)
Terpenes
• Geraniol
• Terpineol
• Nerolidol
• Linalool

Mesocarp (fleshy fruit)

Norisoprenoids
• β-Damascenone
• β-Ionone

Organic acids
• Malic acid
• Tartaric acid

Thiols
S-3-(Hexanol-cysteine)

Sugars
• Glucose
• Fructose

FIGURE 12.18   Anatomy of a grape. The various compounds in wine as found in grapes.

FIGURE 12.19 *Red and white* **grapes for a Shiraz or Chardonnay.** Red or white wine has the color due to the time the skin color is extracted from the wine.

the "must," which is Latin for young wine. The crushed solid portion is called the pomace. The longer the pomace remains mixed with the must, the more of the color and other compounds like tannins that will be extracted into the liquid must.

The grape varietal, location and environment of the vine, condition of the fruit, and maturation of the grape berry are all key to making a good wine. Prior to maturation, there is less glucose and fructose; instead the fruit produces organic acids, tartrate and malate. These two bitter tasting acids are produced as a defense mechanism against foraging animals and birds until the seed develops. As the fruit reaches maturity and the seed is ready, plant hormones switch metabolism to stimulate the production of pigments (anthocyanins in red grapes) and decrease glycolysis to allow the buildup of glucose and fructose. Malate is used to produce other compounds and several volatile organic compounds are created. This results in a sweet, tasty, and attractive smelling berry for animals to eat and spread the seeds (Fig. 12.19).

The maceration of grapes and stems is the beginning of wine. The distinction between a white, rose, or red wine is in large part due to the color compounds in the pomace where the skin and crushed seeds lie. When must and pomace are in contact, tannins and other compounds are extracted from the solid pomace into the liquid must. These compounds impart color and give body and flavor to the wine. The pale color of white wines are a product of a short time mixing the must and the pomace, while heavy-bodied reds like the Italian Amarone and Chianti will be fermented in the presence of the pomace. In this case the heat from fermenting yeast and the ethanol will help to extract the flavor and color compounds into the water (aqueous) phase. Of course, red grapes will provide more color than white grapes but much depends on the temperature and time of maceration.

An important factor in growing wine grapes is "terroir," a French concept widely embraced by viticulturists. Terroir is the effect of local soil and climate on the development of the complex nature of flavor and aroma molecules in grapes. The basic idea is that exact clones of a grape will produce different tasting wine depending on the environment, geography, and local climate of the vineyard. These factors are suggested to impact the metabolic composition of grapes including the levels of flavonoids, polyphenols, sugars, organic acids, and amino acids. One in-depth chemical study using a technique called mass spectrometry focused on four different vineyards located in the Burgundy region in France. In this work, the chemists found a significant, discernable difference between the Pinot Noir grapes before and after maceration. They also found that after several years in the bottle, a number of flavorants were at different levels even though the vineyards were only 40 km from each other [6]. Polyphenolic compounds seemed to have the greatest differences in these studies.

12.9.2   Fermentation

At this point, as the must is mixed with pomace or separated from the solids, winemaking has begun. Environmental (or more excitingly named "wild") yeast and bacteria have already been introduced to the fruit juice and begun to ferment. As in beer and rising bread, alcoholic fermentation is conducted by the yeast genus *Saccharomyces*. There are over 3000 yeast species and strains available for biomedical research that are nearly as complex as the hundreds of yeast variants available from wine and beer supply houses. The two most common species are *S. cerevisiae* (although there are many strains of this yeast) and *S. bayanus*. A strain is made from small differences within a species of an organism. For yeast, changes in the genetic material (i.e., DNA) lead to altered proteins, metabolism, and characteristic of a new strain. Over time, both species of yeast have mutated slightly to produce subtle but very important differences in growth rate and optimal growth temperature, metabolism, and the production of important flavor compounds. For example, there are two closely related strains of *S. cerevisiae*, but one strain produces a low concentration of isoamyl acetate (responsible for a pear fruit taste) of 1.2–3.5 mg/l of wine, while a second closely related *S. cerevisiae* strain produces 9.0–16 mg of the ester per liter of wine. To fully appreciate the different characteristics in wine yeast strains, there are several prominent and trusted suppliers of yeast strains for beer and wine production.

There are key characteristics for any yeast strain to be considered for winemaking. First is the ability to metabolize most of the sugar into ethanol and the other products. A yeast strain inhibited by lower levels of ethanol (<10–14%) will leave considerable sugar in the end product, resulting in a very sweet wine. Lack of sugar is what makes a dry wine. A good strain should also ferment at a reasonable and predictable rate to compete with other environmental microbes. However, a high fermentation rate will result in higher temperatures, which can cause the evaporation of desirable volatile esters and other volatile aroma compounds. Strains of yeast for white wines are fermented at lower temperatures (12.14°C) while red wines are fermented at higher temperatures to assist in the extraction of color and aroma flavonoids (35–42°C). Strains that do not produce "off-flavors" are also an important consideration. For example, hydrogen sulfide (H_2S) produces a rotten egg smell, while some strains may produce acetic acid giving a vinegar smell. Like fish in an aquarium, if the winemaker mixes yeast strains or bacteria, the cells must play together nicely. Some yeast produce toxic proteins or peptides called "killer factors," which are lethal to other strains of yeast or bacteria. If an advanced winemaker plans to use a mixture of yeast or bacteria (see malolactic fermentation in the following), it is important to use a strain of yeast resistant to the "killer yeast" toxins. Many wild yeasts produce and are resistant to the killer toxins, while *Saccharomyces* does not produce toxins yet is resistant to their effects.

Wild yeasts (indigenous to the grapes vs. added or inoculated yeast) are found near the pedicels and stomata of grapes and flourish in sites on damaged or broken skin. *Saccharomyces* is not typically found on grapes in the field; rather, the two classes Ascomycetes and Basidiomycetes are mostly commonly found on grapes. Molds, such as *Aspergillus*, *Penicillium*, *Rhizopus*, and *Mucor*, and bacteria (*Bacillus*, *Pseudomonas*, *Micrococcus*, and *Acetobacter*) are also part of the wild flora of grapes. *Saccharomyces* does not tolerate the harsh conditions of the vineyard and is considered a domesticated strain of yeast.

The question of whether the winemaker should produce wine from the existing "wild" yeast or instead add a precise strain of *Saccharomyces* remains a topic of discussion. The predominant argument for using wild yeast to ferment the wine is that these unique strains of yeast create a complex and unique character in the finished wine, not repeatable with *Saccharomyces*; however, these strains can also produce a diverse set of flavor molecules, some of which may not be acceptable in the wine, for example, higher amounts of acetic acid and ethyl acetate. Depending on the number of yeast starting on the grape, the initial cell seeding may be too low to ferment before bacteria or other microbes take over the fermenting wine. Because most wild yeasts cannot tolerate high levels of ethanol, winemakers may mix their favorite strain of *Saccharomyces* to the culture or add back a small reserved volume of wine fermented with *Saccharomyces*. However, like handcrafted unique beer brewers, some winemakers stay with the tradition to accept the benefit of native yeast fermentation to provide notes and flavors difficult to achieve and unique from mass produced wine.

A typical yeast growth curve has three or four phases (Fig. 12.20). The culture begins with adding a portion of living yeast cells—called the inoculum—to the larger culture. The original inoculum grows, divides, and multiplies (Fig. 12.21). This is a crucial step for the production of wine. The yeast cells must have sufficient oxygen and micro- and macronutrients in order to build sufficient biomass before converting their metabolism over to ethanol production. Winemakers also carefully measure the sugar concentration of the must (in winemaking, units of sugar concentration are measured in Brix, where 1°Bx is 1 g of sucrose in 100 g of the liquid), because high sugar concentrations are toxic to the yeast cells. In this beginning phase, the winemaker will add either dried yeast, rehydrate the yeast with nutrients, or if available add an active liquid culture purchased from a supplier. As mentioned, if the starting

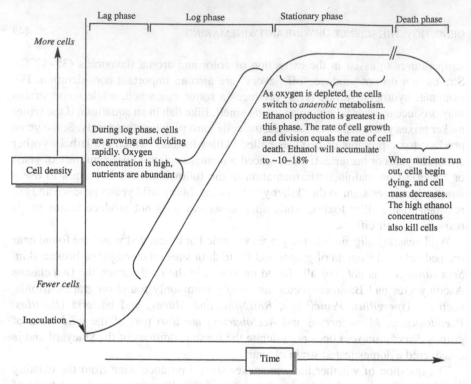

FIGURE 12.20 Cell growth of wine yeast.

Labels within the figure:

Lag phase   Log phase   Stationary phase   Death phase

More cells

Cell density

Fewer cells

Inoculation

Time

During log phase, cells are growing and dividing rapidly. Oxygen concentration is high, nutrients are abundant

As oxygen is depleted, the cells switch to *anaerobic* metabolism. Ethanol production is greatest in this phase. The rate of cell growth and division equals the rate of cell death. Ethanol will accumulate to ~10–18%

When nutrients run out, cells begin dying, and cell mass decreases. The high ethanol concentrations also kill cells

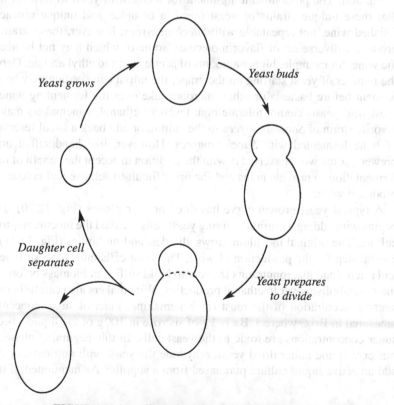

Labels within the figure:

Yeast grows

Yeast buds

Daughter cell separates

Yeast prepares to divide

FIGURE 12.21 Growth cycle of a yeast cell.

inoculum is small, bacteria and molds may dominate the fermentation. A typical starting concentration of yeast is 10^8–10^6 yeast cells per liter of juice. Dry active yeasts typically have 25×10^9 cells/g. Starting with higher numbers of cells in the starting inoculum can produce more fusel alcohols and esters, which, depending on the wine and grapes, may be desirable.

After the lag phase, where cells adapt to the new environment and begin to grow and divide into new daughter cells, they begin to grow at a higher rate called the log phase. Temperature and oxygen are critical in the early phases of cell growth. The numbers of cells will double depending on temperature. At 10°C cells double every 12 h, while at 20°C they divide every 5 h and every 3 h at 30°C. White wines fermented at lower temperatures produce fewer colloids (particles of insoluble protein and carbohydrate), which cloud during cold temperatures. Furthermore, cold growth temperatures produce a brighter, fruity bouquet as more esters are retained in the wine (the colder temperatures prevent evaporation of volatile compounds), where the tannic and heavy-bodied red wines need the higher temperatures to extract the color and flavor compounds from the pomace.

In both lag and log phases, cells are using nutrients to create the biological building blocks (proteins, lipids, DNA, RNA, complex carbohydrates) needed to make new cells. Little ethanol is produced as the must contains dissolved oxygen. Most of the sugar is metabolized using the oxygen to produce CO_2, water, and ATP via glycolysis, the tricarboxylic acid cycle, and the electron transfer chain in mitochondria (see Chapter 2 for a review). As long as oxygen is present, most of the sugars will be metabolized to CO_2 producing up to 38 ATP molecules per molecule of glucose, or less for fructose. Remember that yeasts are facultative anaerobic organisms and in the absence of oxygen, they will produce ethanol as a way to regenerate NAD^+ from NADH to allow continued glycolysis and ATP production; the coupling of glycolysis to ethanol production ultimately produces 2 ATP and 2 ethanol molecules for every molecule of glucose. While ethanol production produces fewer ATP molecules per glucose, it is *faster* than the aerobic version that produces 38 ATP. So in the early stages of fermentation when sugar concentration (Brix) is high, yeast will use a mix of anaerobic and aerobic metabolism and produce low levels of ethanol.

In the stationary phase, cells are no longer actively growing and dividing—they are stationary. Yeast cells also die and divide at the same rate, so there is no net accumulation of cell mass. It is in this phase that the bulk of the ethanol is produced. After the flurry of metabolism that took place in the log phase, there is little dissolved oxygen left— the conditions are becoming increasingly anaerobic. If the must is left still, escaped carbon dioxide gas will produce a barrier layer of gas at the surface of the liquid, reducing further oxygenation of the liquid. The result is anaerobic metabolism as cells convert pyruvate to acetaldehyde and ethanol. Of course, not all of the acetaldehyde is converted to ethanol, and some is shuttled to make other compounds (aldehydes, ketones, and esters) which we will learn about later in this chapter. As long as sugar and other nutrients (nitrogen from proteins, phosphate, and vitamins) are available, viable cells will continue to metabolize sugar into pyruvate and ultimately ethanol.

Eventually waste products overcome the fermentation and cells will no longer grow and continue to die. Ethanol inhibits its own production from acetaldehyde in a term

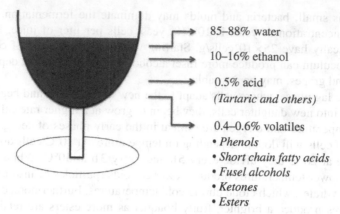

→ 85–88% water

→ 10–16% ethanol

→ 0.5% acid
(Tartaric and others)

→ 0.4–0.6% volatiles
• *Phenols*
• *Short chain fatty acids*
• *Fusel alcohols*
• *Ketones*
• *Esters*

FIGURE 12.22   Chemical composition of a typical wine.

called feedback inhibition. Fermentations that are held too long will begin to produce more and more acetaldehyde once ethanol concentrations rise. In addition ethanol will disrupt the way proteins interact with cell wall lipids and the proteins themselves will unravel and denature in higher concentrations of ethanol. An interesting note is that ethanol allows H^+ (protons) to enter through the membrane acidifying the cell and causing cell death.

The final result of fermentation will be from 10 to 16% ethanol depending on the wine, yeast strain, and fermentation conditions. The carbons from glucose and fructose will have mostly been converted (95%) to carbon dioxide and ethanol. Other final products of wine include pyruvate (organic acid) acetate, acetaldehyde, esters, ketones, glycerol, phenols and other volatiles, tartaric acid, and lactate (Fig. 12.22).

12.10   SULFUR, SORBITOL, AND OAKING: ADDITIVES IN FERMENTATION

Both the home and larger professional winemakers use a number of additives to ensure a safe, noncontaminated final product and to produce the full spectrum of finished wine flavor. Sulfur dioxide is a strong antioxidant (reducing unwanted reactions between oxygen and polyphenols and other compounds) and microbial growth inhibitor. This explains why Romans burned candles (producing sulfur compounds) inside wooden casks to prevent a vinegar smell. The sulfur compounds would inhibit the vinegar produced by bacteria contaminating the wood; thus the ancients practiced a form of sterilization without realizing.

The sulfur dioxide (SO_2) produced by burning candles or matches has a sharp smell, easily detected but different from another sulfur compound, hydrogen sulfide (H_2S). To prevent mildew growth, winemakers can dust their grapes with sulfur to inhibit growth of the mold. However, when yeasts work on these sulfur-dusted grapes, the product hydrogen sulfide (H_2S) smells like rotten eggs. Sulfur dioxide easily dissolves in water where it reacts to form bisulfite (HSO_3^-), which will further react to form sulfite (SO_3^{2-}):

$$H_2O + SO_2 \rightarrow H^+ + HSO_3^{-1} \rightarrow 2H^+ + SO_3^{2-}$$

Sulfites can react irreversibly with other compounds in the wine by forming covalent bonds or reversibly by binding noncovalently to other molecules—in either case, the sulfite is considered "bound." Remaining unbound and unreacted sulfite molecules are considered free sulfites. The free sulfite has antioxidant and antimicrobial properties. Many winemakers measure both bound and free sulfites, which sum to total sulfite.

In red wines, polyphenolic compounds can react with oxygen reducing the color and astringency of the wine. Phenolic compounds in white wines will react with oxygen to form darker, polyphenolic pigments. Addition of sulfite to wine will compete for such reactions. Sulfur dioxide is added to must to inhibit the browning enzymes released during the grape crush. Sulfur dioxide also reacts with the products of polyphenol oxidation producing acetaldehyde as a side reaction. Unfortunately, sulfur dioxide gas (SO_2) is most soluble at very low pH ($pH < 1$). At the pH of wine, approximately 3–4 pH units, the SO_2 converts to bisulfite (HSO_3^-) from reaction with water. Therefore, in must sulfur dioxide is added as a liquefied gas or added as a potassium metabisulfite that will convert back to sulfur dioxide in the water.

Wild yeasts are typically inhibited by low concentrations (20–50 ppm free SO_2) of sulfur dioxide, while *Saccharomyces* growth rate is only slightly inhibited at these levels of sulfide dioxide. Resistant yeasts express higher levels of protein pumps in the cell membrane (SSU1), which help to transport the sulfur compound out of the cell where it will not impact cell growth. Wild yeasts do not produce these transporters and are more susceptible to SO_2. Bacteria are also very susceptible to sulfur inhibition, and if a winemaker plans to perform malolactic fermentation (MLF), she should avoid the compound altogether. Because of the sensitivity of bacteria to sulfites, it is used to inhibit most nonyeast microorganisms. By the end of fermentation, most of the added sulfite will be in the bound form and not effective to inhibit bacterial growth and second batch is often added for longer-term storage. This helps to block unwanted contaminant bacteria from turning wine to vinegar. Sodium metabisulfite is often used at higher concentrations (50 g/l of water ~1250 ppm) as a sanitizing agent. At these levels a small amount will revert to SO_2 gas where it will act as a strong antimicrobial and can sanitize equipment. However, since SO_2 is most soluble at acidic pH, some will add citric acid to decrease the pH and shift the equation to maximize the concentration of the SO_2 gas form. This can be dangerous in confined conditions, where the gas can be harmful and short-term exposures induce asthma and other lung problems.

There are those who are sensitive to sulfites and avoid wine for fear of headaches or induce asthma attacks. The FDA estimates that 1 in 100 people have some sort of sulfite sensitivity and within this group 5% have asthma. Most of these reactions are the result of breathing in SO_2 gas, which is very low in wine. Yeast will produce small amounts of sulfites and the USDA allows up to 350 ppm (Table 12.3).

TABLE 12.3 Sulfur Compounds in Wine.

Molecular SO₂	Bisulfite HSO₃⁻	Sulfite SO₃²⁻	Bound versus Free Sulfur
• Antioxidant and inhibitor of browning oxidizing enzymes • Gas, soluble in water • Only found in appreciable levels in pH < 1	• More prevalent at the low pH of wine • Binds aldehydes, sugars, and anthocyanins (bound SO₂) • Prevents browning enzymes (quinones) and chemical oxidation (reacts with R—C=O groups)	• Major form at pH greater than 7.5 • Strong antioxidant • Slowly reacts with oxygen and is one of the effects of aging	Bound—sulfite bonds or interacts with carbohydrates, polyphenols, and aldehydes Free—unbound sulfur, in wine bisulfite, in more pH neutral water a mix of sulfite and bisulfite. Only free forms of sulfites act as an antimicrobial

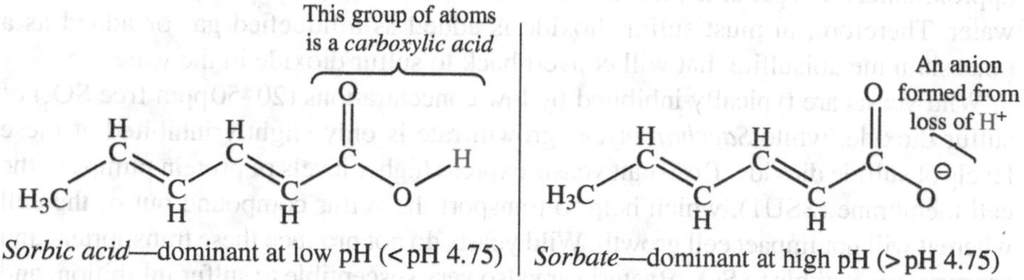

This group of atoms is a *carboxylic acid*

An anion formed from loss of H⁺

Sorbic acid—dominant at low pH (< pH 4.75) | *Sorbate*—dominant at high pH (> pH 4.75)

FIGURE 12.23 Sorbic acid is the dominant form at the low pH of wine.

12.10.1 Sorbate

Once the wine has finished fermenting, some residual yeast can continue to grow. If the wine is not dry—that is, it still contains sugars—existing or wild yeast can begin to grow and ferment as the wine ages. If the wine has already been bottled, this is particularly troublesome. Carbon dioxide gas will build up as the yeasts metabolize residual sugars, eventually bursting the glass bottles. Potassium sorbate is a simple organic acid (a short-chain fatty acid originally isolated from the Mountain Ash Tree) commonly added to wines to inhibit remaining yeast. Sorbate is also often used as a food preservative to inhibit mold and yeast growth. While not typically required for dry, red wines, sweeter reds, fruit, and white wines will have enough sugar to support yeast growth—these wines benefit from the addition of sorbate. In the acidic pH of wine, sorbate is present as the acid form—sorbic acid (pK_a of 4.75; Fig. 12.23), which is 20 times more inhibitory to yeast than the ionized sorbate form.

12.10.2 Oak

Traditional aging of wine in oak barrels is not common in modern winemaking, as industry home winemakers most often use stainless steel. However, the impact of oak on wine aroma and flavor is significant and wood is often included in the fermenting process. There are a number of compounds that are extracted from wood that positively impact the flavor and aroma of wine. Vanilla-like compounds and specialized esters called lactones are a few examples of important compounds brought to wine by oak (Fig.12.24). These molecules will also react with the grape tannins and anthocyanins and other phenolics providing an additional spectrum of flavors while reducing some of the astringency of a young wine. If barrels are not used for aging, adding wood to the fermentation is a critical step for achieving depth and complexity of flavor.

For wine production, oak is divided into French and American oak (see Table 12.4). French oak is found in Limousin Burgundy, the Central and Vosges regions of France,

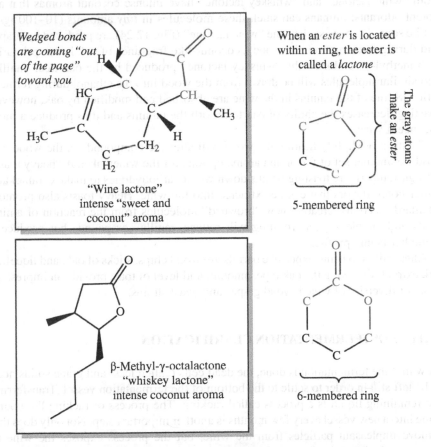

FIGURE 12.24 Lactones provide some of the flavors of wine and whiskey. A compound with a ring structure containing a carbon double bonded oxygen (aka ester) is a lactone.

TABLE 12.4   Types of Oaks and their Aromas Used in Winemaking.

Vanillin	Vanilla aroma
Fresh oak	*trans*-Lactones
Coconut	*cis*-Lactones
Spice and clove	Eugenol and isoeugenol
Caramel, butterscotch, and sweet	Furfural and 5-methylfurfural
Cinnamon and spice	Coumarin
Charred and smoky aromas	Guaiacol and 4-methylguaiacol

and is now also grown in Eastern Europe (*Quercus robur* and *Quercus sessilis*). American white oak is grown in Kentucky, Missouri, Arkansas, and Michigan (*Quercus alba*). In general the French oaks have a higher polyphenol (tannin) content, while American oaks have a lower phenol, higher lactone, and a very high aromatic content—the lactones are responsible for the oak and coconut aromas. Both "wine lactone" and "whiskey lactone" have intense coconut aromas that are potent odorants; humans can smell these molecules in *tiny* amounts (10–100 ng/l). While some lactones such as the "wine lactone" (Fig. 12.24) are produced by grapes and during fermentation, the woody, coconut-like fragrance of oak is due primarily to β-methyl-γ-octalone—the "whiskey lactone" produced by the oak tree. Vanillin and similar molecules will be drawn from the wood into the ethanol during fermentation. Some of the tannins in the wine are absorbed and modified by oak; however, toasting decreases the ability of oak to absorb the tannins and thus produce a more astringent wine.

Toasting or slightly heating the wood will alter the compounds in the wood and produce another set of flavor and aroma options for the winemaker. A "heavy toast" or high-temperature toasting breaks down wood carbohydrates to make caramel and butterscotch flavors that can be extracted into the wine. Heavy toasts also promote Maillard reactions, creating new "browned" molecules from the reaction of amino acids and simple sugars. Toasting will reduce vanilla compounds but produce a smoother tannin profile.

Alternatives to using expensive barrels are wood chips, sticks of oak, and liquefied oak extract. Choosing the oak type, amount, and level of toast provide an impressive array of diversity of wine beyond grapes and yeast strains.

12.11   POSTFERMENTATION CLARIFICATION

Now that the fermentation is done, the dead cells (called lees) and grape solids need to be left still in order to settle to the bottom of the fermentation vessel. Transferring the remaining liquid is a process called racking. The process of racking the young wine into a new vessel every few months is another important step. Not only does this remove unpleasant particles from the wine, but the process exposes the wine to oxygen, which can lead to unwanted oxygenation reactions with flavor molecules. Red wines with more polyphenols to react with the oxygen (and are typically treated with higher sulfites) have a greater capacity to react with the oxygen introduced during racking. Whereas white wines with less of the phenols will be more affected by oxygen. An additional trick winemakers use to reduce cloudiness is to rack at lower temperatures. This decreases the solubility of several components and reduces the need for multiple racking transfers.

Some wines like champagne or Chardonnays are left to "sit on the lees" (sur lie in French) for weeks with periodic stirring. The bready and smooth mouthfeel of these wines are due to the aging of the dead yeast (lees), and the process involves gentle stirring to mix the lee with the wine without adding oxygen. The entire process is called battonage and results in the release of mannoproteins—cell wall *glycoproteins* that are a combination of protein (10%) and carbohydrate (90%). During battonage these glycoproteins (the classification of protein–carbohydrate hybrids) are released from dead, broken yeast debris. The mannoproteins stabilize the wine, keeping proteins and tartaric acid in solution, but they also interact with other flavor compounds including tannins. Through an unknown mechanism, these complex carbohydrates and protein–carbohydrates increase the perception of a creamy denseness to the wine that is attractive to some.

Once the cell debris has been removed by racking, there may still remain an unattractive haze or cloud in the wine. Over time, the compounds causing the haze will precipitate as crystals or a fluffy solid at the bottom of your wine bottle that will alter the aging and flavor of wine. Hazing is due to either complex polysaccharides or proteins remaining partially in solution. Proteins from the grapes or broken cells may polymerize into a semisoluble cloudy solution. Polysaccharide hazes are the result of cell wall carbohydrates of the grape and yeast. This is a significant issue when using fruit to produce wine and is solved by adding the enzyme pectinase to degrade and remove the haze.

Bentonite is a fining agent used to bind proteins and precipitate to the bottom of the vessel. Bentonite is a clay that readily absorbs water and will bind positively charged proteins. At the low pH of wine, most proteins will be positively charged with the $R-NH_3^+$ groups and a significant fraction of the carboxylic groups neutral $R-COOH$. The balance will be a positive charged protein, which will bind to the bentonite particles and through coprecipitation will be removed from solution. The clay will also remove flavorants and in one study up to 13% of the volatile compounds were removed by bentonite. Polysaccharides also cause hazing but will not be removed by clay. Enzyme digestion of this haze is often the first choice producing smaller more soluble pieces of carbohydrate.

Red wine will produce a phenol polymer haze that will leave a fine solid precipitate in aged bottles of wine. Proteins that bind to the polymerized phenols creating a larger insoluble mixture often remove such particles. Isinglass is a collagen protein made from swim bladders of fish often used in white wines. Gelatin is another type of collagen derived from bone that is used to remove tannins and phenol hazes in red wine. Two commercial products, Sparkolloid and Kieselsol, are also used. Sparkolloid is a mix of algae extract with positive charged components. Kieselsol is a suspension of silicon dioxide prepared as a side product in glass production. These are negatively charged particles that will bind to positive charged proteins and tannins and encourage settling. Chitosan is the ground exoskeleton of crustaceans and is positively charged. Combinations of charged fining agents with an understanding of the source of haze are a good application of chemistry and biology.

After racking and fining, wine is often given one more treatment to stabilize it and avoid precipitation or haze during aging. Cold stabilization is achieved by cooling wine to 81°F/27°C for two or more weeks until acids like tartrate precipitate. Heat stabilization is used to remove unfined proteins.

12.12 FLAVOR AND AROMA

The flavor and aroma of wine starts but does not finish with the maceration. There are over 1000 different aroma and flavor compounds identified in various grapes and wines. The volatile compounds that escape the wine into the air where we can detect levels as low as 2 ng/l (about two drops of a compound in a swimming pool) provide a major component of our sensory wine pleasure. Flavor compounds are different in wine than in living plants or plant material. In the plant material, many of the flavorant molecules are covalently bonded to sugars or amino acids. This helps maintain the compound in solution, as the adducts are more water soluble and less volatile. Enzymes from the grape or yeast will cleave this bond releasing the aroma compound as the wine matures.

Red wines are described as bitter and astringent, while white wines may be bitter but rarely astringent. Astringency is the taste, flavor, or aroma that causes a puckering of the mouth or dryness after drinking. This is due to the level of tannins in the wine. Lower molecular weight tannins are both bitter and astringent and found in fairly high levels in young wine that lack body and flavor. The bouquet of flavor and aroma for wines is complex. The white wine Sauvignon Blanc is high in 2-methoxy-3-isobutylpyrazine, a largely unpleasant compound produced by most plants as a deterrent from herbivores. In grapes, the methoxypyrazine compound is produced in the young immature berry and slowly metabolized as the fruit matures. The level of remaining methoxypyrazine depends on sun exposure and local climate. A hint of the compound is desirable but too much will ruin the batch of wine. Thus, a good understanding of the compounds that provide the flavor and aroma to wine is a powerful tool for producing, appreciating, and, of course, consuming wine.

As already described, there are thousands of flavor and aroma compounds involved in a good wine. These compounds can be classified into small organic molecules, many metabolically produced by grape and by yeast during fermentation, and other compounds (polyphenols and anthocyanins) originating from the skin, seed, and stems of the grape as well as from oak used in storage or added during fermentation. Complicating matters are the vast numbers of these compounds, how much are extracted during maceration, the kind of yeast strain used, and the reactions between these molecules during aging. Your taste perception of flavor depends on the numbers and slight variations of receptors in your mouth and nose. Let's look at a few examples of the molecules involved with flavor and aroma in wine.

12.13   SMALL ORGANIC FLAVOR AND AROMA COMPOUNDS

12.13.1   Organic Acids

Wine is basically an acidic aqueous (water) solution with ethanol. The pH of wine is fairly acidic with a pH between 3.3 and 3.6 for red wines and 3.1 and 3.4 for white wines. Acids are organized into soluble or fixed acids that are responsible for most of the acid content of wine (tartaric, malic, and several other acids) and volatile acids (including acetic, butyric, and propionic acids).

The bulk of the acid in wine is tartaric acid, which imparts a sour flavor. In aged wine, tartaric acid reacts with oxygen in the presence of Cu^{2+} or Fe^{2+} to produce glyoxylic acid. This new product binds phenolics like tannins resulting in a mature, less astringent taste. Tartaric acid is less soluble at cold temperatures; the acid will precipitate as crystals with calcium or potassium ions. Some winemakers take advantage of this chemistry by cold treating the wine prior to bottling and removing any precipitated tartaric acid. The L racemic form of tartaric acid is the dominant version found in grapes and spontaneously and slowly isomerizes (converts) to the D form. A mixture of D and L tartaric acid is very insoluble compared with either of the pure isoforms (Fig. 12.25).

Malic acid in mature grapes is much less than half of the total acid content but in immature grapes can be equal to tartaric acid. Grapes grown in warm temperatures metabolize much of this acid, but grapes picked early or grown in cool conditions will have much higher malic acid. The sour, green apple flavor from malic acid is undesirable. One way to remove the offending acid is to initiate a second fermentation using bacteria to convert malic acid to lactic acid. This process called malolactic fermentation is carried out by *Oenococcus oeni*, a lactic acid bacterium that can grow in the acidic conditions of wine. Bacteria convert malic acid to lactic acid and carbon dioxide, thus reducing the overall acidity (lactic acid has one less H^+ producing carboxylic group) and help create a smoother, creamy mouthfeel due to the lactic acid (Fig. 12.26).

Fermenting yeasts produce several volatile short-chain fatty acids. The principal volatile acid is acetic acid, which provides a vinegary flavor and smell. Smaller levels of propionic acid give wine a fatty taste, while butyric acid has a rancid butter aroma. Typical levels of acetic acid are low; however, bacterial contamination of wine will convert ethanol into acetic acid and this is how wine vinegars are produced.

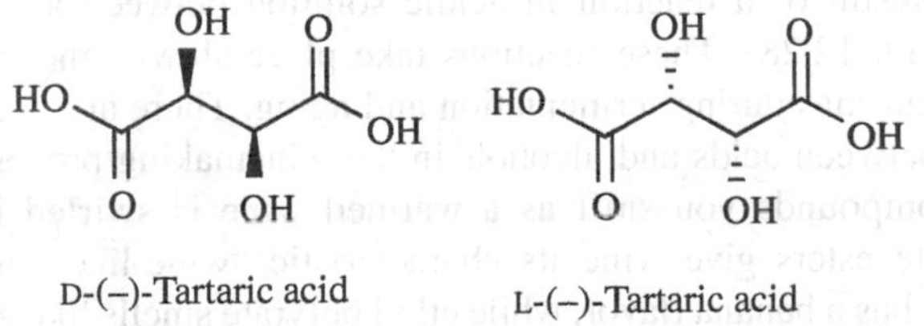

D-(−)-Tartaric acid   L-(−)-Tartaric acid

FIGURE 12.25   Stereoisomers of tartaric acid. Notice the difference between D and L stereoisomers is the spatial arrangement of the two —OH groups. The darkened dart indicates the OH is coming out of the page and the dashed darts show the OH is behind the page.

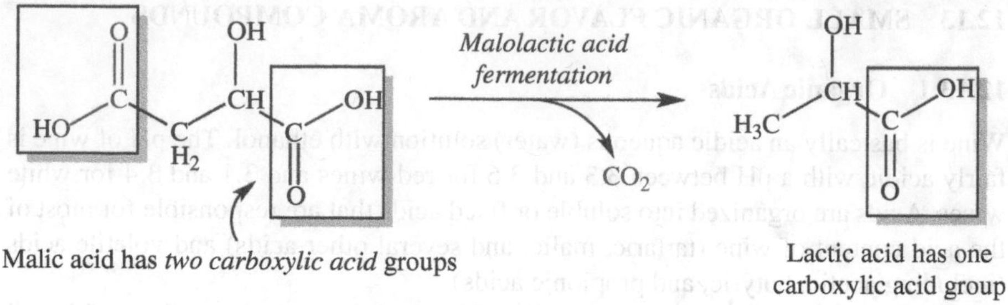

Malic acid has *two carboxylic acid* groups

Lactic acid has one carboxylic acid group

FIGURE 12.26 Malolactic fermentation. A secondary fermentation of bacteria performed after yeast fermentation to produce lactic acid from malic acid.

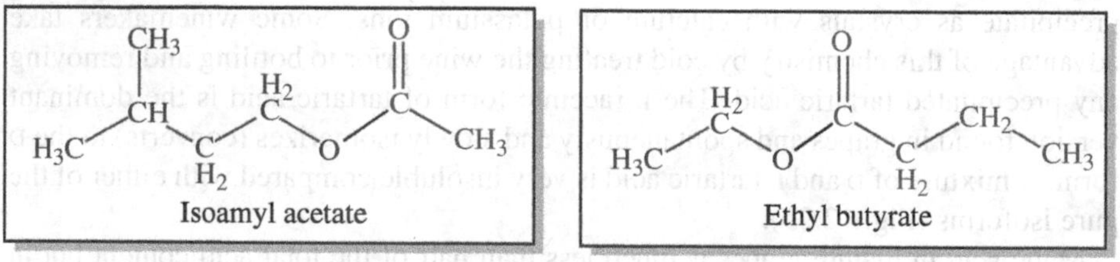

Isoamyl acetate

Ethyl butyrate

FIGURE 12.27 Fusel alcohols.

12.13.2 Higher, Fusel Alcohols

While ethanol is the primary alcohol produced by yeast, there are several other longer carbon chain alcohols produced along the way (Fig. 12.27). The most important of these are propanol, isobutanol, and isoamyl alcohol, which account for nearly 50% of the aromatic constituents of wine after ethanol but only about 0.2% of the total mass of wine. In addition to the direct impact of these higher alcohols on the aroma of wine, they play an important indirect role in developing aged wine bouquet. Alcohols will react with organic acids found in wine to produce a new class of volatile compounds, the esters.

12.13.3 Esters

Esters are the result of a reaction in acidic solution between an alcohol and an organic acid (Fig. 12.28). These reactions take place slowly and generate a wide range of fruity aromas during fermentation and aging. There are over 300 different esters created between acids and alcohols in the winemaking process. These products are the compounds you sniff as a warmed wine is swirled in a glass. For example, acetate esters give wine its characteristic "wine-like" or vinous scent. Isoamyl acetate has a banana flavor, while ethyl butyrate smells like apples and ethyl hexanoate has a pineapple odor. As the carbon chain of the ester gets longer, the ester becomes less volatile and the flavor turns soapy and lard-like [7]. Because the alcohols, acids, and enzymes that react to make these esters are specific to the type

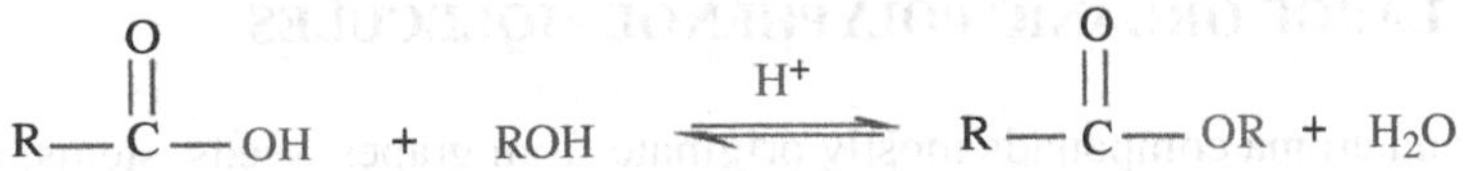

FIGURE 12.28   Conversion of an acid and alcohol to an ester.

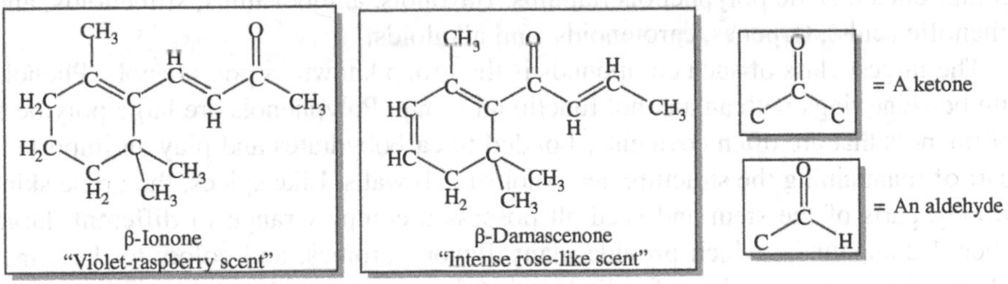

FIGURE 12.29   A ketone (Ionone) and an aldehyde (damascenone).

and strain of yeast and the fermentation conditions, the bouquet of esters can vary drastically from wine to wine and the ester concentrations and ratios can change during wine aging.

12.13.4   Aldehydes and Ketones

Aldehydes and ketones are both types of molecules that contain a C=O double bond; they only differ in where the C=O is placed within the molecule. Aldehydes have a C=O on the terminal carbon, whereas ketones have the C=O in a "middle carbon." Acetaldehyde is the dominant aldehyde in wine. At lower levels this aldehyde gives a fresh cut apple aroma. At higher levels (100–125 mg/l) acetaldehyde is an off-odor and flavor (also known as a wine fault or defect) and is considered to be a contributing factor in hangovers. There are few ketones in grapes. β-Ionone and β-damascenone are two examples of grape-derived ketones that survive fermentation and have powerful scents (Fig. 12.29). Both β-ionone and β-damascenone can be detected by the human nose at very low levels, and due to this fact, they contribute to the wine bouquet even in very small amounts. The diketone diacetyl is most commonly produced during malolactic fermentation where higher levels can even give a caramel-like, buttery aroma to the wine.

12.13.5   Thiols/Mercaptans

Organic compounds that have a sulfur–hydrogen group (R–SH) are collectively termed thiols and mercaptans. Thiols, which may bind mercury, are given the term mercaptan. There is several sulfur-containing compounds produced by the grape or by yeast. Many of these thiols are chemically bound to grape proteins through the cysteine amino acid. During fermentation, yeast enzymes cleave the bond producing a volatile thiol. The flavor of blackcurrant and grapefruit in Sauvignon Blanc and Semillon wines is due to such thiols.

12.14   LARGE ORGANIC POLYPHENOL MOLECULES

Flavor and aroma compounds mostly originate from grapes, seeds, stems, and, to a lesser extent, the yeast. In addition to the smaller organic molecules described previously are much larger, more complex carbon compounds. A few classes of these compounds include polyphenols (tannins, flavonols, anthocyanins, stilbenoids, and phenolic acids), terpenes, carotenoids, and alkaloids.

The largest class of such compounds is the group known as polyphenols. Phenols are benzene rings with an alcohol functional group. Polyphenols are large polymers of phenols that are often covalently bonded to carbohydrates and play an important part of maintaining the structure and color of cell walls. Like spices, the grape skin, woody parts of the stem and seed all possess a complex range of different, large phenol compounds, which provide many flavors, aromas, and colors to the wine. Flavonoids are a large class of polyphenols (phenols) to which tannins, anthocyanin, stilbenes, and flavonol belong. The types of polyphenols found in skin versus stem or seeds are different. Remember that red wine pomace is left in contact with the must so that more polyphenols are found in red than white wine.

The astringent and some of the bitter character of wine is due to a class of polyphenols called tannins. When these molecules hit the tongue, they give a sensation of dryness (i.e., the opposite of salivation), and at high concentrations, tannins taste bitter. These substances are found bound to proteins, carbohydrates, free in solution and as polymers with other phenolic compounds. Grape and oak tannins are placed in two categories: condensed (also called proanthocyanidin) and hydrolyzable. Bonding of catechin and/or epicatechin monomers together forms condensed tannins. As shown in Figure 12.30, catechin and epicatechin are stereoisomers of one another. When part of a condensed tannin polymer, the A ring of one monomer is joined to the C ring of another by covalent bonds between the 4 and 8 positions (numbers from standard flavonoid numbering, also shown in Fig. 12.31). Condensed tannin polymers can also release an anthocyanidin pigment molecule from the end of the polymer during an *oxidative cleavage* reaction, and it is this reaction that gives condensed tannin polymers the name "proanthocyanidin" (Fig. 12.32). Condensed tannins polymerize with anthocyanins and are mostly found in seeds. These tannins are derivatives of the flavonoid catechin or its stereoisomer, epicatechin. Generally, the smaller tannin monomers have more flavor than the larger tannin polymer, since the larger polymers are simply too big to fit into a taste receptor protein! But the larger tannin polymers are responsible for the astringency or "dryness" that wine creates in the mouth. It has been proposed that the polymeric tannins bind to proteins in saliva, causing them to denature and "precipitate" or become *insoluble*. While this has not been conclusively demonstrated, it would explain the "dry" or "rough" feeling that red wines have in the mouth and on the tongue. Condensed tannins can undergo an oxidative cleavage reaction to produce anthocyanidin molecules. Anthocyanidin pigments are known for their purple and reddish hues. If the anthocyanidin is modified by a sugar (sometimes called a glycone), the molecule is called an anthocyanin. If the anthocyanidin reacts with the end of the condensed tannin polymer, it *terminates* the polymer. The carbon 4 of the anthocyanidin is doubly bonded to another carbon—this prevents the addition of any other monomers.

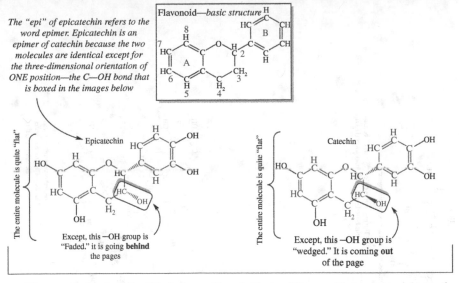

Flavonoid—*basic structure*

Epicatechin

The entire molecule is quite "flat"

Except, this —OH group is "Faded." it is going **behind** the pages

Catechin

The entire molecule is quite "flat"

Except, this —OH group is "wedged." It is coming **out** of the page

These two molecules only differ in the position of a single —OH group. They are a special type of isomer called a *stereoisomer*. Stereoisomers have the same number and types of atoms, all connected in the same way—but they differ in how the atoms are arranged in three-dimensional space

FIGURE 12.30 Epicatechin and catechin. Basic structures of polyphenols/flavonoids.

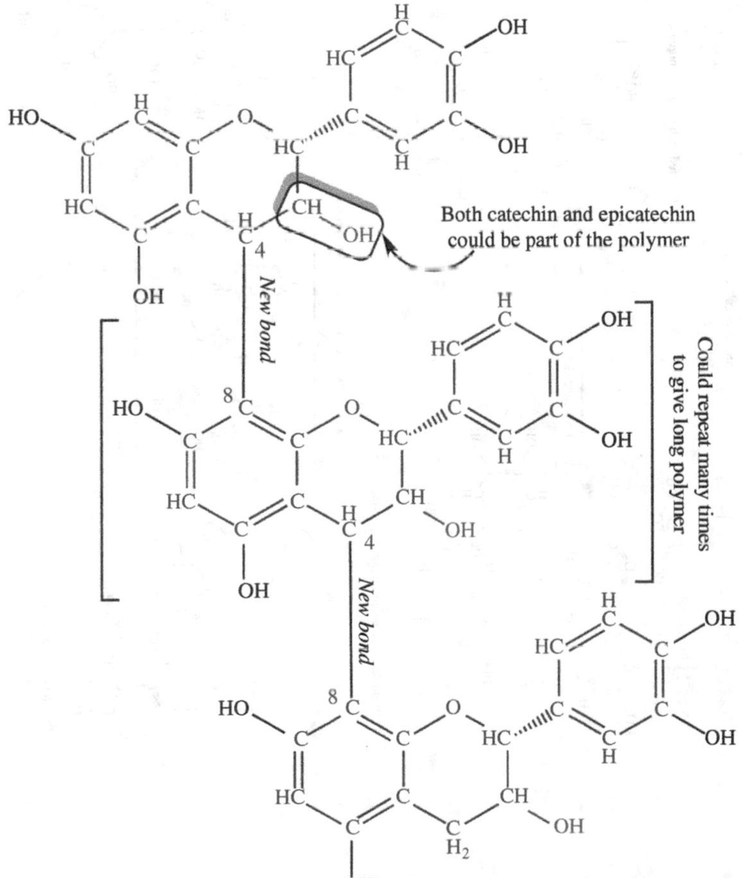

Both catechin and epicatechin could be part of the polymer

New bond

New bond

Could repeat many times to give long polymer

FIGURE 12.31 An example of a condensed tannin. Notice the building blocks of tannin and the numbering system.

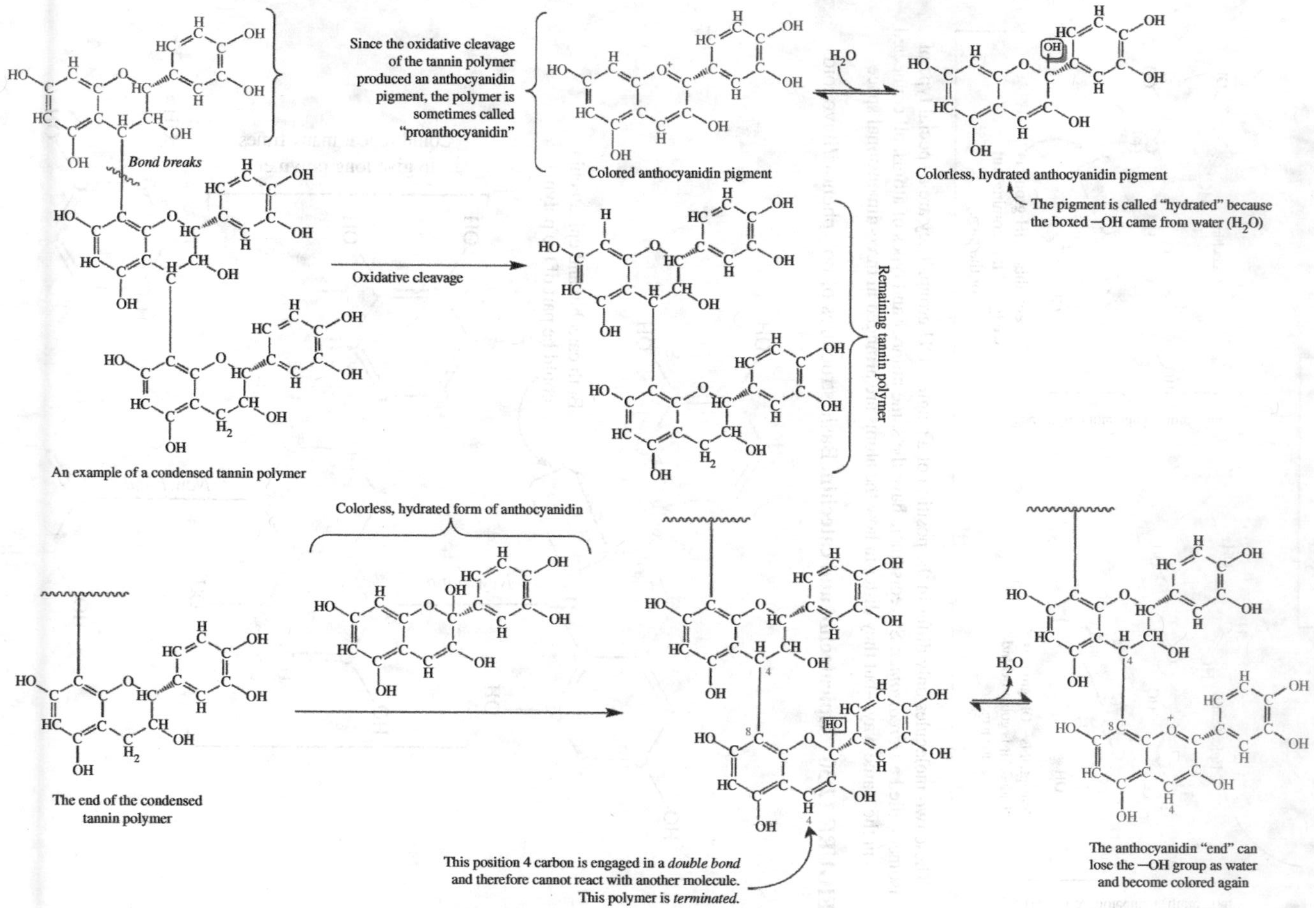

Since the oxidative cleavage of the tannin polymer produced an anthocyanidin pigment, the polymer is sometimes called "proanthocyanidin"

Remaining tannin polymer

Colorless, hydrated anthocyanidin pigment

The pigment is called "hydrated" because the boxed —OH came from water (H_2O)

H_2O

Colored anthocyanidin pigment

H_2O

The anthocyanidin "end" can lose the —OH group as water and become colored again

Oxidative cleavage

An example of a condensed tannin polymer

Bond breaks

Colorless, hydrated form of anthocyanidin

The end of the condensed tannin polymer

This position 4 carbon is engaged in a *double bond* and therefore cannot react with another molecule. This polymer is *terminated*.

FIGURE 12.32   The complex reaction of tannin.

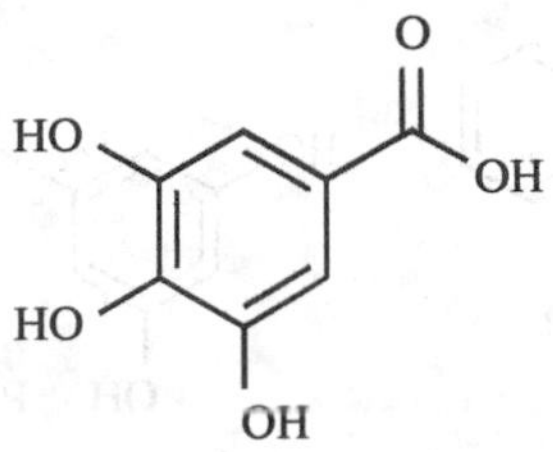

Gallic acid

FIGURE 12.33   Gallic acid is used to form hydrolyzable tannins.

The anthocyanins are considered a different class of polyphenols, and they are largely responsible for the color and organoleptic character of wine. Anthocyanins (and anthocyanidins) change color with pH and react with oxygen to lose their color. They are also antioxidants (by their nature of reacting with dissolved oxygen sparing the reaction from compounds) and have antimicrobial and reported anticarcinogenic activity.

Hydrolyzable tannins are derivatives of gallic acid and found bound to carbohydrates. Thus, the term "hydrolyzable tannins" comes from the enzymatic removal of tannins from the core sugar by using a water molecule to cleave the connecting bond. There are over a thousand variants of these tannins. The two core forms are the gallotannins and ellagitannins. Both are larger water-soluble tannins that are found in spices, oak, as well as wine grapes. Grapes have low quantities of hydrolyzable tannins. However, oaking a wine (or aging in a wooden barrel) will extract some tannins into the wine. These tannins are considered soft as they have an astringent feel but are not bitter. This differs from the smaller molecular weight tannins, which are called hard tannins, because they are both bitter and astringent (Fig. 12.33).

12.14.1   Nonflavonoid Phenols

Stilbenes represent a predominant class of nonflavonoid phenol compounds found in wine; the more simple is resveratrol (Fig. 12.34). These compounds originate from the stems and skin rather than the pulp or seed of the grape vine. Several studies indicate a significant role for stilbenes in cardiovascular disease and some types of cancer [8]. Caftaric acid is an example of a second nonflavonoid phenol class known as hydroxycinnamates. These compounds are easily found in most plants and react with tartaric acid. Caftaric acids, like stilbenes, have little or no flavor component.

Terpenes are a large class of compounds created by building polymers of isoprene units (Fig. 12.35). These are not considered polyphenols. Several of the terpenes found in wine are the same as those found in spices and along with esters provide spicy notes in wine. Terpenes accumulate in the skin and pulp of the berry and are therefore prevalent in white and red wines. Most terpenes are modified with an alcohol making the compound volatile and therefore part of the wine fragrance. The most common terpenes in wine give the aromas of floral, rose-like (geraniol and nerol), coriander (linalool), and citrus (citronellol and limonene).

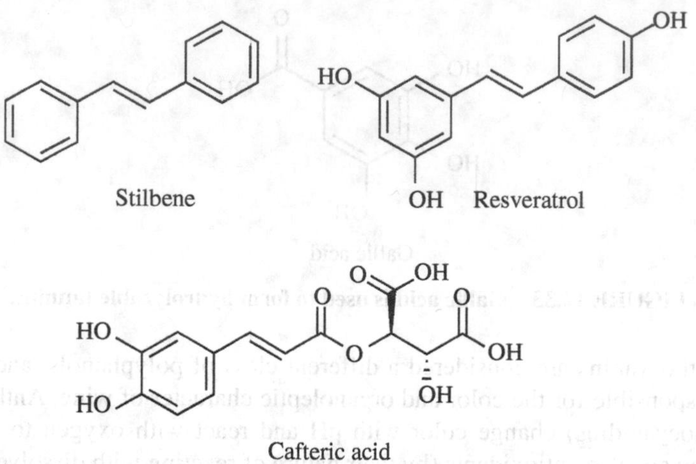
Stilbene
HO
OH
OH
Resveratrol
HO
HO
O
O
O
OH
O
OH
OH
Cafteric acid

FIGURE 12.34 Nonflavonoids.

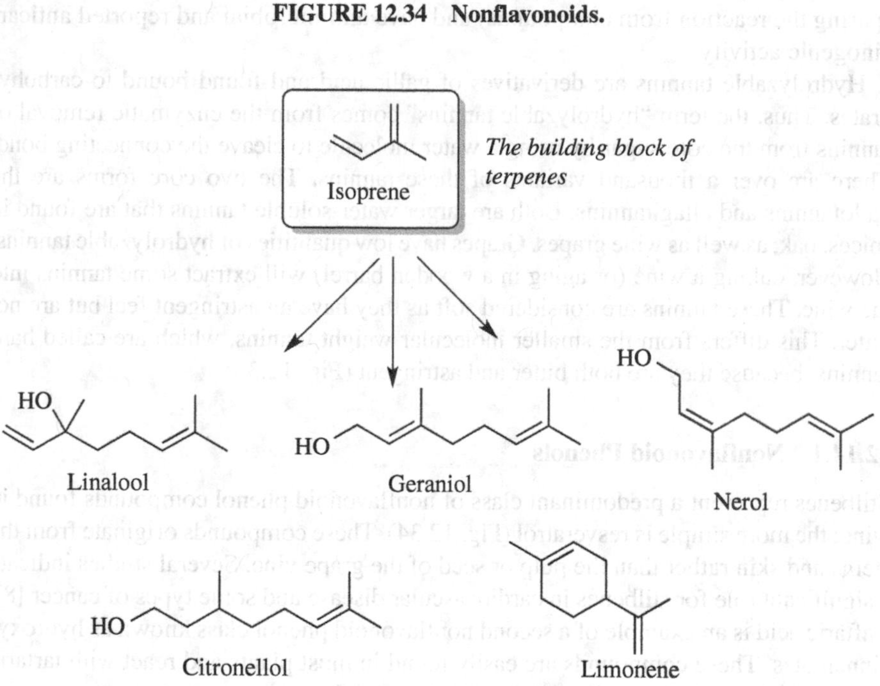
Isoprene
The building block of terpenes
HO
Linalool
HO
Geraniol
HO
Nerol
HO
Citronellol
Limonene

FIGURE 12.35 Common terpenes created from units of isoprenes.

12.15 AGING AND REACTIONS

Dr. Ann C. Noble, a Professor at the University of California Davis Viticulture and Enology Department, created the aroma wheel to characterize and appreciate the complexity of flavor and aroma in white and red wines [9]. This wheel is a fantastic demonstration of the diversity of flavorants in wine. While many (but, by far, not all)

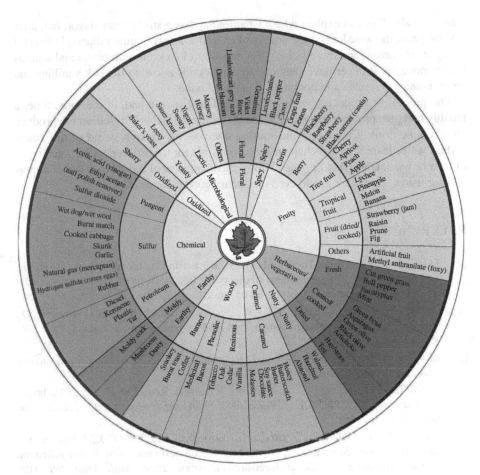

FIGURE 12.36 Wine aroma wheel. Copyright A C Noble 1990, 2002. www.winearomawheel. com. The smallest most inner circle is the general wine bouquet terms. The next circle describes the simple terms, while the outer circle groups the individual flavors of that category.

of the important components have been described in this chapter, we have yet to account for reactions between many of these molecules as a wine ages. Interaction of wine with proteins, fining agents, and oxygen will result in a loss or change in the flavor and aroma molecules. Consider the impact of adding bentonite to remove fines. Tannins will bind to proteins and grow to a large insoluble polymer. Acetic acid and acetaldehyde will react with several of the large molecules enhancing and changing their impact on wine flavor and aroma (Fig. 12.36).

Young red wine is often very astringent with excess tannins and bitter compounds. Aging will lead to loss of some of these and other aromatic compounds, mellowing out the wine. Tannins will react with sugars in the wine reducing their bitterness and softening the wine's astringency. Esters of fatty acids change during aging, diminishing the volatile notes. Young white wines will lose a few types of esters while other more stable esters remain. This can explain why a Chardonnay has a strong pear flavor, but, after several years, the wine loses those notes and gains the notes of buttery diacetyl (diacetyl is higher in content and more stable). The reaction of oxygen with esters and tannins will remove the stronger flavor and aroma compounds allowing the oak vanillins and softer tannins to remain and dominate in the aged wine.

The process of producing wine and beer is fairly simple and, at the same time, a terribly complex process. Sugar and yeast with a few nutrients will ferment to produce an alcoholic drink. Conditions and starting components make a world of difference and, as we've learned in this chapter, are responsible for the differences between types of beer or wine. Understanding the biology and chemistry of these processes is important for using, appreciating, and even producing good wine or beer.

REFERENCES

[1]   *The Code of Hammurabi*. Translated by L.W. King with Commentary from C.F. Horne. The Avalon Project. Yale University. Retrieved 12 July 2015. http://avalon.law.yale.edu/ subject_menus/hammenu.asp (accessed on November 17, 2015).

[2]   Caspeta, L., Chen Y., Ghiaci, P., Feizi, A., Buskov, S., Hallstrom, B.M., Petranovic, D., Nielsen, J. 2014. Altered sterol composition renders yeast thermotolerant. *Science* 346: 75–78.

[3]   Lam, F.H., Ghaderi, A., Fink, G.R. and Stephanopoulos, G. (2014) Engineering alcohol tolerance in yeast. *Science* 346: 71–75.

[4]   USDA Table of Nutrient Retention Factors. Release 6. Retrieved 12 July 2015. http:// www.ars.usda.gov/SP2UserFiles/Place/80400525/Data/retn/retn06.pdf (accessed on November 17, 2015).

[5]   Libkind, D., Hittinger, C.T., Valerio, E., Goncalves, C., Dover, J., Johnston, M., Goncalves, P. and Sampaio, J.P. (2012) Microbe domestication and the identification of the wild genetic stock of lager-brewing yeast. *Proc. Natl. Acad. Sci.* 108: 14539–14544.

[6]   Roullier-Gall, C., Boutegrabet, L., Gougeon, R.D. and Schmitt-Kopplin, P. (2014) A grape and wine chemodiversity comparison of different appellations in Burgundy: vintage vs terroir effects. *Food Chem.* 152: 100–107.

[7]   Jackson, R. (2008) *Wine Science: Principles and Applications*. 4th edn. Elsevier.

[8]   Tome-Carneiro, J., Larrosa, M., Gonzales-Sarrias, A., Tomas-Barberan, F.A., Garcia-Conesa, M.T. and Espin, J.C. (2013) Resveratrol and clinical trials: The crossroad from in vitro studies to human evidence. *Curr. Pharm. Des.* 19(34): 6064–6093.

[9]   Ann Nobles and the Aroma Wheel. Retrieved 12 July 2015. http://winearomawheel.com (accessed on November 17, 2015).

13

SWEETS: CHOCOLATES AND CANDIES

Guided Inquiry Activities (Web): 1, Elements, Compounds, and Molecules; 3, Mixtures and States of Matter; 7, Carbohydrates; 12, Emulsion and Emulsifiers; 13, Flavor; 31, Chocolate Properties; 32, Chocolate Tempering; 33, Sugar

13.1   INTRODUCTION

Few people can walk by a candy shop without being drawn in by the sight of brightly colored rock candies and lollipops and the aromas of chocolate and caramel. And then, once you try a sample of smooth, creamy fudge or a piece of chewy, sweet salt-water taffy, you might yearn for more, want to prepare some at home, or wonder about the history and science behind candymaking. If you resonate with any of these ideas, this is the chapter for you! In this chapter, we are going to learn about some of the history and science behind different sweeteners, chocolate, and candy.

13.2   SUGARS AND SWEETENERS

We have all used a variety of sweeteners during baking and food preparation: white granulated sugar sprinkled on strawberries, molasses in gingersnap cookies, and powdered sugar in a frosting. All of these sweeteners fulfill the criteria of being sweet; however, if you have ever replaced brown sugar with white sugar in a cookie recipe, you know that this small change makes a big difference in the final product. The white sugar cookie is crisp and crunchy, while the brown cookie is chewy and

The Science of Cooking: Understanding the Biology and Chemistry Behind Food and Cooking,
First Edition. Joseph J. Provost, Keri L. Colabroy, Brenda S. Kelly, and Mark A. Wallert.
Companion website: www.wiley.com/go/provost/science_of_cooking

FIGURE 13.1   Granulated sugar. Table sugar (sucrose) in granular form.

flexible. What is the difference in the sugars and sweeteners that we use in cooking? Why might you choose to use one to serve a particular culinary purpose but not another? The answer lies in understanding the structure and chemistry of carbohydrates (Fig. 13.1).

13.2.1   Sugars

13.2.1.1   Granulated Sugar   Ah ... granulated white sugar. We use it for making a pitcher of cold lemonade in the summer, we sprinkle it on our morning cereal, and we use it in our favorite brownie recipe. When someone refers to "sugar," this is the white crystalline solid that we visualize. Granulated sugar is produced from sugarcane, a member of the tallgrass family that grows in tropical regions, and sugar beets, a parsnip-looking vegetable that grows in the more temperate, cooler northern climates (Fig. 13.2a). About 75% of the world's production of sugar comes from sugarcane; Brazil is the leading producer of the product and generated 35.75 million tons in 2011/2012. Sugar beets produce 55–60% of the sugar used in the United States, with the rest produced from sugarcane (Fig. 13.2b).

At this point, you already know about the molecular structure of granulated white sugar; it is over 99% pure sucrose; granulated or table sugar is pure sucrose, the disaccharide that contains one molecule of glucose linked to a molecule of fructose via an oligosaccharide bond (Fig. 13.2c). The crystalline structure and molecular components of sucrose give it some really interesting properties in the kitchen. Let's learn more about table sugar.

First, how does granulated sugar become a crystal? The crystalline structure of sugar comes from its production process. To produce sugar, juice that contains sugar is extracted from the cane or beet. Some of the water in the liquid juice is evaporated and impurities removed; the liquid is boiled further to make a supersaturated sugar

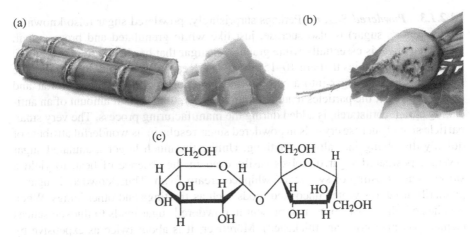

FIGURE 13.2   Sucrose. Two of the most common sources of sucrose are (a) sugarcane, (b) sugar beet, and (c) Sucrose is a disaccharide of glucose and fructose.

syrup. Then, as the supersaturated sugar syrup cools (we will discuss this further later in this chapter when we look at chocolate), the individual sucrose molecules align themselves to interact with one another via intermolecular interactions. Eventually this gives rise to the solid crystals that you know as sugar forms. Each sugar crystal is comprised of many sucrose molecules that interact with one another via intermolecular forces. The intermolecular forces that hold sugar crystals together are noncovalent, hydrogen bonds, formed between –OH groups of different sucrose molecules.

The size of the crystals that form gives us different types of granulated sugars. Typical granulated sugar crystals are 400–600 µm in size, superfine or "caster sugar" crystals are 200–450 µm, and sanding sugar granules that you might use on top of a muffin are 600–800 µm in size. However, even though the crystal size varies, all of these types of sugars are comprised of crystals of sucrose.

13.2.1.2   Brown Sugar   Brown sugar is a sucrose sugar, just like white granulated sugar, that has a distinctive brown color and texture due to the presence of molasses. Brown sugar contains from 3.5 (light brown sugar) to 6.5% molasses (dark brown sugar). Its soft, moist texture is due to the hygroscopic nature of molasses; it attracts water even more readily than sucrose and causes baked goods to retain moisture and chewiness that cannot be matched by granulated sugar. Molasses is a thick solution made by heating syrup of crushed sugarcane or sugar beets. The liquid is heated to remove water, condense the liquid, and promote Maillard reactions between proteins and sugars and caramelization reactions between sugars to produce some of the browned/blackened components of molasses. The crystallized sugars that result from this process are removed, and the resulting liquid is molasses. The additional flavor molecules present within the molasses due to the Maillard and caramelization reaction products impart a rich, caramel-like aroma and taste to the food. If you replace brown sugar with white sugar in a recipe or vice versa, you will notice a difference. However, the myth stating that brown sugar is healthier than white sugar is not true.

13.2.1.3   *Powdered Sugar*   Perhaps surprisingly, powdered sugar (also known as confectioner's sugar) is also sucrose, just like white granulated and brown sugar. Powdered sugar is essentially white granulated sugar that has been ground or milled to very small particles that are 10–15 μm in size. Really! You could do this at home by putting normal sugar into a coffee grinder or pulverizing it with a mortar and pestle. However, the particles tend to clump together, thus a small amount of an anti-caking agent, cornstarch, is added during the manufacturing process. The very small particle size of sucrose crystals in powdered sugar results in its wonderful attribute of quickly dissolving into almost anything. Unlike the much larger granulated sugar crystals, powdered sugar dissolves easily, even in the absence of heat, to yield a smooth cake frosting or sweetened whipped cream foam. Thus, powdered sugar is primarily used to sweeten uncooked foods such as frostings and other icings. When it is heated, the 2–3% cornstarch present in powdered sugar tends to thicken sauces (which you may not want thickened). Moreover, it is about twice as expensive by weight, as granulated sugar, which is another limitation on its use.

13.3   PROPERTIES OF THE SUCROSE-BASED SUGARS AND USE IN THE KITCHEN

What are some of the properties of sucrose that are important in cooking? As you have learned, like dissolves like. Sucrose has a lot of polar, hydroxyl (–OH) groups in it and so does water (Fig. 13.3). Thus, when you put a teaspoon of sugar into a cup of water, the sugar crystals dissolve. Why? Each molecule of sucrose disperses away from its sucrose partners within the crystal and becomes surrounded by and forms new intermolecular interactions with water because there are so many more molecules of water than molecules of sucrose (Fig. 13.3). Because of these interactions, sugar dissolves in your glass of iced tea and, as long as you stir it and maintain a 2:1 ratio of water to sugar, doesn't sit on the bottom of the glass.

Intermolecular interactions between sucrose and water are not only important for solubility but also in the preparation of baked goods and candy. Due to the intermolecular interactions between the sucrose and water, sugar helps to retain the moisture in a baked good. Did you ever think about how you can bake a pan of brownies at 350°F/177°C and not lose all of the moisture out of them? How on earth do brownies remain soft and chewy when water evaporates at 212°F/100°C? The intermolecular interactions between sucrose and water are so strong that some of the molecules of water remain bound to the sucrose and do not evaporate away, even when baked at this high temperature.

However, you also may have figured out that sucrose is not the only thing that is important to retention of moisture in and appearance of baked goods. Remember the differences between a brown sugar cookie and a white sugar cookie? A brown sugar cookie is soft and chewy but kind of dull looking. A white sugar cookie is dry and crisp, with a glossy or shiny appearance. There must be something else about these two sugars that explains why you use white sugar in a brownie and brown sugar in a chocolate chip cookie. The "something else" is an invert sugar. And, no, there is not an extraverted sugar.

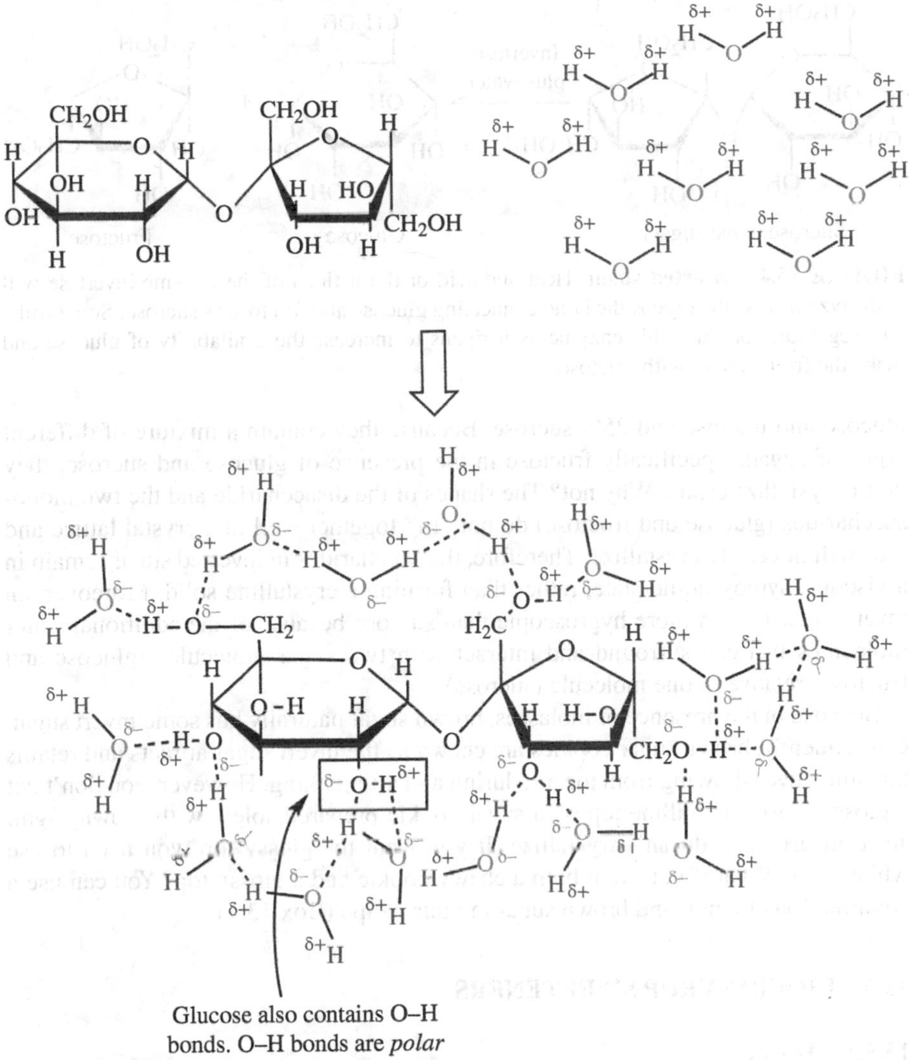

Glucose also contains O–H
bonds. O–H bonds are *polar*

FIGURE 13.3 Like dissolves like. The polar characteristic of water allows for it to form
hydrogen bonds with the polar –OH bonds of sucrose as water dissolves the sugar.

13.4 INVERTED SUGARS

When sucrose is heated in the presence of acid (like the citric acid from a lemon) or
an enzyme called invertase, its disaccharide bond is broken, and the free glucose and
fructose molecules are released (Fig. 13.4).

This process is called inversion; the resulting mixture containing a mixture of
unbroken sucrose, glucose, and fructose is called an invert sugar or invert syrup.
What does it have to do with the properties of sugar? Invert sugars are about 75%

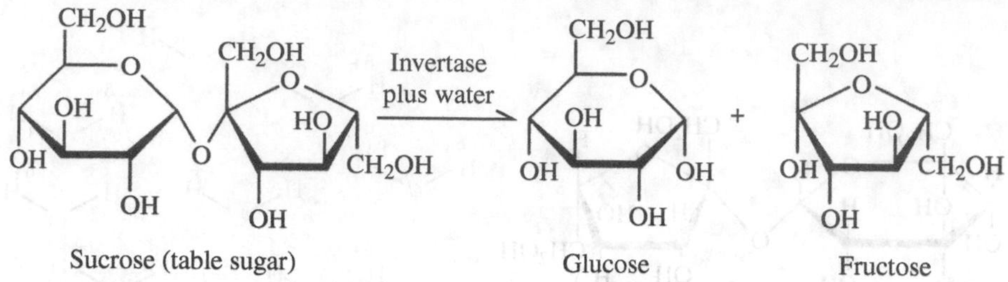

FIGURE 13.4   Inverted sugar. Heat and acid or the actions of the enzyme invertase will hydrolyze (break) the glycosidic bond connecting glucose and fructose as sucrose. Some fruits and vegetables produce this enzyme as it ripens to increase the availability of glucose and make the fruit sweeter with fructose.

glucose and fructose and 25% sucrose. Because they contain a mixture of different types of sugars, specifically fructose in the presence of glucose and sucrose, they don't crystallize easily. Why not? The shapes of the disaccharide and the two mono-saccharides (glucose and fructose) do not "fit" together well in a crystal lattice and thus will not easily crystallize. Therefore, the saccharides in inverted sugar remain in a viscous, syrupy liquid state, rather than forming a crystalline solid. Moreover, an invert sugar is even more hygroscopic than sucrose because of the additional water molecules that can surround and interact with two sugar molecules (glucose and fructose) relative to one molecule (sucrose).

Because of the presence of molasses, brown sugar naturally has some invert sugar. Consequently, brown sugar cookies are chewy as the invert sugar attracts and retains moisture, even drawing from the air, during and after baking. However, you don't get a glossy, crisp, crystalline top on a sugar cookie prepared solely with brown sugar, since invert sugar doesn't crystallize. If you want the glossy top, you need to use white sugar. What if you want both a chewy cookie and a glossy top? You can use a combination of white and brown sugar in your recipe (Box 13.1).

13.5   LIQUID SYRUP SWEETENERS

13.5.1   Honey

Honey is a sweet syrup made by bees using the nectar of flowers. The sweetening power of honey comes from the presence of fructose (25–44%), glucose (24–36%), small amounts of sucrose (0.5–2%), and trace amounts of approximately 25 differ-ent oligosaccharides. However, because honey is the product of bee-collected pollen, there are also a number of other nonoligosaccharide components including proteins, enzymes, amino acids, trace minerals, and polyphenol compounds (providing aroma). The source of pollen, climate, and type of bee will change the ratio and type of the sugars and these important trace components; thus, the aroma, taste, and sweetness of honey varies quite widely. Given the high concentration of fructose within all honeys, it is very sweet, was the most important sweetener in Europe until the sixteenth century when cane sugar became more widely available,

---

BOX 13.1   INVERTING SUGAR FOR JAM

The inversion of sucrose is an essential part of the process of making jams and preserves. During the jam-making process, a fruit and a large amount of added sugar are boiled together for a period of time. During boiling, the fruit softens, its pectin (plus any added pectin) is dissolved into the mixture, and much of the fruit's water is evaporated away. The mild acidity of the fruit promotes the hydrolysis of up to half of the added sucrose to glucose and fructose (i.e., inversion). This chemistry has various beneficial effects in jam making. First, the total number of sugar molecules is increased by up to 50%. This boost gives a modest increase in the sweetness of the mixture, but, most importantly, it increases the proportion of the water that is bound to sugar molecules; this water is therefore no longer available to support the growth of microorganisms. Moreover, the invert sugar is much more soluble in water than the original sucrose and represses crystallization. Because of these characteristics, it is easy to reach the very high sugar levels (and low water levels that are essential for long-term stability of a jam).

---

and does not crystallize easily (remember that fructose doesn't crystallize in the presence of glucose).

In cooking, honey is often used as a spread on bread and in teas or other hot drinks where its sweetness and nuanced floral flavors and aromas can be enjoyed. In baking, it effectively retains moisture in breads and cakes because it is hygroscopic like brown sugar and contributes to the leavening of quick breads and other baked goods that use baking soda as a leavening agent due to its acidity, which lies between pH 3.1 and 6.1 depending upon the source. The sugars in honey facilitate caramelization and Maillard browning reactions (with proteins) in the crusts of baked goods and meat glazes (think of a honey-baked ham) (Box 13.2).

13.5.2   Corn Syrup and High-fructose Corn Syrup

Corn syrup and other syrups made from starches were not developed until the early 1800s but are now major players within the food industry due to their inexpensive production costs and useful culinary characteristics. As you know, all fruits and vegetables contain starch, which are long chains of glucose molecules (see Chapter 1).

Starch can be broken into smaller pieces in the presence of acid or enzymes, which catalyze cleavage at the glycosidic bond between glucose molecules. As this degradation occurs, you are left with some smaller chains of starch and some individual glucose molecules. If you have eaten a "starchy" russet potato, you know that it does not taste particularly sweet. However, when some of the starch is broken down into individual molecules of glucose, it becomes sweeter and has a thick, viscous consistency. If the original starch comes from corn, you have a corn syrup. If it comes from potatoes, you have a potato syrup. Basically, you can generate syrup from any

BOX 13.2   THE HEALTH BENEFITS OF HONEY

The antibacterial properties of honey are fairly well known. The low water content of honey makes the sugar inhospitable for microbial growth. In fact, there is so little water in honey, the sugar solution will cause water to move out of most microbes into the sugar solution (via osmosis) effectively dehydrating and killing the bacteria. In addition to the low water content, the growth of over 60 different species of bacteria has been found to be partially or fully inhibited by honey. Ancient Sumerians used honey as a drug on wounds, and Aristotle told of the healing power of honey on sore eyes and wounds. What are the molecular components of honey that provide this antimicrobial effect? The enzyme glucose oxidase converts glucose to hydrogen peroxide (H_2O_2) and gluconolactone, which is further transformed to gluconic acid. Hydrogen peroxide modifies the proteins and lipids of bacteria, stopping their growth, while gluconic acid acts as an antifungal and antibacterial agent. Several other compounds, including methylglyoxal and methyl syringate, which are produced by the partial digestion of plant material by the bee, also block microbial growth.

high-starch fruit or vegetable. However, although you have a sweet syrup, the resulting syrup only has 30–40% of the sweetness of sucrose. Thus, you would need to add a lot of the syrup to a recipe to yield the same sweetness as sucrose. How is sweetness measured? The sweetness of a particular molecule or food depends on the type of sugar and how the sugar molecule binds to the taste receptors (see Chapter 2). Understanding the differences in how sugars taste and their sweetness can help a candymaker create interesting confections. Lactose, the sugar found in milk, has the lowest sweetness value, while table sugar, high-fructose corn syrup (45 and 55%), and fructose all have similar sweet intensity. Glucose is about 30% as sweet as table sugar (sucrose).

In the 1960s, there was a breakthrough in the starch syrup industry that led to the discovery of high-fructose starch syrups. Starch syrup is typically prepared by extracting the starch out of a fruit or vegetable (oftentimes from "field" or "dent" corn, which is high in starch and low in glucose, making it unpalatable for humans) and is broken down into smaller starch fragments and individual glucose molecules using enzymes called amylose (found in your saliva to break down starch into smaller chunks). After the long starch polymers are broken into smaller pieces, another enzyme called a xylose isomerase is added, which converts (isomerizes) some of the glucose into fructose. Since fructose is sweeter than glucose, a high-fructose corn syrup has more sweetening power than a regular corn syrup. In fact, a high-fructose corn syrup that contains approximately 53% glucose and 42% fructose will provide the same sweetness as the equivalent weight amount of sucrose. If more of the glucose is isomerized to fructose to reach a 53% fructose and 42% glucose ratio, you have a syrup that is as sweet as honey. Because some longer polysaccharides are still present within the mixture, the long molecules tangle up with one another, interfere with crystallization of the individual glucose and fructose molecules, and slow down the motion of all molecules; this gives corn syrup a much thicker, more viscous consistency than any sucrose syrup.

13.5.3 Molasses

You already know a little bit about molasses and how it contributes to the properties of brown sugar, but molasses, by itself, can also be used as a sweetener. Molasses is essentially a by-product of extracting sugar from sugarcane. After the juice is extracted from the sugarcane (by crushing or mashing the cane), the juice is boiled to concentrate the sugar. A raw form of granulated sugar crystallizes out of this boiled solution. However, the remaining liquid syrup (called cane syrup or first molasses) still has a very high sugar content. The cane syrup is mixed with some uncrystallized sugar syrup and boiled again; more raw sugar crystallizes. This is the syrup from which molasses is derived. This process can continue a third time to yield "blackstrap" molasses. With each boiling, the liquid syrup gets darker and darker due to the caramelization of the sugars that occurs during the repeat boiling.

The darker the molasses, the more its sugars have been transformed by caramelization and Maillard browning reactions. Thus, blackstrap molasses is very bitter and not particularly sweet, while molasses made from the first and second boiling are readily used in cooking and baking when a complexity of flavors, including hints of caramel, butter, green tones, and sweetness, are necessary. Not only do molasses provide a sweetness and moisture retention to foods, but also it adds a rich flavor to baked goods such as gingerbread and spice cakes and a savory appeal to barbeque sauces and baked beans.

13.6 CHOCOLATE

Chocolate is arguably one of the world's most loved foods. Over 600 different types of molecules that contribute to its acidity, bitterness, astringency, sweetness, creaminess, and chocolate flavor and aroma make chocolate one of the most complex, flavorful foods that is known. Interestingly, however, chocolate comes from a bean that is itself quite unpalatable; it is crunchy, astringent, bitter, and essentially aromaless. Thus, the realization of its full potential as a delectable food has taken hundreds of years by numerous peoples. The details of the very early history of chocolate are not well established. As far as we know, the cocoa tree, *Theobroma cacao*, first appeared in South American tropical rain forests, and its potential as a food was realized by Mayan, Inca, and Aztec civilizations (Fig. 13.5).

Around 700 AD, the tree was carried northward (toward Mexico) by the Mayans and was cultivated and utilized as a food source throughout that region, as the beans contained fat, starch, and protein. The tree only flourishes in areas that are 20° north and south of the equator, so cultivation of the tree was (and still is) quite limited. However, during this time, the beans were exported northward into what is now the United States and were so highly valued that they were used as a form of currency.

FIGURE 13.5   Early use of chocolate. The cover of a famous book *Traités nouveaux &
curieux du café du thé et du chocolate*. Dufour first described the beneficial pharmacological
effects of chocolate including counteracting drunkenness, upset stomach, and menstrual
disorders.

The word chocolate is derived from the Aztecs as "xocolatl"; xocolatl was the name
of a drink in which roasted beans were simmered in hot water, flavored with red
pepper and vanilla, and thickened with ground corn. The beans were also used as a
spice to flavor meat dishes in the Aztec civilization, similar to moles that are used in
Mexican cuisine still today.

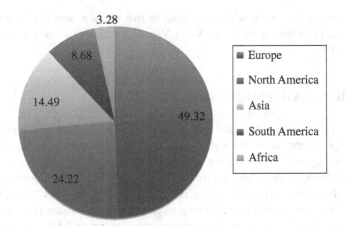

FIGURE 13.6   Global chocolate consumption.

The detailed history of chocolate becomes more documented when the first cocoa beans were brought to Europe in 1502. The beans were seized during the fourth and last voyage of Christopher Columbus near an island off the coast of what is now Honduras, unimpressively mistaken for almonds. However, the Europeans had little understanding of how to use the beans and primarily prepared a drink similar to that of the Aztecs, with less red pepper. Eventually, the chocolate drink was modified and sweetened by mixing the cocoa beans with milk, sugar, and eggs in Spain, and it gradually spread throughout Europe, arriving in England in 1650 where it became particularly popular and was the basis for the famous chocolate houses of London.

Chocolate drinks continued to be developed and adored throughout the seventeenth and eighteenth centuries in Europe; however there was an undesirable layer of fat within the drink due to the high-fat content of the beans.

The first report of the use of a press to remove fat from the beans occurs in a French treatise, published in 1678 [1]. However, the credit for the invention of the process for extracting cocoa butter from the cocoa bean to make cocoa powder is given to the Dutchman, Coennraad van Houten, in 1828. The ability to separate the cocoa butter from the cocoa solids was a breakthrough in the culinary world and allowed the production of solid chocolate by an English company in 1847, and 2 years later, milk chocolate was developed by a Swiss firm. Interestingly, the love of chocolate continues to live on in both Switzerland and England. The Swiss ranked first in annual chocolate consumption, at 11.9 kg per person in 2012, while the Irish and English ranked second and third, respectively, consuming 9.9 and 9.5 kg per person (Fig. 13.6). The average American eats 5.5 kg of chocolate annually, which corresponds to 128 Hershey's milk chocolate bars. We love our chocolate!

Fortunately, chocolate is not only delicious to eat, but there is also a myriad of cool science that is fundamental to its flavor, aroma, and behavior. In this chapter, you will learn about the steps and science involved in the production and manufacture of chocolate from cacao beans, the favorable or unfavorable transformations that occur when you work with chocolate in the kitchen, and the different types of chocolate. Let's take a walk into Willy Wonka's factory to learn more about the science of chocolate.

13.7 CHOCOLATE PRODUCTION

Chocolate is a natural product of the cacao tree, *T. cacao*, which now grows within a narrow range of conditions throughout the wet, lowland tropics of Southeast Asia, South America, and West Africa but was native to South America. The Swedish botanist, Carl Linnaeus, classified the tree as *T. cacao* in 1728. The name was perhaps appropriately given as *Theobroma* is Greek for "food of the gods," even though its use at the time was still limited largely to hot drinks. *T. cacao* is a broad-leaved evergreen tree that flourishes in damp places, growing up to 7.5 m tall. At the age of 3 or 4, it produces white flowers; once it flowers, it flowers year round, undergoing pollination by tiny flies called midges. Pollinated cacao flowers produce fruit (i.e., seed pods) in approximately 40 days, which becomes melon shaped (i.e., seed pods) with a leathery shell of 20 cm in length and 10 cm in diameter in another 5–6 months. Each pod contains 20–40 seeds (i.e., cacao beans; Fig. 13.7). Each bean consists of two cotyledons and an embryo surrounded by a seat coat (testa) and is enveloped in a sweet, slightly acidic, juicy white pulp that comprises approximately 40% of the seed flesh weight.

Both the bean and the pulp are key components to obtaining the chocolate that we love to eat, although at this early stage it neither tastes, smells, nor looks like chocolate. How, then, do we get chocolate? Chocolate is produced through a few steps some of which we have already talked about: (i) fermentation of the beans and pulp, (ii) roasting of the fermented beans, (iii) grinding and more mixing, (iv) conching, and (v) tempering.

FIGURE 13.7   Cocoa pod.

13.8 FERMENTATION

Fermentation? You might think that this is a mistake, as chocolate doesn't have any of the flavors or aromas of either cheese or alcohol. However, fermentation is a key step to chocolate production and is carried out by similar yeast and microbes that we discussed earlier. What occurs during fermentation of cacao beans? Fermentation is essential for removing the fleshy, sweet pulp that envelops the beans and developing the precursor molecules that are essential for chocolate flavor. When does fermentation occur? Fermentation occurs on the farms and plantations where the cacao bean seed pods are harvested. Let's take a closer look at the process and science of cacao bean fermentation.

After harvest of the cacao pods, which may occur by a human with a machete (or less ideally an animal with claws and teeth), the pods are placed in piles and left to ferment in the heat and humidity of the tropics for a few days. The resting allows the seed pod shell to weaken. After a few days, workers break open the pods and pile the beans and the pulp together in heaps, covering them with leaves. Exposing the nutrient- and sugar-rich pulp to the environment of warm and humid temperatures supports the growth of a succession of microbes including yeasts and several forms of bacteria. These microbes are in their element; they have food (i.e., the pod flesh), moisture, and tropical temperatures to grow and flourish for 2–8 days. What types of microbes are involved? The sugar-rich and acidic pulp favors growth by yeast first. The yeasts utilize the sugars (i.e., primarily glucose and fructose) from the pulp to make ethanol and CO_2 (Fig. 13.8), generate heat, and metabolize some of the naturally occurring citric acid in the pulp (this is what makes the pulp acidic). As expected, the pH of the pulp increases, and the amount of available oxygen decreases, which favors the growth of our microbe friends from cheese making, the lactic acid bacteria (lactobacillus).

As the lactic acid bacteria take over the fermenting mass, they utilize glucose to first lactic acid and citric acid and then metabolize the citric acid to other compounds, reducing the acids and further increasing the pH of the mass. In order to allow for continued growth, workers turn the pile at least once a day. As its temperature rises above 99°F/37°C, acetic acid bacteria become dominant, oxidizing ethanol to acetic acid

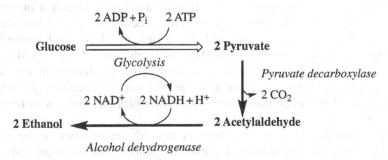

FIGURE 13.8 Ethanolic fermentation. In the absence of oxygen, yeast will metabolize glucose to pyruvate and ultimately produce ethanol to replace NAD for continued metabolism.

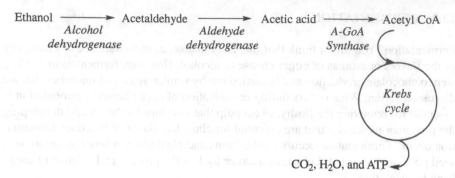

FIGURE 13.9 Ethanol metabolism. Bacteria will convert ethanol produced by yeast to make ATP under higher temperature conditions.

and then to CO_2 and water (Fig. 13.9). As ethanol and lactic acid levels decline, the bacteria shift to producing acetic acid, reducing the pH, and increasing the temperature of the fermenting mass even up to 122°F/50°C or more due to the exothermic nature of the reactions.

You might be thinking, all of this fermentation is happening in the pulp, but what about the bean? Cacao beans are also affected by fermentation. In short, the acetic acid bacteria create a vat of hot vinegar (i.e., acetic acid), which penetrates and produces holes inside the beans. The acidity kills the seed embryo as the compromised cells soak up some of the acids, sugars, and ethanol of the fermenting pulp. Some of the flavor and aroma of chocolate begins to be generated during this stage, as the contents of the cells mix together and start reacting with one another due to the acidic conditions. For example, some of the compounds in the beans react with proteins and oxygen, while the beans' digestive enzymes mix with proteins and sucrose to degrade them into amino acids and simple sugars (like glucose and fructose); these reactions produce some chocolate flavor precursors. Spore-forming bacteria soon take over fermentation, followed by filamentous fungi, which complete the metabolism of acetic acids in the beans (Fig. 13.10).

Although fermentation and metabolism of citric, acetic, and lactic acids involve an extensive amount of chemistry and biochemistry, it is largely a preparation process for the beans, providing them with the precursors that will later develop into molecules that we associate with the character, flavor, and aromatics of chocolate. The presence or absence of these acids also modulates bean pH, which affects the activity of bean enzymes involved in the production of free amino acids, peptides and reducing sugars, and Maillard reaction chemistry, all of which impact the final chocolate flavor. See Figure 13.19 for an overview of fermentation of cocoa [2].

Once fermentation is complete, farmers dry the beans, often in the sun for 1–4 weeks. Once they are dried to 7% moisture content, the beans are resistant to spoilage by microbes (there is not enough water present for a microbe to thrive) and are shipped to manufacturers for further processing.

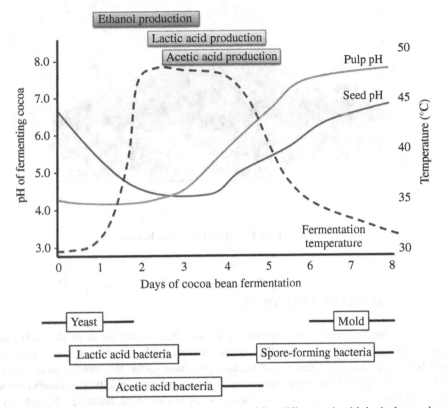

FIGURE 13.10 Fermentation of cocoa. A series of five different microbiological organisms are responsible for producing conditions to extract and begin the production of chocolate from cocoa beans.

13.9 CACAO BEAN ROASTING: THE PROCESS

Dried, fermented cacao beans are less astringent (due to a lower polyphenol content) and more flavorful than unfermented beans, but their flavor has yet to fully develop. In fact, at this point, the flavor is dominated by a vinegar odor and taste from the acetic acid. By roasting the beans for 30–60 min at 250–350°F/121–177°C, the glue that holds the cells together (i.e., the cell wall) continues to break down, and the reported 200–600 molecules that we associate with chocolate flavor are produced (Fig. 13.11).

You may be surprised that a relatively short-lived and mild heating process can transform a vinegar-tasting cacao bean into something that smells, tastes, and looks more like chocolate. What is happening at a chemical level? The gentle heat drives off the sour-tasting acetic acid and ethanol, while Maillard reactions of sugars, free amino acids, and oligopeptides yield a complex array of molecules that we associate with the flavor of chocolate.

FIGURE 13.11   Roasted cocoa beans.

13.10 FLAVORS OF CHOCOLATE

The cacao bean contains many natural molecules that contribute to its flavor. If you have ever (probably unintentionally) eaten unsweetened chocolate or cocoa powder and had your mouth go into a sour pucker, you have tasted the molecules that contribute to the bitterness and astringency of a cacao bean, specifically theobromine and caffeine (found in tea) and polyphenols such as catechins, flavan-3-ols, anthocyanins, and proanthocyanidins (Fig. 13.12).

Polyphenols undergo oxidation by enzymes called polyphenol oxidases to produce a range of tannin compounds we've investigated in wine, vegetables, and spices. Like in wine, tannin-derived polyphenols play an important role in flavor and aroma for chocolate. Another source of aromatics and volatiles found in roasted beans comes from the Maillard reactions that occur between the sugars (i.e., sucrose, glucose, and fructose) and free amino acids or small peptides that were produced during fermentation via acid hydrolysis reactions.

The fermented pulp also provides flavors and aromas of fruits, sherry, and vinegar to chocolate in the form of acids, esters, alcohols, and acetaldehyde, the most notable being phenylacetaldehyde, benzaldehyde, phenylethanol, 3-methyl-1-butanol, phenylethylacetate, and 2-heptanone. One class of molecules, pyrazines, is responsible for a roasted nutty chocolate note. Dutch processing, which decreases the acidity of chocolate, increases the concentration of the pyrazines. Some chocolate manufacturers will add butyric acid during the chocolate-making process. Butyric acid is associated with the rancid flavor present in aged milk; in chocolate, it contributes a tangy taste. As a common additive to American-made chocolate, thought to stabilize milk chocolate from fermentation, the compound is so easily detected by humans that it is one of the distinguishing characteristics between European chocolate and American chocolate (Box 13.3).

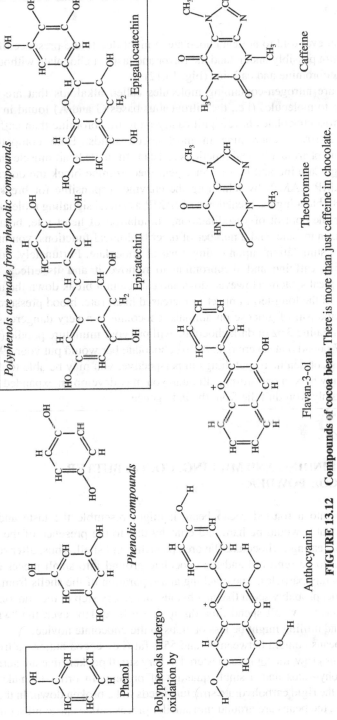

FIGURE 13.12 Compounds of cocoa bean. There is more than just caffeine in chocolate.

BOX 13.3

Were you ever told to not eat chocolate at night due to the presence of caffeine? You cannot possibly understand the flavor and taste of chocolate without thinking about theobromine and caffeine (Fig. 13.12).

Both are nitrogen-containing molecules called alkaloids that are similar in structure to molecules (i.e., the nitrogenous bases G and A) found in DNA and RNA. Most chocolates have significantly more theobromine than caffeine, containing 2–10% theobromine in most cocoa solids. Both compounds have significant pharmacological functions and can stimulate heart muscle and nervous signaling. Caffeine and to a lesser extent theobromine block the degradation of cyclic AMP (cAMP) by inhibiting the enzyme responsible for breaking down cAMP. By blocking the breakdown of cAMP, a critical signaling molecule responsible for the control of vasodilatation, stimulation of heart rate, breakdown of glycogen in muscle, and a number of other biological functions, you may have a sense of "stimulation" upon eating a bar of chocolate. Fortunately, humans can metabolize caffeine and theobromine to remove it and its effects from our physiological system. However, dogs and cats cannot break down the molecules. Because of the long-term impact of increased heart rate, blood pressure changes, and metabolism of glucose on animals, chocolate is a very dangerous food for your pet. Eating 3 oz of dark chocolate will put your miniature poodle (i.e., a dog of 15–17 pounds) at severe risk. A 7 oz chocolate bar would put your 190 lb. large dog at risk of death. To put things in perspective, you may be able to eat several large candy bars in one sitting; although you may develop an expanded waistline, you certainly wouldn't die from the indulgence.

13.11 GRINDING AND MILLING: COCOA BUTTER AND COCOA POWDER

If you bit into a roasted cacao bean, it might resemble the taste and aroma of chocolate, but it would be hard and crunchy due to the presence of the bean shell, perhaps like eating a chocolate-flavored pistachio nut shell. Thus, after roasting, the shells need to be removed and the chocolate ground into small pieces and further milled into tiny particles. The cracking and separation of the shells from the chocolatey product probably seem like an obvious, necessary step of the chocolate-making process; thus we won't spend time discussing it here. However, the "why" behind grinding and milling might be less obvious to the chocolate novice.

Cacao beans contain between 50 and 55% fat (i.e., cocoa butter—a triglyceride). As the beans grow, the fat is sequestered in very small pockets that are surrounded by rigid carbohydrates and a small quantity of protein. In order to make a smooth chocolate, the rigid carbohydrate structure needs to be broken down. In the first stage of grinding, the beans are ground into smaller pieces called cacao nibs, during which the carbohydrate walls surrounding the pockets of fat are broken down. As the nibs are ground further, the ground-up cocoa solids (mainly protein and carbohydrates) become suspended in melted cocoa butter (i.e., fat) yielding a thick, dark chocolate paste called the chocolate liquor. Pure cocoa liquor has a very concentrated chocolate taste, but its flavor is quite bitter, astringent (due to the presence of the polyphenols discussed earlier), and acidic (due to the presence of acetic, lactic, and citric acids).

To turn the bitter, pasty chocolate liquor into something that is similar to the smooth milk chocolate bar that you might want to eat, a manufacturer has to add a few more ingredients, such as sugar, milk solids (fats, proteins, and carbohydrates from dehydrated milk), vanilla, and an emulsifier. The addition of sugar is obvious to anyone who has bitten into a bar of unsweetened chocolate, while vanilla adds flavor to the final product. But why would you add more cocoa butter and an emulsifier? One reason is flavor but another is to keep various components including sugars (which are polar and water soluble) in suspension—or emulsified.

As discussed earlier, the chocolate liquor contains cocoa solids suspended in melted fat. There is enough cocoa butterfat to completely surround all of the more polar protein, and carbohydrate cocoa solids; you can imagine a sea of butter that contains some protein and carbohydrate molecules. However, as sugar is added to the chocolate liquor, the proportion of the fat in the liquor drops; thus there is no longer enough butter to coat both the cocoa solids and sugars. The resulting effect is a heterogeneous mixture, due to lack of fat coating on the sugar, carbohydrate, and protein molecules, that is a pasty mass rather than a smooth, flowing liquid. In order to obtain a smooth texture, additional cocoa butter is added in an amount that is proportional to the amount of added sugar. The extra fat ensures that the cocoa solids and finely ground sugar remain suspended in the liquid fat, thus allowing the liquor to flow. How much fat is necessary? Generally speaking, a smooth chocolate requires a minimum total fat content of 30%, but there is a way to add less cocoa butter. By adding an emulsifier that has amphipathic properties like lecithin, a phospholipid also known as phosphatidylcholine, the cocoa solids and sugar particles can be coated with the hydrophobic end of that molecule, which helps to bind all of the particles together and allows the chocolate to have good flowing properties. Even if a manufacturer is not striving to make a lower-calorie chocolate, lecithin is often still added to help the hydrophilic sugars and proteins adhere to the hydrophobic fat molecules (Box 13.4).

13.12   CONCHING

Conching, whose name was derived from the shell-like shape of the first machines that carried out this step, kneads the chocolate liquor, along with other added ingredients against a solid surface at temperatures of 115–190°F/46–88°C for a period of time between 8 h and 5 days. Wait a minute, this sounds a little like grinding and milling. Why is conching necessary? Although an emulsion is created following grinding and milling of the chocolate liquor, it is not yet a stable smooth emulsion. By breaking the solid particles (cocoa proteins, cocoa carbohydrates, sugar crystals)

BOX 13.4   CHOCOLATE AND EMULSIFIERS

Many liquids contain a suspension of small solid particles; these mixtures are called emulsions. The solid particles have a tendency to stick to one another and "fall out" of the liquid solution; thus an amphipathic molecule, called an emulsifier, is often added to ensure that the solids remain suspended in the liquid solution. How do emulsifiers work? One end of an emulsifier molecule has favorable interactions with and coats the surface of a solid particle, while the other end of the emulsifier dangles into and has favorable intermolecular interactions with the liquid. This interaction with both components of the mixture (solid particles and solution) allows the solid to remain suspended in the liquid. The viscosity of the liquid, as well as the amount and size of the solid particles in the emulsion, will govern the flow properties of an emulsion; the smaller the particles, the thicker the emulsion. In chocolate, the properties of the emulsion lead to the different textures of different types of chocolates. Generally speaking, the solid particles found in English chocolate are larger; thus they flow more quickly around the mouth (less viscous), and the consumer detects the feel and taste of the chocolate quickly. American chocolates are typically a thicker emulsion; thus they stay around in the mouth longer and give a more lingering taste.

---

and remaining sugar crystals into even smaller, fine particles during conching, the emulsion is no longer gritty. Moreover, conching separates the hydrophilic particles from one another, allowing for an even coating of all the particles with cocoa butter such that when the finished chocolate melts, it flows smoothly. Finally, due to the presence of heat, conching mellows the flavor of the chocolate. Aeration and heat cause approximately 80% of the volatile aromatic compounds, including acetic acid and some of the remaining water, to evaporate. This makes the chocolate product more basic or alkali, and importantly from a consumer standpoint, the chocolate no longer smells or tastes like vinegar. Additionally, a number of other volatile and distinctive aromas and flavorants of chocolate are generated, particularly those associated with roasted, caramel, and malty aromas (pyrazines, furaneol, and maltol).

You might be thinking, is the chocolate finished after conching? Essentially yes, it is finished if you are interested in liquid dark chocolate. After conching, dark chocolate you will have a warm liquid of cocoa butter that contains molecules of the original cacao beans, the flavorful and aromatic molecules produced during fermentation and roasting, and sugar. Milk chocolate will contain addition additives from dehydrated milk, which will add milk solids including fat, casein and whey proteins, and lactose. These will add body, emulsification, and smoothness to the mixture. However, it is difficult to sell and eat liquid chocolate; thus, the last step of the chocolate-making process is to cool the liquid to room temperature and allow it to solidify in the appropriate form (e.g., chocolate bars, chocolate-covered cherries). On the surface, tempering, a term that describes cooling down a material to harden it, seems pretty straightforward; cool the chocolate, it hardens, and you eat it. However, in chocolate making, this cooling step is the trickiest step of the entire process. To obtain a glossy chocolate that snaps when you break it and melts with a smooth creaminess in your mouth, a manufacturer must pay very close attention to the liquid as it is cooled and rewarmed. The care required in chocolate tempering is necessary because of the crystallization properties of cocoa butter.

13.13   TEMPERING

13.13.1   Crystallization

You have lots of personal experience with crystals in the kitchen in the form of salts, sugars, and maybe even crystallized ginger. Each of these crystalline forms has particular physical properties and characteristics that lend themselves for some purposes but not others. What is a crystal? Crystals are solids that form by a regular repeated pattern of molecules connecting together. In some solids, the arrangements of the building blocks (atoms and molecules) can be random or very different throughout the material. In crystals, however, the structure of the atoms/molecules is repeated in exactly the same arrangement over and over throughout the entire material. Each crystalline arrangement leads to different physical properties of the substance.

What does this have to do with chocolate? In order to obtain desirable physical properties of chocolate (i.e., a good crisp snap when you break it and a smooth texture upon melting), the fat molecules need to crystallize in a particular manner during tempering. You want the crystals to be stable enough to hold the chocolate together in a solid bar form that snaps when you break it. However, you also desire that the melting point of the crystals is above room temperature, but below body temperature, so that the chocolate truly does melt in your mouth but not in your hand. Finally, you need small crystals so that the chocolate has a smooth, glossy finish that doesn't look (or taste) gritty.

Cacao fat molecules are unusually regular, monounsaturated triglycerides (Fig. 13.13).

The uniformity of the molecules allows them to pack together tightly and form a dense network of stable crystals involving all of the fat molecules. However, this packing organization is only created when crystallization of the fat is carefully controlled. Cocoa butter can actually solidify into six different kinds of fat crystals, I–VI called polymorphs (Table 13.1), depending upon the crystallization temperature. A crystallization temperature is the temperature below which the fat molecules (or other types of molecules of interest) crystallize into a particular structure and above which melt into a liquid. The fats can stack in various ways forming double or triple fatty acid chain overlaps. The more stable and tighter the contact, the higher the melting point (Figs. 13.14 and 13.15). Of the six forms, it is polymorph V that has the favorable visual and textural characteristics that we associate with a great

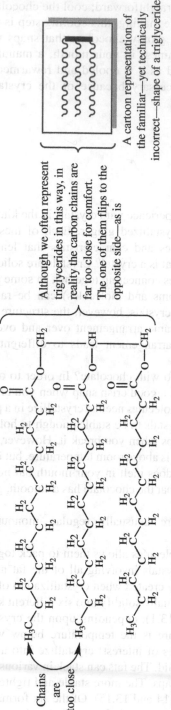

A cartoon representation of the familiar—yet technically incorrect—shape of a triglyceride

Although we often represent triglycerides in this way, in reality the carbon chains are far too close for comfort. The one of them flips to the opposite side—as is

Chains are too close

FIGURE 13.13 Fats of chocolate. Triacylglycerol (a backbone glycerol bound to three fatty acids or acyl chains) is a major component of chocolate.

TABLE 13.1   Chocolate Polymorphic Forms and Melting Point.

Form	Melting Point (°C)	Chain Packing
I	16–18	Double
II	21–22	Double
III	25.5	Double
IV	27–29	Double
V	34–35	Triple
VI	36	Triple

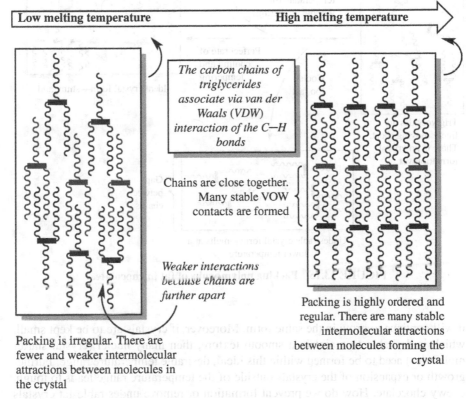

Low melting temperature   **High melting temperature**

The carbon chains of triglycerides associate via van der Waals (VDW) interaction of the C—H bonds

Chains are close together. Many stable VOW contacts are formed

Weaker interactions because chains are further apart

Packing is highly ordered and regular. There are many stable intermolecular attractions between molecules forming the crystal

Packing is irregular. There are fewer and weaker intermolecular attractions between molecules in the crystal

FIGURE 13.14   Fatty acid packing and melting point. The more contact the higher the melting point.

bar of chocolate. Polymorphs I–IV have an unstable, less organized, and looser network of fat molecules, which includes some of the fat molecules within the crystalline structure, while others remain in a liquid, unpacked form. These "liquid" fats ooze away from the solid, yielding a greasy and soft, less appealing chocolate. How do you get chocolate to crystallize as polymorph V? By controlling the temperature of fat crystallization. The temperature at which it begins to crystallize primarily governs the particular crystalline form of cocoa butter. Once a crystal starts growing,

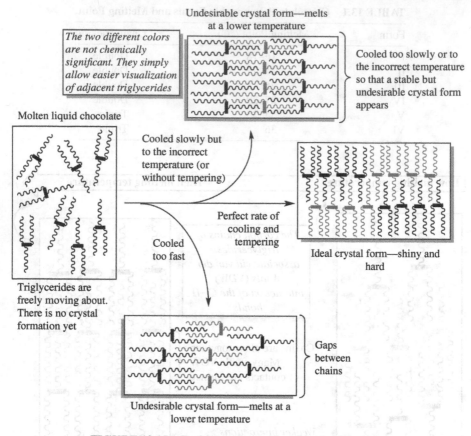

The two different colors are not chemically significant. They simply allow easier visualization of adjacent triglycerides

Molten liquid chocolate

Undesirable crystal form—melts at a lower temperature

Cooled too slowly or to the incorrect temperature so that a stable but undesirable crystal form appears

Cooled slowly but to the incorrect temperature (or without tempering)

Perfect rate of cooling and tempering

Cooled too fast

Triglycerides are freely moving about. There is no crystal formation yet

Ideal crystal form—shiny and hard

Gaps between chains

Undesirable crystal form—melts at a lower temperature

FIGURE 13.15   Packing polymorphs of fat in chocolate.

it will continue to grow in the same form. Moreover, if crystals are to be kept small, which is desirable to maintain a smooth texture, then many nuclei (crystal starter molecules) need to be formed within this ideal, desirable temperature range. Further growth or expansion of the crystals outside of the temperature range leads to tough, chewy chocolate. How do we prevent formation or remove undesirable fat crystals and promote formation of favorable fat crystals? Tempering. Other than the creation of flavor, tempering chocolate is where the magic happens.

13.14   TEMPERING CHOCOLATE

Tempering consists of three steps. Step one involves heating the chocolate to approximately 120°F or 49°C. At this temperature, all of the fat crystals melt, but the chocolate emulsion that was created during conching is not destroyed. In step two, the chocolate is cooled to approximately 80°F (27°C) and stirred. The unstable, less organized and less dense crystals (I–IV) will remain "melted" at this lower temperature due to the lack of organization of the individual fat molecules within these crystalline forms. However, polymorph forms IV, V, and VI will crystallize at 80°F/27°C. When combined with stirring, a set of small, starter crystals, primarily of forms IV and V, are generated. Why not polymorph VI? Polymorph VI forms/ crystallizes on a much slower timescale (i.e., days or weeks) than the other crystalline forms; thus very little is formed within the relatively short time frame of chocolate production. In the final step of tempering, the crystals of polymorph IV need to be removed from the system. By heating the chocolate to approximately 90°F (~32°C), the undesirable polymorph IV crystals melt, while crystals V remain intact. When tempered correctly and carefully, the desirable starter crystals of polymorph V will direct the development of the chocolate crystalline network as the chocolate is cooled and solidifies into its final form, with fat crystals of about 40 millionths of an inch in size.

Have you ever noticed the difference in melting properties and chocolate "snap" in dark chocolate versus milk chocolate? These differences are due to a difference in the crystal structures of milk and dark chocolate. At a crystalline level, the higher relative fat content of dark chocolate (because of the lack of milk solids and less sugar) allows the fat to crystallize in a more organized network than found in milk chocolate. The stronger crystalline network means that more heat will be required to "break" the network or melt the chocolate. You can hear and feel this when you break a bar of dark chocolate relative to a milk chocolate bar. Perhaps you have even carried out the experiment of comparing the effect of placing a bar of dark chocolate versus a bar of milk chocolate in your hand. Because milk chocolate contains a more diverse mixture of fats (i.e., the butterfats from the milk), it softens or melts over a wider temperature range than dark chocolate. The result: sticky fingers with milk chocolate. The solution: eat milk chocolate as quickly as possible (which isn't a problem for most of us). Not only are these crystallization properties felt when you eat a bar of chocolate, but the exact tempering temperatures that a cook uses when working with any chocolate will vary depending upon the fat content and type of chocolate that is being prepared.

13.15 CHOCOLATE BLOOM

Most people have left an unopened chocolate bar in a car at least once, where the chocolate melted and resolidified in the sealed package. Once opened, you probably noticed that the chocolate developed some white or gray streaks, had a soft texture (even in the solid form), and may have even tasted gritty (if you ate the bar at all). Even though the streaks look like mold, you can eat this less appealing chocolate without putting yourself into harm's way. What happened? Your chocolate experienced a condition called "chocolate bloom," a beautiful example of what occurs when chocolate is not tempered properly (Fig. 13.16).

Bloomed chocolate will not hurt you, but it looks unappetizing and loses the texture and flavor release that you expect when you bite into a piece of

FIGURE 13.16   Chocolate bloom.

well-tempered chocolate. If you followed the discussion on fat crystallization in the previous section, you may have figured out that the ideal fat crystals found in the original chocolate bar melted away in the heat of the car and were replaced with the less desirable fat crystals. Let's think more about what happened in the car example.

The desirable form of chocolate has fat particles that crystallize in the 29–32°C (84.2–89.6°F) temperature range depending on the type of chocolate. During the summer, the temperature in a car can reach temperatures as high as 49°C/120°F or even higher. As expected, all of the fat within the chocolate melts at these temperatures; this isn't problematic, as this is exactly what we do during the first step of chocolate tempering. It is the cooling phase that is problematic, as the cooling of a chocolate bar is completely uncontrolled in a car. In short, the chocolate is cooled to whatever temperature is dictated by the weather on a particular day or evening, which is typically a temperature that is less than 82°F/28°C during a summer evening. Thus, the chocolate is being put into a temperature condition that favors the formation of fat crystals that do not have the desirable properties of good chocolate. Specifically, the less stable polymorph forms crystallize first, and then the more stable polymorph crystals develop over a period of hours to days. Remember that polymorph V has a more ordered crystalline structure than polymorphs I–IV; thus the crystal matrix becomes more compact, which pushes the cocoa solids and sugar particles to the surface. It is these particles that lead to the whitish/grayish appearance at the surface but not in the interior of the chocolate. Moreover, the fat crystals are "softer" overall because there is less organized and dense packing of the fat molecules themselves due to the presence of different fat crystalline states. Thus, although you still have the same molecules that were present in your original chocolate bar, the molecules are organized in a fashion that leads to less desirable physical properties. Can you convert your chocolate back to its original, appealing form? Absolutely, you just have to subject the chocolate to the entire tempering process.

13.16 CHOCOLATE BLOOM IN CHOCOLATE CHIP COOKIES

Wait a minute! You may be thinking about the chocolate chips in your favorite cookies, recalling the lovely smooth, velvety texture after being subjected to temperature conditions that are even more drastic than those in your car. Why doesn't bloom happen to chocolate chips in cookies? After all, the chocolate chips melt completely (just like in the car) and as the cookie cools, the chips resolidify, so the cocoa butter recrystallizes, just like in the car. Hmmm. What is different about the chocolate in a cookie and the chocolate in a candy bar? Your experience and scientific research shows that if you have a chip that is embedded in a cookie (or another fat-containing baked good), the chip will not undergo bloom. However, if you have a chip that sits at the very edge of or is not deeply embedded within the cookie (such that it doesn't have significant contact with the dough), it will experience bloom. Thus, it is the presence of or exposure of the chip to the dough that reduces the instance of bloom; additional research has shown that the key factor is the fat.

In cookies with a certain amount of fat, chocolate chips don't bloom, as long as the cookies are made with a noncocoa butterfat that is present at a certain level. For example, in cookies containing 14% palm oil, bloom was observed, but bloom did not occur in cookies made with 20% palm oil. What is causing the bloom inhibition? Fat migration. When two fats are mixed, the resulting fat combination crystallizes differently than either of the two individual fats. Due to the change in crystallization behavior, chocolate in a cookie doesn't bloom. Milk fat (from butter) and cocoa butter are considered compatible fats. When the two are mixed, the milk fat prevents the bloom of cocoa butter and makes the chocolate less snappy. Thus, milk chocolate is less prone to bloom than dark chocolate. Milk fat, found in butter-based cookies, that migrates from the cookie dough into the chocolate chip during baking protects the chips from bloom. Interestingly, the vegetable fats present in shortenings are typically not compatible with cocoa butter and may promote bloom. Thus, the surface of a peanut butter cup is very prone to bloom, but shortening-based cookies are less prone to bloom due to the high-fat content of the cookies.

13.17 COOKING WITH CHOCOLATE

As you have probably surmised from the discussion about chocolate bloom and tempering, chocolate can be difficult to work with in the kitchen. Once you have figured out how to deal with bloom and tempering, another challenge that you may face is seizing. During seizing, a beautiful, smooth liquid chocolate turns into a grainy, muddy mass within seconds. The most common cause of seizing chocolate is a small amount of water. However, a large amount of water gets incorporated into a liquid chocolate without any problems. Why the difference?

Let's think about well-tempered chocolate at a molecular level. You have millions of microscopic, mostly hydrophilic cocoa and sugar particles surrounded by a sea of fat. Remember conching? The way that we made a chocolate liquor in the first place was by adding cocoa butter to ensure that the hydrophilic cocoa and sugar molecules remain separated from one another. If water is added to this mixture, the cocoa and sugar particles are naturally more attracted to the water than the cocoa butter and begin to clump together with the water. The result: a clumpy chocolate ball sitting in cocoa butter. A little bit of water affects a whole lot of particles; the more that you mix, the more seizing that occurs.

How does a large amount of water prevent seizing? If a larger amount of water (or other hydrophilic liquid like alcohol or vanilla extract) is added, there is enough water to keep the sugar and cocoa particles separate from one another and surrounded by water molecules; the resulting chocolate has a smooth, silky texture.

13.18   CHOCOLATE-COATED CANDIES

The preparation of chocolate-coated candies is not trivial and involves some interesting science that cannot be overlooked. Specifically, how is the soft center of a candy sealed or kept separate from the chocolate coating? The answer is invertase. Invertase is an enzyme, obtained from yeast, that catalyzes the hydrolysis or breakdown of sucrose into glucose and fructose. The soft centers of chocolates, which contain a high concentration of sucrose, suspended in glucose syrup, appropriate flavorings, colorings, and invertase, have a moldable, stiff pasty texture when first prepared by the candymaker. This texture is necessary so that the center can be pressed or stamped into a shape; however it is not appealing for the consumer. After stamping the shape of the center, it is coated with melted chocolate. During the following days or weeks (before the candies are sold), invertase hydrolyzes a proportion of the sucrose to its monosaccharide components, glucose and fructose. The glucose and fructose are more water soluble than sucrose and dissolve a bit, converting the paste into a softer, creamy textured center. This process might take up to 2 months for the desired texture to be achieved.

13.19   DIFFERENT TYPES OF CHOCOLATE AND CHOCOLATE-LIKE PRODUCTS

13.19.1   The World of Chocolate

As you go to the grocery store to buy chocolate for baking, cooking, or eating, you have probably noticed a great diversity in the types of chocolate that you can purchase: dark, milk, bitter, or unsweetened chocolate bars, regular, or Dutch-processed cocoa powder. On top of that, you can find chocolate from different countries with different percentage numbers advertised on the wrapper. What are the types of chocolate, what do these numbers mean, and does the type of chocolate used matter when you cook?

Each country has its own definition of chocolate. The different components, along with requirements for the chocolate-making process (i.e., all Swiss chocolate must be made in Switzerland), lead to different textures and tastes that one might associate with chocolate from a particular country or area. In the United States, the FDA regulates the naming and ingredients of cocoa products based on the percent of chocolate liquor and milk solids.

13.20 DIFFERENT TYPES OF CHOCOLATE

You know that there are many different types of products that are called "chocolate." Some products, like white chocolate, aren't really chocolate at all! What distinguishes one form of chocolate from another are the ingredients, the preparation of the cacao beans, and the presence of other additives, as well as the final fat crystalline structure. Remember that chocolate liquor is the cocoa butter and cocoa solids that result from the fine grinding of cocoa nibs. In terms of its components, the liquor is made of approximately 55% cocoa butter, 17% sugars, 10% protein, a few percent of tannins, and a small amount of theobromine.

Different chocolates are distinguished mainly by the sweetness and darkness of the chocolate. Generally speaking, the more chocolate liquor, the less sugar and fat and the more bitter the chocolate will taste. Unsweetened chocolate, also called baking chocolate, is used almost exclusively for baking, since it does not have a desirable flavor when eaten by itself. Unsweetened chocolate is almost pure chocolate liquor, with a very small quantity of additional sugar and fat (lecithin). It has the most intense chocolate flavor, since it is almost exclusively chocolate liquor.

Bittersweet and semisweet chocolates are also used mostly in cooking and baking but can also be eaten alone, depending upon your taste preference. Bittersweet contains between 35 and 50% chocolate liquor and more fat and sugar than unsweetened chocolate; thus one serves as a poor substitute for the other. Semisweet chocolate tends to have more sugar than bittersweet, but the terms are not regulated, so the sweetness of one type versus another will depend upon the manufacturer. Because bittersweet and semisweet chocolates do not contain milk solids, they are types of dark chocolate. Dark chocolates can be unsweetened and sweetened chocolate types but are never milk chocolate. When looking for a dark chocolate bar for consumption, most dark chocolates are described by the amount of chocolate liquor in the bar. More bitter dark chocolate bars will be above 80% cocoa/cacao. These bars will be bittersweet in taste, will have little added fat or sugar, and will not have a smooth or soft consistency. Sweet dark chocolate will have about 30% cocoa/cacao; the addition of more sugar and fat will give the dark chocolate a sweeter taste and softer texture.

Milk chocolate is quite distinct from dark chocolate in its preparation, look, texture, and taste. Since it is prepared by adding milk solids to 10–20% chocolate liquor, milk chocolate is a sweet, creamy, lighter-colored chocolate with hints of caramel and butterscotch flavors. You may remember from Chapter 3 that milk solids include the sugars (mostly lactose) and proteins (casein and whey) from milk. Milk fat is the fat from milk, without the sugars, protein, or water. The US FDA defines milk chocolate as containing no less than 10% of chocolate liquor, 3.39% by weight of milk fat, and 12% by weight of milk solids. In fact, chocolates made with other fats, such as vegetable oils, cannot be called chocolate or milk chocolate. If you see a chocolatey product that is not clearly labeled as chocolate or milk chocolate, read the label. If a plant oil like palm or vegetable oil is used as the fat, the product will likely read "made with chocolate."

I'm sorry to tell all of the white chocolate macadamia nut cookie lovers of the world that white chocolate isn't really chocolate at all. White chocolate contains no cocoa solids of any kind. Instead, white chocolate contains at least 20% cocoa butter, sugar (up to 55%), milk solids, fats, and flavorings including vanilla. Cocoa butter, the fat isolated from the cacao bean, is what gives white chocolate its smooth, melting texture. Why? Cocoa butterfat is a mixture of saturated and unsaturated fats from 20 to 16 carbons long, so the fats melt just below body temperature. The fantastic melting qualities of white chocolate make it a favorite for cookies and dessert bars. If cocoa butter is replaced with palm or vegetable oil during production, then the product is no longer called white chocolate but white chips or vanilla chips. The taste and properties of these chips may be indistinguishable from most white chocolates since most of the flavor is derived from the milk and sugars.

Cocoa powder is used for baking and for drinking with added milk and sugar. To make cocoa, unsweetened chocolate (solidified chocolate liquor) is pressed to squeeze out most of the fat, while the resulting cake is ground into an intense chocolatey powder that contains approximately 80% cocoa solids and between 10 and 22% fat. Cocoa powder is quite acidic (remember the low pH of the chocolate liquor due to acetic and lactic acids). To counter its acidity, a basic or alkaline solution can be added to the cocoa powder, creating Dutched cocoa powder. Dutched cocoa powder has a darker brown appearance than natural cocoa powder; due to the neutralization of naturally acidic cocoa powder, it has a milder and detectably more complex flavor. The taste of natural cocoa powder can be astringent and harsh, which overwhelms the presence of other flavor molecules.

13.21 CANDY

All sugar candy, whether creamy, crystalline, soft, or brittle, is made from sugar and water. It sounds pretty simple, doesn't it? If the ingredients are the same, then, what is the difference between a hard, round peppermint and a soft peppermint-flavored piece of taffy? A cook (or confectioner, in the world of candy) creates the different textures by varying the relative proportions of sugar and water, which changes the intermolecular interactions between sugar molecules and sugar and water molecules. Just when you thought that you could forget about intermolecular interactions, they strike again! Additional variation is imparted by cooking and cooling temperatures, the amount of time spent at a given temperature, the rate of temperature changes, and the presence and timing of stirring or lack thereof. Let's start with the first factor: the relative concentrations of sugar and water.

The first step in the confectionery process is the preparation of sugar syrup. In general terms, some amount of sugar and water are combined, and the solution is heated to or allowed to boil at a particular temperature. As the solution boils, water evaporates, increasing the amount of sugar in the mixture relative to water.

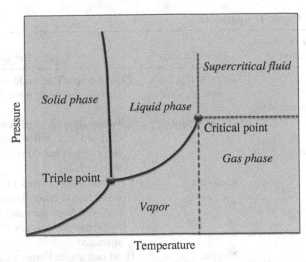

FIGURE 13.17   Phase diagram for water. The diagram describes the relationship between pressure and temperature for water. Water has a critical point where liquid and gas are indistinguishable and forms a supercritical fluid. At the triple point all three phases coexist.

Correspondingly, the solution thickens and becomes more viscous and syrup-like; the more water the syrup contains, the softer the final product. It seems simple; let's talk more about the science, specifically phase diagrams.

Figure 13.17 shows the phase diagram for water. A phase diagram is a graph that shows the state of matter of a particular substance under different pressure and temperature conditions. At an atmospheric pressure of 1.0 atm and a temperature of 212°F/100°C, water changes between a liquid and gas phase. Thus, this is the boiling point of water. As atmospheric pressure decreases, boiling point of water also decreases. Because of this phenomenon, if you are cooking pasta in boiling water at a high altitude (where there is a lower atmospheric pressure), you will need to boil your pasta for more time to achieve an al dente texture.

What does this have to do with candymaking? A phase diagram for a sugar/water solution, with temperature and sugar concentration as the variables (Fig. 13.17), is very valuable for the confectioner. You can see that the higher the sugar concentration (measured as a weight percent sugar), the higher the boiling point of the syrup. Thus, the value of the boiling point is actually indicative of the amount of dissolved sugar found in the syrup. Interestingly though, as a sugar/water solution boils, water molecules begin to evaporate from the solution, and the sugar molecules account for a larger proportion of all of the molecules in the solution. As the syrup gets more concentrated in sugar the boiling point rises. If you are making candy, it is your job to boil the syrup while closely monitoring the temperature with a trusted candy thermometer, as this will tell you the exact sugar concentration of the solution. At 235°F/113°C, you have the optimal sugar concentration for fudge, 270°F/132°C is ideal for taffy, and 300°F/149°C is ideal for hard candies.

TABLE 13.2   Candy Temperature.

Syrup's Boiling Point and Sugar Concentration	Type of Candy	Syrup's Behavior
215–234°F 101–112°C Sugar concentration: 80%	Sugar syrup, fruit liqueur, some icings	Thread stage: The liquid sugar can be pulled into brittle threads between the fingers
234–240°F 112–115°C Sugar concentration: 85%	Fudge, pralines	Soft-ball stage: A small amount of syrup dropped into chilled water forms a soft, flexible ball but flattens after a minute in your hand
242–248°F 116–120°C Sugar concentration: 87%	Caramels	Firm-ball stage: Forms a firm ball in ice water that will not flatten when removed from the water but remains malleable and flattens when it is squeezed
250–268°F 121–131°C Sugar concentration: 92%	Nougat, marshmallows, toffee, gummy candy	Hard-ball stage: Forms a hard ball in ice water that holds its shape on removal
270–290°F 132–143°C Sugar concentration: 95%	Taffy	Soft-crack stage: As the syrup reaches this low moisture content stage, the bubbles on top become smaller, thicker, and closer together. When dropped into ice water, the syrup separates into hard but pliable threads that bend slightly before breaking
300–310°F 148–154°C Sugar concentration: 99%	Lollipops and other hard candies	Hard-crack stage: This is the highest possible cooking temperature for a candy. Because there is almost no water left in the syrup, syrup dropped into ice water separates into hard, brittle threads that break when bent

In the absence of a candy thermometer, there are other methods for knowing when to stop cooking of the candy syrup by carrying out various visual kitchen tests. Perhaps you have heard the terms soft-ball stage or hard-crack stage of candymaking. These terms describe the behavior of the syrup by itself or when a drop of it is placed into ice water. Whether you use a thermometer or conduct a kitchen test, the boiling temperature is key to achieving the desired sugar concentration for whichever candy variety you wish to prepare (Table 13.2).

Based upon the types of candy prepared, you likely recognized a correlation between cooking temperature and candy type/texture. As the cooking temperature increases, the candy gets harder. What is the relationship? A candy texture is determined by the way the sugar molecules in cooled syrup organize into a solid structure. You may be thinking that this sounds a lot like chocolate tempering—you are correct. If the sugar cools and forms a few large crystals, the candy texture will be coarse and grainy. If millions of microscopic sugar crystals develop within a sea of syrup, the candy will be smooth and creamy. If the sugar forms no crystals at all, you will have a big candy mass. As with chocolate, the trickiest stage of candymaking comes as the syrup cools to room temperature. The rate of cooling, movement of the syrup, and presence of the smallest particles of dust or sugar can have dramatic effects on the structure and texture of a candy.

What impacts sugar crystal formation? You already know that the molecular structure of sugar favors crystal formation. You know that sugar crystals form easily based upon the glistening sugar in your sugar bowl or at sanding sugar granules atop a muffin. You also know from our discussion about chocolate that more stirring leads to smaller crystals and crystals develop from seeds of the most stable crystal present within the mixture. Let's try to bring all of these concepts together to better understand sugar crystallization and the candymaking processes.

Prior to cooking a sugar/water syrup, each molecule of sucrose is completely surrounded by water molecules because there is so much water relative to sucrose (Fig. 13.18). However, after evaporating off much of the water, the sucrose molecules have a much greater tendency to bump into and interact with one another. When they do, the sucrose molecules can interact with one another to form a seed crystal. As the water continues to evaporate, the syrup becomes more concentrated in sugar. When the tendency of a dissolved substance to bond to itself is exactly balanced by the water's ability to prevent this bonding, you have something called a saturated solution. The moment at which saturation is reached depends upon temperature and the molecules present. When a sugar/water mixture is boiling, all the molecules (both water and sucrose) are rapidly moving, and intermolecular interactions are constantly being broken and reformed. In this state, the likelihood of crystal formation is low because the molecules are moving so rapidly, even if a large amount of sugar is present relative to water. This type of solution is referred to as a supersaturated solution. However, as soon as the mixture begins to cool (i.e., the molecules stop moving rapidly), the solution contains more sugar than it normally can dissolve at that particular temperature. In this state, even a small disturbance (like a dust particle) will induce the formation and propagation of sugar crystals. As the crystals form, the surrounding solution becomes less concentrated. When the "new" solution reaches a saturating sugar concentration at its new, cooled temperature, the crystal stops growing until the temperature decreases further. You continue to have this balance between sugar molecules that are in solution (surrounded by water) and sugar in the crystalline form sitting in the sugar/water syrup.

Let's talk more about what happens during formation and growth of the crystal. A crystal seed is a surface to which other sugar molecules can attach, increasing the size of the crystal. A seed can be a few sugar molecules that come together during movement of the syrup, which will be facilitated by stirring, since the molecules will be more likely to bump into one another. You may have seen other events that induce crystallization; if the syrup splatters on the side of the pot, it may harden into a crystal. If those crystals are stirred back into the pot, you have a seed! Dust particles or air bubbles can also act as seeds. A sitting, metal spoon acts as a seed due to its ability to conduct heat; it can cool a small area of a syrup, leading to a super-supersaturated

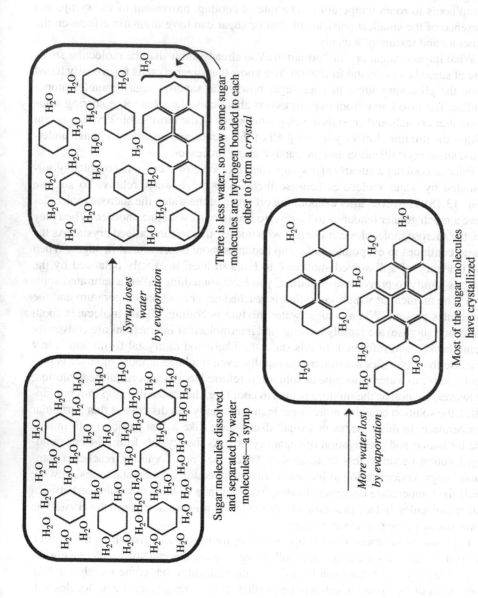

Sugar molecules dissolved and separated by water molecules—a syrup

More water lost by evaporation

There is less water, so now some sugar molecules are hydrogen bonded to each other to form a *crystal*

Syrup loses water by evaporation

Most of the sugar molecules have crystallized

FIGURE 13.18 Crystallization of sugar. Evaporation of water results in the formation of sugar crystal lattice.

solution that will quickly crystallize. Thus, depending upon when you want your sugar to crystallize, you must know when and how much to stir, as well as which type of utensil (typically a wooden spoon is best) should be used for stirring.

What about the effects of temperature? We alluded to it previously, but the temperature at which crystallization is initiated is as important in nonchocolate candies as it was in chocolate. Generally, hot syrups produce large, coarse crystals, while cool syrups produce fine crystals. Why? Remember that molecules are moving faster in a hot solution; thus once a crystal starts to form, more sugar molecules run into and attach to the crystal seed surface, and the crystals grow rapidly. At the same time, stable crystal seeds are less likely to form, as a small complex of molecules is more likely to be broken when the other molecules hit it. Thus, the total number of crystals is smaller in a hot syrup, but the crystals formed are larger. In contrast, in a cooler syrup, more crystal seeds are present, but the slower molecular movement results in fewer molecules attaching to each crystal and slower crystal formation overall. The result: candies that crystallize from hot solutions have a coarse texture, while recipes that require a smooth, creamy texture, like fudge or fondant, call for the syrup to be cooled dramatically before initiating crystallization (by stirring) (Box 13.5).

What else does stirring do? Stirring favors the formation of crystal seeds by pushing sugar molecules into one another. A syrup that is not stirred will develop only a few crystals, while one that undergoes constant stirring will have a large number of crystals. The more crystals there are in the syrup, the fewer free-floating molecules and the smaller the average size of each crystal. Thus, the more a syrup is stirred, the finer the consistency of the final candy. If you want to make smooth and creamy fudge, be sure that your arm is ready to stir constantly until the process is finished.

In some candies (like transparent, hard, butterscotch disks), you don't want crystals to form. How is crystallization prevented? If a syrup is cooled very rapidly, the sugar molecules stop moving before they have a chance to form any crystals at all. In these candies, the water content is only 1–2%; thus the syrup is very viscous, and if cooled quickly, the sucrose molecules never have a chance to order themselves into crystals. Instead, they just settle into a disorganized mass called a glass. Just like window glass, a sugar glass is brittle and transparent. Why transparent? Because individual sugar molecules are too small to deflect light when they are randomly arranged. Crystalline solids appear opaque because the surfaces of even tiny crystals are big enough to deflect light.

What if you want to stop crystallization at a certain point? You can do so by the addition of other ingredients because a unique molecule that is unlike the molecules that make up the crystal will effectively block or interfere with crystallization. These other ingredients, such as corn syrup, milk products, fats, or acid, may also be used to add flavoring or other textural components to the candy (Fig. 13.20).

A variety of candies are made using this general method and are consumed and adored by the masses, including noncrystalline candies, crystalline candies, and candies whose texture is modified with other agents (like gums and gels). The best way to make this process come to life is by looking at some specific examples.

BOX 13.5   SCIENCE OF THE FONDANT

Without starting a confectionery war, both buttercream and fondant are great
icings for a cake or other baking products. For those who don't know, fondant
is either a soft spreadable icing (often used as filling for pasties) or a more
dense, doughlike icing that can be rolled out and shaped to cover cakes and
baked goods (Fig. 13.19). Made from a sugar syrup using more than one form
of sugar to limit the repeating crystals of hardened sugars, fondant is a pliable
and active way to decorate a cake. A basic recipe for rolled fondant includes
confectioner's sugar (finely ground table sugar/sucrose), gelatin (a protein that
holds water to act as a gelling agent and also interferes with crystalline sugar
formation), glycerin (also known as the three-carbon sugar, glycerol), or glucose
(used to limit the crystal sucrose seeds from forming and to maintain pliability
of the final fondant). However, other ingredients can also be added like shortening (fat) to limit the crystallization and add creaminess to the final product
and acids to (vinegar, citric, or tartaric) increase the inversion of sugar. If you
recall a discussion from earlier in the chapter, inversion occurs when some of
the sucrose disaccharide is cleaved or hydrolyzed to the individual building
blocks of glucose and fructose. A very thorough study published by Mary
Stephens Carrick in the American Chemical Society *Journal of Physical
Chemistry* investigated the volume of water, amount of sugar, and tartaric acid
make a consistent fondant. She also found that how the fondant was boiled and
treated is critical for a pliable and well-prepared fondant. The final recipe for
her fondant is 1 cup of sugar, 3/4 cup of water, and 1/8th teaspoon of cream of
tartar. Melt all components and carefully remove the crystals that develop on the
pan at the water level, as the crystals will give your fondant an unwanted crunch.
Continue to heat for approximately 15 min allowing the solution to reach 240°F/
116°C. (The next few steps are critical to avoiding the formation of crystals.)

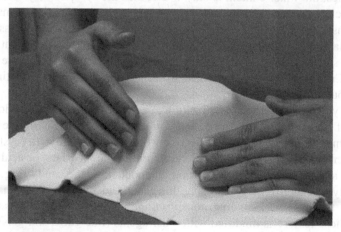

FIGURE 13.19   Fondant. A sheet of fondant applied over a cake.

Remove from the stovetop without jarring and let stand, allowing all remaining bubbles to rise to the surface. Slowly pour the solution into a shallow platter and cool to 104°F/40°C. Beat the cooled mixture for exactly 3 min (she tested the time!) using a wooden spoon using a circular motion (not a cut or folding method), followed by kneading into a soft, moist mass. Kneading, Dr. Carrick found, reduced the stickiness of the fondant. Dr. Carrick's article is a great read, a meticulous approach, and an interesting set of experiments for a new connectionist/baker to carry out their craft [3].

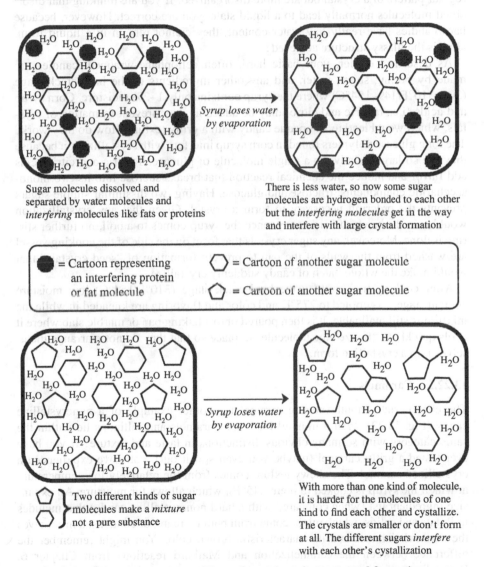

Sugar molecules dissolved and separated by water molecules and *interfering* molecules like fats or proteins

There is less water, so now some sugar molecules are hydrogen bonded to each other but the *interfering molecules* get in the way and interfere with large crystal formation

● = Cartoon representing an interfering protein or fat molecule

⬡ = Cartoon of another sugar molecule

⬠ = Cartoon of another sugar molecule

Two different kinds of sugar molecules make a *mixture* not a pure substance

With more than one kind of molecule, it is harder for the molecules of one kind to find each other and cystallize. The crystals are smaller or don't form at all. The different sugars *interfere* with each other's cystallization

FIGURE 13.20   Impact of fat and protein on sugar crystal formation.

13.22   NONCRYSTALLINE CANDIES: HARD CANDIES AND CARAMELS

13.22.1   Hard Candies

Hard candies, such as butterscotch, clear mints, and lollipops, are the simplest noncrystalline candies in terms of preparation (Fig. 13.21). You may be thinking, but isn't a lollipop one big sugar crystal? These types of candies actually have what is called a glass structure because the molecules are not arranged in the stable, ordered, regular pattern of a crystal but are more disorganized. If you are thinking that disordered molecules normally lead to a liquid state, you are correct. However, because hard candies have really low water content, they cannot remain in a liquid form. How is the glassy structure achieved?

As described earlier, your basic hard, often cellophane-wrapped, candies are made by boiling sugar, water, and any other ingredients to the hard-crack stage (300–310°F/149–154°C) where the syrup contains only 1–2% moisture. Corn syrup and acids (i.e., tartaric and citric) may be added as they prevent formation of crystals, which would lead to an opaque candy with a gritty texture. How do they work? The long glucose polymers found in corn syrup interfere with crystallization because they are so large, relative to a single molecule of glucose. Tartaric and citric acids add flavor and induce the chemical reaction that breaks sucrose into its two monosaccharide components, fructose and glucose. Having two dissimilar simple sugars prevents the ordering necessary to form a crystal. Given that any crystallization would ruin the batch of hard candy, once the syrup comes to a boil, no further stirring is done. Moreover, any sugar crystals that form on the side of the cooking vessel are washed from the walls of the pan to prevent formation of "seed crystals" that would make the whole batch of candy suddenly crystallize.

After cooking the syrup to the hard-crack stage (310°F/154°C) low moisture content stage, it is cooled to 275°F, and color and flavoring are kneaded in while the mixture is still malleable. It is then poured onto a baking pan or marble slab where it cools quickly to "freeze" the molecules in place so that they do not rearrange themselves into a crystalline form.

13.22.2   Caramels

The chewy, rich, mouthwatering caramel is another example of a noncrystalline candy (Fig. 13.22). Wait a minute! How can caramels and lollipops lie within the same category with so many obvious distinctions in taste and texture? If you have ever tasted a gritty caramel (maybe you even spit it out), you have experienced a crystallized caramel. The chewy texture comes from the relatively low temperature at which the syrup is cooked (a mere 245°F), which allows for moisture. An expansive ingredient list (when compared with a hard noncrystalline candy) that includes milk fats and solids, which might come from butter, cream, or condensed milk, gives the caramel richness and its characteristic brown color. You might remember the difference between the caramelization and Maillard reactions from Chapter 6. Depending on how you make the caramel, you will get the color and flavor from just

FIGURE 13.21   Hard candies. Made by boiling sugar water and flavoring.

FIGURE 13.22   Caramels. A different crystalline candy form.

caramelization reactions (if no protein is present) or a combination of the two reactions (if protein is available). Let's talk about the process of making caramels first and then get into some of the science.

If you looked through 10 different recipes that describe how to make caramel candy, you will find slight differences in the milk products used and the exact temperature at which the syrup is boiled. These differences will result in great variation in the softness, color, and richness of the final product. Generally though, many caramels are made by cooking sugar, cream, butter, and a little bit of corn syrup to a temperature between 242 and 248°F/117–120°C. As the syrup is heated to this "firm-ball" stage, it is stirred, but once the desired temperature is reached, stirring ceases to reduce crystallization (as you saw in hard candy production). The thick syrup is poured into a metal pan, allowed to cool, and cut into the bite-sized pieces of caramel candy. Simple, isn't it?

Now, you know about many of similarities between caramel and hard, noncrystalline candy preparation that place the two in the same "class" of candy. However, there are some obvious differences too. The lower boiling temperature used in caramels will increase moisture content; as in baked goods, more moisture means a softer, more chewy texture. However, chewiness also comes from the corn syrup and milk proteins. These ingredients do not only interfere with crystallization but also with the formation of a glass structure. As stated before, the characteristic caramel flavor and brown color come from the Maillard reaction that occurs between the sugar and milk proteins. You have already learned a lot about how the Maillard reaction is important in the browning of meats and that a sugar and a protein are required for the reaction to occur. Guess what we have in our boiling syrup that will eventually become caramel candy? Sugar (sucrose, fructose, and glucose) and milk proteins! The aldehyde group of a sugar molecule reacts with the amine group on a protein, resulting in the characteristic brown caramel color and flavor.

Perhaps as you are reading this section, your mouth is watering, just like it does when you eat a caramel. Caramels are mouthwatering because the process of chewing a caramel releases butterfat from the mass of gooey, chewy sugar goodness.

13.23   CRYSTALLINE CANDIES: ROCK CANDY AND FUDGE

13.23.1   Rock Candy

As a child, you may have begged a parent to buy you a bag of large, coarse, colored sugar crystals, also known as rock candy (Fig. 13.23). Rock candy provides a great example of what happens when you encourage formation of large crystals from a sugar syrup (and is an easy experiment/recipe to carry out in your own kitchen). Basically, you boil sugar and water to the hard-ball stage (250–268°F) and then pour the syrup into a small container (a glass or jar works well) that contains a toothpick or crystal seed around which the crystals can form. Voila! That's it! After a few days of undisturbed sitting, you will have beautiful and large sugar crystals that will

FIGURE 13.23   Brown sugar rock candy.

FIGURE 13.24   Fudge.

impress your friends. You can extend your experiment by stirring or disturbing the solution, adding flavoring or coloring or boiling to different temperatures and noting the differences in crystal size.

13.23.2   Fudge

When you think of fudge, crystalline is probably not a word that immediately comes to mind. After all, good fudge has a creamy and smooth consistency, just like a caramel, doesn't it? Actually, the microcrystals that are present in fudge are key to making it ... fudge (Fig. 13.24). These tiny, very fine crystals give the fudge its firm texture and are small enough not to feel gritty on your tongue. Thus formation of the right type of crystal is key to fudge making. How does the proper crystallization occur? It is all about the cooling, not the cooking. Cooking fudge is quite similar to the preparation of caramel in terms of some added ingredients (i.e., milk solids, milk fat, corn syrup), along with chocolate solids. These ingredients are boiled to the soft-ball stage (234–240°F/112–116°C), which is slightly lower than the boiling temperature required for most caramels. The exact boiling temperature is largely dependent upon the ratio of other ingredients (e.g., cream to sugar), as well as altitude and humidity. If the boiling temperature is too low, you will have runny fudge. If the boiling temperature is too high, your fudge will be hard. Thus, a fudge maker may have to experiment with temperature a little. The next point of distinction comes after boiling; the fudgy hot syrup is allowed to cool UNDISTURBED to a temperature of 100–130°F/38–54°C. This means that the fudge is not stirred, as this would increase the likelihood that the seed crystals (from sucrose, from dust particles, from splashes on the side of your pot) will form. If crystals form at this stage, the crystals will be too large, and your fudge will have a gritty, grainy texture. Hmmm. At the beginning of this section, you were told that fudge has crystals. When do these special fudge crystals form? Crystallization needs to be induced after the fudge has cooled. You stir and beat the fudge continuously for about 15 min until the mixture is thick. The more you stir, the more and smaller crystals you will obtain, which will yield a smoother candy.

FIGURE 13.25   Marshmallows. The familiar combustion of sugar and oxygen catalyzed by the heat of a campfire.

13.24   AERATED CANDIES: MARSHMALLOWS

Jumbo sized, miniature, bunny shaped, and the most beloved Peeps™. You can't really talk about the science of candy (or have a campfire) without talking about soft sticky, spongy, sweet marshmallows (Fig. 13.25). Before we get into some of the science, let's talk about the name.

The marshmallow confection first came from the mallow plant (*Althaea officinalis*), a weedy relative of the hollyhock that grows in marshes, whose roots contain mucilage. Mucilage is a thick, gluey sap produced by some plants and microscopic animals to help with food storage and seed germination. People of ancient cultures used the plant for both culinary and medicinal purposes. The ancient Egyptians, for example, dried the root and mixed it with honey to make mallow treats, while the French, who were introduced to it in the early to mid-1800s, experimented with using its gummy juice to soothe sore throats. The French were the first to make "marshmallows" that resemble the confection that we enjoy today by mixing sap from the marshmallow plant with eggs and sugar and beating the mixture into a foam. Marshmallows were so popular that in the late nineteenth century, candymakers were unable to keep up with the demand. In response, the modern version of the marshmallow was born, which does not rely on the mallow plant at all. Today, manufactured marshmallows are made by combining gelatin (a protein solution) with a sugar syrup (i.e., sugar and corn syrup) and cooking the mixture to 240°F/115°C. The mixture is whipped into a foam that is two to three times the original volume. How does the foam develop? Millions of air bubbles are stabilized and trapped by the protein molecules as the mixture cools and gelatin sets. A marshmallow is, in essence, a solid foam that is only 35–45% as dense as water.

Where does the difference in texture and shape of various marshmallow products, such as marshmallow fluff, a Jet-Puffed™ marshmallow, or a Peep (Fig. 13.26),

FIGURE 13.26   Peeps.

come from? Common types of marshmallows have a 1×1-inch cylinder shape because in marshmallow factories, the liquid foam is piped through a long, 1-inch-diameter tube as it cools. The marshmallow "rope" that emerges from the tube is chopped into pieces that are approximately 1 inch in length. The texture of a marshmallow product can be controlled by adjusting the proportions of ingredients and the amount of whipping. With more whipping, you will incorporate more air bubbles into the foam and have a softer marshmallow. These marshmallows might be great for eating by themselves but don't have a firm enough texture to be covered in chocolate or another candy coating. What makes a roasted marshmallow so yummy? The fire's heat both melts the gelatin and caramelizes the sugar, producing a hot, gooey, caramel-flavored concoction.

REFERENCES

[1]   Philippe Sylvestre, D. (1685) *Traités nouveaux & curieux du café du thé et du chocolate*, Lyon.

[2]   Schwan, R.F. and Wheals, A.E. (2004) The microbiology of cocoa fermentation and its role in chocolate quality. *Crit Rev Food Sci Nutr.* 44(4): 205–221.

[3]   Carrick, M.S. (1919) Some studies in fondant making. *J. Phys. Chem.* 23(9): 589–602.

INDEX

The Science of Cooking: Understanding the Biology and Chemistry Behind Food and Cooking,
First Edition. Joseph J. Provost, Keri L. Colabroy, Brenda S. Kelly, and Mark A. Wallert.
© 2016 John Wiley & Sons, Inc. Published 2016 by John Wiley & Sons, Inc.
Companion website: www.wiley.com/go/provost/science_of_cooking